MAGNETIC COMPASSES
AND
MAGNETOMETERS

Magnetic Compasses
and
Magnetometers

ALFRED HINE, B.Sc.

Late Senior Scientific Officer at the Admiralty Compass Observatory

ADAM HILGER LTD
LONDON

Published by
ADAM HILGER LTD
98 St Pancras Way, London, NW1

Printed by
J. W. Arrowsmith Ltd, Bristol 3

5042

FOREWORD

by

Commander A. V. Thomas, O.B.E., O.ST.J., J.P., R.N.

When one is asked to write a foreword to a book written by a friend who dies before the book is completed, one is torn between the requirements of a valedictory notice and an introduction to the book itself, many of whose readers will never have known the Author.

Alfred Hine and I were friends and colleagues for more than a quarter of a century. His first love, somewhat oddly, was steam rollers, but when he came to the Admiralty Compass Observatory these were duly replaced by magnetism, magnetic compasses and magnetometers and all that they imply. Eventually he acquired a knowledge of these matters which can rarely have been surpassed.

Before, during (especially during) and after the Second World War his unfailing help, coupled with his complete command of his subject, was a source of enormous comfort and support to myself and to all the many people who relied on him. And, with it all, his charm, humanity and kindness made everything seem so much more easy.

He knew great happiness in life, with sorrow and, later, happiness to follow again. And, incidentally, he had a cat called James Watt who, at a ripe age, was run over by a staff car on the drawbridge at Ditton Park. For good measure, he was an accomplished gardener, an addict of choral singing and a pillar of Toc H.

When he retired he wrote this book, mostly at his home in Norfolk, where he could still indulge his love of small sailing boats. Then he fell ill, and before the book was finished he died, to leave its completion in other (but very able) hands.

'Definitive' is a word generally used only by publishers on their dust covers, and no one can say whether this is a definitive work. But it covers a field embracing almost every aspect of compass design and embodies a wealth of fundamental learning and experience which is unlikely to be out-dated in future publications for many years.

In fact, anyone who henceforward needs to study the design of magnetic compasses or magnetometers, or to learn of the history and development of these instruments, will unquestionably include Alfred Hine's book among their sources of reference. And while one pays tribute to the skill and devotion of the Author, one should perhaps (as he would, I think, have wished) include an acknowledgement to that most remarkable and legendary device, the magnetic compass itself, which, given a fair chance, will provide a continual indication of direction and is surely one of the greatest gifts ever bestowed by Providence on the navigators and surveyors of the world.

ACKNOWLEDGEMENTS

Grateful acknowledgement for illustrations and assistance is made to H. Browne & Son Ltd, Barking; Hilger & Watts Ltd, London; Kelvin Hughes, Barkingside; The Sperry Gyroscope Co. Ltd, Bracknell; W. F. Stanley & Co. Ltd, New Eltham; the Controller of H.M. Stationery Office for permission to reproduce Crown Copyright photographs; the Controller of H.M. Stationery Office and the Hydrographer of the Navy for permission to make copies of Admiralty charts; Commander E. M. Penton and members of the Admiralty Compass Observatory; The Rev. Howard Whyntie; and Commander A. V. Thomas.

CONTENTS

CONTENTS

Chapter 1

Introduction
and
Historical Background

The magnetic—or mariner's—compass is, perhaps, one of the oldest of scientific instruments. For centuries it has been the means of directing sailors in their explorations of the oceans and coastlines of the world. In fact its use was well known before the phenomenon of terrestrial magnetism was clearly understood.

The discovery of the property of a lodestone, a variety of magnetic iron ore, of seeking a given direction with respect to the Earth's north pole has been attributed to many and varied sources, none of which is certain. It is probable that the directional behaviour of a lodestone, or a piece of iron that had been stroked or 'touched' by a lodestone, in setting towards north when suspended or floated was known long before the Middle Ages. The Chinese were among the earliest users of a magnetized needle in the form of a compass, but mainly for necromancy and fortune telling; and in the Chinese compasses illustrated in Plate 1.1, the numerous concentric circles of signs include the signs of the zodiac in addition to the points of the compass. Plates 1.2 and 1.3 show other versions of the instrument, one of which incorporates a sundial.

The mariner's compass appears to have been in fairly regular use around the thirteenth century, though it is uncertain whether the Chinese, Arabs, Turks, Persians, Italians or Greeks were the first to use it. Early records of the use of the compass include a twelfth-century work by Alexander Neckham, *De Utensilibus*,* a poem by Guiot de Provins dating from about the thirteenth century, the *De Contemplatione* of Raimon Lull (1272), the encyclopaedia *Livres du Trésor* of Brunetto Latini (1260), and various thirteenth-century Scandinavian records.

The noteworthy *Epistola de Magnete* of Petrus Peregrinus de Maricourt,† addressed in 1269 to Sigerus de Fontancourt, is the earliest detailed work extant on the magnetic compass. The first part deals with early notions of terrestrial magnetism, such as the polarity of the lodestone and the influence upon it of the poles of the heavens; whilst the second part describes a type of azimuth compass using a floated needle with a graduated circle, having 90° to each quadrant (a very

† May, W. E., 'Alexander Neckham and the pivoted compass needle', *J. Inst. Navig.*, 1955, **8**, 3.
* Bertelli, T., 'Sopra Pietro Peregrino de Maricourt e la sua epistola "De Magnete"', *Boll. Bibliogr. Sci. mat.*, 1868, **1**, 1–32, 65–69, 101–139, 319–420.

early example of degree markings) and a movable sight for taking bearings. The author goes on to describe a compass having a needle thrust through a pivoted axis, housed in a box with a transparent cover, and fitted with a graduated circle and alidade or sight for taking bearing. A leaf from this famous manuscript, illustrating what is probably the earliest example of a pivoted compass with degree markings, appears in Plate 1.10.

It is likely that the first form of compass was a lodestone resting on a wooden float in a bowl of water. Artificial lodestones or iron needles 'touched' by a natural lodestone were similarly used and in due course were thrust through straws or reeds to cause them to float, so that they became known as sailing needles. The principal use of sailing needles was to indicate the direction of north in foggy or cloudy weather, when the stars, particularly the pole star, were hidden from sight. The common method of navigation in those early days was by reference to the direction of the wind, and for this purpose the *rosa ventorum*, or wind rose, was used (see Plate 1.4). This device was far older than the compass and bore the eight principal directions marked with the names of the eight principal Mediterranean winds, Tramontano, Greco, Levanter, Scirocco, Ostro, Africo (or Libeccio), Ponente and Maestro. The *rosa ventorum* appears on the early *portulani* or sailing charts of the mediaeval Mediterranean pilots and, for its satisfactory use, required a knowledge of north. Thus, when no stars were visible, the combination of a *rosa ventorum* and sailing needles provided the early navigator with his primary navigational instrument. It is interesting that the French expression for compass card today is *la rose des vents*.

The sailing needle was, in due course, to become a needle pivoted on a pin rising from the bottom of the compass box or bowl. Then, no doubt, compass points or a *rosa ventorum* would have been marked on the bottom and eventually a graduated card would have been attached to the needle itself.

The earliest compass cards or wind roses were marked with a T (Tramontano) and an arrow-head at north, and a cross at east. The arrow-head developed into the customary fleur-de-lys—almost universally used to this day—whilst the cross eventually became a mere ornamentation of the east point, continuing until about the middle of the nineteenth century. The further subdivision of the eight principal points (cardinals and intercardinals) into thirty-two points or rhumbs is attributed to the Flemish navigators. Each point of the compass represented an angle of $11\frac{1}{4}°$ and was further divided later into half points and quarter points. For a long time this was the usual style of compass marking and is still preferred by many fishermen. Apart from the compass described by Peregrinus, the use of degrees as compass marking was not common until about the eighteenth century.

The cards, marked in points, also carried a scale of degrees divided into four quadrants, 0° at N. and S., 90° at E. and W. Compass courses were then identified as, for example, S. 20° E., N. 65° W., S. 10° W., N. 4° E., as an alternative to the point notation.

In this country, the Royal Navy was the first to adopt a 360° card for magnetic compasses, having previously used a degree card marked in quadrants, without any points other than the cardinals and intercardinals. The change took place after the Second World War. Merchant Navy cards eventually came into line with Royal Naval practice, but initially points, half points, and quarter points appeared in conjunction with the degree scale. The tendency has been to simplify the marking of compass cards and the present-day style seems very austere compared with that of the elaborate and decorative cards of an earlier period, as will be seen from the ancient and modern compass cards shown in Plate 1.5 to 1.7.

An example of a mid-seventeenth-century azimuth compass is shown in Plate 1.8. The art of compass-making, however, remained comparatively primitive until the eighteenth century, partly owing to difficulty in magnetizing the needle and to the inability of the iron then available to retain its magnetism, but also because of the complications involved in pivoting the needle; various shapes and combinations of needles were tried so that insertion of the pivot bearing at the centre of the needle was avoided. It was not until 1766 that compass design made any real advance, when Dr Gowin Knight took out his first patent for a compass. He used better steel for the needle than had been used hitherto, to give it increased magnetic strength, and reverted to the use of a single needle with a jewel cap for the pivot in the centre. Knight's method of supporting the compass bowl and his design of the gimbal system represented an advance on previous practice, and his compasses were so successful that they were adopted by the Royal Navy and were also copied in Germany. Plate 1.9 illustrates one of these instruments.

Since then there have been many inventions appertaining to compasses, not all of them effective or successful. A great number have been 'improvements' in the magnet or needle arrangement, e.g. the inclination of the needle at the angle of dip and the introduction of curious and incredible devices such as a cast iron bowl for screening the compass needle from the magnetic effects of a steel ship.

The advent of the iron ship and steam-driven propelling machinery was a serious threat to the efficacy of the mariner's compass, since the magnetic properties of the hull and its contents produced very large deviations of the compass needle from magnetic north. The use of armour plating in warships aggravated this condition. Curiously enough, it was not until well into the nineteenth century that the attractive effect of iron on the ship's compass was appreciated, any imperfections or irregularities in the behaviour of the compass being attributed to faults in its manufacture or in the lodestone which had been used to 'touch' its needles.

Capt. Matthew Flinders, R.N., was the first to investigate the problem of the deviations of the compass on a proper experimental basis, having studied the phenomenon during his voyage to Australia in H.M.S. *Investigator* in 1801–4. He proposed the vertical bar of soft iron, which today is called the Flinders bar, as a means of compass correction.

The numerous marine catastrophes attributable to unreliable compasses and unknown deviations brought into being in 1837 the Admiralty Compass Committee, which was set up to investigate the whole question of compass design, positioning, and correction. This Committee laid down rules for the correct placing of the standard compass in H.M. ships. Moreover, this body was responsible for the design of a new compass, the Admiralty compass, Pattern 1 (described in detail in Chapter 5), which was so successful that it was not only used by the Royal Navy, but was adopted also by the principal foreign navies.

In 1824, Denis Poisson* had published a mathematical analysis of the deviations of the compass and his theories were further developed by Archibald Smith and

* Poisson, D., Memoir to the Institut de France, 1824.

Capt. Frederick Evans, R.N., later Sir Frederick Evans*, between 1851 and 1865. The results of their work appeared in numerous papers read before the Royal Society and the Institute of Naval Architects—and in the *Admiralty Manual for Ascertaining and Applying the Deviations of the Compass* (1862).

As a result of experiments carried out in the years 1838–9, under Admiralty auspices, in the iron steamship *Rainbow* and subsequently in the iron sailing ship *Ironsides*, Sir G. B. Airy,† then Astronomer Royal, proposed a method of correcting the deviations of the compass by permanent magnets and soft iron masses, establishing the principle of correcting like with like. This process, which remains in use today as the fundamental method of compass correction, at first fell into disrepute in the Royal Navy, probably owing to a lack of understanding of the method.

All the compasses so far described were 'dry-card' compasses, which were subject to errors and unsteadiness caused by the vibration of the steamships and by gunfire, and attempts were made to damp out the oscillations of the needle by filling the bowl with liquid. In 1813, Francis Crow of Faversham produced a liquid-filled compass which, though crude, incorporated many of the features of present-day compasses. Some of the problems that had to be solved were the leakage of liquid, expansion troubles, discoloration of the paint and difficulty in repairing a damaged pivot. Considerable advances were made by E. S. Ritchie in 1862, when he patented a magnet system incorporating a float, thereby relieving the pivot of much of the weight of the system and so lengthening its life; and in 1866 by W. R. Hammersley, who invented an expansion chamber, which largely prevented leakage of the liquid. Although Ritchie's compass was adopted by the U.S. Navy, the Royal Navy continued to use the Pattern 1, except in bad weather and during gunfire, when liquid compasses were used.

The dry-card compass was to remain in use by the Royal Navy for several more years and to last longer still in the Merchant Service, owing to the work of Sir William Thomson, who eventually became Lord Kelvin. Thomson's dry-card compass and binnacle were two major achievements of 1876. The former superseded the Admiralty Pattern 1 in H.M. ships and also became a traditional compass for the Merchant Service, only now giving place to the liquid compass. The features which made it so popular were its very light (paper) card and the ease with which the card could be replaced and new pivots and jewels fitted. Spare parts were available at all the world's principal seaports and the Thomson compass was regarded as the mariner's best aid to navigation. Even its degree of unsteadiness due to low damping was held to be in its favour, since the master mariner of those days liked to see a 'lively' compass.

Owing to the lightness of the card, a large diameter could be adopted without subjecting the jewel and pivot to an excessive loading. The liquid compass, on the other hand, became unwieldy with cards larger than about 7 in. diameter. It is a

* Smith, A., and Evans, F. J., 'On the effect produced on the deviation of the compass by the length and arrangement of the compass needles and on a new mode of correcting the quadrantal deviation', *Phil. Trans. roy. Soc.*, 1861, 161.

Admiralty Manual for Ascertaining and Applying the Deviations of the Compass Caused by the Iron in a Ship, 7th edition (reprinted 1920).

Evans, F. J., *Elementary Manual for the Deviations of the Compass* (1870).

† Airy, G. B., *Phil. Trans. roy. Soc.*, 1839, **3**; 'On the connection between the mode of building iron ships and the ultimate correction of their compasses', *Trans. Inst. nav. Archit.*, 1860, **1**, 105.

curious fact that, in the middle of the last century, it was considered that the bigger the ship, the bigger the necessary compass; the *Great Eastern*, for example, had 12-inch compasses. The compass establishment of ships of that time was lavish by present-day standards. One barque, the *Brier Holme*, owned by the author's grandfather, though modest in size (206 feet long and of 894 net register tons) had seven magnetic compasses, 'one in a binnacle, one standard, one brass on tripod on mizen mast, one hanging in Cabin, one Azimuth, one liquid and one boats, all 10 in. diameter'. The use of a liquid compass in this ship, built in 1876, is interesting.

Thomson's other invention, the binnacle, systematized the methods of compass adjustment and greatly facilitated the application of the theories and methods of Airy, Smith and Evans. Not only did the binnacle support the compass and its gimbals, but it was adapted to contain groups of fore-and-aft, athwartship and vertical magnets for the correction of the ship's permanent magnetism and also soft iron spheres and the Flinders bar for the correction of induced magnetic effects. Although light alloys have often replaced wood in current binnacle design, the methods of correction and the arrangement of the correctors devised by Thomson remain to this day.

The liquid compass was introduced into the Royal Navy in 1906 by Capt. L. W. P. Chetwynd, R.N., and its design is very little different from that used today. It was only during the Second World War that the liquid compass was adopted, hesitantly, by the Merchant Navy.

Shortly after the introduction of the liquid compass, there appeared an instrument which challenged the magnetic compass and bade fair to banish it from the larger vessels afloat, especially warships. This was the gyroscopic compass, developed almost simultaneously by Brown, Sperry and Anschütz in England, America and Germany respectively. The combined advantages of remote indication, by several repeaters, of the ship's head, of being a true north-seeking compass and of being unaffected by magnetic materials and electrical appliances seemed so overwhelming that the magnetic compass became relegated by degrees to the status of a secondary or emergency compass, although Queen's Regulations and Admiralty Instructions state that all H.M. ships must carry a properly corrected magnetic compass. In spite of its initial defects, the gyro-compass was so popular that, at any rate in the Royal Navy, a generation of officers grew up that scarcely appreciated the virtues of the magnetic compass and were comparatively unskilled in its use and correction. Merchant ships, however, continued to use the magnetic compass, since it was uneconomic to install gyro-compasses except in the larger vessels.

This state of affairs continued until the Second World War, when the Admiralty introduced the transmitting magnetic compass for use in small craft. There had been previous attempts by Siemens, Holmes and other inventors to produce a magnetic compass that would operate repeaters, but these earlier inventions were not adopted because an unacceptable load was imposed on the magnetic needle in causing it to operate some form of pick-up device or because there was a lack of sensitivity in the transmitting system or because troubles developed owing to the motion and vibration of a ship. It was probably due as much to electronic advances as to the adherence to the fundamental principles of compass design that the Admiralty transmitting magnetic compass developed into a tried and useful navigational instrument. It used the resistance-bridge method of Holmes and a follow-up system with an a.c. servo-amplifier—a comparatively new technique. Designed primarily for small fast coastal craft of the Royal Navy, such a compass challenged the seeming supremacy of the gyro-compass and the magnetic compass began to come into its own once again with the fitting of transmitting compasses in destroyers and frigates, as well as in a few commercial vessels. Work on the transmitting compass had been going on in other countries too, notably by Plath and Siemens in Germany and by various Italian designers. Photoelectric systems were popular and the Plath compass was widely used by the German Navy.

Not only ships but also aircraft need compasses, and though the early aircraft used simple adaptations of the mariner's compass, there was later a distinct trend in compass design aimed at meeting the special conditions arising in high-speed flight. The notorious 'northerly turning error' became evident, and thereafter aircraft compasses were characterized by light-weight, low-strength magnet systems with heavy damping. The need for transmitting compasses in aircraft soon arose. These in some instances were similar to the marine instruments in being designed merely to repeat the indications of the magnetic needle, but owing to the high accelerations encountered the simple magnetic compass soon proved to be inadequate. The solution was found in the gyro-magnetic compass, notably the Royal Air Force DR compass, in which the stable properties of a gyroscope as a heading reference are combined with the magnetic north-seeking properties of a compass needle. In brief, the gyroscope provides high stability for a short time, for example during turning of the aircraft, whilst the magnetic compass, though unstable under accelerations, provides an accurate average indication of magnetic north. The gyroscope smooths the perturbations of the compass needle which, in turn, corrects any drift of the gyroscope and renders the system north-seeking.

The gyro-magnetic compass was soon developed for use at sea. An instrument was designed at the Admiralty Compass Observatory which assumed equal importance with the gyro-compass in ships of the Royal Navy. In fact, in some circumstances, the gyro-magnetic compass has proved more suitable and more accurate than its gyroscopic counterpart. Thus, after a lapse of over half a century, the magnetic compass has once more taken its rightful place aboard ship and is no longer regarded as a secondary instrument.

Modern invention in the field of electromagnetic induction has devised alternative methods of determining the direction of magnetic north, and recent years have seen the development of the inductor compass. A device known as the saturable inductor is a valuable alternative to the pivoted needle. Since it is compact, requires no delicate pivot and jewel, and provides, by the very nature of its arrangement, an electrical signal, it is an ideal sensitive element for aircraft compasses, being frequently sited at the wing tip in order to be far removed from electrical and magnetic interference. Moreover, it is readily adapted for use in a gyro-magnetic system; contemporary aircraft compasses, both civil and military, are of this kind. The saturable inductor has also found application in the marine world and a recent type of gyro-magnetic compass has enabled the problem of fitting a magnetic compass in large warships to be solved.

Ideally, the magnetic compass in its modern form and the gyroscopic compass provide a partnership of complementary instruments of comparable accuracy capable of dealing with the problems of navigation in high magnetic and geographic latitudes.

The authority responsible for compasses used by the Royal Navy is the Admiralty Compass Observatory at Slough. First established at Charlton in 1842, it transferred to Deptford in 1870 and moved to its present site in 1917. Its original purpose was to test compasses, jewels and pivots, and to oversee the repair of compasses belonging to H.M. ships. Today the staff of the magnetic test room at the Observatory test all compasses for H.M. ships, and the establishment gives 'type approval' to many commercial compasses and issues certificates for the standard compasses of all ships fitted under Ministry of Transport rules. The correction of the compasses in naval vessels is administered by the Observatory, which is also responsible for the inspection of these vessels as to their magnetic properties and the correct siting of their compasses. Correction and inspection are carried out by naval officers. In addition, a staff of civilian scientists undertakes research and development on compasses, and several new designs of instrument have been produced for use by both the Royal and Merchant Navies as well as by other services. Workshops are available at the Observatory for small-scale experimental manufacture and repairs, while the supervision of outside contracts is carried out by the production department.

The Admiralty Compass Observatory is recognized throughout the world as a unique authority on compass matters. The international collection of compasses in its fine museum admirably illustrates the development of the magnetic compass down the ages.

Chapter 2

The Earth's Magnetic Field
and its
Measurement

Although the mariner's compass was in general use in the Middle Ages, little was known about terrestrial magnetism, that property of the Earth which, among other things, gives the lodestone and the magnetic needle their directive character-istics. There were many theories and notions, but few bore any resemblance to the facts and explanations of the Earth's magnetism as set forth today. It was soon discovered that the magnetic needle did not always point to the true pole of the Earth and this divergence between the direction of the needle (magnetic north) and the pole (Polaris, or true north) was found to vary at different places on the Earth's surface. This fact was known to Columbus and it is said that the apparently inaccurate behaviour of his compass when crossing the Atlantic led to trouble with his ship's company.

Some compass-makers in Europe, being aware only of the divergence of the needle from true north but not of its variability, constructed instruments with the needle offset from the north point of the card, so that the card indicated true north. This divergence of the magnetic needle from true north became known in due course as magnetic variation and will be discussed further in this chapter.

The first serious attempt to study terrestrial magnetism was made by William Gilbert,* who published his famous treatise *De Magnete* in 1600. This laid down the basis of future studies of the science. Gilbert invented a device called the *terrella*, a sphere of lodestone in a brass case engraved with lines of latitude and longitude and suspended by brass chains. This model of the Earth served as a very good approximation when investigating the behaviour of a magnetic needle on the Earth's surface. A few of these historically interesting devices still exist.

Thus, as Gilbert demonstrated, for most practical purposes the Earth may be considered to be a great magnet surrounded, like all other magnets, by a field of magnetic force. It is this field which directs the compass needle. A useful convention names the end of the compass needle that points north the 'red' end or pole; the south-pointing end is designated 'blue'. Since unlike poles of magnets attract each other, the north magnetic pole of the Earth to which the compass needle points is termed a blue pole and the south magnetic pole a red pole.

* Gilbert, W., *De Magnete*, 1600. Translations: P. Fleury Mottelay (John Wiley and Sons), 1893; S. P. Thompson (Gilbert Club, London), 1900, with notes.

A simple approximation to the magnetic condition of the Earth is obtained by supposing it to have a magnetic dipole near its centre, unsymmetrically placed with respect to its axis of rotation. Whilst this conception is necessarily incomplete and unsatisfactory from many aspects, it can serve to demonstrate the behaviour of magnetic needles on the Earth's surface, and suffices to show the significance of the direction in which a magnetic needle points.

The magnetic field, represented by lines of force emerging from the Earth's south magnetic or red pole and entering the Earth's north magnetic or blue pole is shown diagrammatically in Fig. 2.1. In regions near to, but not coincident with, the

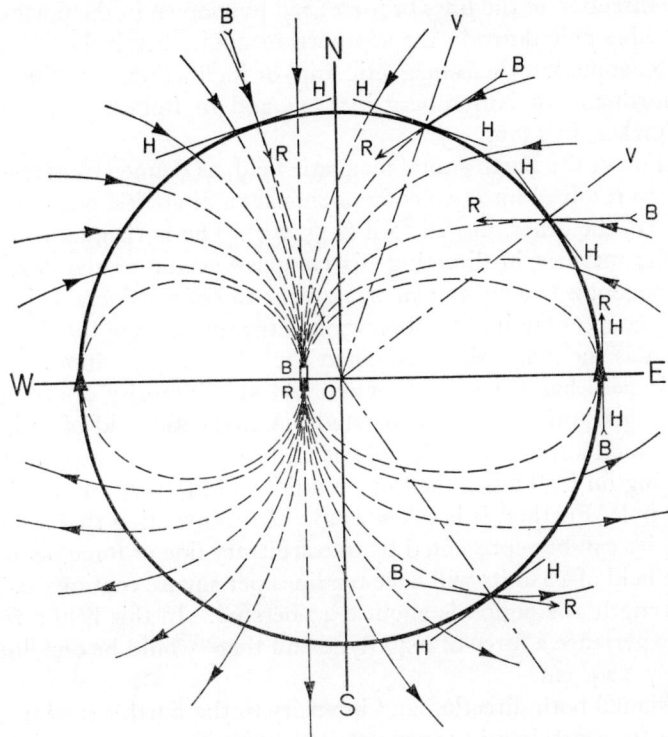

Fig. 2.1. Simplified section through the Earth's magnetic field

B and R signify blue and red poles. The lines marked V and HH show the vertical and horizontal directions in various latitudes.

true or geographic poles, the lines of force are normal to the Earth's surface, and since a freely suspended magnetic needle will lie along the line of magnetic force at its locality, such a needle would point vertically at either of these places—red down at north and blue down at south. These are the magnetic poles, frequently termed the 'dip' poles,* and they cover fairly large areas. The north magnetic pole

* Ian Cameron writes of James Clark Ross: 'His career seemed to have reached its climax when he became the first man to set foot on the north magnetic pole.' This would have been in 1831. We then learn that in 1841 'below in his cabin was the Union Jack which he had unfurled ten years previously at the north magnetic pole, and to raise the same flag at the south pole was his most cherished dream.' *Lodestone and Evening Star* (Hodder and Stoughton), 1965, pp. 221 and 227.

is situated in Viscount Melville Sound, approximately 75·5° N. Lat., 100·5° W. Long., and the south magnetic pole in King George V Land, approximately 66·5° S. Lat., 139·5° E. Long., these positions being for the epoch 1965.0.

It will be seen from Fig. 2.1 that the lines of force are tangential to the Earth's surface, or horizontal with respect to the observer, only in the neighbourhood of the equator. Here a needle suspended at its centre of gravity will rest horizontally— red pole to the north. The irregular line encircling the Earth where this condition obtains is sometimes called the magnetic equator. At other places on the Earth's surface a similarly suspended needle will be inclined to the horizontal, as the lines of force are no longer tangential to the Earth's surface. The needle will therefore lie along the direction of the lines of force: red pole down in the northern magnetic hemisphere, blue pole down in the southern magnetic hemisphere. The discovery of this phenomenon, known as magnetic 'dip' or inclination, is variously attributed to Georg Hartmann of Nuremberg in 1544 and to Robert Norman, a London instrument-maker, in 1576.

The direction of the Earth's total magnetic field, as defined by a freely suspended needle, may be resolved into two components at right angles, one tangential to the Earth's surface, the other normal to it (Fig. 2.2). The horizontal component is of interest to the mariner in directing his compass needle to the north, while the vertical component is usually an embarrassment and of considerable inconvenience. The vertical plane containing the observer and the direction of the Earth's magnetic field is known as the magnetic meridian at the observer's position.

A magnetic field has not only direction but also intensity or strength. This is measured in c.g.s. units known as oersteds. A magnetic field of unit intensity is that field in which a free magnetic pole of unit strength will experience a force of one dyne acting on it. Thus a field of 10 units would exert a force of 10 dynes on the free unit pole. Further, it is taken as a useful convention that a magnetic field of unit intensity can be represented by one arbitrary line of force per square centimetre, so the field of 10 units will have ten lines per square centimetre. The Earth's total field strength at London is about 0·47 oersted. In this field a free magnetic pole would experience a force of 0·47 dyne and there would be one line of force in approximately 2 sq. cm.

Having assigned both direction and intensity to the Earth's total magnetic field, it is possible to appreciate in more detail the significance of the horizontal and vertical components. The inclination or dip of the lines of force with respect to the horizontal plane varies from zero at the so-called magnetic equator to 90° at the magnetic poles. Hence at the magnetic equator the horizontal component is at a maximum value and the vertical component is zero, whereas at the magnetic poles the force is entirely vertical with no horizontal component. In between, the horizontal and vertical components are determined by the local inclination of the lines of total force. They are related thus:

$$T = (H^2 + Z^2)^{1/2}$$

$$Z/H = \tan D$$

where H is the horizontal component, Z the vertical component, T the total force, and D the dip or inclination of the total field to the horizontal. In London, $H = 0·188$ oersted, $Z = 0·436$ oersted, $T = 0·474$ oersted, and $D = 66° 40'$, so that the useful part of the Earth's field for directing a magnetic needle or compass, namely H, is a comparatively small proportion of the total field. If the needle were freely supported at its centre of gravity it would rest inclined to the horizontal at an angle of 66° 40', as will be seen from Fig. 2.2.

The directive force of the Earth's magnetic field, compelling the compass needle to point north, is greatest, therefore, at the magnetic equator, falls off as the observer moves north or south, and is zero at the magnetic poles. The vertical force, which can be such an inconvenient factor in marine compass design and installation, is greatest at the magnetic poles, decreasing to zero at the magnetic equator.

As the early navigators discovered, a freely suspended magnetized needle does not point, except in certain parts of the Earth, to the true or geographical north pole. Moreover, the departure of the direction of the magnetic needle from true north can be easterly or westerly and differs according to the observer's position on the Earth's surface. This departure is known as variation (less frequently as declination).

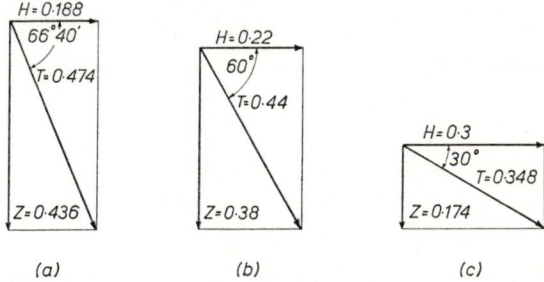

FIG. 2.2. Horizontal and vertical components of the Earth's magnetic field

H, Z, and T are respectively the horizontal component, the vertical component, and the total force, all expressed in oersted units. The field is shown for (a) London, (b) Northern Spain, and (c) the Sahara Desert.

Places where the variation is zero, i.e. where the magnetic needle points to true north, lie along what is known as the *agonic line*. Between a true or geographic pole and its neighbouring magnetic pole, a line may be drawn through places where the variation is 180°; on crossing this line, the variation changes from easterly to westerly (or vice versa).

Not only does the value of variation differ according to position on the Earth's surface, but it is also subject to certain more or less periodic changes. There is the daily change, discovered in 1722 by a London instrument-maker called Graham. This exhibits, at the present time, a range of about 15 minutes of arc during the summer in the neighbourhood of London, between morning and afternoon values of variation, the turning points being about 08 00 hrs and 13 30 hrs local time.* The afternoon value is more westerly than that of the morning. In equatorial regions the change may be very small, whereas in places near the magnetic poles it may amount to many degrees.† Abnormal amounts of daily change in any given locality are usually evidence of magnetic storms and other phenomena associated with sunspot activity.

* In the winter the daily change is smaller and may be only 4 minutes of arc.
† The values of the horizontal and vertical components of the Earth's field and also the angle of dip undergo a daily change similar to that shown by variation.

There is also the annual change, occurring simultaneously but in opposite directions in the northern and southern hemispheres, and having a range in this country of about $2\frac{1}{4}$ minutes of arc. The most easterly turning point occurs in August and the most westerly in February.

Finally, there is the secular change (also manifest in other elements of the Earth's magnetic field); at the present time, the value of variation at London is becoming more easterly at a rate of about 8 minutes of arc per annum. The secular change, in so far as available data show, appears to be cyclic, the variation at London having a maximum easterly value in 1580, becoming zero in 1700, and having a maximum westerly value in 1800. The prevailing value of variation in London is approximately $7\frac{1}{4}°$ west, so that, if the above rate of change continues, the magnetic needle there will point to almost true north in about 54 years' time. After that, variation will become increasingly easterly and, on once more reaching a maximum, will then tend to decrease in value, passing through zero and becoming westerly after a lapse of several centuries. However, the rate of change is not necessarily constant over a long period.

The following values of variation were obtained from maps of Ditton Park, near Datchet, Buckinghamshire.

A.D. 1607	10° east
1718	6° west
1742	14° west
1834	$23\frac{1}{2}°$ west
1960	$8\frac{1}{2}°$ west

When plotted on the terrestrial globe, the lines of equal variation have not at all the symmetrical pattern that might be expected from a development of Fig. 2.1. In fact the pattern of such lines in some areas is irregular in the extreme. It is true that one agonic line traverses the American continent in a fairly uniform way, but its counterpart in the eastern hemisphere is distorted by a great magnetic anomaly or irregularity in Siberia, equivalent almost to a subsidiary magnetic pole. Another large but less imposing anomaly is to be found in the vicinity of South Africa.

Besides the more or less regular changes of the Earth's magnetic components, sudden, sometimes unpredictable changes take place. For instance, variation may alter in the London area by as much as ten or fifteen minutes of arc in a quarter of an hour, and in extreme cases, may suddenly change by as much as one degree.

These irregular disturbances are known as magnetic storms and their study and forecasting are carried out systematically by both magnetic and ionospheric observatories. The close relationship between magnetic storms and radio fade-out and allied disturbances has led to the conclusion that the mechanism causing both lies in the ionosphere, and lends support to the theory that the Earth's magnetic field is affected by electric currents circulating within and without the terrestrial sphere. Moreover, the recurrence of magnetic storms at 27-day intervals led the great English magnetician, C. Chree,[*] to establish the connection of these outbreaks with

* Chree, C. 'An inquiry into the nature of the relationship between sunspot frequency and terrestrial magnetism,' *Phil. Trans. roy. Soc.*, 1904, **203**A, 151–187; 'The 27-day period in magnetic phenomena,' *Proc. roy. Soc.*, 1914, **90**A, 583–599.
  Chree, C., and Stagg, J. M., 'Recurrence phenomena in terrestrial magnetism,' *Phil. Trans. roy. Soc.*, 1927, **227**A, 21–62.

solar activity. The apparent rotational period of the sun being 27 days, Chree had observed the coincidence of sunspots and magnetic storms and their common reappearance after that period as the sunspot once more faced the Earth. Solar storms are now accepted as the prime cause of magnetic and radio disturbances, auroral displays and similar ionospheric and upper atmospheric effects. Sunspot numbers, density of ionization of the ionospheric layers, critical frequency of radio waves, solar flares and the like are now regularly recorded, classified and analysed.

Chree's 27-day recurrence period is not always strictly followed. It seems that small and average storms may recur at 27, 54, 81 . . . days fairly systematically, but great storms do not usually recur, and it may be that there is something in the mechanism behind a great storm that causes it to differ from the lesser outbreaks.

A magnetic storm usually occurs about two days after the central meridian transit of a sunspot and may be accompanied by a world-wide radio disturbance. The delay is accounted for by the time taken by electrified particles ejected from the sunspots to reach the Earth. They then become influenced by the magnetic field of the Earth and cause magnetic storms and, on entering the ionosphere, produce auroral displays and radio interference. Another form of solar activity, frequently associated with sunspots, is the solar flare which may cause a simultaneous temporary 'radio black-out' on the sunlit side of the Earth. At the same time, a transient magnetic effect, known as a crotchet, may occur, followed by a storm two days later. The flare emits ultra-violet radiation which travels towards the Earth at the speed of light. The 11-year cycle of sunspot activity is also reflected in the occurrence of magnetic disturbances.

The components of the Earth's magnetic field may conveniently be represented by Fig. 2.3, showing three orthogonal axes in X, Y and Z; the X axis is directed towards true north, the Y axis towards the east, and the Z axis in the gravitational vertical. H is the horizontal component of the Earth's magnetic field and lies in the plane XY, making an angle V with X. The total force, T, is inclined to the horizontal at an angle D and is in the same vertical plane as H. The vertical magnetic component, Z, lies along the Z axis.

Thus

$$X = H \cos V \quad \text{(the northerly horizontal vector)}$$
$$Y = H \sin V \quad \text{(the easterly horizontal vector)}$$
$$Z = Z$$
$$\tan V = \frac{Y}{X}$$
$$H^2 = X^2 + Y^2 \quad \text{or} \quad H = (X^2 + Y^2)^{1/2}$$
$$\tan D = \frac{Z}{H}$$
$$T^2 = X^2 + Y^2 + Z^2 \quad \text{or} \quad T = (X^2 + Y^2 + Z^2)^{1/2}$$

Consequently, the relationship between the geographical axes of reference and the magnetic components is completely defined.

The elements of the Earth's magnetic field of immediate interest to the mariner and compass-maker are measured, recorded and displayed as maps of lines of equal variation (isogonals, Fig. 2.6), lines of equal horizontal force, lines of equal vertical force lines, of equal total force (isodynamic lines, Fig. 2.7), and lines of equal angle of dip (isoclines, Fig. 2.8). For a given locality, one requires to know horizontal force H and angle of dip, and hence vertical force, Z, and total force, T, or vice versa, and variation. The first isodynamic chart was published by Hansteed in the nineteenth century.

The earliest variation charts are said to have been made by Alonzo de Santa Cruz about 1536, but Edmond Halley* produced in 1701 the first world variation chart of the form used today, illustrated in Plate 2.1. Since then, others have been published by Dodson (1746), Mountaine (1756), Churchman and Dunn (late eighteenth century), Yeates (1817), Barlow (1833) and Evans (1858). The earliest of these

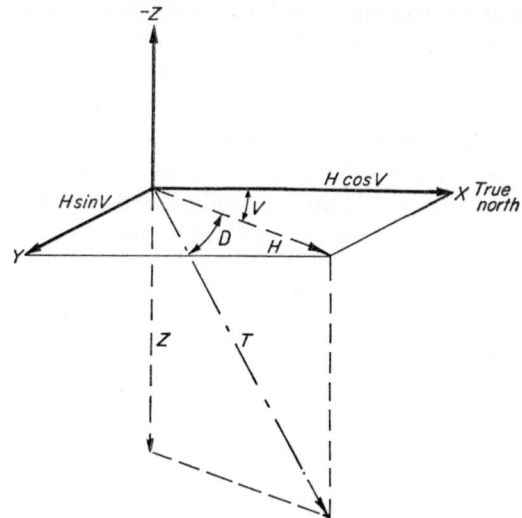

FIG. 2.3. Components of the Earth's magnetic field

V is the magnetic variation or declination. D is the dip or inclination. H, Z, and T, which have the same significance as in Fig. 2.2, are shown in a co-ordinate system in which the X axis is directed towards true north and the Y axis towards the east point; the Z axis is vertical, with its positive direction downwards.

were primarily intended for determining longitude from variation; nowadays, however, variation charts are aids to navigation, whereby magnetic north can be converted to true north, or the deviations of the compass obtained by subtracting variation from observed total compass error. Modern variation charts are published in this country by the Admiralty.

Magnetic data are systematically gathered by magnetic surveys carried out wherever possible over the whole surface of the Earth by land, sea and air. Airborne surveys, using modern techniques such as the magnetic anomaly detector (M.A.D.) make the collection of data comparatively rapid over land and sea. Fixed magnetic observatories on land carry out continual measurements of the strength and direction of the Earth's magnetic field. In this country magnetic observatories are

* Bauer, L. A., 'Halley's Earliest Variation Chart', *Terr. Magn. atmos. Elect.*, 1896, **1**, 28–31.

operated by the Royal Greenwich Observatory at Hartland in Devon and by the Science Research Council at Lerwick and Eskdalemuir. In the United States of America, the U.S. Coast and Geodetic Survey operates a series of stations covering the continent; other nations have similar establishments, so that world-wide magnetic information is readily obtainable. Two examples of magnetograms, or magnetic

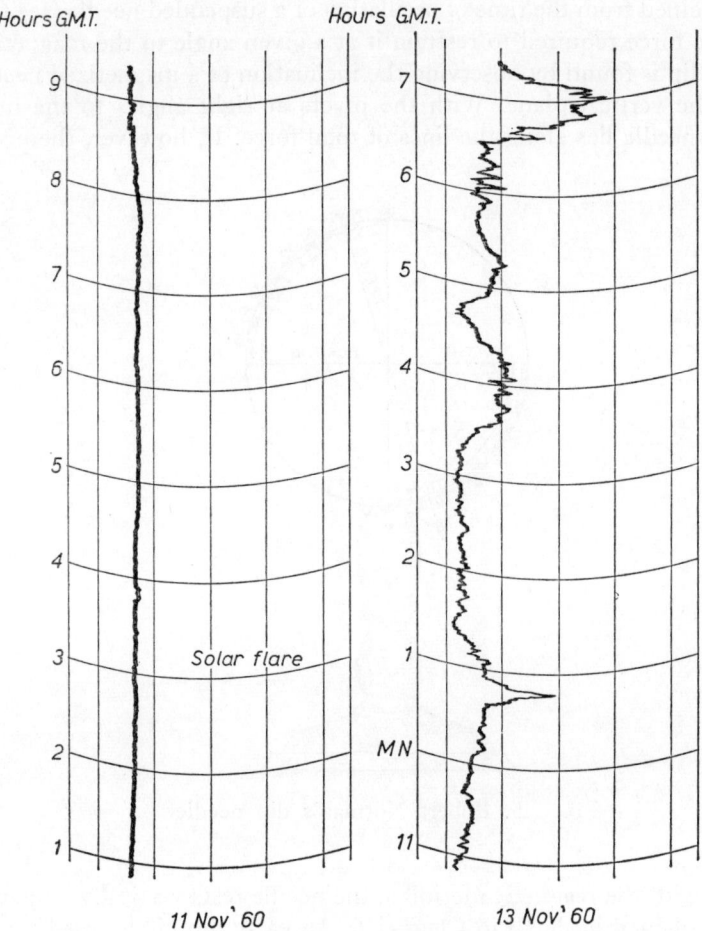

FIG. 2.4. Magnetograms

The left-hand recording was made on a magnetically quiet day, during which a solar flare occurred. The right-hand recording, made approximately two days later, shows the development of a magnetic storm as charged particles from the sun reached the Earth.

records, are shown in Fig. 2.4. Surveys at sea are carried out in special non-magnetic ships, fitted with marine versions of the land instruments for magnetic measurements, and routine determinations of variation are made by H.M. ships. The exploration of the Earth's magnetic field has now been extended beyond the immediate neighbourhood of the terrestrial sphere by measuring devices carried in space rockets and satellites.

Instruments for measuring the Earth's magnetic field are commonly known as magnetometers or variometers; various types of them are described in Chapter 6. Magnetometers are frequently adaptations of a freely suspended magnetic needle. Variation is found by comparing the direction taken up by such a needle in the horizontal plane and comparing it with that of true north, found by astronomical observations or from known survey marks. The strength of the Earth's field may be determined from the time of oscillation of a suspended needle (see Chapter 3) or from the force required to restrain it at a given angle to the magnetic field. The angle of dip is found by observing the inclination of a magnetized needle free to swing in the vertical plane. With the pivots at right angles to the magnetic meridian, the needle lies along the lines of total force. If, however, the pivots are

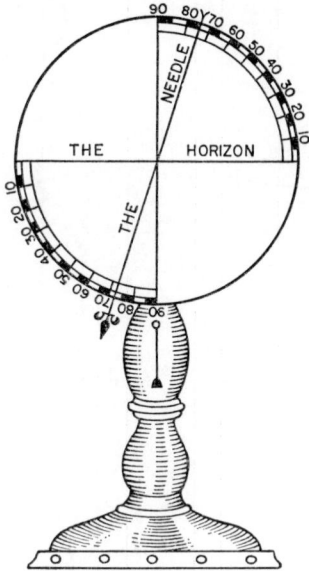

FIG. 2.5. Robert Norman's dip needle

placed in line with the magnetic meridian, the needle rests vertically. The use of a *dip circle* is explained in detail in Chapter 6. An early dip circle, used by Robert Norman in 1576, is shown in Fig. 2.5.

The mariner's compass is a particular form of magnetometer. Its prime use is to determine the direction of the Earth's horizontal magnetic field by means of a magnetic needle free to swing in the horizontal plane, thus indicating magnetic north. Given the prevailing value of variation, true north can be found. Conversely, a mariner's compass may be used to determine variation.

In addition to magnetometers of the suspended or pivoted-needle type, new instruments are being widely used in fixed and mobile observatories employing recently discovered techniques in the field of electromagnetic induction and electronics. These are described in Chapters 6 and 11. Many of them can be readily adapted for continuous recording of magnetic phenomena.

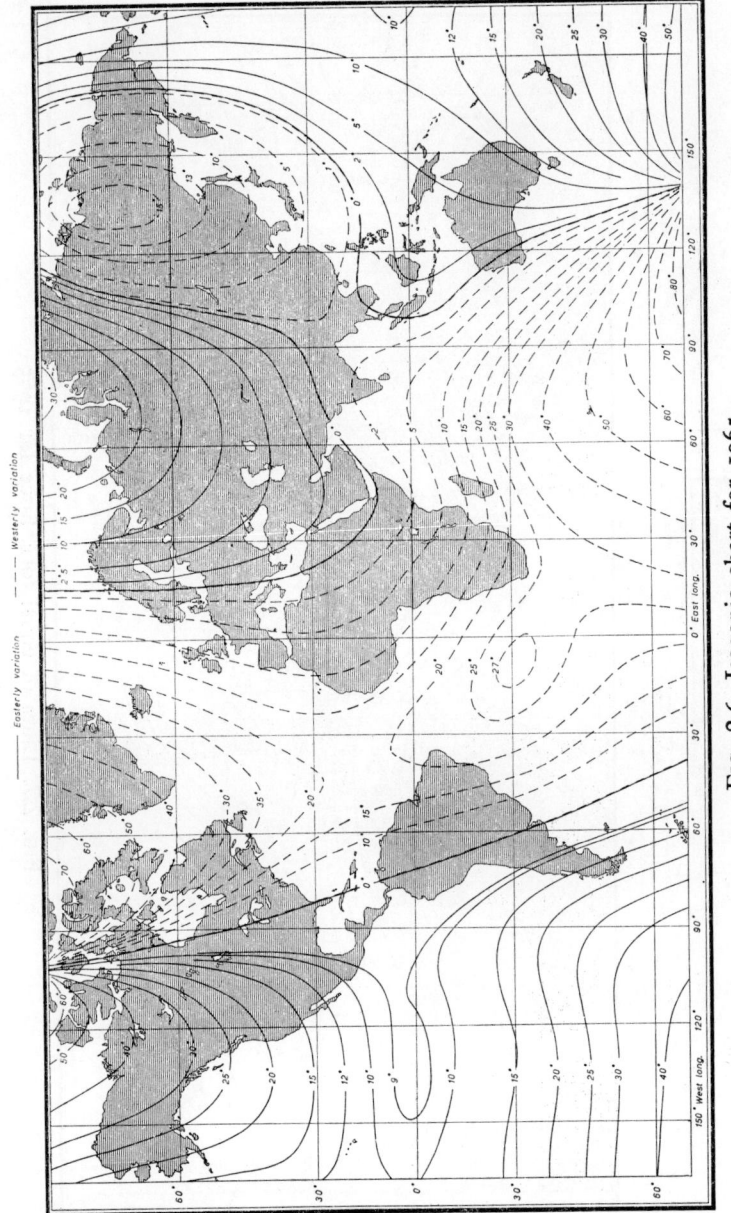

Fig. 2.6. Isogonic chart for 1965

The isogonal lines join places of equal magnetic variation.

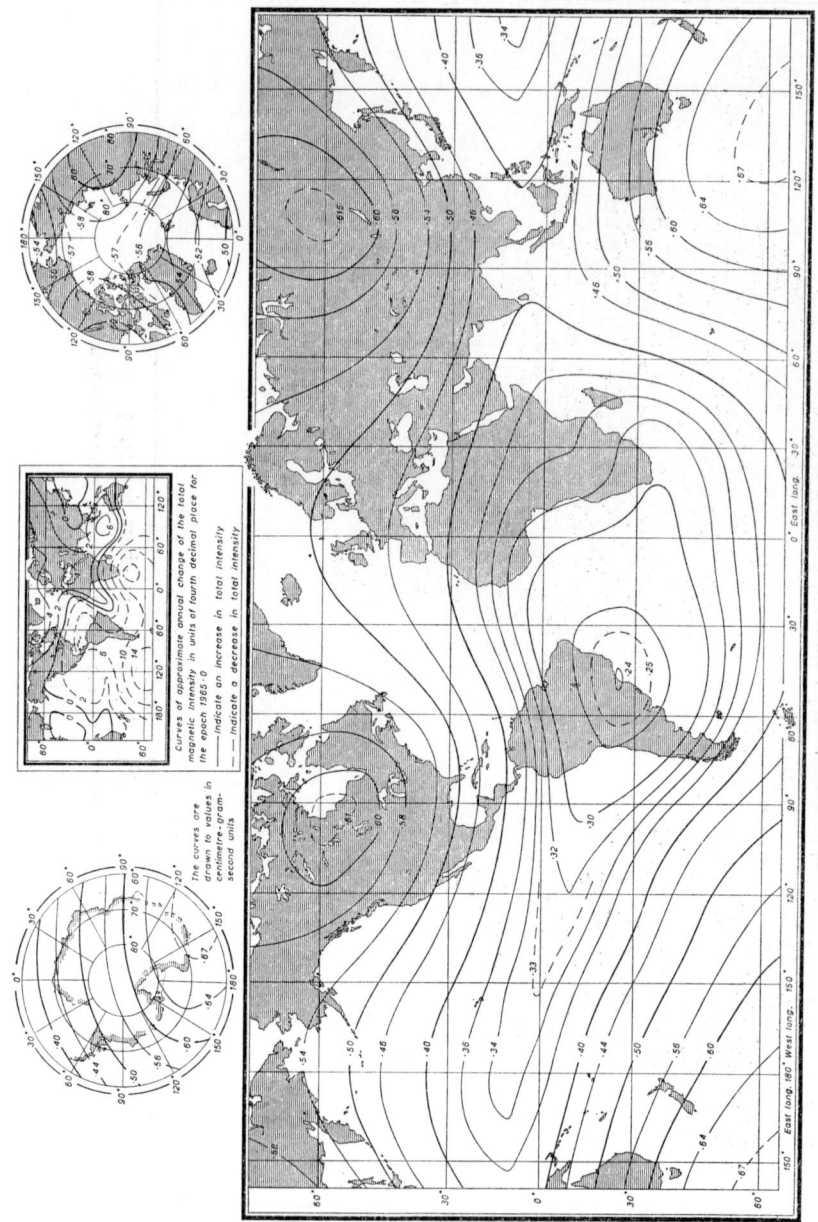

FIG. 2.7. Lines of equal total force for 1965

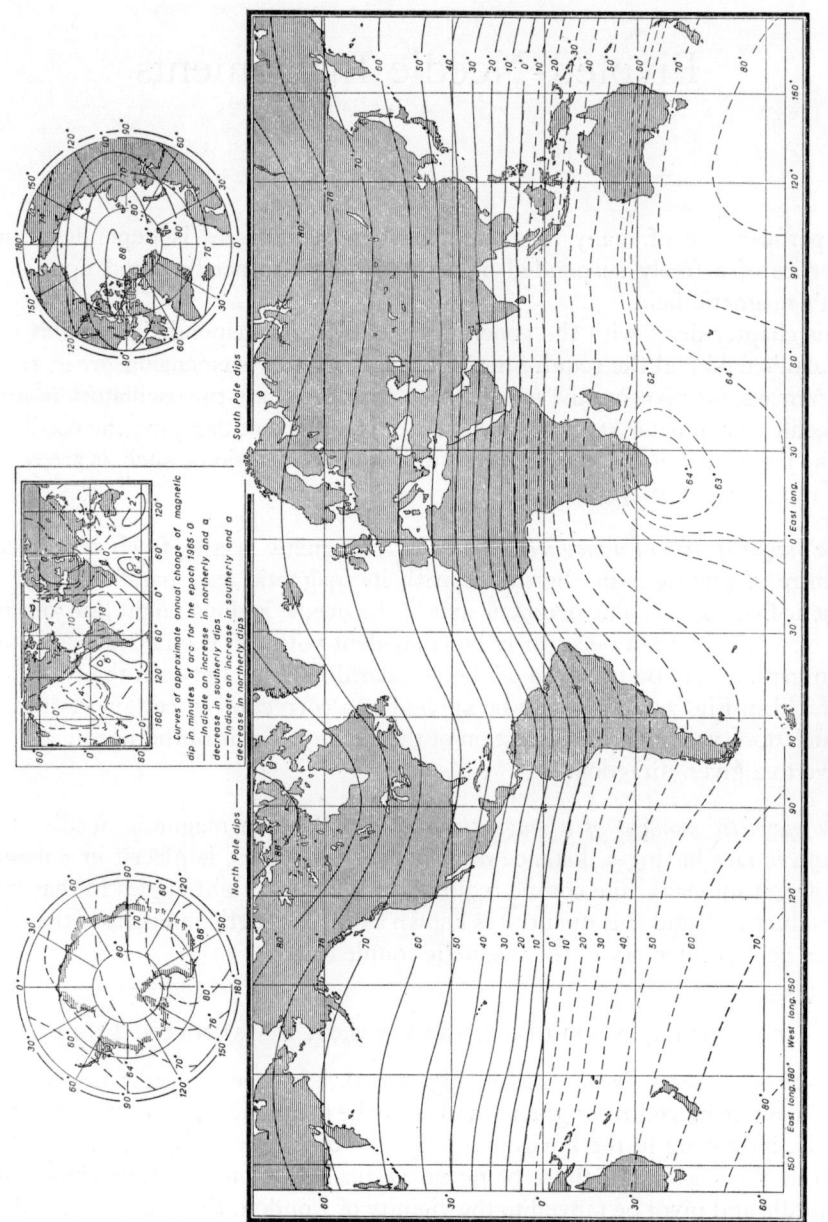

FIG. 2.8. Lines of equal dip for 1965

19

Chapter 3

Pivoted-Needle Instruments

The performance of many magnetometers and compasses is dependent on the properties of a freely suspended or pivoted magnetic needle acted upon by the Earth's magnetic field.

This chapter deals with the general case of a freely suspended needle at rest in the Earth's field and the requirements for a satisfactory suspension or, in the case of a compass, for pivoting the needle. Also considered are the oscillatory motion of the needle or magnet system when disturbed, methods of damping the oscillations, and the behaviour of the needle when subject to accelerations, such as are encountered in ships and aircraft.

The magnetic axis of a magnetized needle. A magnetic needle, freely suspended at its centre of gravity, will align itself with its magnetic axis parallel to the total magnetic field vector. The magnetic axis of the needle is the straight line joining its two poles and may not necessarily be coincident with the geometric axis. Thus, a determination may be involved of the angular difference between the two axes, as illustrated in Fig. 3.1. This may be applied as a correction in certain instruments, in which observations of the direction of the geometric axis of the needle are made relative to a given direction.

The magnetic moment of a magnetized needle. When a magnetic needle of pole strength m and having a distance of $2l$ between its poles, is placed in a magnetic field of unit intensity, the couple required to hold it at right angles to that field is the product $2ml$, and this quantity is known as the magnetic moment of the needle, M. In a field of intensity F, the magnetic couple is MF.

THE PIVOTED NEEDLE IN THE EARTH'S MAGNETIC FIELD

Single-pivoted needle. Now if, as in Fig. 3.2(*a*), a magnetic needle is poised on a single pivot at its centre of gravity and is at the magnetic equator, it will rest horizontally, indicating in the horizontal plane the direction of the Earth's total field, which in this case is H, there being no vertical component. If, as shown at (*b*), this needle and pivot be moved to the vicinity of London, the needle will still set in the direction of the total magnetic field and will therefore rest at an angle of 66° 40′ to the horizontal, the red pole inclined downwards. In a corresponding locality in the southern magnetic hemisphere, the needle will be inclined with the blue pole downwards.

Thus a needle, simply pivoted or suspended at its centre of gravity, has limitations as a magnetometer and more so as a compass. Controlling forces have now to be introduced so that, for example, the direction of the horizontal field may be clearly indicated. This may be achieved as at (c) by so placing a balance weight of mass w that the couple wgx is equal and opposite to the magnetic couple MZ. The needle will rest horizontally, but only for a given value of Z, and this arrangement is usual only for instruments which are kept set up at a given locality. Some types of compass directors are provided with a sliding weight on the needle which is adjusted according to the local value of Z.

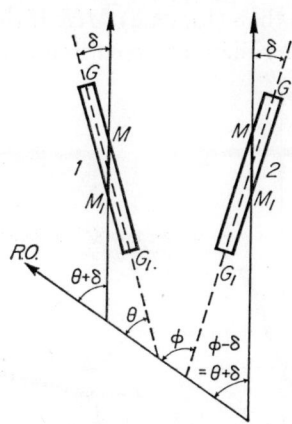

FIG. 3.1. The magnetic and geometric axes of a magnet

GG_1 is the geometric axis, MM_1 the magnetic axis, and δ the (much exaggerated) angle between them. R.O. goes to a reference object. The angle between the geometric axis and the direction of the reference object is θ in one position of the magnet and ϕ after the magnet has been turned over (i.e. rotated through 180° about its geometric axis). The angle between the magnetic meridian and the direction of the reference object is $(\theta+\phi)/2$. The angle δ is $(\phi-\theta)/2$.

A more elegant arrangement of balancing, and one which is almost universally adopted in magnetic compasses, is to arrange the needle system so that its centre of gravity is below the pivot point, i.e. the system is pendulous (d). The needle will be horizontal at the magnetic equator since, in the absence of a vertical magnetic force, the centre of gravity will lie vertically below the point of suspension. Elsewhere, it will rest with a small tilt β in order to give effect to a pendulous couple that will balance the magnetic couple. Therefore

$$MH \sin \beta + wgR \sin \beta = MZ \cos \beta$$

and the constants can be adjusted so that for any terrestrial situation, the angle of tilt can be kept to an acceptably small value, enabling a compass needle to be used in all parts of the world without recourse to rebalancing.

For small values of β

$$wgR\beta = MZ - MH\beta$$

In order to avoid extra weight on the pivot other than that imposed by the magnetic needle, the needle itself may be used as a balance weight with the pivot point situated above it as at (e). This is the common arrangement in magnetic compasses.

Double-pivoted needle. Alternatively, the magnetic needle could equally well be supported on a vertical axle with jewelled bearings at each end as shown in Fig. 3.3. This is a practicable system to a certain extent, but in addition to the weight of the system on the bottom jewel, there is a side thrust on the upper bearing which will impose an additional frictional load, and which is not experienced with a single pivot.

Friction at the pivot or in the suspension

The directive force of the Earth's magnetic field on a magnetometer or compass needle is extremely small even when the needle is deflected through 90°. The couple acting on the needle is then (Fig. 3.4) MH. If the needle is deflected by only a small amount θ from the meridian, the resulting couple is $MH \sin \theta$.

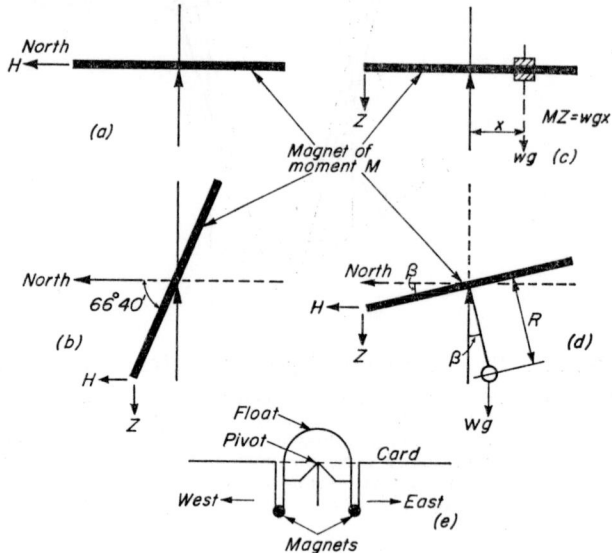

Fig. 3.2. Single-pivoted magnetic needle

Thus if the friction is excessive there will be a limit to the precision with which the needle will return to the meridian after being deflected. If τ is the frictional torque, the limiting condition by which the meridian is defined is given by

$$\tau = MH\delta$$

where δ is the deflection when friction balances the magnetic restoring couple, 2δ being the included angle about the meridian which commonly defines the friction error of a pivoted needle.

It will be seen that a great deal depends on the quality of the pivot and jewel; a cheap pocket compass with its elementary pivot cannot compare with an expensive prismatic compass.

Theory of the pivoted needle

Certain fundamental equations may be derived to describe the behaviour of a magnetic needle, freely moving or acted upon by certain controlling forces in the Earth's magnetic field.

22

In Fig. 3.5, X', Y' and Z' are three orthogonal components of the Earth's magnetic field, OX' defining the direction of true north. The horizontal plane in which X' and Y' lie is $CX'BY'D$. FBE is any plane tilted about OB where the angle COF is the angle of tilt, β. The angle $X'OB$ (α) is the angle measured to the intersection B of the tilted plane from true north, X', and defines the direction of the tilt axis. A positive tilt angle is that which arises from a clockwise rotation of the plane FBE about OB when viewed along the direction of the tilt axis. Conversely,

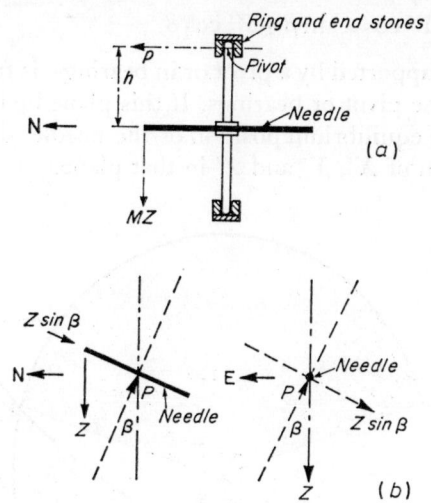

FIG. 3.3. Double-pivoted magnetic needle

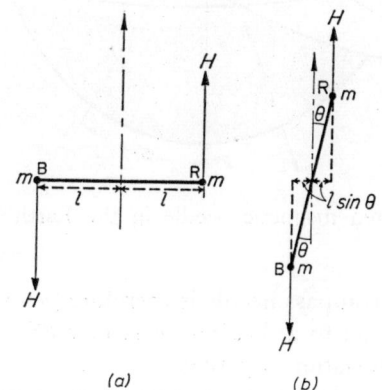

FIG. 3.4. The couple acting on a pivoted magnetic needle

a counter-clockwise rotation gives rise to a negative angle. In the diagram, the direction of the tilt axis makes β positive. The axis OB is always horizontal. DEZ'CF is the vertical plane through O normal to OB. In the tilted plane lies a magnetic needle represented by ON, which lies in the x axis, x, y and z being three orthogonal components related to the needle. The angle BON (η) defines the direction of compass north measured from the intersection B of the tilted plane and the horizontal plane.

We may therefore write

$$x = X'(\cos \alpha \cos \eta + \sin \alpha \sin \eta \cos \beta) + Y'(\sin \alpha \cos \eta - \cos \alpha \sin \eta \cos \beta)$$
$$- Z'(\sin \eta \sin \beta) \tag{3.1}$$

$$y = X'(\cos \alpha \sin \eta - \sin \alpha \cos \eta \cos \beta) + Y'(\sin \alpha \sin \eta + \cos \alpha \cos \eta \cos \beta)$$
$$+ Z'(\cos \eta \sin \beta) \tag{3.2}$$

$$z = X'\sin \alpha \sin \beta - Y' \cos \alpha \sin \beta + Z' \cos \beta \tag{3.3}$$

A magnetic needle, supported by a pivot or in bearings, is free to rotate in a plane normal to the axis of the pivot or bearings. If this plane be tilted and represented by FBE in Fig. 3.5, the equilibrium position of the needle, defined by η, is determined by the resolution of X', Y' and Z' in that plane.

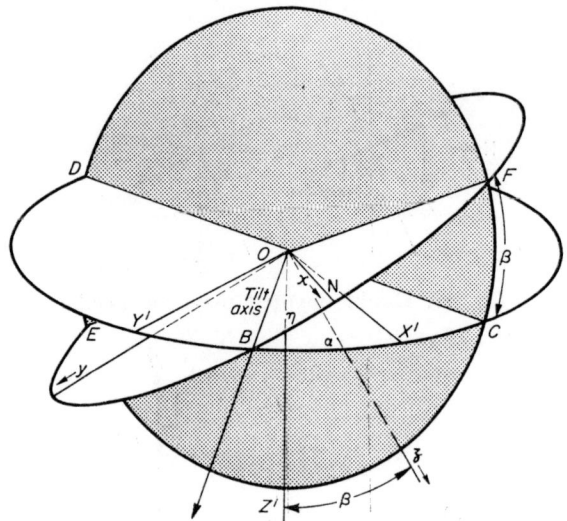

FIG. 3.5. A pivoted magnetic needle in the Earth's magnetic field

The deviation δ of the compass needle is therefore $(\alpha - \eta)$. An easterly deviation (i.e. when the needle points to a direction east of magnetic north) is considered positive and a westerly deviation negative.

If the coordinate system is rotated so that α becomes the magnetic azimuth instead of the true azimuth,

$$X' = H, \quad Y' = 0, \quad Z' = Z$$

The component of H along OB is $H \cos \alpha$. Since Z acts normal to OB, it has no component along it.

The component along OF is $H \sin \alpha \cos \beta - Z \sin \beta$. For equilibrium, it is necessary that the field along Oy be zero; thus

$$H \cos \alpha \sin \eta = (H \sin \alpha \cos \beta - Z \sin \beta) \cos \eta$$

Therefore

$$\tan \eta = \frac{H \sin \alpha \cos \beta - Z \sin \beta}{H \cos \alpha} \qquad (3.4)$$

i.e., equation (3.2) takes the form

$$H(\cos \alpha \sin \eta - \sin \alpha \cos \eta \cos \beta) + Z \cos \eta \sin \beta = 0$$

Let δ be the deviation of the needle from magnetic north. Then

$$\delta = \alpha - \eta$$

and

$$\tan \delta = \frac{\tan \alpha - \tan \eta}{1 + \tan \alpha \tan \eta}$$

$$= \frac{H \dfrac{\sin 2\alpha}{2}(1 - \cos \beta) + Z \sin \beta \cos \alpha}{H\left[1 - \dfrac{1 - \cos 2\alpha}{2}(1 - \cos \beta)\right] - Z \sin \beta \sin \alpha} \qquad (3.5)$$

Equation (3.5) may be expressed as a Fourier series of the form

$$\sin \delta = \bar{C} \cos \eta + \bar{D} \sin(2\eta + \delta)$$

where

$$\bar{C} = \frac{2 \tan D \sin \beta}{1 + \cos \beta}$$

and

$$\bar{D} = \frac{1 - \cos \beta}{1 + \cos \beta}$$

For small values of β, write $\cos \beta = 1 - \tfrac{1}{2}\beta^2$, $\sin \beta = \beta$ and $(1 - \cos \beta) = \tfrac{1}{2}\beta^2$. Then

$$\frac{H \dfrac{\beta^2 \sin 2\alpha}{4} + Z\beta \cos \alpha}{H\left(1 - \beta^2 \dfrac{1 - \cos 2\alpha}{4}\right) - Z\beta \sin \alpha} \qquad (3.6)$$

Neglecting terms in β^2 in equation (3.6)

$$\tan \delta = \frac{Z\beta \cos \alpha}{H - Z\beta \sin \alpha} \qquad (3.7)$$

or

$$\tan \delta = \frac{\beta \tan D \cos \alpha}{1 - \beta \tan D \sin \alpha} \qquad (3.8)$$

The positive sign indicates an easterly deviation of the needle.

25

B

Equation (3.8) is a useful and important expression and appears frequently in the analysis of compass deviations due to tilt of a double-pivoted magnet system and to accelerations acting on a pendulous system.

EFFECT OF TILT ON A PIVOTED MAGNETIC NEEDLE

A single pivot behaves as a universal joint and a magnetic needle balanced and poised on it remains in the horizontal plane unless an acceleration other than gravity acts on it. Therefore tilting of the mounting does not affect the attitude of the needle and causes no deviation.

A pendulous system behaves similarly although it is horizontal only at the magnetic equator. In other localities there is a small inherent tilt of the needle due to the vertical component of the Earth's field, being always in the north-south plane, but this causes no deviation.

In a double-pivoted system, the needle no longer rests in the horizontal plane when its mounting is tilted [see Fig. 3.3(b)], and the resulting deviation is derived from equation (3.5), or from (3.6) and (3.7) if the tilt is small.

If the tilt is in the north-south plane,

$$\alpha = \frac{\pi}{2} \quad \text{or} \quad \frac{3\pi}{2}$$

so

$$\delta = 0$$

If, however, the tilt is in the east-west plane,

$$\alpha = 0 \quad \text{or} \quad \pi$$

and

$$\tan \delta = \pm \sin \beta \tan D$$

Consequently, a double-pivoted system in the vicinity of London suffers a deviation of $2°\cdot32$ for one degree of east-west tilt and, as the system is moved nearer the magnetic pole, so the deviation increases. In the vicinity of Hudson's Bay, for example, the deviation for one degree of east-west tilt is of the order of $9°$.

From equation (3.5), it will be seen that, unless the angle of tilt is small, there is a deviation that is independent of Z, so that at the magnetic equator, i.e., when $Z = 0$,

$$\sin \delta = \frac{1 - \cos \beta}{1 + \cos \beta} \sin (2\eta + \delta)$$

The dip circle is an example of a double-pivoted needle with the pivot axle horizontal, so that

$$\alpha = 0 \quad \text{and} \quad \beta = \pi/2$$

From equation (3.4)

$$\tan \eta = -Z/H$$

Since $\tan D = Z/H$

$$-\eta = D$$

or

$$\delta = D$$

Limiting value of tilt angle: the unstable condition

The angle of tilt β is called the critical angle when the plane of rotation of the compass needle or magnet system is normal to the total field vector, at which time $\beta = \pi/2 - D$. The needle then loses all directional control and, at this angle and beyond, becomes unstable and its deviations largely unpredictable. For small values of tilt angle, the critical state occurs in regions where the angle of dip is large.

EFFECT OF GRAVITATIONAL FORCES

If the effect of gravity* be taken into account, and if a and b are the distances along the x and y axes respectively of the centre of gravity of the magnetic needle from the centre of rotation, additional couples are applied.

The magnetic couple due to OB is $MH \cos \alpha \sin \eta$ and that due to OF is $MH \sin \alpha \cos \beta \cos \eta - MZ \sin \beta \cos \eta$.

The gravitational couple is $wg \sin \beta (a \cos \eta + b \sin \eta)$. Therefore

$$MH \sin \alpha \cos \beta \cos \eta - MZ \sin \beta \cos \eta - MH \cos \alpha \sin \eta$$
$$= wg \sin \beta (a \cos \eta + b \sin \eta)$$

so

$$\tan \eta = \frac{MH \sin \alpha \cos \beta - MZ \sin \beta - wga \sin \beta}{MH \cos \alpha + wgb \sin \beta} \tag{3.9}$$

When $b = 0$, as in a suspension magnetometer or single-pivoted compass,

$$\tan \eta = \tan \alpha \cos \beta - \frac{Z \sin \beta}{H \cos \alpha} - \frac{wga \sin \beta}{MH \cos \alpha} \tag{3.10}$$

and from this it may be shown that when the tilt is in the north-south plane, as in a pendulous magnet system, no deviation occurs.

If, as in a dip circle, $\alpha = 0$ and $\beta = \pi/2$,

$$\tan \eta = - \left(\frac{Z}{H} + \frac{wga}{MH} \right)$$

$$= -\frac{1}{H} \left(Z + \frac{wga}{M} \right) \tag{3.11}$$

UNIFILAR SUSPENSION

A unifilar suspension of unspun silk, quartz fibre or tungsten wire takes the place of pivots and jewels in various precision instruments, such as the Kew magnetometer. The upper end of the suspension is held in a torsion head and the lower end is attached to the magnetic needle or other form of magnet. The torsion head is adjusted to take the twist out of the wire or thread, so that there is no torsional couple tending to turn the needle out of the magnetic meridian. For small deflections, a unifilar suspension may be regarded as frictionless.

* Fig. 3.5 and equations (3.1) to (3.11) describe the behaviour of a magnetic needle in the northern magnetic hemisphere. In the southern magnetic hemisphere, Z acts upwards, i.e. its sign is reversed, with consequent alterations to the sign of the deviations of the magnetic needle.

The needle or magnet is attached to the suspension at a point above its centre of gravity and thus is pendulous. If the pendulous couple is great enough, it will tend to rest near the horizontal, like the single-pivoted compass needle (see Fig. 3.6). Any extra couple to bring the magnet horizontal in the northern magnetic hemisphere can be applied by the addition of a small balance weight towards the south-seeking end (blue pole) of the magnet or by moving the point of attachment towards the north-seeking end (red pole), and vice versa for the southern magnetic hemisphere.

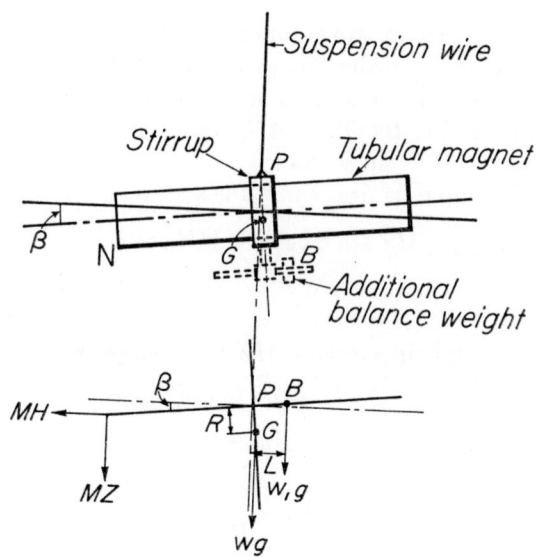

FIG. 3.6. Unifilar magnetometer suspension

The effects of torsion in the wire and of a small twist are illustrated in Figs. 3.7 and 3.8.

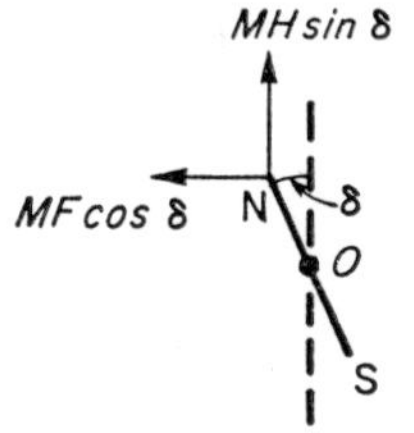

FIG. 3.7. A needle deflected by torsion in its unifilar suspension

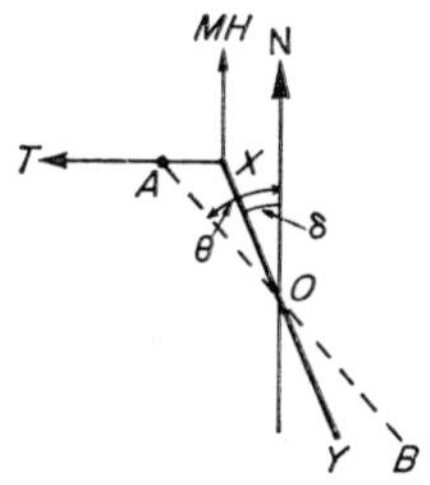

FIG. 3.8. The effect of a small amount of twist in a unifilar suspension

In Fig. 3.7, let a magnetized needle be suspended at O; let the small angle δ be the deflection due to a field F; let M be the moment of the magnet and H the Earth's field. The couples present are due to F, H, and the twist in the suspension due to the deflection angle δ. Thus

$$MF \cos \delta = MH \sin \delta + \tau$$

Now, for small angles, where C is the modulus of rigidity of the material of the

28

wire, L the length of the wire, and d the diameter of the wire,

$$\tau = \frac{\pi d^4}{32} \cdot \frac{\delta}{L} \cdot C$$

Thus

$$MF = MH\delta + \frac{\pi d^4}{32} \cdot \frac{\delta}{L} \cdot C$$

or since $\pi d^4 / 32 = I$, the moment of inertia of the wire,

$$\delta = \frac{MF}{MH + \dfrac{I}{L} \cdot C}$$

In Fig. 3.8 let ON be the direction of the magnetic meridian, AOB the direction of the magnet before magnetization (i.e. the free position of the suspension), XOY the direction of the magnet acted upon by the Earth's field. The angle OAN is the twist θ in the suspension, the angle OXN is the deviation δ of the magnet from the meridian owing to the twist, M is the magnetic moment of the magnet, H is the Earth's field strength, τ is the torsional couple, $(\pi d^4/32)[(\theta - \delta)/L]C$.

Equating couples (when angles are small),

$$MH\delta = \frac{\pi d^4}{32} \cdot \frac{(\theta - \delta)}{L} \cdot C$$

$$MH\delta + \frac{\pi d^4}{32} \cdot \frac{C}{L} \cdot \delta = \frac{\pi d^4}{32} \cdot \frac{C}{L} \theta$$

or

$$\delta \left(MH + \frac{\pi d^4}{32} \cdot \frac{C}{L} \right) = \frac{\pi d^4}{32} \cdot \frac{C}{L} \theta$$

Dividing by

$$\frac{\pi d^4}{32} \cdot \frac{C}{4}$$

$$\delta \left(\frac{32 MH . L}{\pi d^4 . C} + 1 \right) = \theta$$

If δ_1 is error per degree of twist θ,

$$\delta_1 = \frac{\pi d^4 . C}{32 MHL + \pi d^4 . C}$$

OSCILLATING NEEDLE

A suspended magnetic needle, once it is disturbed, will oscillate about the magnetic meridian. In Fig. 3.9, NS is a magnetic needle pivoted at O and oscillating about XOY, the amplitude of the oscillation being small.

Let I be the moment of inertia of the needle about O and M its magnetic moment. Let the field along OX be H.

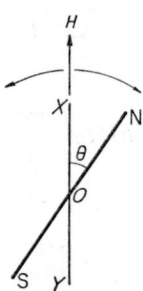

FIG. 3.9. An oscillating magnetic needle

The couple about O due to the magnetic field is $MH\theta$ and the inertial couple is $I\ddot{\theta}$. Assuming simple harmonic motion,

$$\theta = A \sin \omega t$$

$$\ddot{\theta} = -\omega^2 A \sin \omega t$$

Since

$$MH\theta + I\ddot{\theta} = 0$$

we have

$$MHA \sin \omega t = I\omega^2 A \sin \omega t$$

or

$$MH = I\omega^2 = 4\pi^2 I/T^2$$

where T is the period of oscillation.

Therefore,

$$T = 2\pi \left(\frac{I}{MH}\right)^{1/2} \tag{3.12}$$

With a unifilar suspension the only forces which eventually bring the magnet system to rest are those due to the resistance of the air and to the twist in the suspension; thus it is usual for the oscillations to persist for an appreciable period of time before dying away. In the Kew magnetometer, for instance, owing to the continually varying value of the Earth's magnetic field, the magnet system is hardly ever at rest.

DAMPING

It is customary in many magnetic instruments to introduce some means of damping the oscillations and bringing the needle or magnet to rest. Damping vanes moving in a dashpot of oil are sometimes used in certain types of suspension magnetometers.

A magnetic compass of the dry-card type, however, relies almost entirely on air resistance to bring the card to rest. Moreover, since the magnets generally used in such compasses are of low moment, the period of oscillation and the time taken for the card to settle are comparatively great. However, this type of compass was once much in demand by ship masters owing to its lively behaviour and their fear of a 'sticky' compass.

Eddy-current damping

An effective and useful form of damping for a dry-card compass is that known as eddy-current damping. With a freely swinging or oscillating magnet system, the equation of motion is of the form

$$I\ddot{x} + MHx = 0 \qquad (3.13)$$

representing an undamped or free oscillation. In the absence of any restraining force—friction or air resistance, for example—the magnet continues to oscillate.

It is desirable that a damping force should be large when the magnet is moving fast and small when it is moving slowly. Moreover, the resistance to motion, when the magnet is started from rest, should be approximately zero. This is important, since any force tending to apply restraint or 'drag' to the magnet system of a compass, when the vessel in which it is mounted makes a turn, is undesirable.

The equation of motion for a damped oscillation is of the form

$$I\ddot{x} + \Delta + MHx = 0 \qquad (3.14)$$

where Δ is a damping term.

Since the requirement is for a retarding force proportional to the speed of oscillation of the moving magnet, we can rewrite the equation

$$I\ddot{x} + k\dot{x} + MHx = 0$$

Such a force may be very simply provided by the generation of eddy currents (or Foucault currents) in a conductor placed adjacent to the poles of the magnet. This device is familiar in electricity meters, where a rotating copper disk between the poles of a permanent magnet applies to the turning motor a torque proportional to the power to be measured. In a magnetometer or compass the process is reversed—it is the magnet that moves and the conductor that remains at rest. A convenient form of conductor is a copper bowl surrounding the magnet system.

Now, when a magnetic pole is placed adjacent to the surface of a conductor such as copper, the lines of force from the magnet will pass into (and through) the conductor as shown in Fig. 3.10. When the magnet is moved in the direction shown, there will be relative movement between the conductor and the magnetic field, resulting in the generation of an e.m.f. This can most easily be seen by supposing that the conductor immediately opposite the magnetic pole is made up of a number of small strips δx lying at right angles to the direction of movement of the magnet.

Applying Fleming's rule, it will be seen that the induced current will flow in the direction of the arrows along the strips δx. In the area of the surface ahead of the magnet and into which the field is being introduced by the movement of the latter, a field of the opposite direction to that of the magnet would be produced, according to Lenz's law, by currents flowing in circular paths as shown, i.e., a north (red) pole would be produced, whereas in the area behind the magnet we find currents tending to maintain the field and giving a south (blue) pole in the conducting sheet.

Thus it will be readily appreciated that a red pole moving as shown will produce opposition to its movement by creating another red pole ahead of itself and a blue pole behind. These induced poles occur only when movement takes place and, since the induced e.m.f. (and therefore the current and field strength) is proportional to the speed at which the moving field passes the conducting strips, the desired type of retarding force or damping is produced. The magnitude of the retarding force, being dependent on current, also depends on the resistance of the conductor, and the desired amount of damping can be achieved by adjustment of the thickness and area of the conducting member and of the gap between the magnet and the conductor.

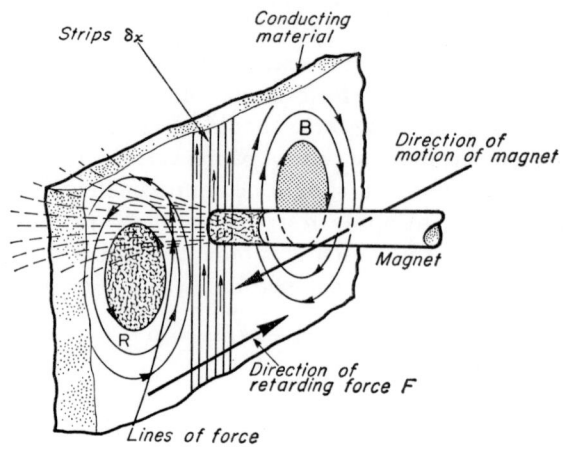

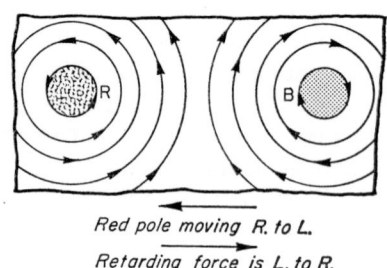

Fig. 3.10. Eddy-current damping

The retarding force F is proportional to the induced field, which is proportional to the current, which is proportional to the e.m.f., which is proportional to the rate of change of the field, which is proportional to the speed of the magnet.

Examples of the use of this form of damping are shown in Figs. 3.11 and 3.12. Fig. 3.11 is a section of the Sucksmith* magnetometer, in which a small but powerful magnet is surrounded by a thick copper pot with a conical bore. The thickness of the copper provides a low resistance path for the eddy currents and the conical bore allows for adjustment of the damping by raising or lowering the magnet. The damping of this instrument is very heavy and produces, for all practical purposes, 'dead-beat' operation. Fig. 3.12 shows a German (Askania) design of compass, similar to that used in the notorious flying bombs. A double-pivoted needle system

* Sucksmith, W., 'An improved magnetometer', *J. sci. Instrum.*, 1945, **22**, No. 7.

is mounted between two thick copper disks, in which the eddy currents are produced by the movement of the magnet poles adjacent to them.

Liquid damping

A more familiar method of damping is provided in the liquid-filled compass. The liquid is used for two reasons. Firstly, it enables a large proportion of the weight of the magnet system to be taken off the pivot by the incorporation of a float in the system. By this means much larger and stronger magnets may be used without damaging the pivot and introducing friction. Secondly, the liquid, by virtue of its viscosity, provides a convenient method of damping. This is not quite so exact a method as that provided by eddy currents, since surface tension and other effects tend to produce restraining forces and to introduce 'drag' on the magnet system independently of its rate of movement; consequently the damping term is not proportional to dx/dt.

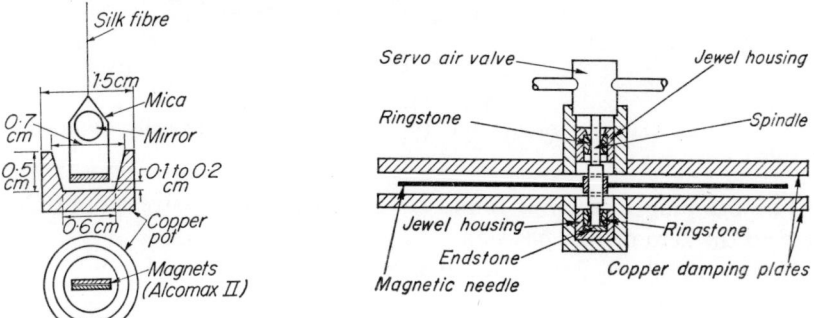

FIG. 3.11. Detail of the Sucksmith magnetometer

FIG. 3.12. A double-pivoted needle with eddy-current damping

This is the Askania compass that was used in the V1 flying bombs during the 1939–45 war.

The effect of swirl can be much more pronounced than with-eddy current damping since rotation of the whole compass imparts an angular momentum to the liquid which does not cease when the compass bowl comes to rest. There is then relative movement between the liquid and the bowl, and between the liquid and the magnet system, with the result that the latter may be displaced by a large angle for an appreciable period of time. To reduce the effects of swirl, the magnet system is made as free as possible from unnecessary projections and a large annular gap is allowed between the magnet system and card and the bowl.

In the usual type of marine compass, the damping afforded by the 'friction' between the liquid and the surfaces of the magnet system is sufficient. The system is neither aperiodic nor critically damped and displays the characteristics of a simple damped oscillation, as illustrated by curve (*a*) in Fig. 3.13. The aperiodic system is shown by curve (*b*), and curve (*c*) represents a critically damped oscillation.

Compasses intended for use in aircraft are required to have much heavier damping and experience has shown that the aperiodic system is to be preferred. To this end, extra projections, such as filaments of wire, are added to the magnet system to provide additional resistance to the motion of the system through the liquid. It must be borne in mind that any such expedient enhances the effect of swirl.

Damping factor

The degree of damping of an oscillating system is sometimes expressed as the damping factor. A heavily damped system will show a rapidly decreasing amplitude of successive swings or oscillations, whereas light damping causes a succession of oscillations whose amplitudes decrease slowly. An aperiodic system cannot be said to have a damping factor.

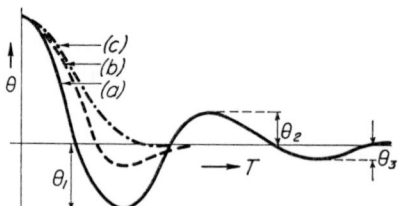

FIG. 3.13. Damping curves
(*a*) Damped oscillation. (*b*) Aperiodic. (*c*) Critically damped.

The damping factor f is found by comparing successive amplitudes of swing beyond the zero or rest position.

If these are θ_1, θ_2 and θ_3 . . . as in curve (*a*),

$$f = \frac{\theta_2}{\theta_1} = \frac{\theta_3}{\theta_2} \quad \text{etc.} \tag{3.15}$$

or, in terms of compass headings,

$$f = \frac{\zeta'_3 - \zeta'_2}{\zeta'_1 - \zeta'_2} \tag{3.16}$$

BEHAVIOUR OF A PIVOTED-NEEDLE COMPASS IN A MOVING VEHICLE

The effect of acceleration

When a single-pivoted compass of conventional design and having a pendulous system as described on page 21, is mounted in a ship, aircraft or any other vehicle, its performance and accuracy are much affected by the movement of the vehicle. The most notable effects are those due to acceleration of the pivot or suspension. This arises from changes of speed, turning, pitching, rolling and yawing. In aircraft, the accelerations are frequently so great as seriously to limit the usefulness of a simple magnetic compass, since circumstances arise when the magnet system of the compass becomes unstable and its deviations completely unpredictable.

East-west (linear) acceleration

If, for example, the craft is moving eastwards at a constant speed, the magnet system will be at rest on its pivot as shown in Fig. 3.14(*a*). If the craft then increases

34

speed, the magnet system, being pendulous, tilts as shown in Fig. 3.14(*b*). The amount of tilt depends on the magnitude of the acceleration *a*. Tan β, or β for small angles of tilt, is equal to a/g, where g is the acceleration due to gravity.

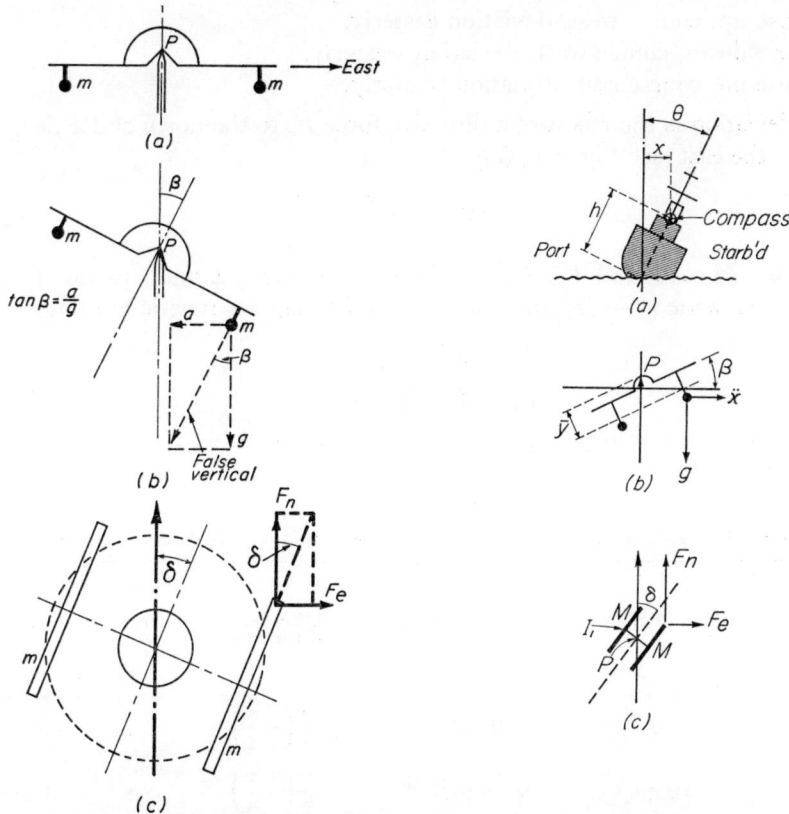

FIG. 3.14. Magnetic compass subject to east-to-west acceleration

P is the pivot, *mm* the magnets, *a* the acceleration, *g* the acceleration due to gravity, and β the resulting angle of tilt.

FIG. 3.15. Magnetic compass in a rolling ship

In (*a*) the view is in the direction of the vessel's forward motion.

If the direction of the tilt axis is to the north (i.e. $\alpha = 0$) the tilt angle, β, is positive, and from equation (3.8),

$$\tan \delta = \beta \tan D$$

$$= \frac{a}{g} \tan D$$

and the positive sign indicates an easterly deviation (the needle or magnet system being deflected towards a point east of north). A deceleration on a westerly course also causes an easterly deviation, whereas an acceleration to the west or a deceleration to the east causes a westerly deviation. This effect of east-west acceleration is frequently experienced in fast marine craft and is very marked in aircraft.

35

The action of pulling up an aircraft's nose in flight decreases its effective forward speed. Putting the nose down increases the speed and deviations will appear owing to the acceleration acting on the compass needle.

Nose down, course east, deviation easterly.
Nose up, course west, deviation easterly.
Nose down, course west, deviation westerly.
Nose up, course east, deviation westerly.

If the deviation is the result of a directive force F_n to the north and a deviating force F_e to the east [see Fig. 3.14 (c)],

$$\tan \delta = \frac{F_e}{F_n} \qquad (3.17)$$

where $F_e = Z\beta \cos \alpha$ and $F_n = H - Z\beta \sin \alpha$ [see Fig. 3.4 and equation (3.7)]. Further, if we write $(\zeta - \pi/2)$ for α, where ζ is the ship's magnetic heading,

$$F_e = Z\beta \sin \zeta$$
$$F_n = H + Z\beta \cos \zeta$$

Therefore, from $\zeta = 0$ to $\zeta = 2\pi$, the values shown in Table 3.1 may be assigned to F_e and F_n.

TABLE 3.1 *Relative deviating and directive forces acting on a tilted compass system*

Ship's magnetic heading ζ	Deviating force F_e	Directive force F_n	Direction of tilt axis	Acceleration § Direction of tilt	Deviation
0 (N.)	0	$H + Z\beta$*	$\frac{3\pi}{2}\left(-\frac{\pi}{2}\right)$	+	0
$\pi/4$ (NE.)	$0.707\,Z\beta$	$H + 0.707\,Z\beta$*	$\frac{7\pi}{4}\left(-\frac{\pi}{4}\right)$	+	E. (+)
$\pi/2$ (E.)	$Z\beta$	H	0	+	E. (+)
$3\pi/4$ (SE.)	$0.707\,Z\beta$	$H - 0.707\,Z\beta$†	$\pi/4$	+	E. (+)
π (S.)	0	$H - Z\beta$†	$\pi/2$	+	0
$5\pi/4$ (SW.)	$-0.707\,Z\beta$	$H - 0.707\,Z\beta$†	$3\pi/4$	+	W. (−)
$3\pi/2$ (W.)	$-Z\beta$	0	π	+	W. (−)
$7\pi/4$ (NW.)	$-0.707\,Z\beta$	$H + 0.707\,Z\beta$*	$5\pi/4$	+	W. (−)
2π (N.)	0	$H + Z\beta$*	$3\pi/2$	+	0

* Directive force increased.
† Directive force decreased.
§ For deceleration, signs in the last two columns will be reversed.

This table is for the northern magnetic hemisphere (see footnote on p. 27 for southern magnetic hemisphere).
If $Z\beta$ should be equal to or greater than H, a state of instability arises (see p. 27).

Accelerations due to rolling and pitching

If the compass position in a ship is distant from the centre of roll or pitch, accelerations will be experienced at that position due to the motion of the vessel about the roll or pitch axis. A pendulous compass system will therefore be tilted in a plane normal to this axis, the orientation of which depends upon the ship's course. Consequently, deviations will occur in the same manner as described for linear accelerations of the ship, except that the motion will be oscillatory and the amplitude of the deviations will depend on the period of roll or pitch, on the pendulous period of the compass, and, since the deviation is manifest as an azimuthal angle, on the rotational period of the compass needle system about its pivot.

Consider a ship that is rolling [Fig. 3.15(a)]. Let the roll be $\theta = A \sin pt$, a roll to starboard being regarded as positive, where $p = 2\pi/T$, T being the period of roll and A its amplitude. The lateral displacement of the compass is x and h is the distance from the centre of roll to the compass position.

Now

$$x = h \sin \theta$$

and if θ is small enough,

$$x = h\theta = hA \sin pt$$

Let I be the moment of inertia of the pendulous magnet system about its pivot P. It is assumed that the magnet system is of the form described in Chapter 5, in which the moments of inertia about all horizontal axes through the centre are equal. Then

$$I = mk^2$$

where m is the mass of the system and k the radius of gyration.

In Fig. 3.15(b) let β be the angle of tilt. The couples acting on the system are

$$I\ddot{\beta} + mg\bar{y}\beta = m\bar{y}\ddot{x} \tag{3.18}$$

Writing mk^2 for I,

$$mk^2\ddot{\beta} + mg\bar{y}\beta = m\bar{y}\ddot{x}$$

Therefore

$$\ddot{\beta} + \frac{g\bar{y}}{k^2}\beta = \frac{\bar{y}}{k^2}\ddot{x} \tag{3.19}$$

Assuming simple harmonic motion, $\ddot{\beta} = p^2\beta$. Solving for β,

$$\beta = \frac{1}{1 - (p^2k^2/g\bar{y})} \cdot \frac{\ddot{x}}{g} \tag{3.20}$$

Now $p^2 = 4\pi^2/T^2$ and $2\pi(k^2/g\bar{y})^{1/2} = T'$, the period of the magnet system in tilt. Therefore

$$\beta = \frac{T^2}{T^2 - T'^2} \cdot \frac{\ddot{x}}{g} \tag{3.21}$$

Since

$$x = hA \sin pt$$

$$\ddot{x} = hAp^2 \sin pt$$

$$= -\frac{4\pi^2}{T^2} . hA \sin pt$$

Thus

$$\beta = -\frac{1}{T^2 - T'^2}4\pi^2\frac{hA}{g}\sin pt \qquad (3.22)$$

The attenuating effect of having T and T' of widely differing magnitudes will be noticed. If $T^2 \gg T'^2$ as is usual and desirable,

$$\beta = -\frac{4\pi^2}{T^2}\cdot\frac{hA}{g}\sin pt \qquad (3.23)$$

This is an oscillatory tilting motion of the compass needle system at the period of roll (or pitch), the negative sign indicating that the system oscillates in tilt in anti-phase with the rolling (or pitching) of the ship.

For small angles of deviation, δ, and from equations (3.8) and (3.23)

$$\delta = -\frac{\dfrac{4\pi^2}{T^2}\cdot\dfrac{hA}{g}\cdot\tan D\cos\alpha\sin pt}{1+\dfrac{4\pi^2}{T^2}\cdot\dfrac{hA}{g}\cdot\tan D\sin\alpha\sin pt} \qquad (3.24)$$

provided the compass needle system has an infinitely short rotational period about its pivot.

In the case of a rolling ship, $\alpha = \zeta$, as the tilt axis may be regarded as lying in the direction of forward motion of the ship; and in the case of a pitching ship, the direction of the tilt axis may be taken as $\pi/2$ in a counter-clockwise direction from the direction of forward motion. Hence in this instance, $\alpha = (\zeta - \pi/2)$. Therefore,

$$\delta_{\text{Roll}} = -\frac{\dfrac{4\pi^2}{T^2}\cdot\dfrac{hA}{g}\cdot\tan D\cos\zeta\sin pt}{1+\dfrac{4\pi^2}{T^2}\cdot\dfrac{hA}{g}\cdot\tan D\sin\zeta\sin pt}$$

$$\delta_{\text{Pitch}} = -\frac{\dfrac{4\pi^2}{T^2}\cdot\dfrac{hA}{g}\cdot\tan D\sin\zeta\sin pt}{1-\dfrac{4\pi^2}{T^2}\cdot\dfrac{hA}{g}\cdot\tan D\cos\zeta\sin pt}$$

In order to reduce the magnitude of the deviations, the compass needle system is designed to have an azimuthal (rotational) period which is different from, and preferably greater than, the period of roll or pitch. The couples acting on a compass needle system in the rotational plane are provided by the inertial property of the system, the magnetic field component to the magnetic north (directive) and the magnetic field component to the magnetic east (deviating) [see Fig. 3.15(c)].

By way of example, consider the case of a ship heading magnetic north and rolling ($\zeta = 0$). Referring to Fig. 3.15(c),

$$I_1\ddot{\delta} + MF_n\delta = MF_e$$

for small values of δ, where I_1 is the moment of inertia of the compass needle system and F_n and F_e are the directive and deviating magnetic forces respectively.

$$\ddot{\delta} + \frac{MF_n}{I_1}.\delta = \frac{MF_e}{I_1} \tag{3.25}$$

Assuming simple harmonic motion, $\delta = -p^2\delta$, and solving for δ,

$$\delta = \frac{1}{1-(p^2I_1/MF_n)} \cdot \frac{F_e}{F_n}$$

$$p^2 = \frac{4\pi^2}{T^2} \quad \text{and} \quad \frac{I_1}{MF_n} = \frac{T_1^2}{4\pi^2}$$

where T_1 is the period of the compass needle system in the field F_n. Therefore

$$\delta = \frac{T^2}{T^2-T_1^2} \cdot \frac{F_e}{F_n} \tag{3.26}$$

From equations (3.7) and (3.17)

$$\frac{F_e}{F_n} = \frac{Z\beta \cos\alpha}{H - Z\beta \sin\alpha}$$

and writing $\alpha = \zeta = 0$

$$\frac{F_e}{F_n} = \frac{Z\beta}{H}$$

Inserting the value of β given by equation (3.23)

$$\delta = -\frac{T^2}{T^2-T_1^2} \cdot \frac{4\pi^2}{T^2} \cdot \frac{hA}{g} \cdot \frac{Z\sin pt}{H} \tag{3.27}$$

It is obviously undesirable to have a compass needle system with an azimuthal period near in value to the period of roll or pitch.

By making $T_1 > T$, as is usual in ships' compasses, $T_1^2 \gg T^2$.

$$\delta = \frac{4\pi^2}{T_1^2} \cdot \frac{hA}{g} \tan D \sin pt \tag{3.28}$$

If the ship is heading magnetic south ($\alpha = \zeta = \pi$)

$$\frac{F_e}{F_n} = -\frac{Z\beta}{H}$$

$$\delta = -\frac{4\pi^2}{T_1^2} \cdot \frac{hA}{g} \tan D \sin pt \tag{3.29}$$

If the ship is heading magnetic east ($\zeta = \pi/2$) or west ($\zeta = 3\pi/2$)

$$\frac{F_e}{F_n} = 0$$

and so

$$\delta = 0$$

If the ship is pitching,

$$\alpha = \zeta - \pi/2$$

When heading magnetic north or south,

$$\delta = 0$$

When heading magnetic east,

$$\delta = \frac{4\pi^2}{T_1^2} \cdot \frac{hA}{g} \tan D \sin pt \tag{3.30}$$

When heading magnetic west,

$$\delta = -\frac{4\pi^2}{T_1^2} \cdot \frac{hA}{g} \tan D \sin pt \tag{3.31}$$

For any value of ζ a similar attenuation occurs, but the solution is more complex than for the cardinal courses, and it is not strictly correct to assume simple harmonic motion.

However, provided $\beta \tan D$ remains small the attenuating term $T^2/(T^2 - T_1^2)$ is approximately correct for the general case given by equation (3.24).

Equation (3.24) shows that the deviations when $pt = \pi/2$ and $pt = 3\pi/2$ are not equal and that there is asymmetry in the deviation curve except when $\alpha = 0$, π, $\pi/2$ or $3\pi/2$.

Write the equation in the form

$$\delta = -\frac{a \cos \alpha \sin pt}{1 + a \sin \alpha \sin pt}$$

where

$$a = \frac{4\pi^2}{T^2} \cdot \frac{hA}{g} \cdot \tan D$$

This may be expressed by the series

$$\delta = \bar{A} - \bar{B} \sin pt - \bar{E} \cos 2pt$$

including terms up to second order in a and where

$$\bar{A} = \frac{a^2 \sin 2\alpha}{4}$$

$$\bar{B} = a \cos \alpha$$

$$\bar{E} = \frac{a^2 \sin 2\alpha}{4}$$

$\bar{A}$ indicates a small constant deviation whilst the ship is rolling or pitching, so that the average direction of the ship's head indicated by the compass is not correct. However, this deviation is very small and amounts in practice to only a few minutes of arc, provided a is small compared with unity. The predominantly sinusoidal character of the deviation curve is indicated by $\bar{B}$.

Further examination of equation (3.24) shows that the maximum value of the deviation occurs when $\sin \alpha = -a$.

Centrifugal acceleration

The deviations of the compass due to the motion of a vehicle or craft are particularly marked in aircraft flying in a banked turn. The same effect occurs to a small extent in ships, but to any important degree only in fast vessels.

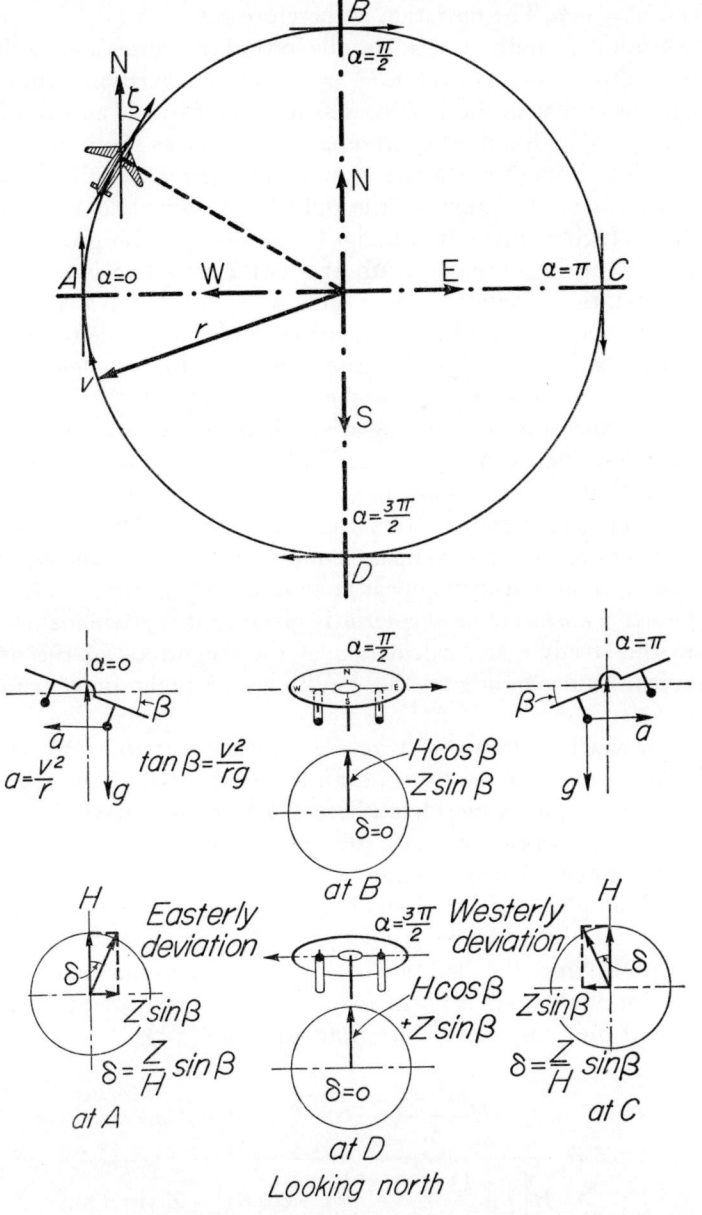

FIG. 3.16. Centrifugal acceleration and northerly turning error

Consider an aircraft flying in a clockwise sense along the circumference of a circle of radius r and at a speed v. The centrifugal acceleration will be v^2/r (Fig. 3.16). This acts upon the pendulous magnet system of the compass, causing it to tilt inwards.

At A, the aircraft is heading north (magnetic) and the tilt is clockwise or positive about the direction of flight, which may also be regarded as the direction, α, of the tilt axis. The deviation of the compass needle will be easterly [see equation (3.5), when $\alpha = 0$].

At B, the tilt is inwards and of the same magnitude as at A, but the heading is now east and $\alpha = \pi/2$. The deviation is therefore zero.

At C, the heading is south, $\alpha = \pi$, and the inward tilt causes a westerly deviation.

At D, where the heading is west, $\alpha = 3\pi/2$, and the deviation is once more zero.

The net result is that in the left-hand semicircle, there is an easterly deviation (i.e. to lower apparent headings by compass). This gives the impression that the aircraft is turning through north at a slower rate than is actually the case. When flying east, there is no deviation. In the right-hand semicircle, there is a westerly deviation (i.e. to higher apparent headings by compass). This gives the impression that the craft is turning through south at a faster rate than is actually the case. When flying west, the deviation is again zero. When turning in a counter-clockwise sense, the above sequence still applies; for example, in the left-hand semicircle, both the aircraft and the compass indicate movement towards lower readings and therefore a faster turn to south than is the case.

It will be seen that a condition may arise when the deviation on approaching a northerly course is increasing at a rate equal to the rate of turn and hence the compass will indicate no turn at all. In an extreme case, the compass may indicate a turn in the opposite direction from that being made. On approaching south, however, nothing prevents the compass from indicating the same sense of turn as that being made, though it may appear at an exaggerated rate.

Hence the name *northerly turning error* is given to this phenomenon.

Except in the steady state, calculation of the magnitude of the deviation involves complexities depending on the conditions of flight immediately prior to the turn.

In Fig. 3.17, which is drawn to correspond with Fig. 3.16, AWPNBEQS is the horizontal plane containing the fore-and-aft line of the craft AOB, the athwartship line POQ and the magnetic meridian NOS. EOW is the east-west line.

As the craft moves in a circular path, AOB is tangential to the path and centrifugal force will cause a pendulously mounted compass needle or magnet system to tilt about AOB. The plane of rotation of the needle or magnet system is thus AP'N'B Q' and the tilt is the angle QOQ' (β). This is clockwise when viewed in the direction AOB. Since the direction of the tilt axis, α, is the same as the heading, ζ, of the craft, the angle of tilt is positive. For a counter-clockwise turn it is negative. The deviation of the compass is δ and from equation (3.5)

$$\tan \delta = \frac{H\dfrac{\sin 2\alpha}{2}(1 - \cos \beta) + Z \sin \beta \cos \alpha}{H\left[1 - \dfrac{(1 - \cos 2\alpha)}{2}(1 - \cos \beta)\right] - Z \sin \beta \sin \alpha}$$

In terms of ζ

$$\tan \delta = \frac{H\dfrac{\sin 2\zeta}{2}(\text{I} - \cos \beta) + Z \sin \beta \cos \zeta}{H\left[\text{I} - \dfrac{(\text{I} - \cos 2\zeta)}{2}(\text{I} - \cos \beta)\right] - Z \sin \beta \sin \zeta} \qquad (3.32)$$

for a clockwise turn, and

$$\tan \delta = \frac{H\dfrac{\sin 2\zeta}{2}(\text{I} - \cos \beta) - Z \sin \beta \cos \zeta}{H\left[\text{I} - \dfrac{(\text{I} - \cos 2\zeta)}{2}(\text{I} - \cos \beta)\right] + Z \sin \beta \sin \zeta} \qquad (3.33)$$

for a counter-clockwise turn.

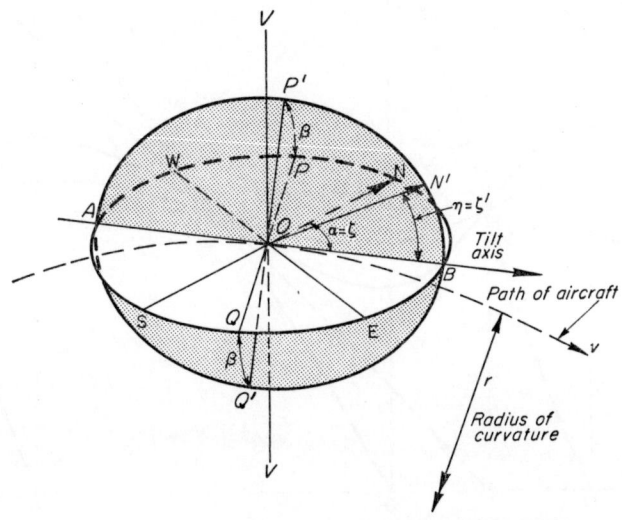

FIG. 3.17. The effect of centrifugal acceleration on a magnetic compass

The unshaded plane is horizontal and the shaded plane tilted. The tilt axis AOB is the fore-and-aft line of the aircraft. NOS is the magnetic meridian, ON' the deviated direction of the compass needle, VOV the vertical, QOQ' ($= \beta$) the angle of tilt, NOB ($= \alpha$) the magnetic heading (i.e. ζ), and N'OB ($= \eta$) the compass heading (i.e. ζ'). The angle $\alpha - \eta = \zeta \pm \zeta'$ is the deviation δ.

The above expressions for $\tan \delta$ are general and the deviations curve plotted against magnetic heading is regular in form as long as the angle of tilt is less than $\pi/2 - D$. This is the critical condition described on page 27.

The heading at which the maximum deviation occurs moves from approximately north and south for small angles of tilt towards east (or west according to the sense of the turn and whether the manoeuvre takes place in the northern or southern magnetic hemisphere) as the angle of tilt increases.

The compass loses all directive force when the tilt equals $(\pi/2 - D)$, and the rotational plane of the needle becomes normal to the total force vector. The deviation curve shows a sudden discontinuity as the deviation changes from $+\pi/2$ to $-\pi/2$. Instability of the system is inevitable and as the tilt increases beyond the critical angle, discontinuities in the curve become more marked; also apparent turns in the wrong direction are evident with increasing instability, the form of the deviation curve having altered completely. This is a serious limitation on the use of a simple magnetic compass in aircraft.

Fig. 3.18 shows the deviation curves plotted against magnetic heading for motion in a circular path for angles of tilt less than, equal to, and greater than the critical angle, the sign of the deviation depending on whether the turn is clockwise or counter-clockwise or executed in the northern or southern magnetic hemisphere.

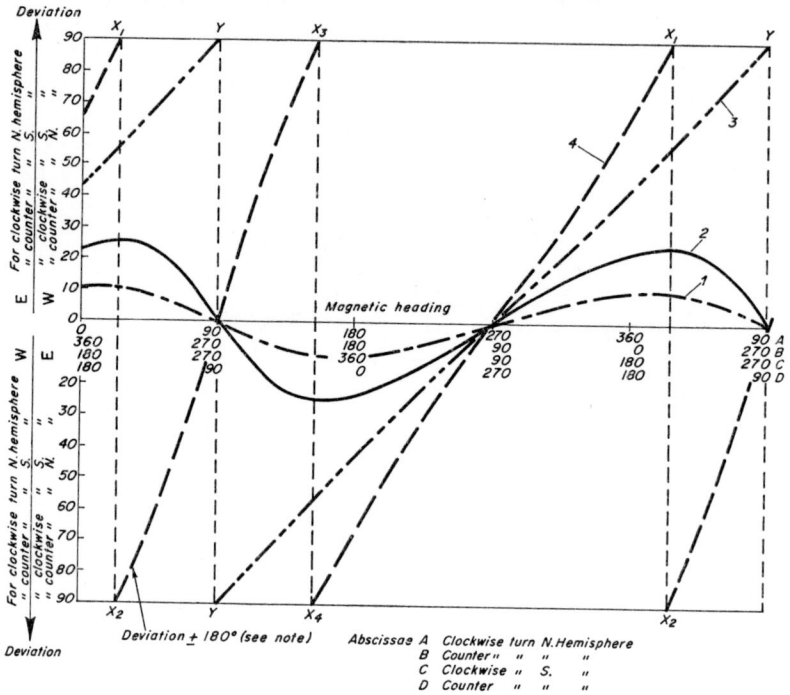

FIG. 3.18. Northerly turning error

If the aircraft is correctly banked, the 'false vertical', i.e. the resultant of the centrifugal and gravitational accelerations, is perpendicular to the deck. Although there is a tilt of the compass needle system with respect to the gravitational vertical and therefore with respect to the vertical component Z of the Earth's magnetic field, there appears to be none with respect to the aircraft. It will be appreciated that, with fast flying aircraft, large values of tilt may be encountered in turns. Methods of combating compass instability are discussed in Chapters 5 and 9.

In the case of ships, where the accelerations involved do not cause large angles of tilt, it is sufficiently accurate to write δ for $\tan \delta$, β for $\sin \beta$ and unity for $\cos \beta$.

44

From equations (3.32) and (3.33)

$$\delta = \frac{Z\beta \cos \alpha}{H - Z\beta \sin \alpha}$$

$$= \frac{\beta \tan D \cos \zeta}{1 - \beta \tan D \sin \zeta} \qquad \text{for a clockwise turn} \qquad (3.34)$$

or

$$\delta = -\frac{\beta \tan D \cos \zeta}{1 + \beta \tan D \sin \zeta} \qquad \text{for a counter-clockwise turn} \qquad (3.35)$$

Curve (1) in Fig. 3.18 is typical for a small fast vessel.

Since the deviation is a periodically changing function as the circular motion continues, and therefore for ζ we can write $2\pi t/T$ where T is the time taken for a complete circle, the magnitude of the deviation will be modified by the dynamic properties of the compass needle system.

The acceleration is always in the same direction relative to the fore-and-aft axis of the craft (that is to say, to the tilt axis). Consequently, the tilt β is constant relative to the tilt axis. Compare the case of a rolling or pitching ship, where the direction of the fore-and-aft axis is constant and the acceleration is oscillatory.

As in this latter case, the azimuthal (rotational) period of the compass magnet system has an effect on the magnitude of the deviations experienced (page 38).

A period, T_1, that is well removed from the time, T, for a complete turn will prevent oscillations from building up, though the solution for the value of the attenuating factor is complex owing to the nature of the directive and deviating forces involved.

In ships' compasses, the period T_1 is usually small compared with T and it is reasonable to assume that the attenuation that may be applied to equations (3.34) and (3.35) is unity. Now

$$\tan \beta = a/g$$
$$= v^2/rg$$

where r is the radius of the circular path and v is the speed of the ship. If T is the time for a complete circular turn,

$$\tan \beta = 2\pi v/gT$$

Thus for small values of δ and β,

$$\delta = \frac{\dfrac{2\pi v}{gT}\tan D \cos \zeta}{1 - \dfrac{2\pi v}{gT}\tan D \sin \zeta} \qquad (3.36)$$

for a clockwise turn, and

$$\delta = \frac{-\dfrac{2\pi v}{gT}\tan D \cos \zeta}{1 + \dfrac{2\pi v}{gT}\tan D \sin \zeta} \qquad (3.37)$$

for a counter-clockwise turn.

If v is in knots, T is in seconds, g is 32·2 ft/sec² and δ is in degrees,

$$\delta° = \frac{\pm\, 18\cdot9 \dfrac{v}{T} \tan D \cos \zeta}{1 \pm 18\cdot9 \dfrac{v}{T} \tan D \sin \zeta} \tag{3.38}$$

Deviation in terms of ship's head by compass

By writing $\tan \delta = \sin \delta/\cos \delta$ and cross-multiplying, equation (3.32) may be written in the form

$$\sin \delta = \bar{C} \cos \zeta' + \bar{D} \sin (2\zeta' + \delta)$$

where

$$\bar{C} = \frac{2 \tan D \sin \beta}{(1 + \cos \beta)} \quad \text{and} \quad \bar{D} = \frac{(1 - \cos \beta)}{(1 + \cos \beta)}$$

Similarly, equation (3.33) may be written

$$\sin \delta = -\bar{C} \cos \zeta_1 + \bar{D} \sin (2\zeta' + \delta)$$

When δ and β are small, as in equations (3.36) and (3.37),

$$\delta = \bar{C} \cos \zeta' \quad \text{for a clockwise turn}$$
$$\delta = -\bar{C} \cos \zeta' \quad \text{for a counter-clockwise turn}$$

where

$$\bar{C} = \beta \tan D$$

$$= \frac{2\pi v}{gT} \tan D$$

Thus

$$\delta = \frac{2\pi v}{gT} \tan D \cos \zeta' \quad \text{for a clockwise turn}$$

and

$$\delta' = -\frac{2\pi v}{gT} \tan D \cos \zeta' \quad \text{for a counter-clockwise turn}$$

[When v is in knots and δ is in degrees, $\bar{C} = (18\cdot9/T)\, v \tan D$.]

Chapter 4

Inductor Instruments

Faraday, in demonstrating the connection between electrical and magnetic phenomena, showed how current will flow in a conductor moving in a magnetic field.

EARTH INDUCTOR

The simplest form of earth inductor is a coil, suitably mounted in bearings and rotating so that the turns of wire cut the Earth's magnetic field. An alternating e.m.f.
is generated at the terminals of the coil and is at a maximum when the rate of change
of magnetic flux through the coil is greatest and is zero when the rate of change of
magnetic flux is nil.

By suitably mounting the spinning coil so that its axis of rotation can be orientated
as required, it is possible to find the direction of the total field of the Earth by
observing zero e.m.f., in which position the axis of the coil lies at right angles to the
direction of the total field, and the axis of rotation lies along it. Hence the angle of
dip may be ascertained. (The earth inductor is a more precise instrument for this
purpose than the dip circle.) By now orientating the coil perpendicular to this first
position, the total intensity of the Earth's field may be determined. The earth inductor is shown diagrammatically in Fig. 4.1 and described in detail in Chapter 6.

The instrument may be used as a compass by indicating on a suitable scale or
bearing plate the appropriate orientation for zero output, in which case the axis of
rotation is in the plane of the magnetic meridian. Fig. 4.2 shows this arrangement
diagrammatically with a galvanometer or similar instrument as an indicator. To
steer with this device, the scale would be set to the appropriate course against
the fore-and-aft line of the craft with the plane of the axis of rotation containing the
0°–180° diameter of the scale. Adjustment for the angle of dip would be necessary.
The craft would then be manoeuvred until the indicator showed zero, when the
desired course could be steered. Although this system is crude and is little more
than a simple demonstration of a principle, an instrument of this type was used
as an aircraft compass in an early transatlantic flight.

Various modifications of the rotating coil have been adapted as direction instruments and a few of these have been chosen to illustrate some of the ideas that have
contributed to the development of the inductor compass. These include the addition within the coil of cores of a material having high magnetic permeability, in
order to enhance the flux density.

Gunn's compass

Gunn's compass* is a variety of the rotating-coil compass in which a long thin coil or solenoid is wound on a rod of highly permeable material. This coil-and-rod assembly is rotated rapidly in the horizontal plane. The rate of change of flux is least as the rod passes through the north–south direction and greatest as it passes through east–west. The instantaneous e.m.f. generated is a function of the angle between the magnetic meridian and the rod.

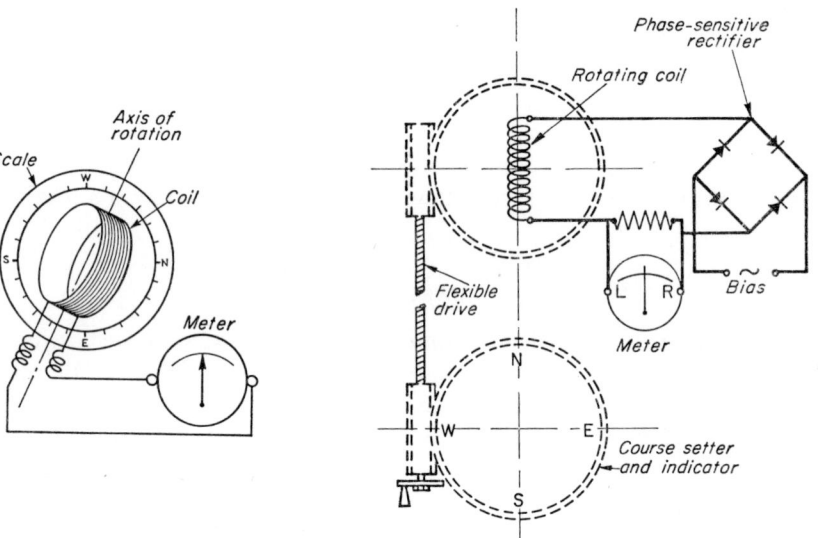

FIG. 4.1. Rotating-coil Earth inductor

FIG. 4.2. A simple course indicator using a rotating-coil Earth inductor

In Fig. 4.3, XY is the axis of the rod which is rotating in the plane of the diagram. NS is the magnetic meridian and FA, PS are the fore-and-aft and athwartship axes respectively of the craft or vehicle in which the compass is installed. The angular velocity of the rod in the horizontal plane is ω and the instantaneous e.m.f. is given by $E = K \sin \alpha$, where α is the angle between the rod and the magnetic meridian, K being a constant.

Considering the angular position of the rod relative to the axes of the craft,

$$\alpha = \omega t + \theta$$

where θ is the compass course. Thus $E = K \sin (\omega t + \theta)$ and the phase angle of the e.m.f. is the course θ. This angle can be ascertained by the use of a rotating contact coupled to the coil-and-rod assembly, whereby an indicating circuit is completed at a predetermined angular position of the rod with respect to the axes of the craft in which it is mounted. When the circuit is completed, the instantaneous value of the e.m.f. is indicated, corresponding to the angular position of the rod with respect to the Earth's magnetic field. By adjusting the contactor, an angular position may be found where the e.m.f. is zero for any given course. This corresponds to the

* Gunn, R., U.S. Pat. 2054318.

instant when $\alpha = 0$. Thus by means of a suitable scale and method of rotation or adjustment, the phase or course angle θ can be determined.

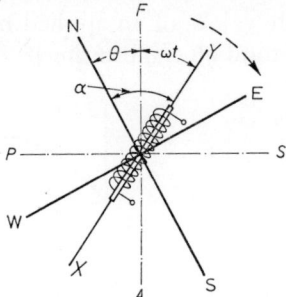

FIG. 4.3. Gunn's compass

Siemens inductor compass

A modification of Gunn's instrument is the Siemens inductor compass,* which entails the use of an alternating current to energize the coil wound round the highly permeable rod. As the rod rotates in the Earth's field, a sinusoidally varying flux is induced in the rod and is at a maximum when the rod is directed north-south and at a minimum when it is directed east-west. By a suitable choice of effective permeability and amplitude of excitation current, the impedance of the winding is a maximum twice per revolution and a minimum twice per revolution. The current in the winding is at the supply frequency but modulated in accordance with the curve shown in Fig. 4.4(a). By supplying a superimposed polarizing magnetic field, either by means of a permanent magnet or by a d.c. component in the excitation supply, the flux in the rod can be made to have only one maximum and one minimum in a revolution, as shown in Fig. 4.4(b). On removing the d.c. component from the effective current in the winding, the modulation is as shown in Fig. 4.4(c) and may be related to the fore-and-aft axis of a craft or vehicle by the phase angle θ [Fig. 4.4(d)], as in Gunn's invention. The system proposed by Siemens and shown in Fig. 4.5, uses two inductors connected in the form of a bridge circuit with metal rectifiers providing an a.c. excitation with a d.c. component. The bridge connection permits only the sinusoidally modulated voltage to appear at the output.

Aga-Baltic compass

Another interesting type of inductor is that developed by Aga-Baltic A.G.† In this instrument, two horizontally opposed, fixed soft-iron cores, each enclosed in an excitation winding, are spaced with a permanent magnet rotating at a high speed between them, as in Fig. 4.6(a). A rapidly alternating flux is induced in the coils at a frequency of about 4 kc/s. This, combined with the excitation supply at about 48 c/s, will give rise to an effective current consisting of a 4 kc/s signal modulated at 48 c/s.

When an external magnetic field, for instance that of the Earth, is applied axially to the system, and when the inductor is directed N. or S. as shown in the vector

* Siemens-apparate und Maschinen Gesellschaft, Berlin, Brit. Pat. 520826.
† Aga-Baltic Aktiebolag-Sweden. Brit. Pat. 569839.

diagram, Fig. 4.6(b), the effective current is as shown diagrammatically in Fig. 4.6(c)·
When the inductor is placed in the direction corresponding to X or Y, the modulated
current assumes the form shown in Figs. 4.6(d) and (e), which are diagrams corresponding to equal and opposite values of an applied magnetic field. The envelope
contains components of 48 c/s and 96 c/s, but the 48 c/s component changes phase

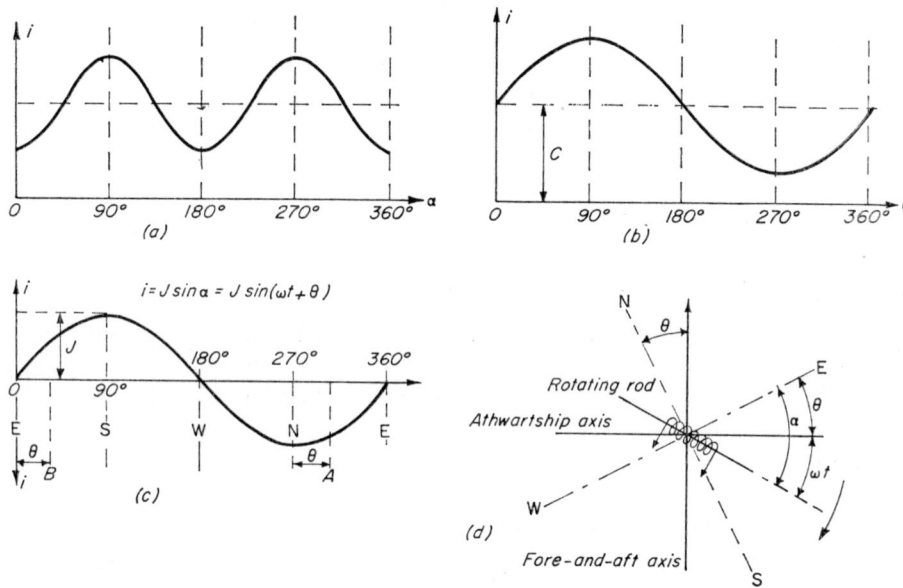

FIG. 4.4. Siemens compass

The curves show the relationship between angular position and current.
Curve (a), the unpolarized condition, with maxima at 90° and 270°; (b) shows
the polarized condition, with a single maximum at 90°; (c) shows the polarized
condition with the d.c. component removed. Diagram (d) illustrates the
geometry of the compass.

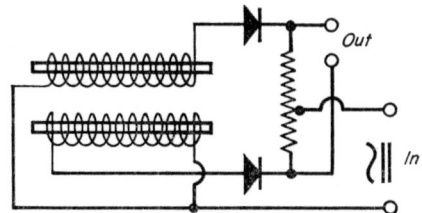

FIG. 4.5. Circuit diagram of the Siemens compass

with change of direction of the field [Fig. 4.6(f)]. An indication of this can be seen
by using a phase-sensitive meter, and Fig. 4.6(g) shows an arrangement in which an
oscillator provides the modulation and reference signals. When the inductor system
is orientated east–west, the envelope of the modulated signal consists of 96 c/s only
whilst the 48 c/s component is zero, giving a null indication at the phase-meter.
Hence, by suitable means of control, the inductor may be adapted to indicate heading.

Of the many types of high-permeability inductors that have been devised, the most successful and most commonly used is the saturable inductor. It is known also by the proprietary trade names of Flux Gate (Pioneer Bendix) and Flux Valve (Sperry). The basic patent is generally attributed to Antranikian* and there have been many modifications and variations on the fundamental principle. The most common forms are single-core and twin-core inductors depending for their operation on a.c. excitation of such magnitude that the cores are periodically driven into saturation on either side of the unmagnetized or zero field state. The application of an ambient magnetic field along the axis of the core or cores alters the relative positions in the magnetizing cycle where saturation is reached in either direction of magnetization.

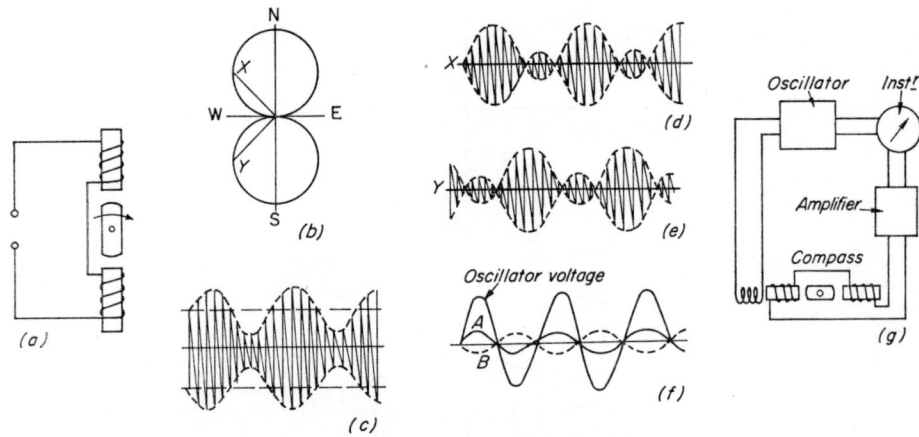

Fig. 4.6. Operation of the Aga-Baltic inductor compass

The corresponding distortion of the flux can be detected as an e.m.f. in the excitation windings (or, better still, in a separate secondary winding), this e.m.f. being in certain conditions proportional to the axially applied field. Four forms of saturable inductor are commonly used, namely:

    (i) fundamental-frequency inductor;
    (ii) second-harmonic inductor;
    (iii) peak-output inductor;
    (iv) pulse-difference inductor.

The complete analysis of these systems is complex, but a good approximation is obtained both to the theory and to the operation in practice by assuming an idealized magnetizing or B–H curve of a highly permeable material, such as Permalloy or Mu-metal. Such a curve is shown in Fig. 4.7 as being constructed of three straight lines, ab, bc and cd. Of these, ab and cd represent the region where the material is saturated ($|\mu| = 1$) and bc is the part of the curve where the permeability is high. H_s is the value of magnetizing field at which saturation occurs.

* Antranikian, H., Magnetic field direction and intensity finder, U.S. Pat. 2047609, 1936.

Alternatively, rather than the B–H curve, the μ–H curve may be used. Such a curve, corresponding to *abcd*, is shown at *pqrs*. A more refined approximation is one in which the permeability changes gradually from its maximum value to $|\mu| = 1$, in which case the figure *pqrs* is no longer a rectangle but is trapezoidal or even bounded by two curved lines p_1q and rs_1.

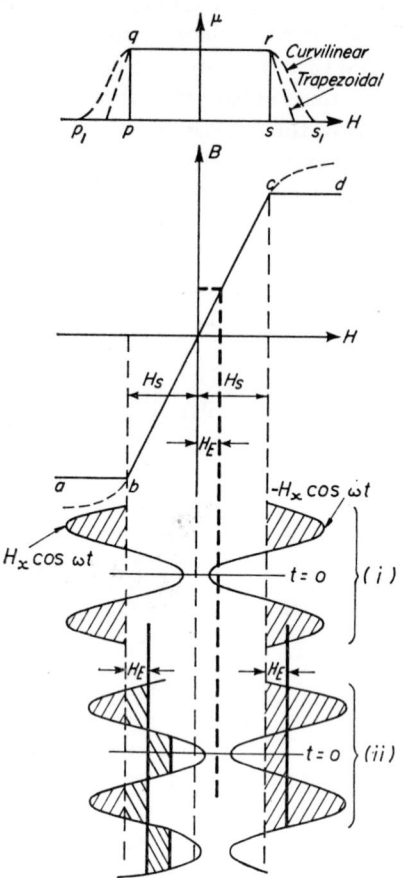

FIG. 4.7. The saturable inductor

Idealized B–H and μ–H curves and fundamental-frequency operation. (i) Zero axial field; the shaded half-cycles are beyond saturation and the unshaded half-cycles cancel out, so that the resultant flux and e.m.f. are zero. (ii) With the application of an axial field H_E; unshaded areas no longer cancel, leaving a residual field in the opposite direction, shown by shading; hence there is a resultant flux change and an e.m.f.

Fundamental-frequency inductor

The operation of the fundamental-frequency inductor can be illustrated by reference to Figs. 4.7 and 4.8. Two parallel or collinear Permalloy cores 1 and 2 are each surrounded by suitable excitation windings which are fed in anti-phase; 3 and 4 are the input terminals, 5 and 6 are equal impedances and 7 and 8 are the output terminals. The cores are in equal and opposite polarizing fields applied either electrically or magnetically. An a.c. supply is connected to the terminals 3 and 4. For the most effective operation,

$$H_p = H_s$$

and in Fig. 4.7 (i) this condition is shown, with zero axial field and with a magnetizing field of $H_x \cos \omega$. The diagram is symmetrical, half of each excitation cycle being greater than H_s and the remaining half cycles cancelling out. The resultant flux, and therefore the voltage at 7 and 8, is zero.

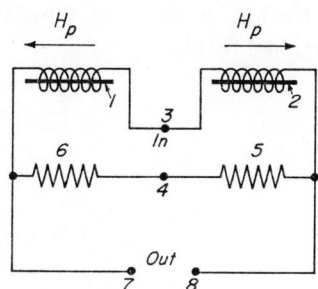

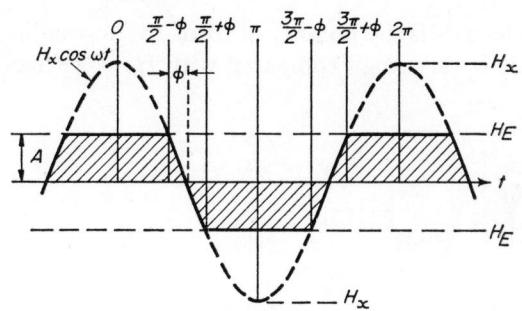

FIG. 4.8. Circuit diagram of a fundamental-frequency saturable inductor

FIG. 4.9. Derivation of fundamental-frequency output from a saturable inductor

The fundamental term is $(2A/\pi)(\cos \phi + \phi/\sin \phi)$. Since $A = H_E$, $\sin \phi = H_E/H_x$.

Now apply an ambient axial field H_E as in Fig. 4.7(ii). In one core, $H_x \cos \omega t$ is superimposed on $H_p + H_E$ and in the other on $H_p - H_E$. An unbalanced flux now exists since corresponding half cycles do not cancel out and are not in the saturation zone to the same extent as in (i). From considerations of symmetry, the effective magnetizing force that produces flux changes in the cores is as shown in Fig. 4.9. If the direction of H_E is reversed, the phase of the magnetizing force is also reversed. Consequently an e.m.f. at the excitation frequency whose phase reverses with reversal of the direction of H_E is developed at the terminals 7 and 8.

Calculation of the output of the fundamental-frequency inductor. Since a linear B–H relationship during the unsaturated part of the cycle is assumed, the flux–time curve repeats Fig. 4.9, but with μH_x and μH_E instead of H_x and H_E.

The fundamental term of the harmonic series which represents this curve is

$$\frac{2A}{\pi}\left(\cos \phi + \frac{\phi}{\sin \phi}\right)\cos \omega t$$

Thus in terms of H_E,

$$H_1 = \frac{2H_E}{\pi}\left(\cos \phi + \frac{\phi}{\sin \phi}\right)\cos \omega t = \frac{2H_E}{\pi}\left\{\left[1 - \left(\frac{H_E}{H_x}\right)^2\right]^{1/2} + \frac{\sin^{-1}(H_E/H_x)}{H_E/H_x}\right\}\cos \omega t$$

where $\phi = \sin^{-1}(H_E/H_x)$. Since $B_1 = \mu H_1$ and $E_1 = dB_1/dt$, and substituting $2\pi f$

53

for ω, we have

$$E_1 = -4\mu f. H_E\left\{\left[1-\left(\frac{H_E}{H_x}\right)^2\right]^{1/2} + \frac{\sin^{-1}(H_E/H_x)}{H_E/H_x}\right\}\sin \omega t$$

$$= -4\mu f. H_E\left\{\left[1-\left(\frac{H_E}{H_x}\right)^2\right]^{1/2} + \frac{\sin^{-1}(H_E/H_x)}{H_E/H_x}\right\} \times 10^{-8}NA \sin \omega t \text{ volts}$$

$$E_1(\text{max}) = -4\mu f. H_E\left\{\left[1-\left(\frac{H_E}{H_x}\right)^2\right]^{1/2} + \frac{\sin^{-1}(H_E/H_x)}{H_E/H_x}\right\} \times 10^{-8}NA \text{ volts}$$

where N is the number of turns in the winding and A the area of the core. Only if H_E is very small compared with H_x is $E_1(\text{max})$ proportional to H_E.

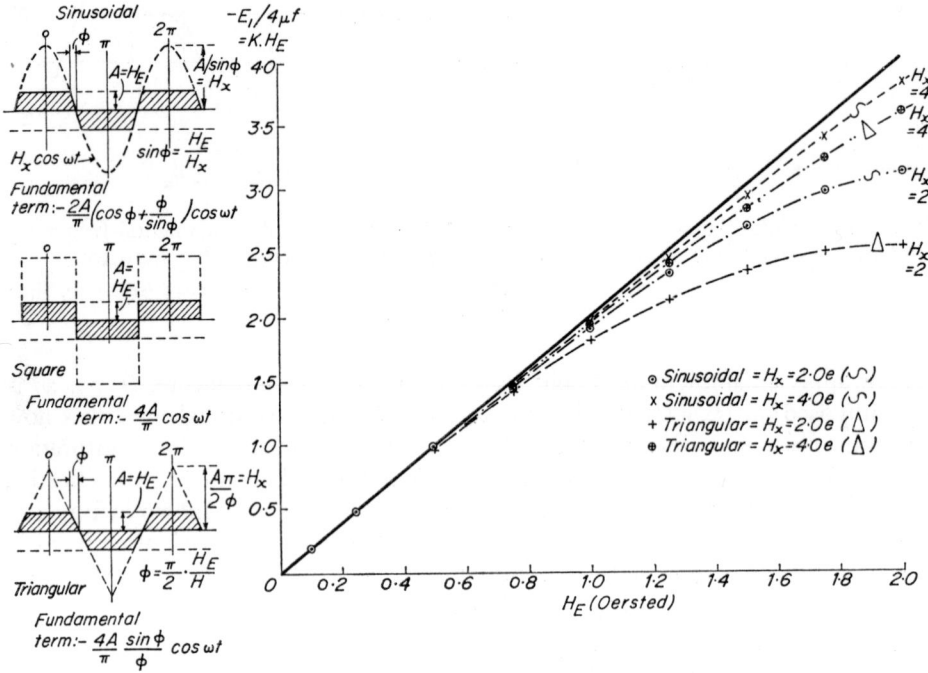

FIG. 4.10. Derivation of fundamental-frequency output from a saturable inductor for three types of excitation, with corresponding output characteristic curves

Fig. 4.10 illustrates this relation for different values of H_x and shows the improved linearity obtained by increasing H_x. There is a limitation to the magnitude of H_x since a condition can be reached when H_E is no longer effective, a greater part of the excitation cycle being in the region of saturation. Consequently a substantially rectangular wave form exists between b and c (Fig. 4.7) and does not effectively alter with changes in H_E.

If $H_E = 0$,

$$E_1 = 0$$

54

If H_E is sufficiently small to say $\sin \phi = \phi$ and to neglect $(H_E/H_x)^2$,

$$E_1 = -8\mu f . H_E \times 10^{-8} NA \sin \omega t \text{ volts}$$

If $H_E = H_x$,

$$E_1 = -4\mu f . H_x \left[\frac{\sin^{-1} 1}{1} \right] \times 10^{-8} NA \sin \omega t \text{ volts}$$

$$= -\mu . \omega H_x . 10^{-8} NA \sin \omega t \text{ volts}$$

This last value for H_E is that due to a sinusoidal excitation operating on the part bc of the B–H curve, whose amplitude is $H_x \cos \omega t$. Moreover, this value of E_1 is a maximum which remains constant until H_E becomes equal to $(2H_s - H_x)$ after which the system moves into saturation and in due course, E_1 becomes zero when $H_E = (2H_s + H_x)$.

For small values of H_E/H_x, Fig. 4.9 suggests that the H–time and B–time curves may be regarded as trapezoidal in form, or even square.

The fundamental component of a trapezoidal wave is $(4A/\pi) [(\sin \phi)/\phi] \cos \omega t$ which in terms of H_E is

$$\frac{4H_E}{\pi} . \frac{\sin \phi}{\phi} \cos \omega t = \frac{4H_E}{\pi} . \frac{\sin (\pi/2) . H_E/H_x}{(\pi/2) . H_E/H_x} . \cos \omega t$$

where $\phi = \dfrac{\pi}{2} . \dfrac{H_E}{H_x}$;

$$B_1 = \mu \frac{4H_E}{\pi} . \frac{\sin \dfrac{\pi}{2} . \dfrac{H_E}{H_x}}{\dfrac{\pi}{2} . \dfrac{H_E}{H_x}} \cos \omega t$$

Therefore

$$E_1 = -8\mu f . H_E . \frac{\sin \dfrac{\pi}{2} . \dfrac{H_E}{H_x}}{\dfrac{\pi}{2} . \dfrac{H_E}{H_x}} \sin \omega t$$

and

$$E_1(\max) = -8\mu f . H_E . \frac{\sin \dfrac{\pi}{2} . \dfrac{H_E}{H_x}}{\dfrac{\pi}{2} . \dfrac{H_E}{H_x}} . 10^{-8} NA \text{ volts}$$

For small values of H_E/H_x where $\sin \phi = \phi$,

$$E_1 = -8\mu f . H_E \times 10^{-8} NA \sin \omega t$$

The above result is given by a triangular excitation wave, and the performance of the system is comparable with that obtained with a sinusoidal excitation, except at higher values of H_E/H_x.

The fundamental component of a square wave is $(4A/\pi) \cos \omega t$ or, in terms of H_E,

$$H_1 = \frac{4H_E}{\pi} \cos \omega t$$

$$B_1 = \mu . \frac{4H_E}{\pi} \cos \omega t$$

Therefore

$$E_1 = -8\mu f . H_E \sin \omega t$$
$$= -8\mu f H_E \times 10^{-8} NA \sin \omega t \text{ volts}$$

and

$$E_1(\max) = -8\mu f H_E \times 10^{-8} NA \text{ volts}$$

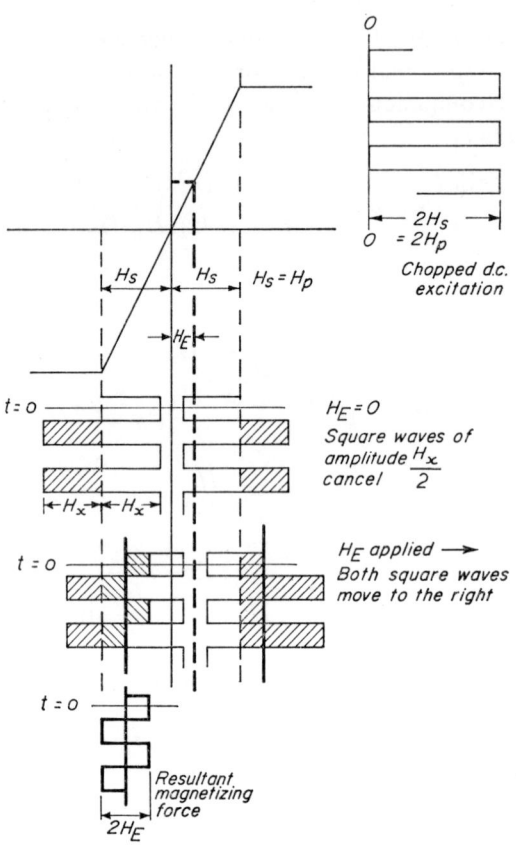

Fig. 4.11. Fundamental-frequency saturable inductor, illustrating square-wave and chopped d.c. excitation

Hence there is a strictly linear relation between H_E and E_1, provided that H_E is not greater than H_x and that $H_E + H_x$ is not greater than $2H_s$. The linear relation then extends from $H_E = 0$ to $H_E = H_x$. Output curves for triangular and square-wave excitations are shown in Fig. 4.10. Square-wave excitation is illustrated in Fig. 4.11 and suggests that the polarizing field H_p can be combined with the exciting field H_x by feeding chopped d.c. to the terminals 3 and 4 of Fig. 4.8. In fact this principle is used in a well-known form of magnetometer in which pulses of d.c. of the appropriate magnitude are applied to the input terminals of the system.

As an alternative to measuring the output at terminals 7 and 8 of Fig. 4.8, a secondary coil of many turns may be wound over the primaries, and the output voltage measured at the terminals of the coil. The secondary may be used as a backing-off coil and fed with direct current until a null indication is obtained on a meter connected across 7 and 8. The current in the secondary coil measures the backing-off field, which is equal and opposite to the ambient axial field. These circuits are shown in Figs. 4.12 and 4.13.

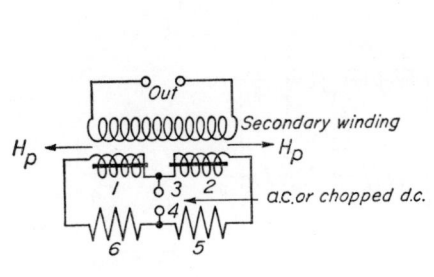

FIG. 4.12. Alternative circuit for fundamental-frequency saturable inductor

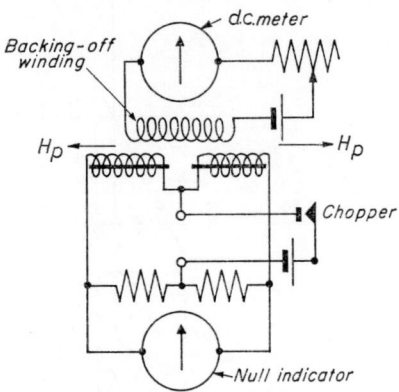

FIG. 4.13. Fundamental-frequency saturable inductor with backing-off circuit and chopped d.c. excitation

Another arrangement of a fundamental-frequency magnetometer in which the coils are excited in series is shown in Fig. 4.14. In this instance, impedances 5 and 6 complete a bridge circuit. As before, equal and opposite polarizing fields H_p are applied and the excitation is such that $H_x = H_s - H_p$. In zero ambient field, the impedances of the two windings are equal and no voltage appears across the output terminals. If an ambient field H_E is applied axially, one core is saturated during part of the magnetizing cycle, as shown by the shaded portion of the right-hand curve, and the impedances of the coils are different. Consequently, the bridge is unbalanced and an output voltage appears. This is derived from a resultant flux whose form is that of the tip of a sine wave.

The fundamental term of the harmonic series that represents this function is

$$H_1 = \frac{H_E}{\pi[1 - \cos(\phi/2)]} \cdot \frac{1}{2}(\phi - \sin\phi)\cos\omega t$$

where $\dfrac{\phi}{2} = \cos^{-1}\left(1 - \dfrac{H_E}{H_x}\right)$, or

$$B_1 = \frac{\mu H_E}{2\pi} \cdot \frac{(\phi - \sin\phi)}{1 - \cos(\phi/2)}\cos\omega t$$

57

C

Thus

$$E_1 = -\mu f H_E \frac{(\phi - \sin \phi)}{1 - \cos (\phi/2)} \sin \omega t$$

$$E_1(\text{max}) = -\mu f H_E \frac{\phi - \sin \phi}{1 - \cos (\phi/2)} \times 10^{-8} NA \text{ volts}$$

which is not a strictly linear function of E_1 with respect to H_E.

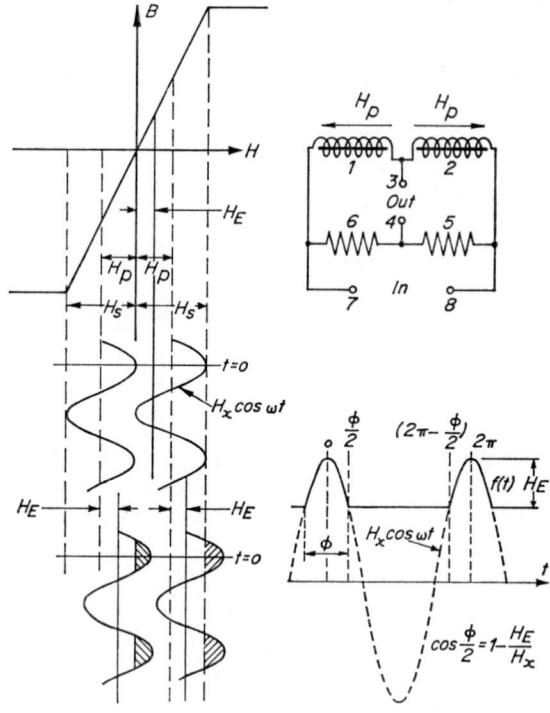

FIG. 4.14. Fundamental-frequency saturable inductor with series excitation

The fundamental term is

$$\frac{H_E}{\pi\left(1 - \cos \dfrac{\phi}{2}\right)} \cdot \tfrac{1}{2}(\phi - \sin \phi) \cos \omega t$$

Second-harmonic inductor

The second-harmonic inductor and the closely-related peak inductor can be discussed on the basis of 'pulse-shifting', i.e. the displacement in time of equal or almost equal and opposite voltage pulses so that a resultant voltage occurs, by the application of an axial magnetic field.

The second-harmonic inductor, being more versatile and sensitive than the fundamental-frequency inductor, is more widely used. Its operation is illustrated in Figs. 4.15 and 4.16 and alternative circuit diagrams are shown in Figs. 4.17 and 4.18.

58

Consider the idealized B–H curve (Fig. 4.15) *abcd* (or its corresponding μ–H curve *pqrs*); in the absence of any external axial field along the cores 1 and 2, the excitation fields, $H_x \sin \omega t$, which are applied to the cores in anti-phase, are symmetrically disposed about $H = 0$, and carry both cores into saturation and beyond.

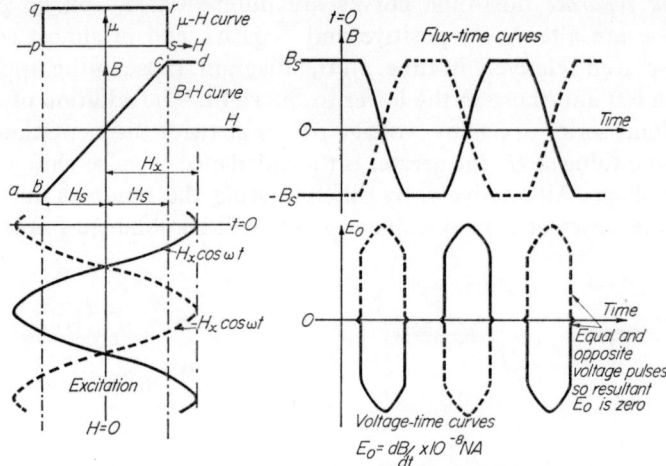

FIG. 4.15. Second-harmonic saturable inductor
Idealized B–H and μ–H curves—zero axial-field condition.

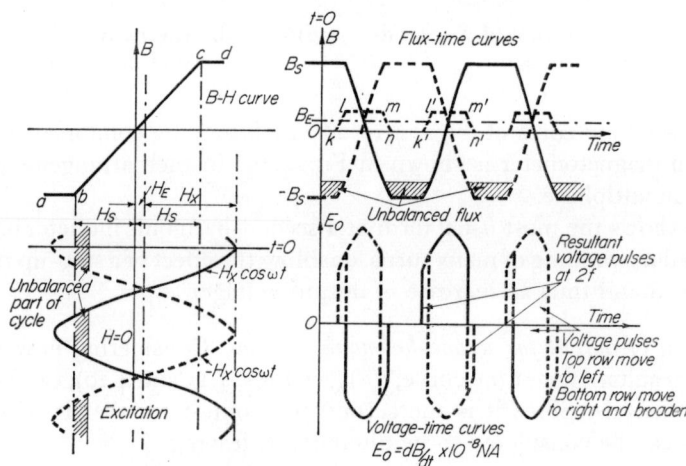

FIG. 4.16. Second-harmonic saturable inductor
Idealized B–H curve with diagrams to illustrate the derivation of output voltage pulses when an axial field is applied.

The flux–time curves, which again are equal and opposite, are shown between $B = 0$ and $\pm B_s$, B_s being the saturation flux. There is consequently no resultant flux and therefore no e.m.f. developed across the output terminals.

By differentiating the flux–time curve, the voltage pulses proportional to dB/dt will be seen to be equal and opposite, and these are shown in the diagram immediately below the flux-time curves. They have a characteristic narrow form during the excursions from $+B_s$ to $-B_s$, with $dB/dt = 0$ during the time spent in saturation.

59

If, however, an external field H_E is applied axially, the excitation field is redistributed and is now symmetrical about $H = H_E$. This causes one core to become saturated before the other, resulting in a period when the cores are unbalanced, with a resultant flux available to produce an e.m.f.

If the new *separate* flux–time curves are differentiated, voltage pulses again appear. These are alternately positive and negative and of almost equal magnitude, but displaced relatively in time. In the diagram, those in the upper row have moved to the left and those in the lower to the right. The addition of these pulses gives a resultant series of narrow voltage pulses at twice the excitation frequency. The greater the value of H_E the greater is the imbalance or pulse-shift and therefore the output voltage. Alternatively, by differentiating the *resultant* flux–time curve *klmn*, the same series of narrow voltage pulses will be obtained (see Fig. 4.16).

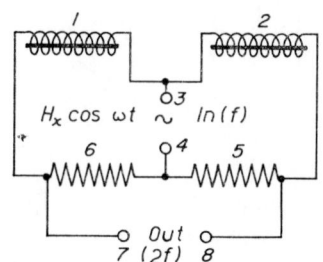

FIG. 4.17. Circuit diagram of a second-harmonic inductor

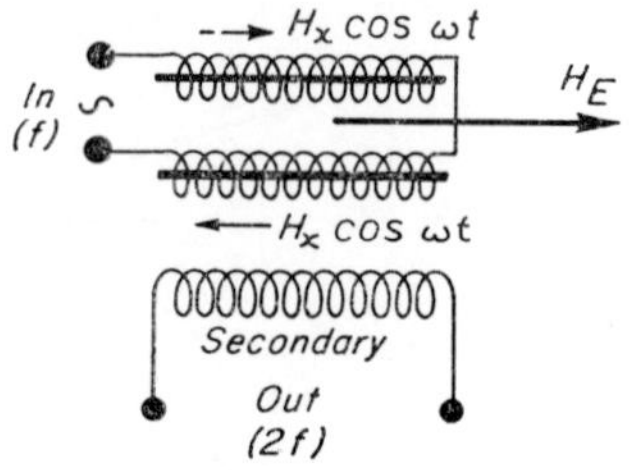

FIG. 4.18. Alternative circuit diagram of a second-harmonic inductor

The two separate cores of Figs. 4.17 and 4.18 may be combined into a single core. Such a magnetometer is shown in Fig. 4.19. In each arrangement the cores are excited in antiphase.

Fig. 4.18 shows the most usual form of a second-harmonic inductor or Flux Gate with a secondary winding of many turns, enabling the effect of a step-up transformer to be achieved and thus an increase of output voltage.

Calculation of output for second-harmonic systems. Considering now the voltage output, the resultant flux–time curve, $f(x)$ (see Figs. 4.15 and 4.20), can be subjected to Fourier analysis. Since it is the second harmonic term that is of interest, all other terms can be considered to be filtered out, leaving

$$- \frac{4}{3}\mu\frac{H_x}{\pi}(\sin^3\theta - \sin^3\alpha)\cos 2x$$

Writing $2x$ as $2\omega t$ and differentiating,

$$E_o = 10^{-8}NA \cdot dB/dt \text{ volts}$$

$$= \frac{4}{3}\mu H_x\frac{2\omega}{\pi}(\sin^3\theta - \sin^3\alpha)\sin 2\omega t \times 10^{-8}NA \text{ volts}$$

where N = turns in the winding and A = area of core.

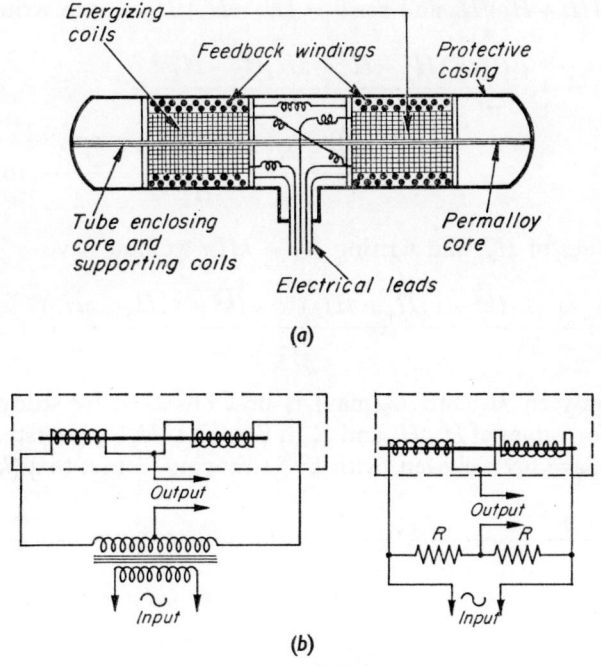

(a)

(b)

FIG. 4.19. Single-core saturable inductor
(a) Probe unit with feedback coils and (b) input circuits for the detector unit.

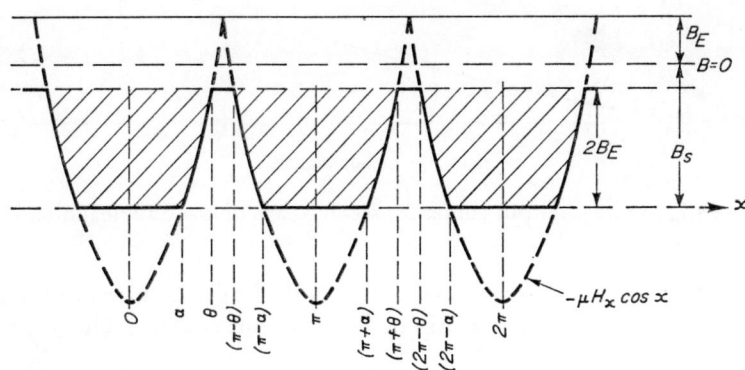

FIG. 4.20. Flux–time curve to illustrate the derivation of output voltage
from a second-harmonic saturable inductor

$$f(x) = \frac{2\mu H_z}{\pi}[\cos\alpha(2\alpha-\pi) - \cos\theta(2\theta-\pi) - 2(\sin\alpha - \sin\theta)]$$

$$+\frac{4}{3}\cdot\frac{\mu H_z}{\pi}(\sin^3\alpha - \sin^3\theta)\cos 2x + \ldots \text{ even cosine terms only}$$

Since $\cos \alpha = (H_s + H_E)/H_x$ and $\cos \theta = (H_s - H_E)/H_x$ we can write

$$E_2(\text{max}) = \frac{8}{3} \cdot \frac{\mu H_x \omega}{\pi} \cdot \frac{(H_x^2 - H_s^2 + 2H_s H_E - H_E^2)^{3/2}}{H_x^3}$$
$$- \frac{(H_x^2 - H_s^2 + 2H_s H_E - H_E^2)^{3/2}}{H_x^3} \cdot 10^{-8} NA \text{ volts}$$

For small values of H_E, and writing $H_x = kH_s$, we now have

$$E_2(\text{max}) = \frac{8}{3} \cdot \frac{\omega}{\pi} \cdot \frac{\mu}{k^2} \cdot \frac{(\overline{k^2 - 1} \cdot H_s + 2H_E)^{3/2} - (\overline{k^2 - 1} \cdot H_s - 2H_E)^{3/2}}{H_s^{1/2}} \cdot 10^{-8} NA \text{ volts}$$

The relation between H_E and $E_2(\text{max})$ is best observed by studying the set of curves plotted for values of H_s, H_E and k. In Fig. 4.21, $H_s = 4$ oersteds and $H_E = 0$ to 0·2 oersteds have been chosen, with $k^2 - 1$ ranging from 0 to 4. ($k = 1$ to 2·24.)

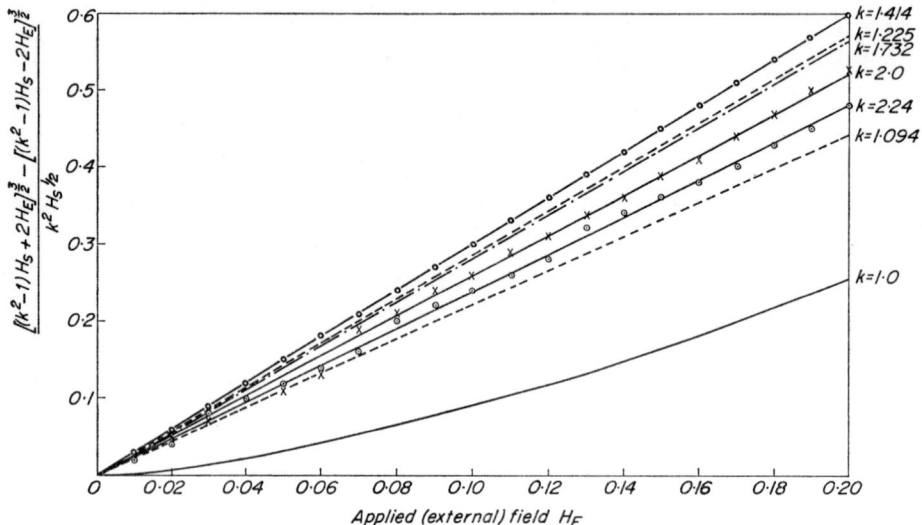

FIG. 4.21. Voltage output curves for second-harmonic saturable inductor

There is an optimum value of $k^2 - 1$, giving both the greatest output for a change of H_E and linearity in the relationship between output and H_E. This is $k^2 - 1 = 1$ or $k = \sqrt{2}$.

An essential requirement for the operation of the inductor is that saturation should occur in each core during each half cycle or

$$H_x > H_s + H_E$$

Therefore

$$k > \frac{H_E}{H_s} + 1$$

In the examples so far considered

$$\frac{H_E}{H_s} + 1 = 1 \cdot 05$$

Therefore k must be greater than $1 \cdot 05$.

When $k = 1$ we have the limiting conditions under which the excitation field is just equal to the saturation field. The requirement that the cores should become saturated during each half cycle of the excitation field is therefore not fulfilled. The curvature of the line $k = 1$ is apparent.

Over-excitation serves only to reduce the output. The relationship between output and k is plotted in Fig. 4.22.

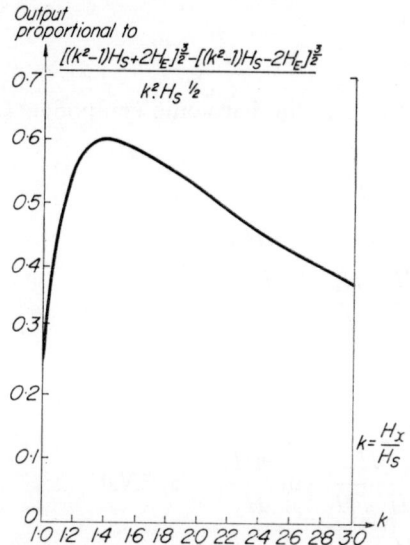

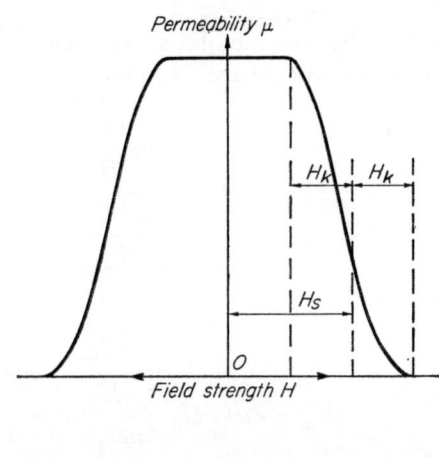

FIG. 4.22. Voltage output curve for second-harmonic saturable inductor

The curve illustrates the optimum value of H_x/H_s.

FIG. 4.23. Permeability versus field strength

The curve illustrates Feldtkeller's method of analysis. H_s is the saturation field strength and H_k the curvature field strength. $H_k = \frac{1}{2}H_s$. The μ-H curve between $H = H_s - H_k$ and $H = H_s + H_k$ is sinusoidal.

The expression for $E_2(\text{max})$ may also be written

$$E_2(\text{max}) = \frac{8}{3} \cdot \frac{\omega}{\pi} \cdot \frac{\mu}{k^2} \left\{ \left[1 + \frac{2H_E}{(k^2-1)H_s} \right]^{3/2} - \left[1 - \frac{2H_E}{(k^2-1)H_s} \right]^{3/2} \right\}$$
$$\times \frac{\{(k^2-1)H_s\}^{3/2}}{H_s^{1/2}} \times 10^{-8}NA \text{ volts}$$

which, on expansion and neglecting terms in H_E^3 and beyond, reduces to

$$E_2(\text{max}) = 32\mu f H_E \frac{(k^2-1)^{1/2}}{k^2} \times 10^{-8}NA \text{ volts}$$

If now we make $(k^2 - 1) = 1$,

$$E_2(\text{max}) = 16\mu f . H_E \times 10^{-8} NA \text{ volts}$$

Feldtkeller's method of analysis. R. Feldtkeller[*] uses the μ–H curve, instead of the B–H curve, and an excitation of triangular waveform in his analysis of the output of inductor systems.

The μ–H curve is such that the permeability does not fall suddenly to zero at the point of saturation but follows, for the sake of simplicity, part of a sine curve from maximum to zero (Fig. 4.23).

A further field-strength term is introduced: H_k, the 'curvature field strength'. For all practical purposes,

$$\frac{H_k}{H_s} = \frac{1}{2}$$

Analysis by this method shows that, for each core, the harmonic components of the output voltage are given by

$$E_n = 8\mu f . H_s \frac{\sin \dfrac{nH_s}{H_x}}{\dfrac{nH_s}{H_x}} . \frac{\cos \dfrac{nH_s}{H_x} . \dfrac{H_k}{H_s}}{1 - \dfrac{2}{\pi} . \dfrac{nH_s . H_k}{H_x . H_s}} \cos \frac{nH_E}{H_x} \times 10^{-8} NA$$

where $n = 1, 3, 5 \ldots$ etc.

$$E_m = 8\mu f . H_s \frac{\sin \dfrac{mH_s}{H_x}}{\dfrac{mH_s}{H_x}} . \frac{\cos \dfrac{mH_s . H_k}{H_x . H_s}}{1 - \dfrac{2}{\pi} . \dfrac{mH_s . H_k}{H_x . H_s}} \sin \frac{mH_e}{H_x} \times 10^{-8} NA$$

where $m = 2, 4, 6 \ldots$ etc.

H_s/H_x is taken as $1/1 \cdot 5$ and corresponds to $1/k$ used in previous expressions. In the conventional arrangement of inductors, where two cores and windings are used, either side by side or collinearly, the odd harmonic components can be shown to cancel one another and the even components to add. Thus E_2 for a second-harmonic inductor is given by $E_2 = 8\mu f \sin (2H_E/H_x)(0 \cdot 92 \, H_x \times 10^{-1} NA)$. For small values of H_E we can write $\sin (2H_E/H_x) = 2H_E/H_x$. Then

$$E_2 = 16\mu f H_E . 0 \cdot 92 \times 10^{-8} NA \text{ volts}$$

If the curvature field strength is not taken into account, i.e., a sharp knee or a sudden discontinuity in the form of the B–H curve is assumed, making H_k equal to zero, and if a value of $k = \sqrt{2}$ is used,

$$E_2 = 16\mu f H_E . 0 \cdot 988 \times 10^{-8} NA \text{ volts}$$

[*] Feldtkeller, R., 'The measurement of magnetic fields by using Förster probes' (Trans. of German military document deposited in Admiralty Compass Observatory library).

In their analysis of the second-harmonic output of a magnetic modulation, Williams and Noble* derive expressions similar to the above.

Useful functions of x that arise in the analysis of the output of saturable inductors are given in Fig. 4.24.

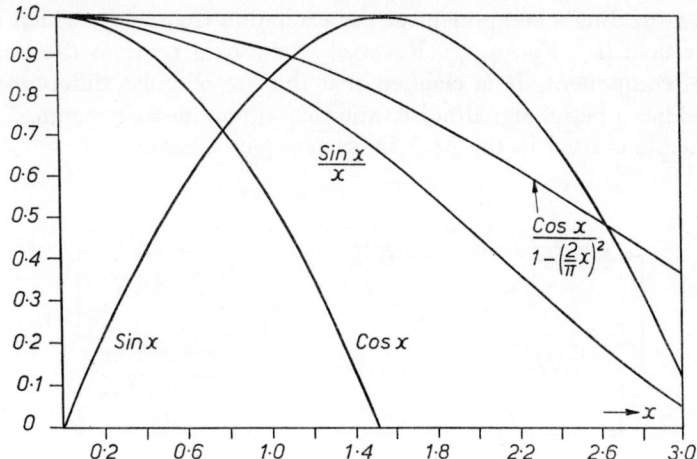

FIG. 4.24. Some useful functions used in saturable-inductor output analysis

The peak inductor

One version of the peak inductor takes the same form as the second-harmonic inductor, except that instead of extracting only the second-harmonic component of the output, the peak value of each voltage impulse is used. J. Squires gives the relative sensitivities as 10–12 $\mu V/\gamma$ for the second-harmonic system and 35–40 $\mu V/\gamma$ for the peak system.† The signal-to-noise ratio, however, is much higher for the former.

An interesting and useful variant is the unbalanced or pulse-difference inductor attributed to Schmitt.‡ This instrument is constructed in the same way as the second-harmonic inductor, but one half is deliberately unbalanced by connecting a suitable impedance in parallel with one exciting winding, as shown in Fig. 4.25. The exciting currents and, therefore, the fields are unequal. It will be seen from Fig. 4.26 that the resulting inequality of the flux–time curve produces a set of unequal and opposed voltage pulses. A salient feature of this system is that it takes advantage of an open hysteresis loop and of the fact that the cores go into saturation faster than they come out, giving rise to quasi-trapezoidal instead of rectangular pulses. The difference is now no longer zero in zero axial field, but a series of equal narrow pulses, alternately positive and negative, the interval between two positive (or two negative) pulses being 2π. For the purpose of measurement, the negative

* Williams, F. C., and Noble, S. W., 'The fundamental limitations of the second harmonic type of magnetic modulator as applied to the amplification of small d.c. signals', *J. Inst. elect. Engrs*, 1950, **97**, Part II, 445.

† 'Airborne magnetic surveys', *Nature*, 13 Feb. 1954 (report of geophysical discussion, Royal Astronomical Society, 27 Nov. 1953).

‡ U.S. Patent 2560132 (1951).

pulses are reversed in the amplifier and a series of positive pulses appears. The fundamental component of this function, i.e. at the excitation frequency, is zero.

When an ambient axial field is applied, the original sets of quasi-trapezoidal voltage pulses are shifted, as in the second-harmonic type of inductor. The resultant alternate pulses are widened and the intervening ones are narrowed, so that the function now exhibits a component at the excitation frequency, which is proportional to the field (see Fig. 4.27). Reversal of the field reverses the phase of this fundamental component. It is claimed that the use of pulse differences is more sensitive and has a better signal/noise ratio than direct measurements.

This principle is used in the M.A.D. system (see Chapter 11).

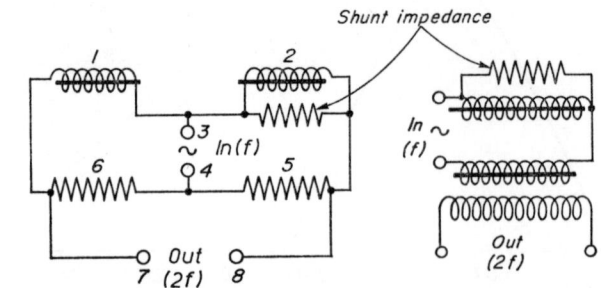

FIG. 4.25. Unbalanced or pulse-difference saturable inductor

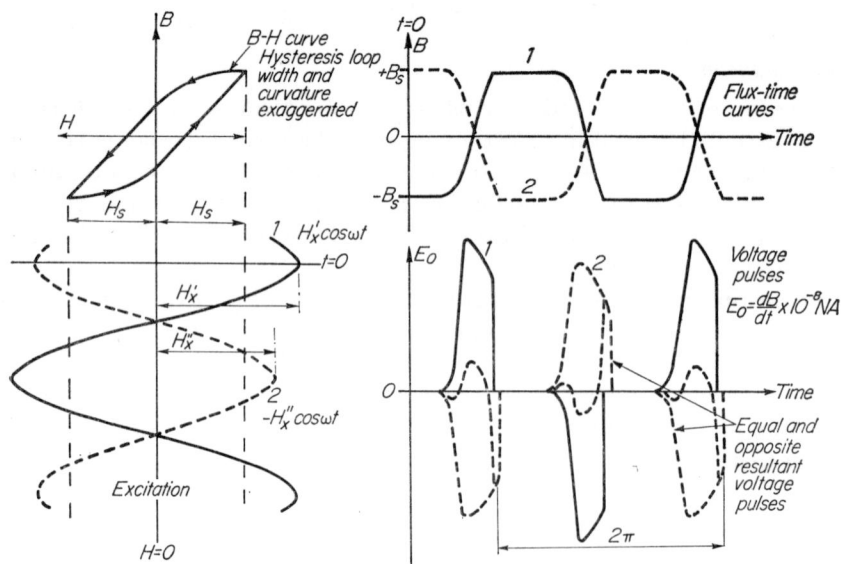

FIG. 4.26. Idealized *B–H* curve and derivation of output voltage pulses of an unbalanced or pulse-difference saturable inductor— zero axial field condition

High-frequency modulated inductor*

When the amplified output of a saturable inductor is applied to a measuring circuit or to a servo motor, it is sometimes necessary to provide a reference signal

* Hine, A., and Hitchins, H. L., 'Apparatus for measuring and detecting magnetic fields', Brit. Pat. 619525, 1946 (Modulated H.F. saturable inductor).

66

with which the phase of the output is compared. With fundamental-frequency and the unbalanced inductors, the reference signal will be at the excitation frequency and can conveniently be obtained from the excitation source, but with the second-harmonic inductor, a reference signal at twice the excitation frequency is needed and therefore a frequency-doubling circuit is required.

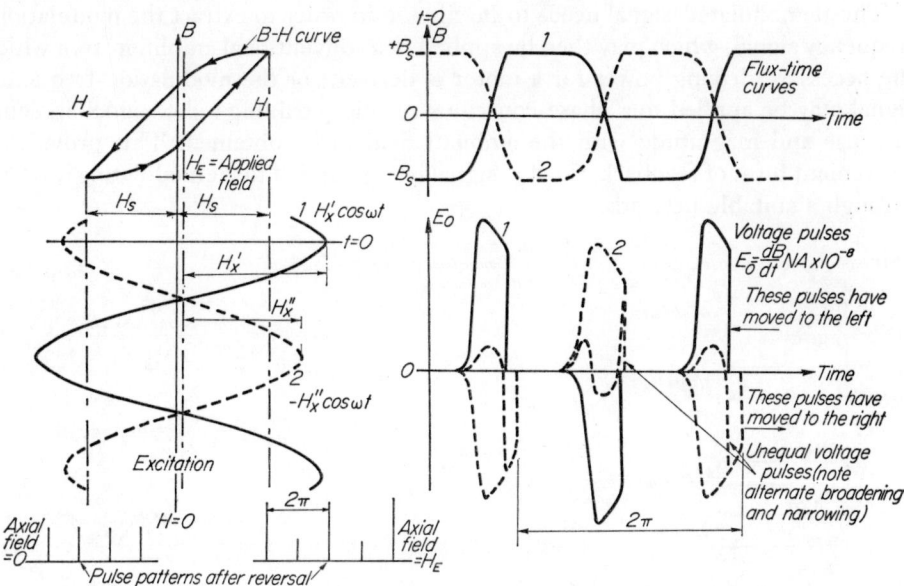

FIG. 4.27. Idealized B–H curve and derivation of output voltage pulses of an unbalanced or pulse-difference saturable inductor when **an axial** field is applied

A method of exploiting the advantages of a second-harmonic saturable inductor with high-frequency excitation, but avoiding the complications of frequency-doubling and a high-frequency output is to apply a modulation frequency to the inductor so that the ambient field is $H_M \sin \omega t + H_E$ instead of H_E. The excitation is $H_x \sin \omega_1 t$ where $\omega_1 \gg \omega$, and the output from the inductor will be at a frequency ω_1/π ($= 2f$) modulated by $H_M \sin \omega t$. The modulation may be applied to the secondary winding through a suitable network or directly to a tertiary winding [Fig. 4.28 (a) and (b)]. The output signal in zero field will resemble a very much over-modulated 'carrier', as shown in Fig. 4.28(c). The phase of the carrier is reversed at each 'loop' of the diagram, as the direction of $H_M \sin \omega t$ reverses at each half cycle of its oscillation.

Now, when a steady ambient field H_E is applied to the inductor, it will add to $H_M \sin \omega t$ and the modulation pattern is now as shown in Fig. 4.28(d). If the direction of H_E be reversed, the pattern undergoes a phase change of 180° [Fig. 4.28(e)].

If this signal is demodulated, the result is, as indicated in Fig. 4.28 (f) and (g), a complex waveform with a marked component in $\sin \omega t$. Thus, by using a high excitation frequency from an oscillator and modulating at any convenient low frequency derived, perhaps, from the mains, an output signal at mains frequency will be obtained, whose amplitude is dependent on the applied ambient field and whose phase is determined by the direction of that field. This system shares the advantage of the fundamental-frequency inductor in that the power supply and output signal have the same frequency, so often desirable when servo-motors have to be driven.

The demodulated signal needs to be filtered in order to extract the modulation-frequency signal, which may then be applied to a conventional amplifier from which the necessary driving power for a motor is derived; or the modulation-frequency signal may be applied to a phase-conscious rectifier enabling a d.c. signal agreeing in sense and magnitude with the ambient field to be obtained. This provides a convenient form of feedback, the d.c. signal being applied to the modulating winding through a suitable network.

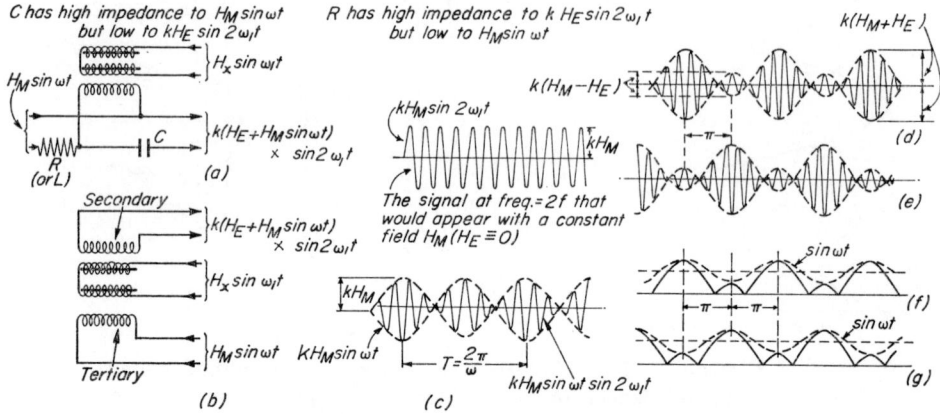

FIG. 4.28. Modulated high-frequency saturable inductor
Circuit and waveform diagrams.

The output from the phase-conscious rectifier, instead of being used for feedback, may be taken directly to an indicator or recorder for magnetic field measurements. Typical schematic arrangements are shown in Fig. 4.29 (a) and (b). Fig. 4.30 illustrates a method of using a common winding for output, modulation and feedback. The system provides an output free from much of the spurious noise that can arise in a simple inductor system, since the modulation tends to clean up the signal and noise is rejected into sidebands, well separated from the output signal frequency. In other words, the detector signal is effectively smoothed and filtered without the use of elaborate circuits. Some of the advantages of the pulse-difference inductor previously described are realized in that, except for the instant when the modulation is zero, the inductor is never in zero field and throughout the greater part of the modulation cycle, it is producing a signal well out of its inherent noise range. Thus the envelope of the modulated output of the inductor is free from disturbance, and so in zero ambient field a noise-free output is possible, since the high-frequency noise components are effectively avoided or suppressed. A sensitivity of one gamma is readily achieved. A recording magnetometer using this method based on Fig. 4.29(b) is described in Chapter 6.

68

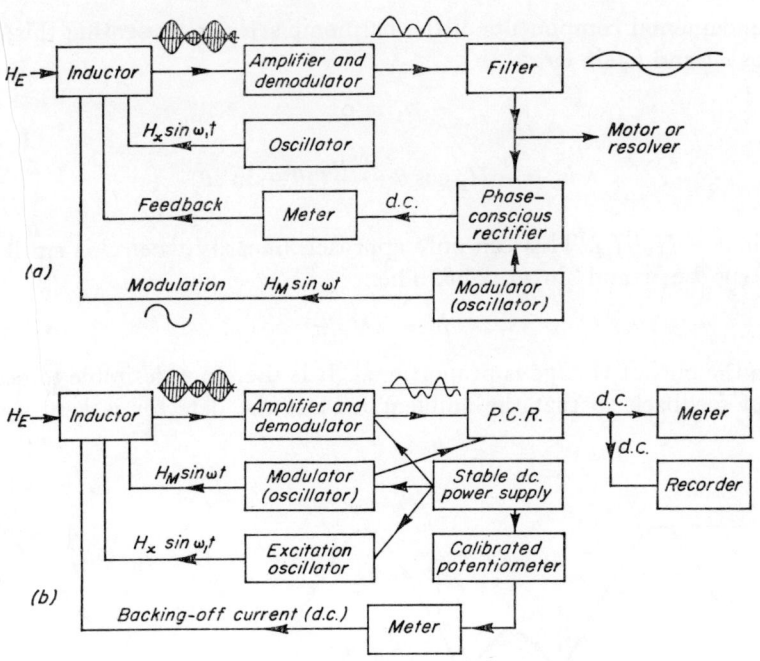

FIG. 4.29. Block circuit diagrams for systems
using the MHF saturable inductor

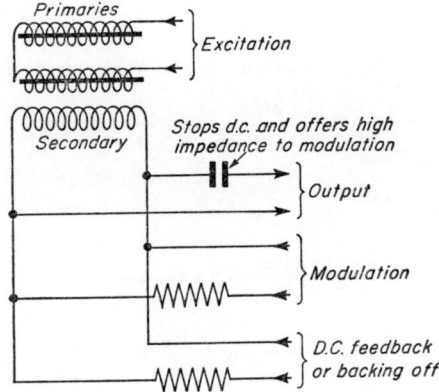

FIG. 4.30. Alternative circuit diagram for the MHF saturable inductor

The diagram shows the connections of the modulation and feedback circuits
to the secondary winding.

Linearity of the output. If it is proposed to use the output as an absolute measure
of H_E, such as in a sine–cosine system or a magnetometer, it is important that
$H_E \ll H_M$. The demodulated waveform (Fig. 4.31) can be expressed as

$$
y = \left(H_M \sin \omega t + H_E \right)_0^{\pi+\theta} + \left(-H_M \sin \omega t - H_E \right)_{\pi+\theta}^{2\pi-\theta} + \left(H_M \sin \omega t + H_E \right)_{2\pi-\theta}^{2\pi}
$$

If the fundamental components of the harmonic series representing this function are $a_1 \cos \omega t$ and $b_1 \sin \omega t$,

$$a_1 \equiv 0$$

and

$$b_1 = \frac{4}{\pi} H_E \cos \theta + \frac{H_M}{\pi}(2\theta - \sin 2\theta)$$

where $\sin \theta = H_E/H_M$. This can only approach linearity when θ is small enough to write $\cos \theta \simeq 1$ and $\sin 2\theta \simeq 2\theta$. Then

$$b_1 = 4H_E/\pi$$

to which the output voltage is proportional. It is therefore desirable to use a high degree of feedback so that the ambient field at the detector is kept near zero.

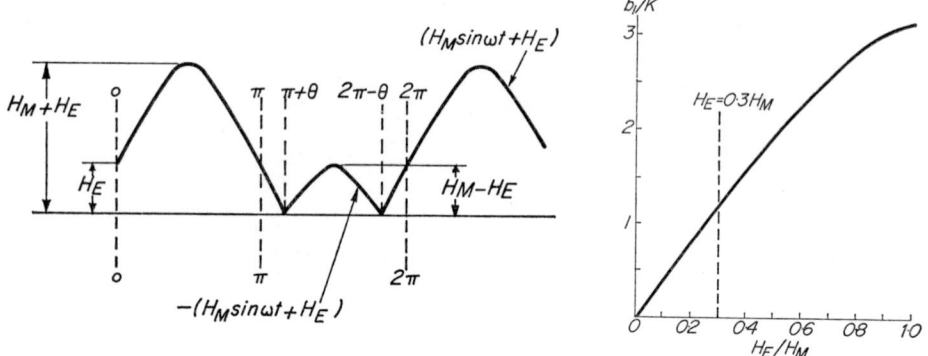

FIG. 4.31. Analysis of output from MHF saturable inductor

$$a_1 = 0 \qquad b_1 = \frac{4H_E}{\pi} \cos \theta + \frac{H_M}{\pi}(2\theta - \sin 2\theta)$$

Put

$$\frac{H_M}{\pi} = k$$

Then

$$\frac{b_1}{k} = \frac{4H_E}{H_M} \cos \theta + 2\theta - \sin 2\theta$$

and

$$\theta = \sin^{-1} \frac{H_E}{H_M}$$

The system reaches its limit when $H_E = H_M$. The carrier is now modulated by exactly 100 per cent and the output is $H_M \sin \omega t$. If H_E is increased further, there is no corresponding increase in the signal. Modulation is now less than 100 per cent. Moreover, $(H_M + H_E)$ must never be so large that the linearity of the inductor output is impaired.

Sperry Flux Valve

A widely used form of second-harmonic saturable inductor is the Flux Valve, illustrated in Fig. 4.32, developed by the Sperry Gyroscope Co. Ltd. This consists of three pairs of radial Permalloy arms, spaced at 120° and connected in the centre by a Permalloy core, P. A primary or excitation winding is placed round the central core, and each pair of arms carries a secondary winding. These windings are star-connected.

Consider one pair of arms and the primary. When excited, the instantaneous magnetic flux is, for example, inwards in the top arm and outwards in the bottom arm, returning across the air-gap. When placed parallel to an ambient field H, the excitation field subtracts from H in the top arm and adds to it in the lower, creating a situation similar to that occurring in the conventional saturable inductor of Fig. 4.18. A secondary winding about the pair of arms will have a second-harmonic output at its terminals. The iron circuit may be closed at the ends of the radial arms and 'flux collectors' in the form of arcuate extensions may be added. The Flux Valve is used in the Gyrosyn compass and certain American equipment.

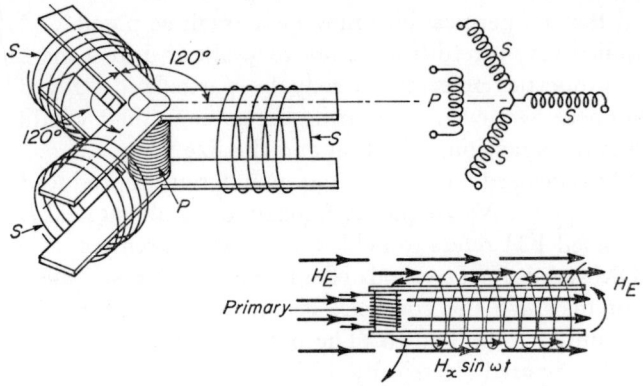

FIG. 4.32. The Sperry Flux Valve

DESIGN OF CORES

The development of a high permeability is obviously important, but it must, of course, be realized that the absolute value of permeability is not applicable. The value required is the effective permeability, which depends on the 'demagnet-izing' factor. That is to say, if a rod of permeable material is placed in an axial magnetic field, poles are produced near each extremity. These poles produce a magnetic field similar to that of a bar magnet placed so as to oppose the ambient axial field, with a corresponding diminution in the flux density within the core material.* This results in an effective permeability of less than the absolute value and approaching the absolute value only for a rod that is very long compared with its diameter. Practical considerations lead to cores in which $l = 200\ d$ approximately, e.g. 4 in. long × 26 S.W.G.

A further aspect is the frequency of excitation. Whilst 400 c/s is a common value, improved results are often obtained at 1 kc/s or higher provided the excitation power is adequate. However, beyond some 7 or 8 kc/s, trouble may be experienced from lack of penetration of the core—or 'skin effect'—in which case the material of the core is not being used efficiently.

* Starling, S. G., *Electricity and Magnetism for Degree Students* (Longmans Green & Co.), 8th ed., 1953, pp. 271–3.

The dimensions of the core may be adjusted to give certain desired characteristics. For instance, a long thin core will provide a high output for a given driving power compared with a shorter core, but the latter will provide a greater range of linearity. In other words, range and sensitivity are not always compatible. This is understood when the relative values of effective permeability are considered.

A very important consideration is the matching of the cores. When small values of ambient field have to be measured, the noise level of the detector needs to be low. Hence unwanted fundamental frequency response and other odd harmonics, which normally do not exist in an ideal saturable inductor, must be kept to a minimum, however good any subsequent filtering circuit may be. Cores should be matched in dimensions and in their magnetic properties, and this latter consideration calls for effective and uniform annealing.

The greatest care must be taken in annealing the material, so that it is rendered stress-free and that its permeability may be as high as possible. After annealing, it must be handled very carefully so as not to strain or distort the core in any way, otherwise the properties referred to may be partially destroyed. Considerable precautions are necessary even in securing the cores in their windings, in order to avoid strain due to expansion, vibration and the like.*

Feldtkeller has devised an interesting nomogram, reproduced in Fig. 4.33, relating excitation power N_1, frequency f, maximum ambient field H_o, and length l. The column headed FM refers to field strength measurement, as described so far in the text. FDM stands for a method of field difference measurement, an adaptation of the inductor in which the cores are separated, each in ambient fields whose difference is required to be known. The outputs from each half of the inductor thus formed are compared electrically.

MOUNTING INDUCTOR ELEMENTS

A single inductor element behaves very much as a double-pivoted compass needle (Fig. 3.3) in that it must be perfectly level before accurate determinations of the Earth's horizontal field can be obtained. The various arrangements of multiple-axis inductor magnetometers measure H or some component of H only when stabilized in the horizontal plane by a gyroscope or when pendulously mounted as, for instance, in a gimbal system. A three-axis system can admittedly be used to measure the three components of field in the deck plane and normal to the deck plane in a ship, but before this can have any meaning in relation to the magnetic meridian, the field strengths measured must be related to the true vertical (or horizontal) plane. Thus, whether they are to be used as magnetometers or as compasses, inductor elements must be mounted with the same stringent precautions as are taken in mounting pivoted magnetic needles.

In determining the direction of the magnetic meridian, an inductor may be orientated in the horizontal plane in such a direction that its output is nil, indicating that it is lying in zero field and therefore normal to the magnetic meridian. If by mischance there is a tilt, β, of the inductor [see Fig. 4.34(a)], a component $Z \sin \beta$

* Annealing is carried out in an atmosphere of dry hydrogen at 1100°C for about 4 hours. The cores may be in the form of wires suspended under slight tension to ensure straightness or cut to length and laid on a refractory bed. The refractory is important—any free silica must be avoided and a magnesia base is generally found to be satisfactory.

will act along the axis of the inductor. Consequently, zero field will be found only by rotating the device through an angle δ from the true east–west direction, where δ is given approximately by

$$H \sin \delta = Z \sin \beta$$

or

$$\sin \delta = \frac{Z}{H} \sin \beta = \tan D \sin \omega$$

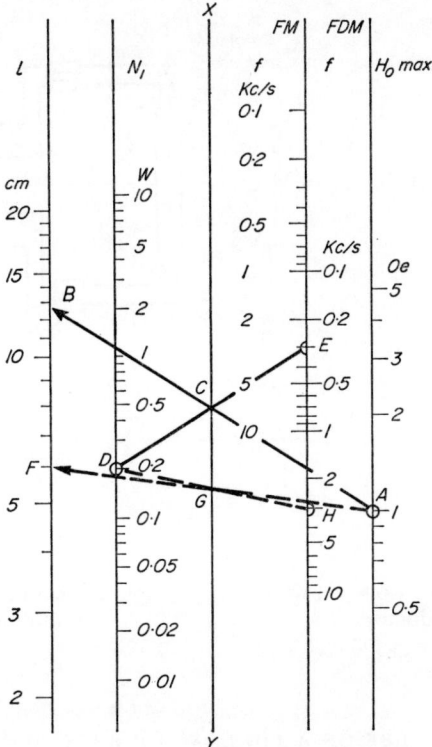

FIG. 4.33. Feldtkeller's nomogram

Examples. Join the value of the maximum field strength (right-hand line) to the length of the core (left-hand line); i.e. draw AB, which intersects the line XY in C. Join the excitation frequency (line FM) to C and produce to D on the scale of N_1, the power output required from the oscillator. The full index lines show that with a maximum field strength of 1 oersted, a core length of 12·8 cm, and an excitation frequency of 3 kc/s, the power output required from the oscillator is 0·2 watt. The broken index lines (AD and FH, which have the common point G on the line XY) show how to find the length of core required to measure field strengths up to 1 oersted with the above excitation frequency and power output. Thus, the nomogram depicts the relationship between the various constants of an inductor.

For small angles, $\delta = \beta \tan D$. In the latitude of London $\tan D = 2\cdot32$. Thus for 1° of tilt, $\delta = 2° 20'$.

In Fig. 4.34(*b*) the inductor is tilted in the meridian. No error in azimuth occurs here in seeking the maximum field strength, which, however, differs from H by approximately

$$H(\cos \beta - 1) + Z \sin \beta$$

Where an inductor element is used simply as a probe for exploring a magnetic field, the field measured is that parallel to the axis of the probe.

For a further discussion of tilt errors due to acceleration, reference should be made to Chapter 3, since the fundamental equations apply equally to an inductor and a needle.

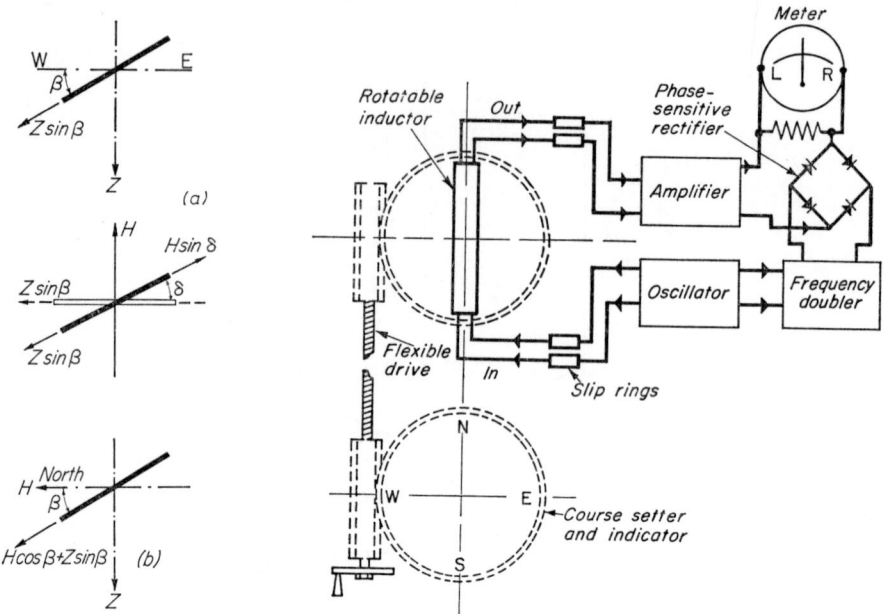

FIG. 4.34. Errors in direction due to tilting of an inductor

For zero output, $H \sin \delta = Z \sin \beta$, so that $\sin \delta = (Z/H) \sin \beta$.

FIG. 4.35. A simple course indicator using a saturable inductor

USE OF A FOLLOW-UP SYSTEM

So far the inductor has been described as a device for producing a voltage proportional to the field acting along its length.

Various techniques suggest themselves whereby a recording meter can be adapted for measuring field strength continuously against time. There are numerous applications of inductors, notably in compass development, where a continuous and automatic indication of direction is required. An arrangement similar to that shown in Fig. 4.2 which could be used is shown in Fig. 4.35, but this hardly satisfies the requirements of a practical compass system. Fig. 4.36 illustrates a better method. Here an output amplifier provides power to drive a motor which is geared to the inductor in such a way that the latter is continually maintained in a state of zero output, i.e. orientated east–west. A compass card attached to the inductor will now indicate direction. The inductor must, of course, be kept horizontal.

The methods of amplification and frequency doubling and the design of the necessary filters to ensure the most efficient use of the desired harmonic component of the inductor output are subjects for text-books on electronics. The circuits, however, are quite straightforward and may include phase-advance systems for the elimination of hunting.

Resolvers

By using several inductors in combination (usually two or three) a follow-up system can be devised without recourse to actual rotation of the inductor elements. Two inductors in the horizontal plane and at right angles to each other will be acted upon by fields equal to $H \sin \theta$ and $H \cos \theta$, as shown in Fig. 4.37. Corresponding signals will be developed, namely $k_1 H \sin \theta$ and $k_2 H \cos \theta$ where k_1 and k_2 are constants depending on the inductor characteristics, and are normally assumed to be equal. If these signals are applied to the stator windings of a resolver consisting

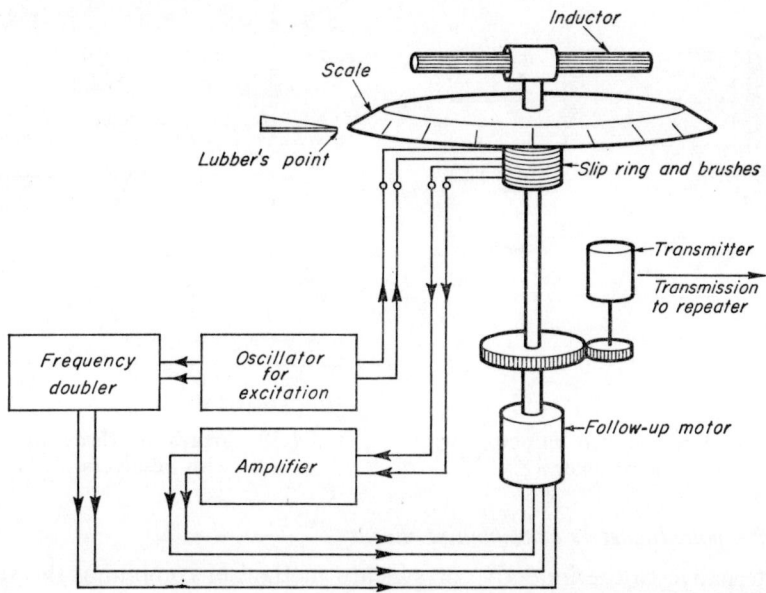

FIG. 4.36. A simple compass incorporating a saturable inductor and a follow-up system

of two coils at right angles to each other, a resultant alternating field will be generated in the space surrounded by the stator windings. The direction of this field ϕ is related to the direction of the stator windings in the same way as the direction of the Earth's field H is to the inductor assembly.

$$\tan \phi = \frac{k_1 H \sin \theta}{k_2 H \cos \theta} = \tan \theta, \text{ if } k_1 = k_2$$

therefore $\phi = \theta$, or in terms of heading, $\zeta' = \zeta$.

The rotor of the resolver which occupies the space within the stator windings will have induced in its windings by the resultant alternating field, a voltage whose phase and magnitude are determined by the angular position of the rotor. There will be two positions 180° apart where this voltage will be zero. Thus, by amplifying the rotor voltage and applying the output to a follow-up motor geared to the resolver rotor, the stable null position can be maintained and a compass card attached to the rotor spindle will provide directional information. It will be necessary to provide a frequency doubler to generate a reference signal for the motor of a second-harmonic inductor.

Since no rotation of the inductor system is required, it is now possible to site it remotely from the rest of the apparatus and thus we have the basic arrangement of a practical inductor compass. A similar system may be devised using three inductors and a synchro-element of conventional form (Fig. 4.38).

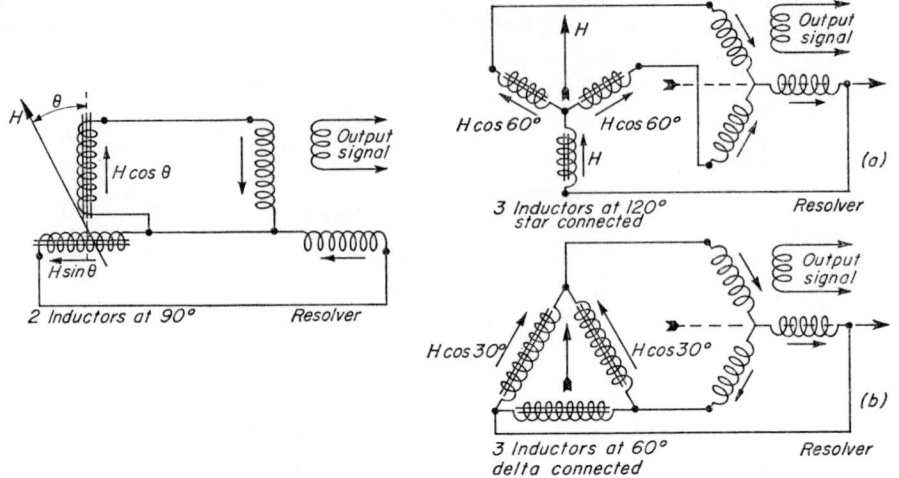

FIG. 4.37. Array of two inductors with a resolver

FIG. 4.38. Arrays of three inductors with resolvers

Sine–cosine potentiometers as combining means

An alternative to the inductive or synchro method of combining the sine and cosine vectors is the use of sine–cosine potentiometers (see Fig. 4.39). These potentiometers are flat slab-wound resistors having a central contact and a radial arm with a second contact which describes a circle about the first one.

If a voltage V is applied across AA, when contact D is at B, a voltage equal to $\frac{1}{2}Vk$ is tapped off, k being a constant depending on the radius of the contact arm and the half-length of the resistance winding. In any other position, the voltage is proportional to the resistance intercepted between C and D, i.e. to the number of turns of wire between C and D, and therefore to the distance x.

Now $x = CB \cos \alpha$, where α is the angle of displacement from line AA. Thus, the voltage tapped off across CD is $\frac{1}{2}Vk \cos \alpha$, or $K \cos \alpha$. Similarly, a voltage proportional to $\sin \alpha$ may be obtained by starting α from a line at right angles to AA. In practice, it is usual to have an L-shaped contact arm, pivoted at the corner of the 'L' so that sine and cosine functions may be derived from the same potentiometer, if required.

Two potentiometers are shown in Fig. 4.40, and voltages proportional to $\sin \theta$ and to $\cos \theta$ are applied across AA and A'A' respectively. The potentiometers AA

76

and A'A' then have their contacts arranged to tap off voltages proportional to cos α and sin α respectively. Thus the outputs will be

$$k' \sin \theta \cos \alpha \quad \text{and} \quad k' \cos \theta \sin \alpha$$

These are subtracted by suitable arrangement of the connections, so that the output at XX is

$$k'(\sin \theta \cos \alpha - \cos \theta \sin \alpha)$$

which is zero when $\theta = \alpha$.

Thus, by driving the potentiometer contacts with a follow-up motor energized by the amplified output from XX, a null position may be sought, as in the case of the inductive resolver.*

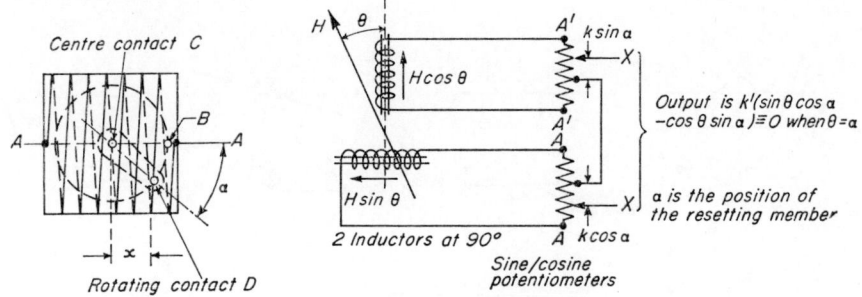

FIG. 4.39. Sine–cosine potentiometer   FIG. 4.40. Sine–cosine potentiometers used with an array of sine and cosine inductor elements

SATURABLE INDUCTORS IN A NON-UNIFORM FIELD

The ambient field acting on a saturable inductor has been assumed to be uniform over the whole length of the inductor. If, however, the inductor is placed in a non-uniform field such as may occur in the proximity of a bar magnet, the question arises as to the strength of field measured by the inductor. It has been demonstrated experimentally, and can be shown theoretically that the field measured by an inductor is proportional to the field at its *centre*, i.e. half-way between the two ends. Thus, for instance, in the case of the 'end-on' position, the field measured is not the mean field but that at the mid-point of the field–distance curve; similarly, in the 'broadside' position, the field measured is different from the mean field. The factor of proportionality is approximately unity.

This makes it possible to describe the behaviour of an array of inductors placed in proximity to a compass needle as a method of obtaining a distant transmission of compass heading. Such 'compass-aided inductors' form the basis of the Magnesyn compass and the Kelvin-Hughes transmitting magnetic compass (see pp. 168 and 194 in Chapter 8).

COMPASS-AIDED INDUCTOR

In its simplest form, an inductor with a follow-up system as in Fig. 4.36 is used to detect and to indicate the direction at right angles to the reverse field from a

* See also section on errors in sine–cosine systems in Chapter 8.

pivoted-needle compass situated vertically above it, thereby enabling compass heading to be transmitted to a distant position. It has certain advantages over a simple Earth's field inductor in that a stronger ambient field is available, the greater part of which, by virtue of its source in a pivoted compass needle, is horizontal only and has no vertical component. This means that, in certain circumstances, effective dip is much reduced and that the deviations due to tilts may be appreciably less than those in a simple inductor system.

Deviation due to tilt

The field acting in the plane of the inductor in Fig. 4.41(a) is $(H_E - H_c)$ in the direction of magnetic north, where H_c is the field from the compass needle and H_E is the horizontal component of the Earth's magnetic field.

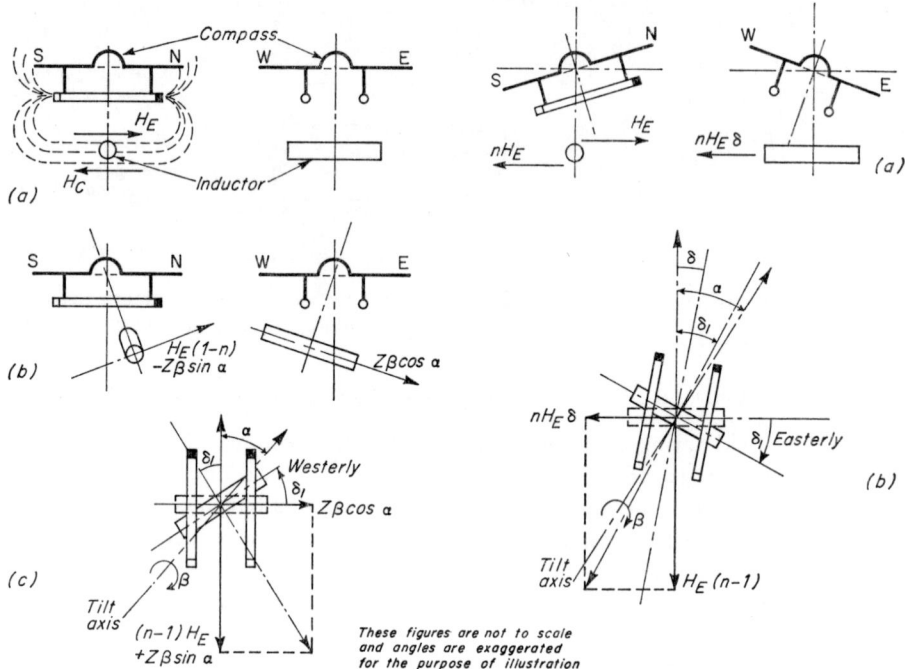

FIG. 4.41. Compass-aided inductor
Vessel heeled

FIG. 4.42. Compass-aided inductor
Compass system tilted

If $H_c = nH_E$ where n must be greater than 2, the ambient field to north in the plane of the inductor is $H_E(1-n)$. The inductor, which is regarded as being fixed to the ship (i.e. not gimballed), by seeking zero axial field, lies at right angles to the ambient field.

Now if the plane containing the inductor be tilted as shown in Fig. 4.41(b) (for example, when the ship is heeled), and the angle of tilt is β about an axis whose direction with respect to magnetic north is α, the northerly field acting in the plane of the inductor is

$$H_E(1-n) - Z\beta \sin \alpha$$

78

and the easterly field acting in the same plane is

$$Z\beta \cos \alpha$$

provided β is small (see also Chapter 3, p. 24).

The direction of the resultant field, at right angles to which the inductor will lie by virtue of its zero-field-seeking property [see Fig. 4.41(c)], is given by

$$\tan \delta_1 = \frac{Z\beta \cos \alpha}{H_E(1-n) - Z\beta \sin \alpha}$$

$$= -\frac{Z\beta \cos \alpha}{(n-1)H_E + Z\beta \sin \alpha}$$

For small angles, $\tan \delta_1 = \delta_1$, so

$$\delta_1 = \frac{-\beta \tan D \cos \alpha}{n - 1 + \beta \tan D \sin \alpha} \tag{4.1}$$

where δ_1 is the deviation from magnetic north. This is appreciably less than the deviation of a simple inductor when subjected to a similar tilt. When $\alpha = 0$,

$$\delta_1 = -\frac{\beta \tan D}{n - 1} \tag{4.2}$$

(a westerly deviation).

Deviation due to acceleration

If the compass system is accelerated (as by a change of speed of the vessel) so that the compass card and needle system are tilted into the false vertical, the inductor meanwhile remaining horizontal, the ambient field acting on the inductor will be composed of the Earth's horizontal field and the field from the compass magnet, which latter has been deviated from magnetic north on account of being tilted (Fig. 4.42). The angle of tilt, β, is given by $\tan \beta = a/g$ ($= \beta$ for small angles) where a is the horizontal acceleration acting on the system.

Again considering only small angles, the northerly field in the plane of the inductor is $H_E(1-n)$ and the easterly field is $-nH_E\delta$ where δ is the deviation of the compass needle system. Thus the resultant field, at right angles to which the inductor will rest, lies at an angle δ_1 to magnetic north, when for small angles,

$$\delta_1 = \frac{n\delta}{n - 1} \tag{4.3}$$

which is greater than the normal deviation of a pivoted-needle compass or an inductor when tilted through an angle β.

Now

$$\delta = \frac{\beta \tan D \cos \alpha}{1 - \beta \tan D \sin \alpha} \tag{4.4}$$

therefore

$$\delta_1 = \frac{n}{n - 1} \cdot \frac{\beta \tan D \cos \alpha}{1 - \beta \tan D \sin \alpha} \tag{4.5}$$

When $\alpha = 0$,

$$\delta_1 = \frac{n}{n-1} . \beta \tan D$$

If the ship's magnetic heading is ζ, $\alpha = (\zeta - \pi/2)$.

If the acceleration is such that the vessel in which the compass is mounted is inclined into the false vertical (as in a banked turn), the compass needle system and the inductor will lie in parallel planes, as shown in Fig. 4.43.

On account of the Earth's magnetic field and taking the case where angles of tilt and deviation are small, the field to magnetic north in the plane of the inductor is $(H_E - Z\beta \sin \alpha)$ and the field to magnetic east in the same plane is $Z\beta \cos \alpha$ (see also Chapter 3, page 24).

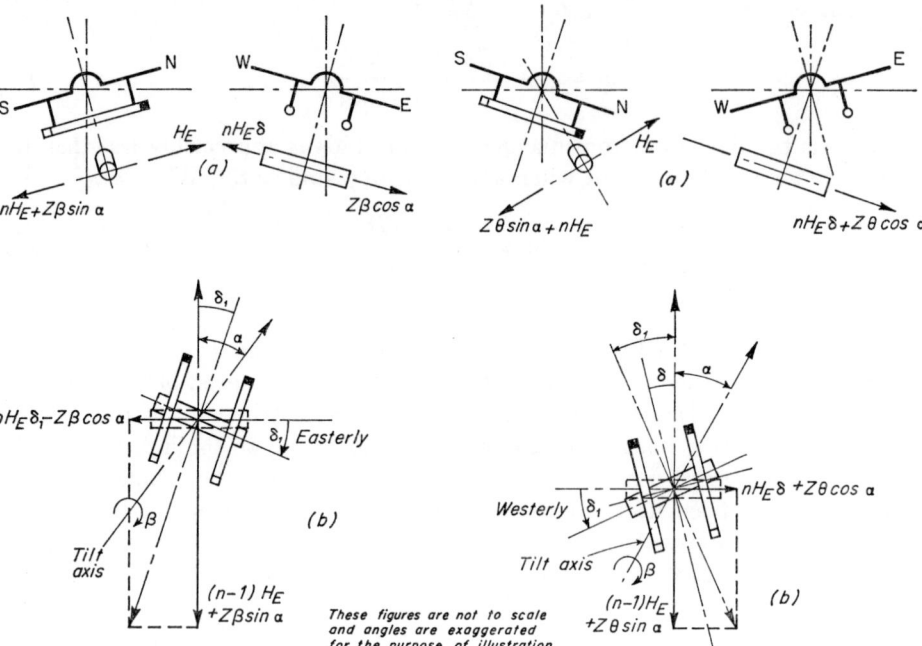

FIG. 4.43. Compass-aided inductor, with the compass system and the vessel inclined into a false vertical owing to acceleration

FIG. 4.44. Compass-aided inductor with the ship rolling or pitching

Owing to the compass magnet, the field to magnetic north in the plane of the inductor is $-H_c = -nH_E$ and the field to magnetic east is $-H_c . \delta = -nH_E . \delta$.

Thus the resultant field lies at an angle δ_1 to magnetic north where, again for small angles,

$$\delta_1 = \frac{-n . H_E . \delta + Z\beta \cos \alpha}{(1-n)H_E - Z\beta \sin \alpha}$$

$$= \frac{n . \delta - \beta \tan D \cos \alpha}{n - 1 + \beta \tan D \sin \alpha} \qquad (4.6)$$

and the inductor is deviated accordingly.

80

$$\delta = \frac{\beta \tan D \cos \alpha}{1 - \beta \tan D \sin \alpha}$$

Neglecting terms containing β^2,

$$\delta_1 = \frac{(n-1)\beta \tan D \cos \alpha}{n-1-(n-2)\beta \tan D \sin \alpha} \tag{4.7}$$

When $\alpha = 0$,

$$\delta_1 = \frac{(n-1)\beta \tan D}{n-1} = \delta \tag{4.8}$$

In a rolling or pitching ship, the inductor, which is fixed to the ship, is tilted through an angle $\theta \, (= A \sin pt)$ in the opposite direction from the compass needle system [see equation (3.23) and Fig. 4.44].

On account of the Earth's magnetic field and in the case of small angles of roll or pitch, the field to magnetic north in the plane of the inductor is $H_E - Z\theta \sin \alpha$ and the field to magnetic east is $Z\theta \cos \alpha$.

Owing to the compass needle, the field to magnetic north in the plane of the inductor is $-H_c = -nH_E$ and the field to magnetic east is $H_c . \delta = +nH_E . \delta$. The resultant field lies at an angle δ_1 to magnetic north, where, for small angles of deviation and tilt

$$\delta_1 = \frac{Z\theta \cos \alpha + nH_E . \delta}{H_E(1-n) - Z\theta \sin \alpha}$$

$$= -\left(\frac{\theta \tan D \cos \alpha + n . \delta}{n - 1 + \theta \tan D \sin \alpha} \right) \tag{4.9}$$

The evaluation of δ is described in Chapter 3, pp. 37–9.

For example, if T_1, the period of the compass needle system, is greater than T, the period of roll or pitch, so that $T_1^2 \gg T^2$, and if $\alpha = 0$,

$$\delta = -\frac{4\pi^2 h A}{g T_1^2} . \tan D \sin pt$$

Therefore

$$\delta_1 = -\left[\frac{1 - (4\pi^2 h / g T_1^2)n}{n - 1} \right] \tan D . A \sin pt \tag{4.10}$$

When both inductor and compass are gimballed, the deviation due to tilt disappears, since heeling the vessel tilts neither the inductor nor the compass system. The deviation due to acceleration is now given by equations (4.6) and (4.7) since, provided the period in tilt of the card and magnet system and the period of the gimbal system are both small compared with the period of roll or pitch (see Chapter 3, pp. 37 and 38), the compass needles and inductor remain in parallel planes. The gimballed arrangement, therefore, is the better form.

Similar deviations due to tilting arise in the compass-aided inductor systems to be described in the next section, where the single inductor, actuated by a follow-up system, is replaced by a fixed array of inductors.

Inductor arrays acted on by a compass system

The single inductor previously considered, driven by a follow-up system, would normally be used to seek east–west. It would therefore rest at right angles to the reverse field of the compass magnet, regardless of the dimensions of the arrangement. When several inductors are used to give, for instance, sine and cosine outputs or 'three-phase' output to a resolver, careful consideration must be given to the dimensions of the inductor array, the compass magnet and the respective distances of the components from each other.

Consider the effect of the field due to a horizontal magnet PQ on an inductor in a horizontal plane at a distance z. If R is the mid-point of the inductor and OR perpendicular to the inductor axis, the field along the inductor in the direction RX and measured thereby (Fig. 4.45) is $ml \sin \theta (\mathrm{PR}^{-3} + \mathrm{QR}^{-3})$.

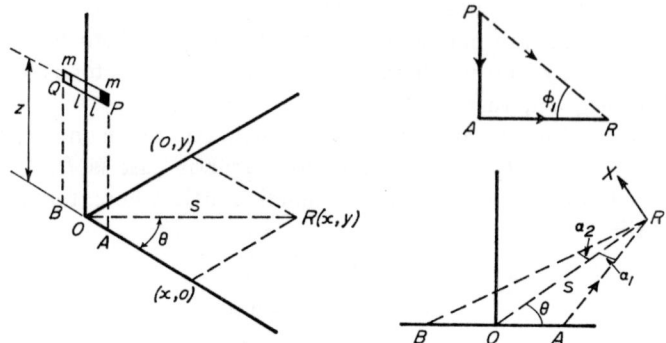

FIG. 4.45. Action on an inductor element of the magnetic force due to a magnet in a parallel plane

Case 1. Square array of inductors. For inductors at right angles (Fig. 4.46), one fore-and-aft and the other athwartships, $\theta = 2\pi - \zeta$, where ζ is the magnetic heading of the system. The field along RX is $-ml \sin \zeta (\mathrm{PR}^{-3} + \mathrm{QR}^{-3})$. The field along

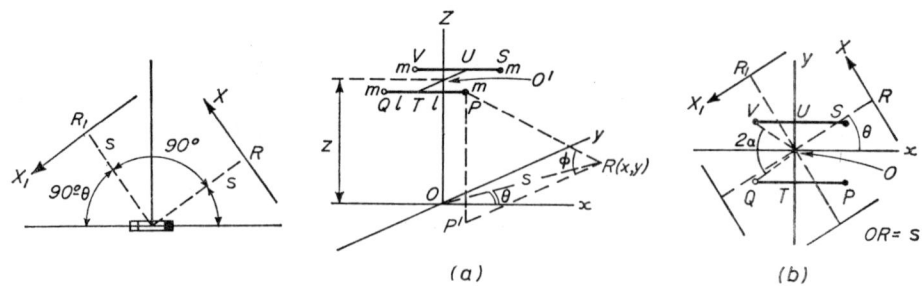

FIG. 4.46. Action on two inductor elements at right angles of the magnetic forces due to a magnet in a parallel plane

FIG. 4.47. Action on an array of inductor elements at right angles of the magnetic forces due to a pair of magnets in a parallel plane

$R_1 X_1$ is $ml \cos \zeta (\mathrm{PR}_1^{-3} + \mathrm{QR}_1^{-3})$. The output of the inductors is proportional to these fields and hence, when applied to a resolving system, an angle $2\pi - \zeta'$ will be determined where

$$\tan \zeta' = \tan \zeta \left(\frac{PR^{-3} + QR^{-3}}{PR_1^{-3} + QR_1^{-3}} \right)$$

Unless $(PR^{-3} + QR^{-3})/(PR_1^{-3} + QR_1^{-3}) \equiv 1$, $\tan \zeta' \neq \tan \zeta$ and errors will occur. It can be shown that

$$\frac{PR^{-3} + QR^{-3}}{PR_1^{-3} + QR_1^{-3}} = \frac{[1 - 2(b/a)\cos\zeta]^{-3/2} + [1 + 2(b/a)\cos\zeta]^{-3/2}}{[1 - 2(b/a)\sin\zeta]^{-3/2} + [1 + 2(b/a)\sin\zeta]^{-3/2}}$$

where $a = z^2 + s^2 + l^2$ and $b = ls$.

Provided $2b/a < 1$, the series are convergent, but only slowly except when $2b/a \ll 1$. Expanding the expression for values of $2b/a \not> \frac{1}{2}$ and ignoring terms in $(2b/a)^4$ and beyond,

$$\frac{PR^{-3} + QR^{-3}}{PR_1^{-3} + QR_1^{-3}} = \frac{1 + 0.27(a^2/b^2) + \cos 2\zeta}{1 + 0.27(a^2/b^2) - \cos 2\zeta}$$

Thus

$$\tan \zeta' = \tan \zeta \frac{1 + 0.27(a^2/b^2) + \cos 2\zeta}{1 + 0.27(a^2/b^2) - \cos 2\zeta}$$

The deviation δ represented by the above expression is mainly octantal, i.e. a term in $\sin 4\zeta$.

Since $\delta = \zeta - \zeta'$

$$\tan \delta = \frac{\tan \zeta - \tan \zeta'}{1 + \tan \zeta \tan \zeta'}$$

$$= -\frac{\sin 4\zeta}{2x - 1 - \cos 4\zeta}$$

where $x = 1 + 0.27(a^2/b^2)$. Thus

$$\sin \delta = -\bar{H} \sin(4\zeta - \delta)$$

where $\bar{H} = b^2/(b^2 + 0.54a^2)$. The deviation can be maintained at a reasonable level by keeping $2b/a$ small. Some typical examples are shown in Table 4.1.

TABLE 4.1 *Octantal deviations in a square array of inductors acted on by a single compass needle.*

	2b/a				
	0·1	0·2	0·3	0·4	0·5
$\bar{H}$	16′	1° 03′	2° 17′	3° 57′	5° 56′

The condition for $2b/a$ being small is met by making z large compared with s and l, but care must be taken to see that the field from the magnet at the inductor is still large enough to make n large.

Case 2. Square array of inductors but with a compass system of two magnets subtending an angle of 2α at the centre. Instead of a single magnet system PQ, a double magnet system PQ, SV is considered (Fig. 4.47). As before, the field along RX can be calculated and is found to be

$$\frac{ml}{\cos\alpha}[\sin(\theta+\alpha)(PR^{-3}+VR^{-3})+\sin(\theta-\alpha)(QR^{-3}+SR^{-3})]$$

which is the same as that due to two crossed magnets as represented by PV and QS, in length equal to $2l/\cos\alpha$ and in orientation to OR, $\theta+\alpha$ and $\theta-\alpha$ respectively.

The field along R_1X_1 is

$$\frac{ml}{\cos\alpha}[\cos(\theta+\alpha)(PR_1^{-3}+VR_1^{-3})+\cos(\theta-\alpha)(QR_1^{-3}+SR_1^{-3})]$$

Writing as before $\theta=2\pi-\zeta$,

$$\tan\zeta'=\frac{\sin(\zeta-\alpha)\,(PR^{-3}+VR^{-3})+\sin(\zeta+\alpha)(QR_1^{-3}+SR_1^{-3})}{\cos(\zeta+\alpha)\,(PR_1^{-3}+VR_1^{-3})+\cos(\zeta-\alpha)(QR_1^{-3}+SR_1^{-3})}$$

If now $z^2+s^2+l^2/\cos^2\alpha=a$ and $ls/\cos\alpha=b$, it can be shown that

$\tan\zeta'=$

$$\frac{\sin(\zeta+\alpha)[(1-K)^{-3/2}+(1+K)^{-3/2}]+\sin(\zeta-\alpha)[(1+L)^{-3/2}+(1-L)^{-3/2}]}{\cos(\zeta+\alpha)[(1+M)^{-3/2}+(1-M)^{-3/2}]+\cos(\zeta-\alpha)[(1-N)^{-3/2}+(1+N)^{-3/2}]}$$

where

$$K=\frac{2b}{a}\cos(\zeta+\alpha)\qquad\qquad L=\frac{2b}{a}\cos(\zeta-\alpha)$$

$$M=\frac{2b}{a}\sin(\zeta+\alpha)\qquad\qquad N=\frac{2b}{a}\sin(\zeta-\alpha)$$

Now by expansion, and ignoring terms in $(2b/a)^4$ and beyond,

$$\tan\zeta'=\tan\zeta\left[\frac{1-\cos2\zeta(15b^2/4a^2)+\cos2\alpha(15b^2/4a^2)+\cos2\zeta\cos2\alpha(15b^2/2a^2)}{1+\cos2\zeta(15b^2/4a^2)+\cos2\alpha(15b^2/4a^2)-\cos2\zeta\cos2\alpha(15b^2/2a^2)}\right]$$

This is a general case.

Now let $\alpha=0$, and

$$\tan\zeta'=\tan\zeta\,\frac{0\cdot27(a^2/b^2)+1+\cos2\zeta}{0\cdot27(a^2/b^2)+1-\cos2\zeta}$$

as before, but let $\alpha=30°$ and $\tan\zeta'=\tan\zeta$. Thus if terms in b^4/a^4 and beyond can be neglected there is no octantal deviation with the conventional compass magnet system whose poles subtend 60° at the centre (see Chapter 5). This arrangement, which was adopted to avoid sextantal deviations in compass correction, is therefore well suited to energizing a square inductor array.

By experiment, a ring magnet also gives a satisfactory result.

Case 3. Equilateral triangular array of three inductors. Consider now in Fig. 4.48 an equilateral array as shown at ABC of inductors in the plane $x'yxy'$ acted upon by a permanent magnet PQ of length $2l$ at a distance z perpendicularly above the centre of the triangle. Let one side BC be parallel to $x'Ox$ and then be rotated clockwise through an angle θ, which may represent a change in compass heading.

The three field vectors are now

$$\mathrm{RX} = ml\sin(30° - \theta)(\mathrm{PR}^{-3} + \mathrm{QR}^{-3})$$

$$\mathrm{R_1X_1} = ml\sin(150° - \theta)(\mathrm{PR_1^{-3}} + \mathrm{QR_1^{-3}})$$

$$\mathrm{R_2X_2} = ml\sin(90° + \theta)(\mathrm{PR_2^{-3}} + \mathrm{QR_2^{-3}})$$

If the inductors are connected to a three-phase resolver of conventional type, the components of the induced voltage may be compounded so that the voltage vector along NL (in the direction 1 to 0) is $3(\mathrm{RX} - \mathrm{R_1X_1})/4$ and along LM (in the direction 2 to 3) is $\sqrt{3}(\mathrm{RX} + \mathrm{R_1X_1} + 2\mathrm{R_2X_2})/4$.

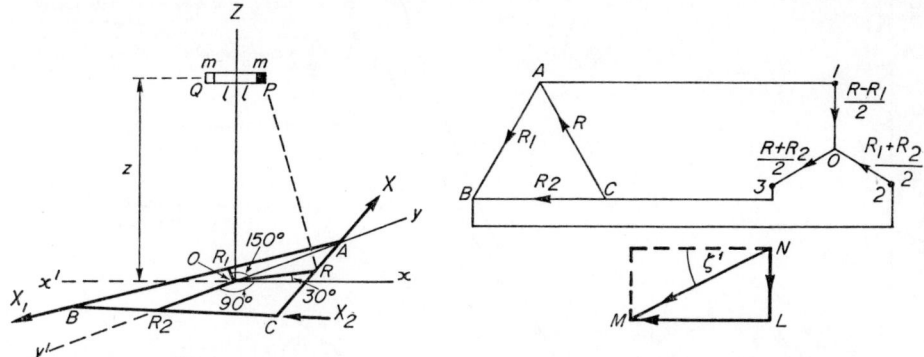

FIG. 4.48. Action on an equilateral triangular array of inductor elements of the magnetic forces due to a magnet in a parallel plane

The field, due to a magnet PQ, along any inductor in the plane $x'yxy'$ is $ml\sin\theta\ (\mathrm{PR}^{-3} + \mathrm{QR}^{-3})$. When BC is parallel to $x'Ox$, the three field vectors along the sides of the triangular array ABC are

$$\mathrm{R} = ml\sin 30°(\mathrm{PR}^{-3} + \mathrm{QR}^{-3})$$
$$\mathrm{R_1} = ml\sin 150°(\mathrm{PR_1}^{-3} + \mathrm{QR_1}^{-3})$$
$$\mathrm{R_2} = ml\sin 90°(\mathrm{PR_2}^{-3} + \mathrm{QR_2}^{-3})$$

Thus the indicated direction of the resultant field from the resolver is given by

$$\tan\zeta' = \frac{\sqrt{3}(\mathrm{RX} - \mathrm{R_1X_1})}{\mathrm{RX} + \mathrm{R_1X_1} + \mathrm{R_2X_2}}$$

$$\tan\zeta' = \frac{3^{1/2}(\mathrm{R} - \mathrm{R_1})}{\mathrm{R} + \mathrm{R_1} + \mathrm{R_2}}$$

Again if $2b/a \ll 1$, neglecting terms in b^4/a^4 and beyond and writing $\theta = \zeta$ = magnetic heading of the craft,

$$\mathrm{RX} = ml\sin(30° - \zeta)\left[2 + \frac{15b^2}{a^2}\cos^2(30° - \zeta)\right]a^{-3/2}$$

$$\mathrm{R_1X_1} = ml\sin(150° - \zeta)\left[2 + \frac{15b^2}{a^2}\cos^2(150° - \zeta)\right]a^{-3/2}$$

$$\mathrm{R_2X_2} = ml\sin(90° + \zeta)\left[2 + \frac{15b^2}{a^2}\cos^2(90° + \zeta)\right]a^{-3/2}$$

85

from which

$$\tan \zeta' = -\frac{3 \sin \zeta[2 + (15/4)(b^2/a^2)]}{3 \cos \zeta[2 + (15/4)(b^2/a^2)]} = -\tan \zeta$$

Hence

$$\zeta' = -\zeta$$

and there is no octantal deviation if b^4/a^4 and higher powers of b/a can be neglected. This is the same result as that obtained when a square inductor array is acted upon by a pair of compass magnets whose poles subtend 60° at the pivot.

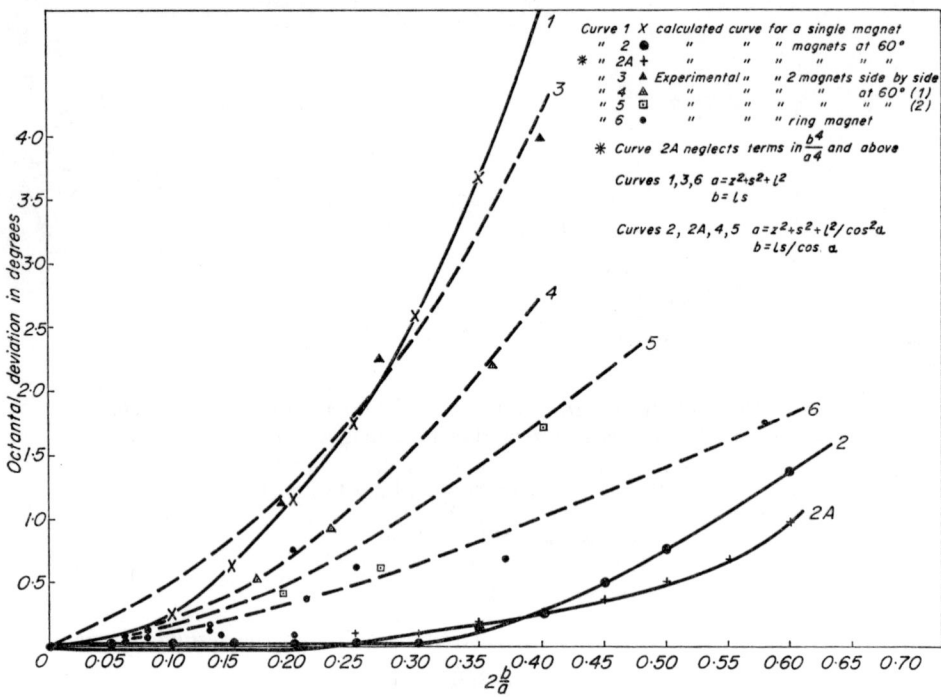

FIG. 4.49. Deviations arising in a square array of inductor elements with different forms of driving magnet

The results of experiments with a square inductor array and using different types of magnet system are illustrated in Fig. 4.49.

Chapter 5

The Pivoted-Needle Compass

The component parts of a pivoted-needle compass (Fig. 5.1) are as follows:

The compass card, engraved around its edge in degrees or points, the sub-divisions being a matter of choice and readability;

The magnetic assembly on which the card is mounted, and which, in liquid compasses, usually includes a float;

The pivot on which the magnetic assembly is poised;

The compass bowl enclosing the card, magnetic assembly and pivot, and possibly including arrangements for damping the oscillations of the system by either liquid or eddy currents;

A lubber's point normally provided to define the compass heading.

An azimuth-taking device, sometimes mounted on the upper surface of the bowl.

Dry-card compass. In the dry-card compass, the magnet and card assembly is not immersed in liquid, reliance being placed on the dynamic properties of the system in air to provide the desired characteristics, and on the air resistance to provide damping of the oscillations of the system. In some cases, eddy-current damping may be used. Such compasses suffer from the weight limitation in loading the pivot, and therefore the magnet and card assembly is, of necessity, light and the magnets comparatively weak. The low degree of damping causes the compass to be lively. On the other hand, the construction is simple and replacement of the card and pivot is easily effected.

Liquid compass. The magnet and card assembly of the liquid compass is immersed in a suitable fluid to improve the damping and to permit the use of a stronger and heavier magnet arrangement than would be allowable in a dry-card compass. Whereas the dynamic characteristics of the latter are readily calculated, they are more intractable as soon as the system is immersed in liquid. Such compasses are more stable under shock and vibration than dry-card compasses.

Design considerations

A measure of compromise is inevitable in the application of the fundamental theory outlined in Chapter 3 to the design of satisfactory pivoted-needle compasses for use in ships, aircraft or vehicles. Examples of such compasses are described in this chapter, together with the general principles underlying their design.

In an instrument intended for use in an observatory or in the field, the prime consideration is the choice of a needle whose magnetic moment results in an acceptable period of oscillation, and which is capable of overcoming any frictional or other restraint due to the suspension and yet is not so unduly heavy as to cause damage to the suspension. Even when instruments of this kind are operated in different magnetic field conditions, the necessary adjustment for balance and the like may conveniently be made on the spot by the user. The choice between torsional and pivotal suspensions, methods of damping and means of reading will depend on the particular purpose for which the instrument is intended.

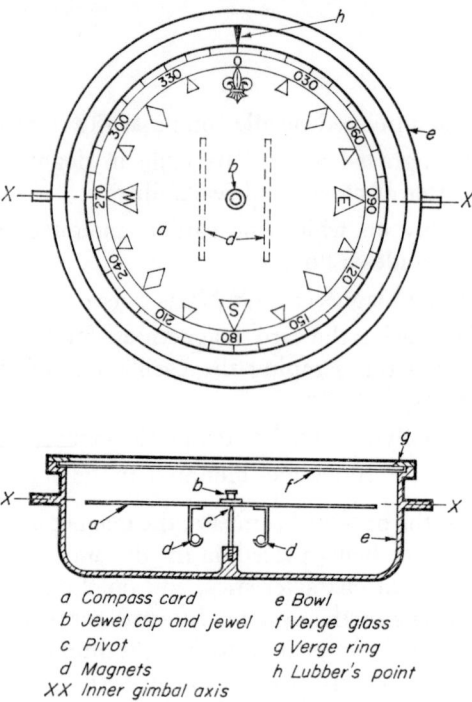

a	Compass card	e	Bowl
b	Jewel cap and jewel	f	Verge glass
c	Pivot	g	Verge ring
d	Magnets	h	Lubber's point
XX	Inner gimbal axis		

FIG. 5.1. Simple pivoted-needle compass

However, when considering the problems of mounting a needle system in a moving craft and of correcting its deviations due to the craft, the static and dynamic requirements of the system impose what sometimes appear to be conflicting limitations. The magnetic moment of the system determines the period of oscillation, in combination with its moment of inertia (and therefore its weight). The directive force to overcome pivot friction is a function of magnetic moment, and when soft-iron correctors are placed in the proximity of the needle, there is a limitation on the moment in order to avoid undesirable induction effects in the correctors.

The magnetic moment also determines the pendulousness of the system and the change of tilt in different vertical field strengths. Moreover, the geometry of the needle system has to be such that with permanent magnetic correctors, spurious deviations of the needle are avoided. The weight of the system must not be so great as to cause damage to the pivot. If eddy-current damping is used, the degree of damping is a function of the magnetic moment.

The features of an instrument of high magnetic moment may be summarized as follows:

Advantages

1. Short period of oscillation and therefore quick settling.
2. Minimal effects of friction and liquid restraints (swirl).
3. Effective eddy-current damping.
4. Avoidance of an excessively long period in regions of low directive force.

Disadvantages

1. Marked induction in soft-iron correctors.
2. High pivot load.
3. Large size of system (usually).
4. Possible change of tilt with change of vertical field.

When using a magnet of low moment, the above characteristics are reversed.

For marine use, the advantage is with a fairly high moment, which reduces the effect of friction and provides a period of oscillation compatible with the characteristics of modern ships. Since present-day compasses are mostly of the liquid-filled variety, on account of their better damping when in a sea-way or under shock or vibration, the high pivot load associated with heavy magnetic systems is largely alleviated by making the system suitably buoyant; although if the effective weight on the pivot is too small, there will be a tendency for the system to centre incorrectly. The addition of a card to the needle system increases the weight and the moment of inertia and modifies the dynamic characteristics. Hence the diameter of the card should, to some extent, be related to the magnetic moment of the system. Although somewhat empirical, this is a convenient point of departure for the design of a compass magnet system, and has found favour among Continental designers of liquid compasses. A graph showing the relation between card diameter and magnetic moment is reproduced in Fig. 5.2.*

Thus the design of the card, magnet system and pivot needs to be viewed as a single problem. The accuracy and performance of the compass largely depend on the quality of the pivot, which is the most vulnerable part of the instrument. Pivots of osmium–iridium are almost universally used, and the forms illustrated in Appendix 3 are generally adopted as being the most suitable for dry-card and liquid compasses. The jewel cup at the centre of the needle system and resting on the pivot point is of artificial sapphire. Jewels appropriate to the pivots illustrated are also shown in the Appendix, which includes extracts from the British Standard Specification for instrument jewels.

For liquid compasses, experience indicates that the weight on the pivot should be between 5 and 10 grams in liquid (about 40 grams in air). In dry-card compasses, a weight of 20 grams is permissible, although in some types this has been as much as 68 grams.

The period of oscillation of the compass magnet system is normally between 23 and 35 seconds at a temperature of 15°C and in a field of 0·18 oersted. The plane of the card must lie within 30 minutes of arc of the horizontal in a vertical field of

* Appendix 2 gives extracts from British Standards 1699 : 1966, in which a similar graph appears.

0·43 oersted in the northern hemisphere, and the change in tilt of the directional system when subject to a change of vertical field of 1 oersted should not be more than 3°. A working temperature range of −30°C to +50°C is normal.

Because of the effect of the proximity of the deviation correctors on a compass needle when placed in the compass position in a ship, the geometry of the magnet system must be considered carefully. Appendix 5 deals with the problem in detail and it need only be said here that short magnets are preferable to long ones and that single magnets are not desirable. Two magnets of equal moment, parallel to each other, and whose poles subtend an angle of 60° at the centre of the array, meet two important requirements: the necessity to approximate to a very short magnet, thereby avoiding the spurious deviations caused by the proximity of correctors, and the need to have a constant moment of inertia about all horizontal axes through the centre (Fig. 5.3 and Appendix 6).

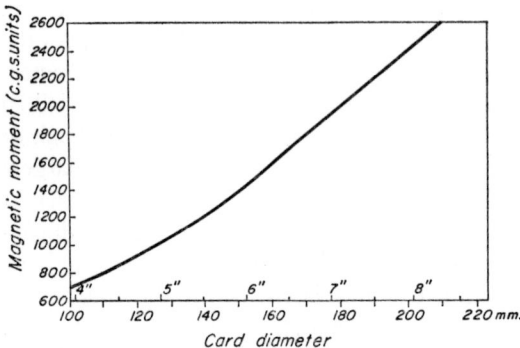

FIG. 5.2. Curve relating magnetic moment and card diameter

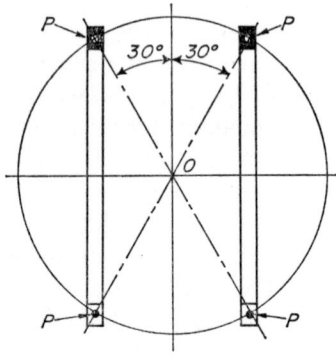

FIG. 5.3. Compass magnet system with two parallel and equal magnets (P = magnet pole)

Other arrangements using more than two magnets satisfy these conditions: for example, a system using four magnets, the outer pair of which subtend 90° at the centre and the inner pair 30°, the array once more being symmetrical about 60°. Other suitable systems using 6, 8 and 10 magnets are frequently employed in dry-card compasses.

As the technique of uniform magnetization is being mastered, the use of the ring magnet (see Fig. 5.4 and Appendix 6) is becoming more general. This type of compass magnet has many advantages. It is perfectly symmetrical in form and dynamics, and in liquid compasses it can be housed within the float, with considerable reduction in swirl errors.

Also, as an approximation to an infinitely short magnet or to the conventional two-magnet system, the ring magnet performs remarkably well. If it is regarded as an infinite number of pairs of parallel magnets, with a uniform distribution of pole strength around each half of the magnet, mathematical analysis shows that, although an improvement on the single needle, it nevertheless gives rise to spurious deviations. To be equivalent to the conventional two-magnet system, the ring should have flats parallel to its magnetic axis, each subtending 60° at the centre.

However, the above conception of a uniform pole distribution may break down owing to the strong demagnetizing effect in the very short elementary magnets constituting the east and west sectors of the ring. Hence it is conceivable that the effect of removing the two 60° segments may, in fact, be achieved fortuitously.

In all the above-described arrays of magnets, the individual magnets are symmetrical about a subtended angle of 60° at the centre. Although short magnets are to be preferred, they are subject to demagnetization owing to the proximity to each other of their poles. Therefore their length should be at least fifteen times their diameter.

The design of a compass usually starts with the choice of the card diameter, which determines the size of the graduations and the accuracy with which the direction of the system can be read. Then comes the process of meeting the various requirements already stated.

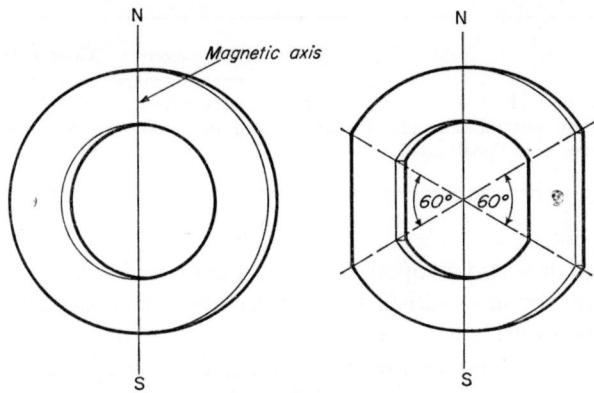

FIG. 5.4. Ring magnets

Expressions used in calculating the parameters of a magnet system

(i) *Period of oscillation*

$$T = 2\pi\left(\frac{I}{MH}\right)^{1/2} \tag{5.1}$$

where T is the period in seconds, I the moment of inertia in g cm² (mk²), M the moment of the magnet(s) and H the ambient horizontal field (0·188 oersted in the United Kingdom).

(ii) *Magnetic moment*

$$M = \bar{M}V \tag{5.2}$$

where M is the magnetic moment in c.g.s. units, $\bar{M}$ the magnetic moment of magnet material per unit volume and V the volume of magnet(s) in cc.

(iii) *Dimensions of magnet.* If $l/d = n$ where l is the length and d the diameter of the magnet in cm, $d = l/n$.

$$V = \frac{\pi l^3}{4n^2} \tag{5.3}$$

$$= \frac{M}{\bar{M}}$$

91

Thus

$$l = \left(\frac{4n^2 M}{\pi \bar{M}}\right)^{1/3} \tag{5.4}$$

(iv) *Pendulous moment*

$$MZ_1 = \frac{mgy\beta_1{}^*}{57.3} \tag{5.5}$$

where M is the magnetic moment in c.g.s. units, Z_1 the change of vertical field strength, usually 1 oersted, m the mass of system in grams, y the distance in cm of pivot point to centre of gravity of the system and β_1 the change of angle of tilt in degrees ($\not> 3°$).

Table 5.1 *Examples of magnet steels*
(Average specific gravity 7·8)

	$\bar{M}$	Remanence	Coercivity
Alcomax II	720	12 400 gauss	575 oersteds
35 per cent cobalt steel	460	9 000 gauss	250 oersteds
15 per cent cobalt steel	410	8 200 gauss	180 oersteds

Card materials

Except in large dry-card compasses where a paper card supported by silk threads is to be preferred, the most satisfactory material for a compass card is undoubtedly ruby mica, since its use ensures clarity, uniformity of thickness and dimensional stability over the working temperature range, all of which are essential. Many seemingly attractive plastic materials fail in these respects and other substances need to be unduly thick and heavy to conform with them. An adequate thickness of mica is from 0·004 in. to 0·008 in. Its specific gravity is 2·8. The other parts of the magnet and card system, e.g. the float, are frequently made of nickel-silver which has a specific gravity of 8·5.

Dry-card compass

In the design of a simple instrument the first step is to decide on a suitable magnet system. The dimensions of the card determine the greatest part of the weight to which the pivot point will be subject and the size, weight and moment of the magnets have to be adjusted so that the permissible pivot loading is not exceeded. The period of oscillation is determined by the magnetic moment and the moment of inertia of the system, the greatest contributor to the latter being the card. A suitable pendulous moment must also be provided.

For example, consider the requirements for a compass with a 3-in. card. If made of 0·005 in. mica, such a card will weigh 1·62 grams and will have a moment of inertia of 12 g cm². Consequently, from equation (5.1),

$$T = 2\pi \left(\frac{12}{M \times 0.188}\right)^{1/2}$$

---

* For more accurate expressions see page 21. The expression given here is the one customarily used in compass design.

If say, a period of 10 seconds is desired, then M = approx. 26 c.g.s. units, and if this is divided between two magnets parallel to each other, their poles subtending 60° at the centre, each magnet will have a moment of 13 units.

If 15 per cent cobalt steel is to be used for the magnet, which has $\bar{M}$ = 410, the length l of the magnet is given by equation (5.4).

$$l = \left(\frac{4n^2 \times 14}{\pi \times 410}\right)^{1/3}$$

Since it is clear that the magnets will be small, n is conveniently made 20, so:

$$l = \left(\frac{1600 \times 13}{\pi \times 410}\right)^{1/3}$$

$$= \sqrt[3]{16}$$

$$= 2\cdot5 \text{ cm} \quad \text{and} \quad d = 0\cdot13 \text{ cm}$$

Two magnets $1\frac{1}{8}$ in. long $\times \frac{1}{16}$ in. diameter would, therefore, be suitable. The weight of such a pair of magnets would be about 1 gram, making the total weight of the card and magnets less than 3 grams. It will be noticed that, in order to obtain the desired period, a low magnetic moment is necessary. It is sometimes preferable to increase the magnetic moment and, at the same time, the moment of inertia of the system. This may be done, for instance, by adding mass in the form of a ring to the circumference of the card, thereby increasing its radius of gyration. This would be possible in the example quoted, since the pivot load so far determined is low, and could with advantage be increased. Too small a load may result in centring difficulties. The addition of a peripheral aluminium ring $\frac{1}{16}$ in. $\times \frac{1}{32}$ in. would increase the moment of inertia to 24 g cm² and allow a magnetic moment of 52 c.g.s. units to be used. The total weight would thereby be increased by about 2 grams.

The distance y of the centre of gravity of the card and magnet system below the pivot point to ensure adequate pendulousness may be determined from equation (5.5):

$$MZ_1 = \frac{mgy\beta_1}{57\cdot3}$$

therefore

$$y = \frac{52 \times 57\cdot3}{3 \times 4\cdot62 \times 981}$$

$$= 0.22 \text{ cm}$$

This is achieved by placing the magnets about one centimetre (0·406 in.) below the pivot.

The Kelvin (or Thomson) compass represents a design that is pre-eminent and is still regarded as a most efficient compass for a great many vessels. It possesses features that are worth considering at this juncture (Fig. 5.5).

By concentrating a large part of the mass of the system in a peripheral ring surrounding a light paper card, the high moment of inertia required to obtain periods of between 23 and 35 seconds is achieved and at the same time the weight of the system is kept low. Magnets of low moment assist in achieving the desired period and their effect on correctors is kept to a minimum. Multiple magnet systems are used consisting of thin needles placed parallel to one another, and conforming to the principle of subtending a mean angle of 60° at the centre. Though the needles may have a large l/d ratio, their low moment limits their length so that the spurious deviations caused by long needles do not arise.

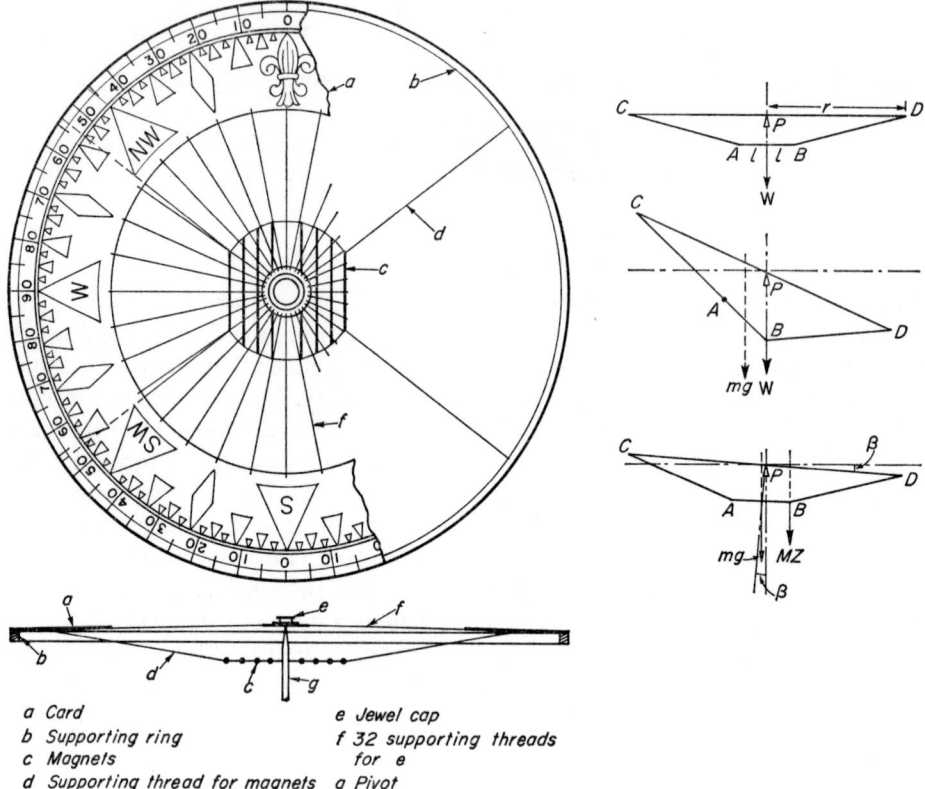

a Card
b Supporting ring
c Magnets
d Supporting thread for magnets

e Jewel cap
f 32 supporting threads
   for e
g Pivot

FIG. 5.5. Kelvin type of dry-card compass

FIG. 5.6. Kelvin type of compass magnet suspension

The paper card is secured to the peripheral ring, and the jewel cup and mounting are retained centrally with the card by a number (usually 32) of radial silk threads. The magnet array is underslung by four silk threads attached to the ring, the pivot stem passing between the two middle magnets. It is interesting to observe that even with the centre of gravity so nearly in the plane of the card, adequate pendulousness is obtained.

With a ring 10 grams in weight and having a moment of the order of 300 c.g.s. units, since

$$MZ_1 = \frac{mgy\beta_1}{57 \cdot 3}$$

$$300 \times 1 = \frac{10 \times 981y \times 3}{57 \cdot 3}$$

94

So

$$y = \frac{300 \times 57 \cdot 3}{10 \times 981 \times 3}$$

$$= 0 \cdot 584 \text{ cm or about } \tfrac{1}{4} \text{ in.}$$

Owing to the slight upward deflection of the card and its threads when resting on the pivot, the pivotal point is almost always raised sufficiently above the centre of gravity of the ring to provide the necessary pendulous moment. The underslung magnets also tend to lower the centre of gravity of the system, as well as making it adequately pendulous and free from excessive tilt with changes of vertical field.

Fig. 5.6 shows the card CD of radius r which, in this instance, may be considered as rigidly connecting the pivotal centre P to the peripheral ring at CD. AC and BD are silk threads of equal length. AB is the magnet of length $2l$.

The whole system can be replaced by an arrangement of rods, which initially are regarded as weightless, joined by pivots at ABCD and poised on a fulcrum P. A load W applied in the centre of AB and therefore vertically below P will not disturb the equilibrium of the system. However, if the load be applied at B, the system will tilt until B falls immediately below P, when equilibrium will be restored.

Now suppose the rod AB has a mass m acting at its centre. Application of W will have moved m to the left and upwards, causing a couple about P and so the equilibrium position will be reached when the resultant of W and mg passes vertically through P. Thus W and mg have equal and opposite moments about P.

If AB is replaced by a magnet, the expression for equilibrium becomes:

$$MZ = \frac{mgy\beta}{57 \cdot 3}$$

as for compass systems in general. AC and BD, being in tension, can be replaced by silk threads and if the centre of gravity of CD is below P, it too produces a couple assisting that of the pendulous magnet.

Consider the design of a dry-card compass of this type. Since much of the mass of the system is concentrated in the supporting ring of the card, the first step is to equate the desired period of oscillation to the estimated moment of inertia and hence to determine the magnetic moment. From equation (5.1)

$$M = \frac{4\pi^2}{T^2} \times \frac{I}{H}$$

Take now the requirements of a 10-in. diameter compass system with eight needles, having a period of not less than 23 seconds or more than 35 seconds, and imposing a load of no more than 20 grams on the pivot. The ring may provisionally be assumed to weigh 10 grams and has a radius of gyration of approximately 5 in.

$$I = 1613 \text{ g cm}^2$$

Let $H = 0 \cdot 188$ oersted and $T = 30$ seconds.

$$M = \frac{4\pi^2}{900} \cdot \frac{1613}{0 \cdot 188} = 377 \text{ c.g.s. units}$$

This total moment is divided between the eight needles. Selecting 15 per cent cobalt steel as the magnet material for which $\bar{M} = 410$, and assuming, since it is evident that the magnets will be comparatively small, that l can be as much as $50d$; from equation (5.4),

$$l = \left(\frac{4 \times 2500 \times 377}{\pi \times 410 \times 8}\right)^{1/3} = 7\cdot 15 \text{ cm (say 2·8 in.)}$$

and

$$d = 0\cdot 143 \text{ cm (say } d = \tfrac{1}{16} \text{ in.)}$$

This is the mean length of the eight needles, which will lie parallel to one another, their poles being on a circle and their spacing conforming to a mean angular distance of 60° (Fig. 5.7). There will be four pairs of needles, whose lengths will be approximately $3\tfrac{3}{8}$ in., $3\tfrac{1}{8}$ in., $2\tfrac{11}{16}$ in. and $2\tfrac{1}{16}$ in. and the set will weigh approximately 9 grams. The total weight of the assembly then is 19 grams. The small increase in the moment of inertia, already calculated above, occasioned by the magnets will have no significant effect on the period of oscillation of the card. The 10-in. supporting ring, weighing approximately 10 grams, could conveniently be made of $\tfrac{1}{8}$ in. $\times \tfrac{1}{16}$ in. aluminium.

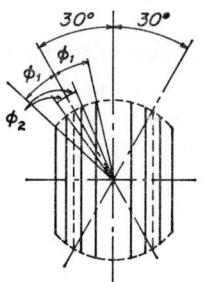

FIG. 5.7. Multiple array of magnets for dry-card compass

From the expression for the pendulousness of the system,

$$MZ_1 = \frac{mgy\beta_1}{57\cdot 3}$$

where M is the magnetic moment, m is the mass of the card and magnet array and y the distance below the pivot point of its centre of gravity, and $Z_1 = 1$.

$$y = \frac{MZ_1 \times 57\cdot 3}{mg\beta_1} = \frac{377 \times 1 \times 57\cdot 3}{19 \times 981 \times 3}$$

since β is limited to 3°. So y = approximately 0·4 cm.

If x is the distance of the centre of gravity of the magnet array below the pivot,

$$9x = 19y$$

therefore

$$x = 0\cdot 85 \text{ cm}$$

Thus if the magnets were suspended $\tfrac{1}{2}$ in. below the pivot, the effect of changes in the vertical field would be within the specified limit.

The assembly of the system upon its pivot within the bowl and the provision of $\pm 10°$ freedom of tilt is common to the more elaborate case of a liquid compass, as is also the geometry of the gimbal arrangement.

Liquid compass

Since the period of oscillation of a liquid compass is not readily determined as in the case of a dry-card compass, the calculation of its period in air being very different from that in liquid, the empirical relation between card diameter and moment is a convenient starting point for the design of a card and magnet system.

As a typical example, consider a compass which is to have a 6-in. card, a period of between 23 and 35 seconds and a weight on the pivot of between 5 and 10 grams.

A moment of 1500 c.g.s. units is suitable for a 6-in. card and the system will be regarded as having two equal and parallel magnets of 750 c.g.s. units each. Since comparatively large magnets are involved a value of 20 for l/d is appropriate.

Selecting 35 per cent cobalt steel as the magnet material, having $\bar{M} = 460$; from equation (5.4)

$$l = \left(\frac{4 \times 400 \times 750}{\pi \times 460}\right)^{1/3}$$

$$= 9 \cdot 4 \text{ cm}$$

$$d = 0 \cdot 47 \text{ cm}$$

Hence magnets $3\frac{3}{4}$ in. long and $\frac{3}{16}$ in. diameter would be suitable. The two magnets would weigh 26·5 grams, and a 6-in. mica card of a thickness of 0·005 in. weighs approximately 6·5 grams; thus the total weight to be supported on the pivot is 33 grams, greatly in excess of that permitted.

The combination of the card and the magnets needs now to be made sufficiently buoyant by immersion in liquid and the provision of a float. It might be presumed that this buoyancy could conveniently be achieved by choosing a dense liquid, but there are severe limitations in this respect. The liquid must be clear and free from sediment, cloudiness or discoloration over the working range of temperature; it must have a reasonably constant viscosity over this range and in no way react with the materials of which the compass is made, nor with the paint used. Thus the choice of liquid is narrow, and mixtures of pure ethyl alcohol and distilled water best fulfil the required conditions. For a temperature range of $-30°C$ to $+50°C$, equal parts of distilled water and alcohol form a satisfactory mixture, with a specific gravity of 0·93.

An extract from an Admiralty Specification for Compass Liquid is shown in Appendix 1, together with boiling-point curves for ethyl alcohol and compass spirit, and freezing-point curves for mixtures of alcohol and water. Other liquids are, of course, used extensively in compasses, namely kerosene, avantine and various silicone fluids, but all should be used with discretion.

A provisional configuration of a float to support the magnets and the card is shown in Fig. 5.8, consisting of a hemisphere $2\frac{1}{4}$ in. diameter, with a cylindrical extension 1·6 in. in diameter between the magnets and a conical insert for the jewel cup. For convenience, the approximate volumes and weights of the various sections of the assembly are tabulated, along with the heights of their centres of gravity above the base of the float and moments of inertia.

The total displacement of the whole assembly of card, magnet and float when immersed in a liquid of sp. gr. 0·93 is 51·9 grams. The weight of the whole system in air being 58·9 grams, this leaves 7·0 grams to be supported by the pivot—on the whole, a satisfactory result.

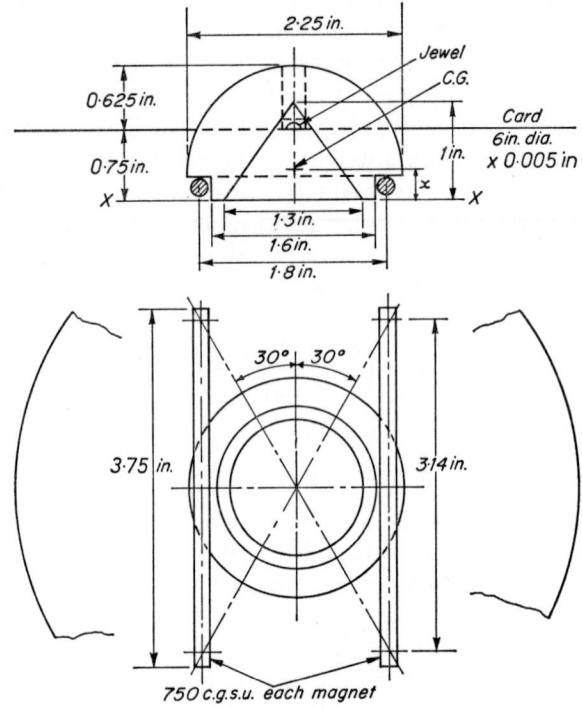

FIG. 5.8. Float and magnet system for liquid-filled compass

Float spun from 30 S.W.G. nickel silver.
By calculation, $x = 0·44$ in.
Tilt for change of $Z = 1$Oe is $1°51'$.
Period of oscillation ~10·0 sec in air
(estimated period in liquid 25–30 sec).

	Displacement cm³	g	Material cm³	g	Height of CG above XX''	k^2 (cm²)	$I = mk^2$
Hemisphere	49·2	45·6	1·64	14	0·966	5·44	75
Cylinder	8·2	7·6	0·26	2·2	0·125	4·1	9
Cone	−7·2	−6·7	0·5	4·3	0·33	1·36	6
Card	2·3	2·2	2·3	6·5	0·75	32·5	211
Magnets	3·4	3·2	3·4	26·5	0·125	13·2	350
TOTAL		51·9		53·5			651
			+10%	5·4			
Weight on pivot 7 g				58·9			

The centre of gravity of the system is determined next, so that its pendulousness may be found. The card is placed at a convenient position on the float, decided more by judgement than by calculation. In this example, 0·75 in. from the bottom of the float is reasonable.

Taking moments about XX,

$$(14 \times 0 \cdot 966) + (2 \cdot 2 \times 0 \cdot 125) + (4 \cdot 3 \times 0 \cdot 33) + (6 \cdot 5 \times 0 \cdot 75) + (26 \cdot 5 \times 0 \cdot 125) + 5 \cdot 4x$$
$$= 58 \cdot 9x$$

where x is the height of the centre of gravity of the assembly above XX.

$$x = \frac{23 \cdot 44}{53 \cdot 5} = 0 \cdot 44 \text{ in.}$$

The distance of the pivotal point above the centre of gravity is $0 \cdot 75$ in. $- 0 \cdot 44$ in. $= 0 \cdot 31$ in. $= 0 \cdot 8$ cm. Inserting this in equation (5.5):

$$1500 = \frac{58 \cdot 9 \times 981 \times 0 \cdot 8}{57 \cdot 3} \beta_1$$

$$\beta_1 = 1° \, 52'$$

which is a satisfactory value.

The period of oscillation in air may now be calculated. The approximate moment of inertia, $I = 651$ g cm^2, so from equation (5.1):

$$T = 2\pi \left(\frac{651}{1500 \times 0 \cdot 188} \right)^{1/2}$$

$$= 10 \text{ seconds (approx.)}$$

The period in liquid, however, is known to be very much greater owing to the adherence of a certain mass of liquid to the system; in general, T in liquid is from $2\frac{1}{2}$ to 3 times greater than in air. Hence, although a practical test is the only means of ascertaining whether the system is satisfactory, it is probable that the above provisional system will have a period in liquid of between 25 and 30 seconds.

The construction of the float must be such that it will withstand an internal pressure of 15 lbs/in.^{2} during the de-aeration of the liquid after the compass is filled, which is essential to prevent the formation of bubbles. Furthermore, the use of only non-corrosive fluxes, such as resin, is permissible in soldering the components.

Although this example of the preliminary design is largely approximate and serves only to indicate the method of dealing with the problem, the result is reasonably typical.

A complete experimental system is constructed at this stage and tested, any further adjustments being made by trial and error. For comparison, the parameters of a known ship's compass are given in Table 5.2.

TABLE 5.2 *Comparison of compass magnet systems*

	Card diameter	Moment	Weight on pivot	Period in air (secs)	Pendulousness
Actual compass	6 in.	1400–1600	5–7g	11·8	≯3°
Calculated example	6 in.	1500	7 g	10	1° 52'

An alternative construction is that using a ring magnet. Such a magnet, $1\frac{9}{16}$ in. outside diameter $\times 1\frac{5}{16}$ in. inside diameter $\times \frac{7}{32}$ in. thick, made of Alcomax II, would have a moment of 1460 c.g.s. units. It would be about 16 grams in weight and could be fitted neatly at the base of and within the float, being adequately protected. Some reduction of buoyancy would be necessary to provide sufficient positive weight on the pivot and the period of oscillation would be reduced owing to the lower moment of inertia of the magnet system.

Having established the parameters of the magnet system, it has now to be mounted within a bowl. In the case of a dry-card compass, this is a simple construction, generally of cylindrical form, closed at the bottom and having a glass disk covering the top. The ring used to secure the cover or verge glass is frequently adapted to take an azimuth circle or alidade. A 10-in. card, for example, may conveniently be enclosed in a bowl of $10\frac{1}{4}$ in. diameter, whose depth must accommodate a tilt of $\pm 10°$:

$$5 \text{ in.} \times \sin 10° = 0\cdot868 \text{ in.}$$

and so the depth would be about 2 in.

In a liquid compass, the bowl needs to be liquid- and, to some extent, pressure-tight and provided with means for dealing with changes of pressure and temperature. In the case of a compass with a 6-in. card, there should be an annular clearance of about 1 in. between the edge of the card and the inside of the bowl; hence an internal diameter of 8 in. is required. To accommodate $\pm 10°$ tilt, a depth of $1\cdot04$ in. is necessary; but the card is $\frac{5}{8}$ in. from the top of the float and $\frac{1}{16}$ in. clearance should be allowed between the top of the float and the underside of the glass, and so $\frac{11}{16}$ in. is already available between card and glass and there is adequate space for 10° tilt.

Space below the magnets is required in order to accommodate 10° tilt, namely $0\cdot325$ in., as well as $0\cdot216$ in. between the bottom of the float and the bridge piece upon which the pivot is mounted; a bowl casting 2 in. deep would therefore be suitable. The top verge glass must be not less than $4\cdot5$ mm thick if of untoughened glass and 3 mm if toughened glass is used.

To accommodate the dilation of the liquid over the temperature range and to prevent an excessive internal pressure, an expansion bellows is fitted on the underside of the bowl. A flexible diaphragm is sometimes used, but this introduces difficulties in adjustment and, since it is equivalent to a spring, in obtaining the proper rate. A far better device is a bellows or sylphon of several convolutions, whose rate in pounds per inch may be accurately determined. The bellows is set to accommodate the maximum expansion of the liquid at the upper temperature limit and to follow its contraction at the lower temperature limit in order to prevent the formation of vapour bubbles. A lower verge glass closes the bottom of the expansion chamber.

Marine compass bowls are usually made of cast brass or bronze which, in common with all the other fittings, must be non-magnetic. Even the presence of particles from a turning tool will cause unpredictable errors because the particles become magnetized in the field of the compass magnet if this happens to have a high moment. The thickness of the bowl should be such as to ensure a sound casting.

Sealing rings or gaskets of a special alcohol-resisting rubber ensure liquid-tight joints between the verge glasses and bowl, the former being firmly held in place by the verge rings. The upper ring may carry a relative bearing scale of degrees and be adapted to take an azimuth circle. The lower verge glass is frequently translucent rather than transparent. Within the bowl is the lubber's point against which the card is read and which should be in the plane of the card. The distance between the lubber's point and the edge of the card should be between 1·5 mm and 3 mm, and the point or mark no wider than 0°·5 of the card graduation. The compass is ultimately mounted in the binnacle or stand which, as well as supporting the compass, contains or supports the deviation correctors.

The connection between the compass and binnacle is made by a gimbal ring, supporting the compass on horizontal pivots in the plane of the compass card and placed athwartships, i.e. at right angles to that diameter of the bowl which contains the lubber's point. The gimbal ring is itself supported in the binnacle on horizontal pivots placed fore-and-aft, i.e. in line with the lubber's point and perpendicular to the athwartship gimbal axes, to within 1°.

The geometry of the bowl and its mounting is important. The vertical plane passing through the lubber's point and the centre of the card must be parallel to the fore-and-aft gimbal axis and perpendicular to the athwartship gimbal axis within a limit of 20 minutes of arc; and the gimbal axes, the lubber's point, the graduated edge of the card and the pivot point must be in the same horizontal plane when the top glass and seating for an azimuth-taking device are horizontal.

In a dry-card compass, the graduated edge of the card may be below the horizontal plane through the pivot point, but the vertical distance must not exceed 0·6 in. and the compass bowl must be gimballed.

The bowl needs to be pendulous, i.e. bottom heavy, and a lead ring encased in brass, attached to the underside of the bowl, is frequently used for this purpose. The pendulous period of the gimballed bowl must not resonate with the roll and pitch of the ship.

Finally, mention must be made of a necessary precaution against the induction produced in the soft-iron correctors associated with the compass (see Chapter 10 and Appendix 5). Not more than one third of the total correction is permitted to be caused by induction from the compass magnet system, and in order to keep within this limit, a minimum distance must be observed in the placing of the correctors, or the moment of the compass magnet must be reduced so that the induction does not exceed the allowable amount when the correctors are placed as close to the compass as the mounting will allow.

Only practical tests will ultimately determine how closely a design calculated on the lines described meets the stated requirements. The calculations can only be empirical and approximate; the final adjustments to dimensions and other parameters must be found by experiment. Many of the other details in a complete compass may vary at the discretion of the designer provided the fundamental principles of design have been met.

In small instruments, eddy-current damping of the oscillations of the compass is a useful alternative to liquid damping. It is commonly used in prismatic compasses, marching compasses and small hand-held compasses, but when applied to large instruments such as ships' compasses, it presents difficulties. These are due to the need for a high magnetic moment in the magnet system to enable sufficient damping to be obtained. This results in a heavy system which has to be supported directly on the pivot.

The great advantage of eddy-current damping, of course, is that the liquid and its attendant complications are avoided. In some liquid-filled instruments, the eddy-current method is used in addition to provide a high degree of damping, as in the A.C.O. datum compass (p. 155). A simple form of compass with eddy-current damping is illustrated in Fig. 5.9.

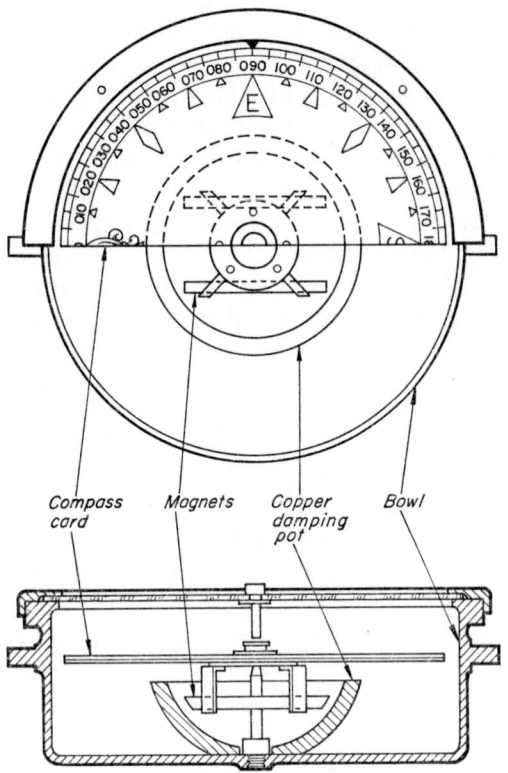

Fig. 5.9. Dry-card compass with eddy-current damping

SHIPS' COMPASSES

Admiralty standard compass

The Admiralty standard compass, or Pattern 1 compass, although marking an exceptional advance in compass design and illustrating many features of good compass construction, is not suitable for present-day use at sea, having been superseded by a number of more recent instruments. Nevertheless, it is of great historical interest. It is also most suitable for use ashore (as a so-called 'landing compass') for magnetic surveys, determination of variation and similar purposes. A detailed description of this compass is given on page 113 of this chapter.

Kelvin compass

The Kelvin compass (already mentioned on page 93), which succeeded the Admiralty standard compass, is illustrated in Plate 5.1. The pivot and jewel were of materials similar to those used in the Admiralty standard compass, namely a sapphire jewel and an iridium pivot. The weight of the system was about one eighth of that of the Admiralty system. Short needles (the longest having a length of 3·5 in.) were used in the 10-in. compass, with a great improvement in correction owing to far lower induction in the soft-iron correctors. The period of the compass was suitable for periods of roll of the ship ranging from 11 to 16 seconds. The 6-in. compass appears to have had an undesirably low period (17·7 seconds) but it was argued that as the compass was for use in battleships' conning towers where the directive force was very low, the effective period would be sufficiently high. In fact, one case is quoted where it was no less than 54 seconds.

The cards were marked at the edge with two sets of degrees, one of which was inverted for use with the azimuth circle. Steering compasses had only one set of figures for direct reading, and points, half points and quarter points were also marked. The bowl rested in the gimbals on knife edges and the gimbal ring was suspended by chains from an elastic wire grummet, which acted as an anti-vibration mounting. At the base of the bowl was a hollow chamber, almost full of castor oil which, being viscous, acted as a brake on undesirable movements of the bowl on the knife edges. If the card became unsteady owing to resonance between the period of the bowl and that of the ship, a shortening of the chains cured the trouble. The azimuth mirror (or circle) (described on p. 109) was mounted on the verge ring. The verge ring on standard compasses was graduated in degrees so that relative bearings could be obtained.

The modern dry-card compass differs only in detail from the Kelvin compass described, as will be seen from a study of the specification for compasses for use in the navigation of merchant ships (see Appendix 2, Extracts from B.S. 1699; Part 1 : 1966). Characteristics of one such compass are given in Table 5.3.

Admiralty compass, Pattern 195

The Admiralty compass, Pattern 195, is a typical example of good modern liquid-compass design. Other contemporary designs may differ in detail but not in the basic features. Many makers have their own preferences for style of card, dimensions and methods of supporting the compass, but the divergences from a basic design are not great; for the requirements for compasses in merchant ships (see Appendix 2) could not otherwise be met.

The Pattern 195 compass is illustrated in Fig. 5.10. It consists of a gunmetal bowl, closed at the top by a verge ring and glass and at the bottom by an expansion disk in which is also fitted a verge ring and glass, both verges being sealed by rubber gaskets. The lower verge glass is frosted and admits light for viewing the compass card at night-time. A bridge supports the stem and osmium-iridium pivot, and the instrument is filled with liquid through a filling plug at the side.

The magnetic system includes a 6-in. card graduated in degrees and mounted on a float in which is fitted a sapphire jewel. When the sapphire jewel is resting on the pivot, the point of the latter lies in the plane of the card. The two magnets are mounted on the float spinning and are below the point of suspension to ensure

TABLE 5.3 *Characteristics of pivoted-needle compasses*

Pivoted-needle compass	Card diameter (in.)	Magnetic moment (c.g.s. units)	Type of magnet steel	Period of oscillation (sec)	Weight on pivot (grams)	No. of needles
Kelvin-Hughes dry-card (1963)	10	370	—	28·5	15·0	8
Kelvin dry-card	10	—	—	38	11·5	8
Kelvin dry-card (1907)	10	—	—	29·7	12·5	8
Hughes dry-card	9	160	—	—	23	10
Kelvin dry-card (1907)	8	—	—	26·5	10	6
Kelvin dry-card (1907)	6	—	—	17·7	5·45	4
Admiralty Patt. I	7½	1700	—	18	68	4
Ship's liquid compass	6	1400–1600	35 per cent cobalt	25–30	5–7	2
Ship's liquid compass	5	750	Alcomax II	25–30	6–8	Ring
Boat's liquid compass	4	800–900	15 per cent cobalt	18–23	5–7	2
Trawler's liquid compass	3 7/8	650–800	—	18–23	5–7	2
Fast motor boat's compass	3 7/8	500–540	15 per cent cobalt	18–23	5–7	2
Submarine's liquid compass	3	250	15 per cent cobalt	18–23	5–7	2
Submarine's projector compass	2	170–200	35 per cent cobalt	11–14	4–6	2

the stability of the system. They are placed parallel to each other and to the 0°–180° diameter of the card, with their poles subtending 60° at its centre.

The moment of the magnet system lies between 1400 and 1600 c.g.s. units, its weight in liquid of 0·93 sp. gr. is between 6 and 8 grams, and its horizontal period of oscillation is between 25 and 30 seconds. The vertical period of oscillation, i.e. about a card diameter, is about 4 seconds. A tilt of ± 10° is allowed.

The lubber's point is inside the bowl, in the same plane as the compass card and pivot tip. The card error, which includes graduation, directional and eccentricity errors, is less than 20 minutes of arc.

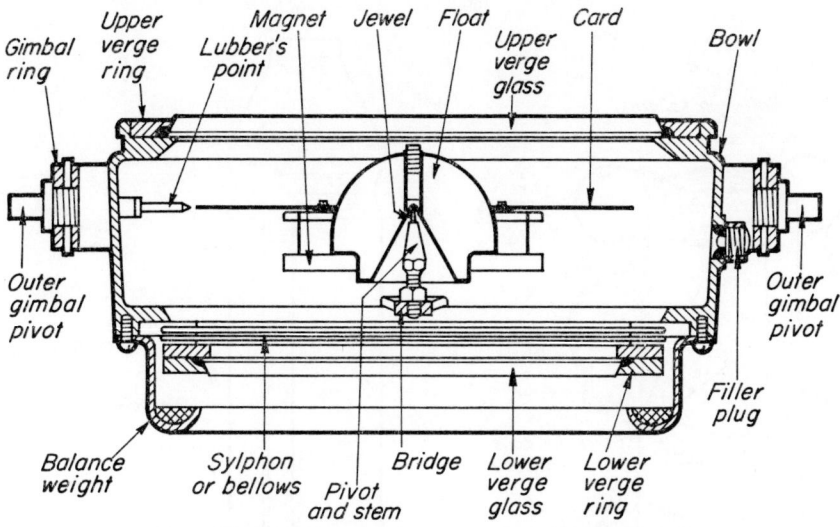

FIG. 5.10. Admiralty compass, Pattern 195

The liquid is in accordance with the specification for compass liquid (Appendix 1) and the required temperature range of the compass is − 30°C to + 50°C. The damping of the system is such that when deflected by 40° and released, 1½ full oscillations are made before it comes to rest. The amplitudes of successive swings are about 11½° and 1½°.

A gimbal ring is pivoted on the compass bowl, the gimbal plane being that of the card, pivot tip and lubber's point. The line through the lubber's point and the pivot is parallel to the outer gimbal axis within 10 minutes of arc. The gimbal ring has pivots which rest in bearings in the binnacle. An azimuth circle may be mounted on the compass verge ring.

In an undisturbed field, this compass is a most accurate instrument, any random errors being represented by a standard deviation of ± 7 minutes of arc. Its design is typical of various smaller Admiralty compasses in present use. Commercial nstruments have very similar performance and characteristics.

The binnacle

The binnacle, illustrated in Fig. 5.11, is both a stand for the compass and a mounting for the correctors. The correctors, described in detail on p. 285, consist of two sets of horizontal permanent magnets inserted in the binnacle, a set of vertical permanent magnets mounted in a container or 'bucket' on the vertical centre line of the binnacle and capable of being raised or lowered, a pair of soft-iron spheres mounted on brackets placed athwartships, and a vertical soft iron bar (the Flinders bar) mounted in a tube provided with brackets normally on the forward side of the binnacle. An arrangement of compass corrector coils, providing adjustable magnetic fields mutually at right angles in the horizontal and vertical directions—the horizontal directions being fore-and-aft and athwartships respectively—encircles the compass. In the latest types of binnacle the coils are enclosed in a cylindrical casting.

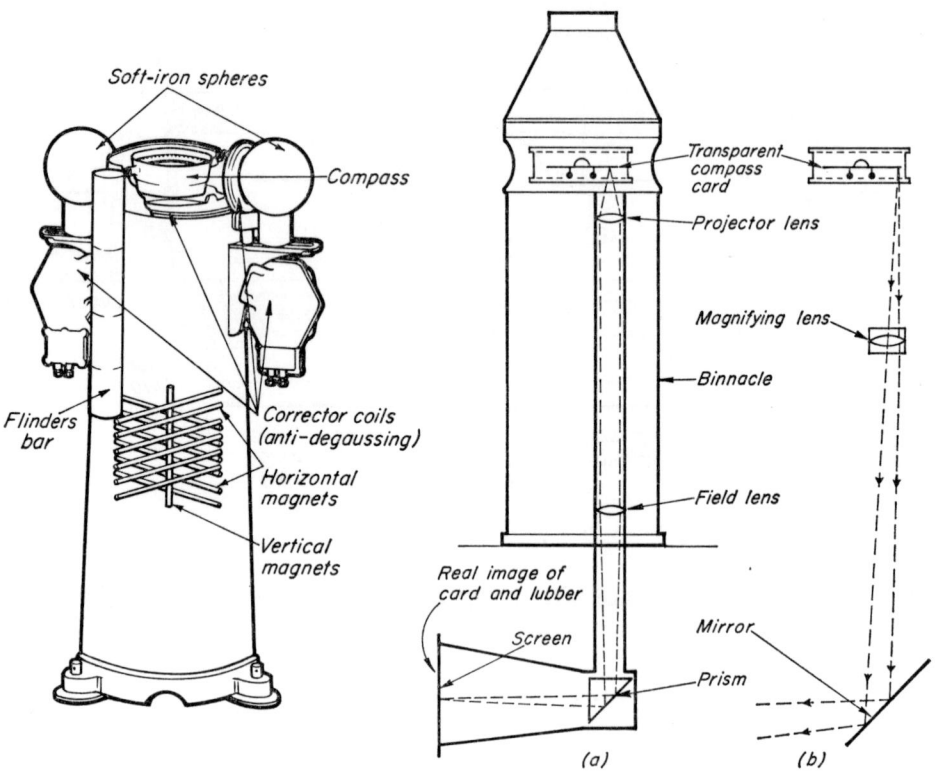

FIG. 5.11. Binnacle   FIG. 5.12. (a) Projector compass.
(b) Reflector compass

An electric lamp inside the binnacle illuminates the compass through the lower verge glass and a brass hood protects it from the weather. The binnacle body is usually made of teak and is mounted so that the outer gimbal axis is in the fore-and-aft line of the ship. Many examples of binnacles for merchant ships could be quoted, but in general they differ little from the Admiralty design. There is no essential difference between the binnacle for a dry compass and that for a liquid compass.

The type of compass and binnacle already described are generally suitable for the navigation of most classes of vessel, but small craft, trawlers and submarines have binnacles specially adapted to their requirements. The compasses appear to differ externally but in principle they conform to normal liquid compass practice.

For small fast motor boats, a spring-mounted compass is used, similar to an aircraft compass (p. 118). The motion of such vessels at high speed or in a seaway is quite different from the rolling and pitching of a large ship, and can well be described as 'slamming'. A brass container with a hood does duty as a binnacle.

Trawlers, with their limited space, frequently use an overhead compass. This may be mounted in a binnacle on the deck immediately over the wheelhouse, with a projector system (similar to that used in a magic lantern) or a simple reflector system to enable the helmsman to see the card from inside the wheelhouse (Fig. 5.12). Alternatively, the compass may be mounted in a special binnacle fitted in the deckhead above the steering position, and is directly visible from above and read from below by means of a mirror. Such an instrument is shown in Plate 5.2. Some trawler fleets have their own preferred types of compass, the Hull ships, for example, favouring a pole compass. This instrument is mounted on a pole forward of the wheelhouse, well clear of magnetic equipment and material, and has a large card readable from inside the wheelhouse.

For steering a submarine, when submerged, a projector type of instrument is sometimes fitted as a stand-by against failure of the gyro-compass. Obviously a magnetic compass would be quite unsatisfactory inside the steel hull of a submarine, and so a position is found outside the pressure hull and on the deck of the conning tower. The binnacle is a pressure-tight tube and passes through the pressure hull where, for safety purposes, a sluice valve is fitted in the tube. Light from a lamp is reflected up the tube past the compass card, and then is turned through 180° by a prism and reflected down *through* the card; and by means of lenses and prisms, a highly magnified image of a sector of the card and the lubber's point is formed on a ground-glass viewing screen at the steering position in the control room. Failure of a lamp is easily remedied, since the lamp is placed inside the submarine and is readily accessible. The arrangement of a submarine projector binnacle is shown in Fig. 5.13.

Foreign compasses

Except for points of detail, there is no fundamental difference between the compasses used by the great maritime powers and those in service in British ships. The principles of compass design are followed by all good manufacturers and spring from the same sources.

American compasses are noteworthy for the fact that it was an American designer, E. S. Ritchie, who first introduced a float into liquid compasses. He also used a magnet system composed of a bundle of thin magnetized wires.

German compasses show a high quality of workmanship but, as in this country, their design often reflects their makers' preferences in matters of detail. Plastic materials have become a feature of modern practice, even for full-size standard binnacles, and the use of a hemispherical verge glass is common. At one time, Bamberg (since absorbed by Askaniawerke) favoured large floats and heavy magnets, which resulted in high moment, quick period and rather large inertia. Strong magnets still seem to be a feature of German instruments. Submarine projector compasses and dry-card compasses follow conventional practice.

Italian and French designers follow mainly conventional lines and have produced some very good instruments. The use of Morel bars (see Chapter 10, p. 290) is a feature of French practice.

Yacht compasses

Most small boats' compasses of conventional design function perfectly well in yachts and other small craft, though sometimes the compass is gimballed with the athwartship gimbal pivots outermost. Alternatively, use may be made of specially designed instruments which follow aircraft compass practice in that they are spring-mounted and have damping filaments. These are not mounted in gimbals. Another form of compass is a completely spherical bowl with the gimbal system inside.

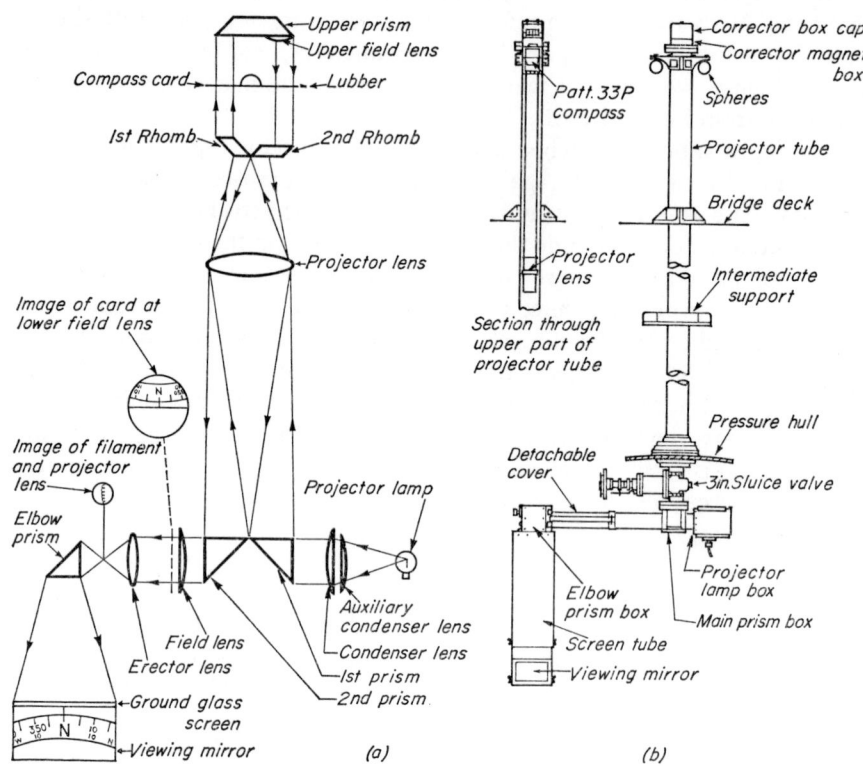

FIG. 5.13. Submarine projector binnacle

Magnifiers, azimuth circles and projection means are used according to the size and arrangement of the vessel. A form of yacht compass, shown in Plates 5.3 and 5.4, is the Sestrel-Moore compass, which has a 5-in. card with both top and edgewise markings, in a transparent bowl. A simple alidade allows bearings to be taken and a half-gimbal bracket supports the bowl. Electric lighting is provided. A variation of this design enables a compass to be mounted either in a gimbal for steering or on a handle for use as a hand-bearing compass.

The azimuth circle (see also *Prismatic compasses*) is mounted on the verge ring of the compass, and enables compass bearings of terrestrial or celestial objects to be observed. It may consist of a simple open sight or alidade (see Fig. 5.14) where a prism is used for reading the card and, on the far side of the circle, a sighting wire which is aligned against the distant object. The prism has a fine slot which acts as a backsight. This slot and the wire must be in the vertical plane which contains the centre of the card, otherwise eccentric errors as explained on p. 116 will arise.

An elaboration of the azimuth circle, and one used extensively in H.M. ships, is an arrangement in which the prism is mounted on the far side of the circle from the observer (see Fig. 5.15). A fine line is engraved on the lenticular front surface of

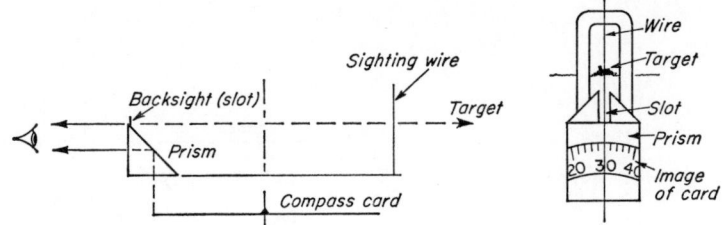

FIG. 5.14. Prism and sighting wire for azimuth circle

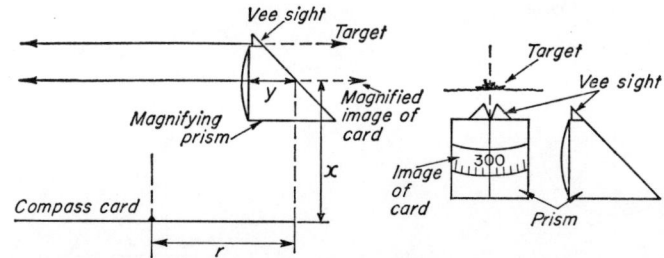

FIG. 5.15. Prism and vee sight for naval azimuth circle
Optical distance $x+y$ = radius of card r

the prism, in which an enlarged image of the card is seen and can be read against the fine line. A foresight and backsight mounted on the prism direct the eye to the distant object; these sights, the line on the prism and the centre of the card all being in the same vertical plane. The foresight can be eliminated without intro-ducing 'ill-aiming error' if the optical distance from the edge of the card to the sighting line on the prism is equal to the radius of the card, and if the vee sight is on this line. Then all angles referred to the vee of the prism are truly degrees, as can be seen from the geometry of the system drawn in Fig. 5.16, and the aim is correct provided the target can be seen in the vee sight. Behind the prism is a reflector which can be tilted, allowing elevated celestial objects to be observed, and glass filters may be used to reduce the light from the sun. The optical plane of sight must be truly vertical and must include the centre of the card; this may be checked by passing a thread along the diameter of the circle parallel to the optical axis of the system and observing the coincidence of its image in the prism with the sighting line and vee. Tilting the prism in altitude must not introduce any bearing error, nor must tilting the reflector cause any apparent change in azimuth of an elevated celestial object.

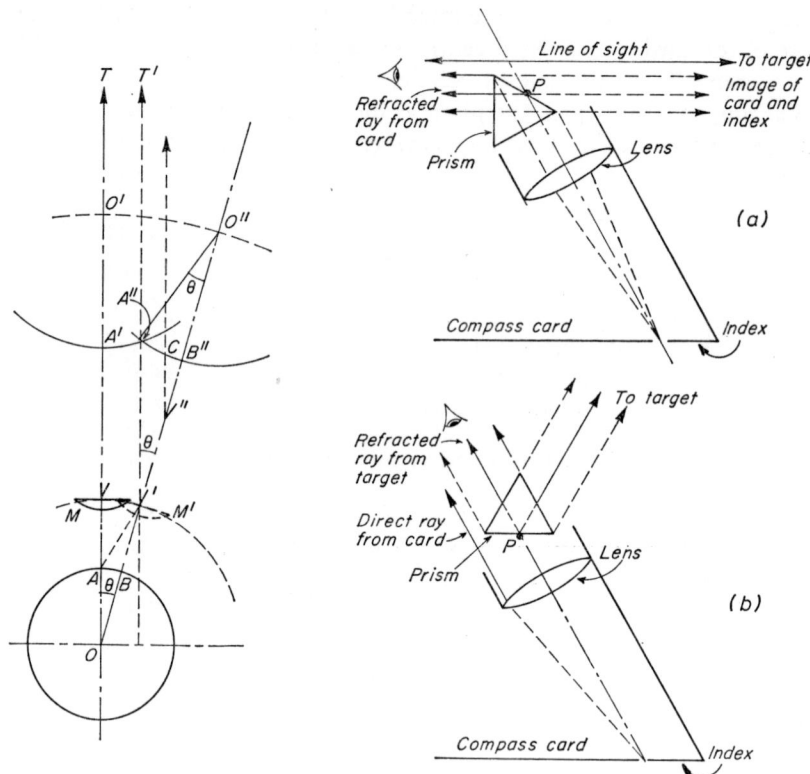

FIG. 5.16. Ill-aiming diagram

FIG. 5.17. Kelvin azimuth circle

O is the centre of the card AB.
M is the reflecting surface with
a magnifying lens. OA = AM.

(a) Arrow down
(b) Arrow up

A′ is the image of A and O′ is the centre of the imaged card. OAVA′O′T is the line of sight to target T through sight V. The bearing of T is given by A′. If the circle is 'ill-aimed' by θ, the line of sight through the vee is now OBV′B″O″. B″ is the image of B.

Triangles V′A″O″ and V′AO must be similar according to the laws of optics. For there to be no error of bearing of T due to 'ill-aiming', the line of sight to T, V′A″T′ must cut the image of the card at A″, so that the angle A″O″B″ = θ.

Hence A″ must be the image of A and if arc A″B″ is small, since O″A″ = A″V′, O″B″ = B″V′ and as the triangles are similar, OB = BV′. This is true, since OA = AM.

If vee were at V″, the 'ill-aiming' error is evident as arc A″C.

$$\frac{O''B''}{OB} = \frac{A''B''}{AB} = \text{magnification of lens}$$

The magnification of the prism should be such that a sufficient sector of the card is visible to ensure correct reading, i.e. at least three figured divisions must be seen. The distance of the image from the observer's eye should be at about the normal limit of close accommodation of vision.

Another form of azimuth circle, often known as the Merchant Navy type, is much favoured in the Mercantile Marine and has certain advantages in ease of accommodation of the eye over the Naval type. It is however a little more complicated. The design was worked out by Lord Kelvin and is illustrated in Fig. 5.17.

A magnifying lens produces a virtual image of the card which is viewed in the rotatable prism. When this prism is set 'arrow down', the distant object on the horizon is viewed over the top of the prism and appears on the rim of the card. For elevated celestial bodies, the prism is set 'arrow up'. The card is viewed along the axis of the lens tube, and the prism reflects an image of the body on the edge of the card. As with other circles, shade glasses are provided.

A shadow pin was at one time used for approximate sun bearings. This was a thin vertical rod mounted at the centre of the circle and immediately over the centre of the card. The position of the shadow cast by the pin on the compass card gave the reciprocal bearing of the sun.

A direct and more elegant method of observing the sun's compass bearing is illustrated in Fig. 5.18. A portion of a cylindrical mirror reflects an image of the sun, in the form of a bright vertical line, to the aperture of a right-angled prism mounted at the opposite end of the card diameter from that at which the mirror is situated.

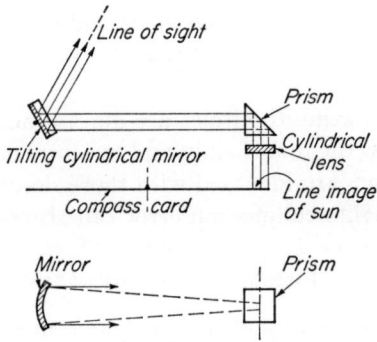

FIG. 5.18. Method of observing compass bearing of sun

This line image is turned through 90° and focused by a cylindrical lens below the prism on the compass card, where again a bright line is formed. Provided the optical plane of the mirror and prism is vertical and includes the centre of the card, the image on the card will be at the compass bearing of the sun.

It cannot be overstressed that with elevated objects, accurate bearings can be obtained only if the circle is level. Any error in cross-level, though not apparent for objects on the horizon, produces errors of up to 1° per 1° of cross-level error when the object is at an altitude of 45°. This applies to all types of alidades and azimuth circles. For this reason it is unwise to take bearings of the sun or stars if their altitude is greater than about 30°.

Telescopic azimuth circles

Telescopic azimuth circles enable accurate bearings to be taken with greater certainty, but the field of vision is limited and the object difficult to find. Moreover, the instrument is apt to be heavy. A well-designed German model is illustrated in Fig. 5.19. The object is viewed through the telescope, whose optical quality is first-class. For elevated objects a reflector is provided, one side being a clear mirror and the other a dark mirror for use with the sun. The image of the object is formed at the focal plane of the eyepiece, which contains cross-wires. A prism brings an image of the compass card into the same focal plane and the distant object is seen superimposed on the edge of the card. It is very important that the optical axis of the telescope should be in the vertical plane that includes the centre of the card. Accuracy of level is also essential.

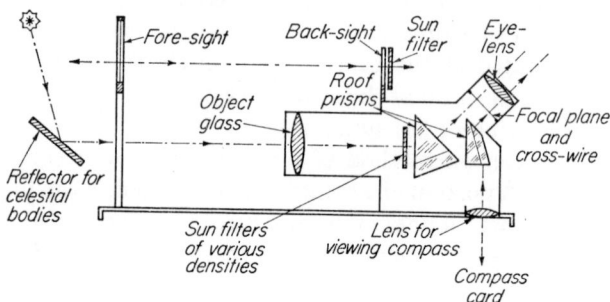

FIG. 5.19. Telescopic alidade
Plath type

The errors that arise in azimuth circles are due to incorrectly mounted prisms and reflectors, to tilt across the line of sight and to eccentricity of the circle with the pivot or bowl. These errors are identical with those described for prismatic compasses. With telescopic sights collimation error can also occur.

PRISMATIC COMPASSES

With a prismatic compass (Plate 5.5), the magnetic meridian is determined by a pivoted magnet system, mounted within a bowl provided with a prism for viewing the card and with a sighting device; thus the magnetic bearings of distant marks may be read directly from the card. This is, in other words, a compass with an integral azimuth circle. The card may be marked in quadrants—0° at north and south, 90° at east and west, or clockwise from 0° at north through 360°. If the latter, the card usually has its graduations reversed to facilitate the direct reading of bearings; hence although it may, in fact, be the south point of the card that is seen in the prism it is marked north and read accordingly.

In an alternative form of the compass, the prism and sight are attached to an azimuth circle which rotates about the verge ring of the bowl. The verge ring may carry a graduated scale of degrees for determining the angular disposition of the sight, as in the Admiralty compass, Pattern 1 (Fig. 5.20). In this last instance, the circle is placed so that the image of the south graduation of the card (which in the Pattern

1 compass is marked 0°, as is also the north point) as seen in the prism is aligned or coincident with the foresight, or sighting wire, seen through the slit in the prism mounting. The reading of the circle index against the graduated scale provides a datum from which magnetic bearings may be measured when the azimuth circle is trained on to a distant object.

A compass with a rotatable circle may give greater precision in reading bearings, especially if the circle carries a vernier, but, as will shortly be shown, random errors can occur owing to any eccentricity of the centre of the card with respect to the centre about which the circle rotates. This error may be avoided by always ensuring that, when determining the magnetic meridian, the circle is trained in a direction that is fixed relative to the bowl. Other errors may be calibrated out. A compass with a fixed sight avoids this difficulty.

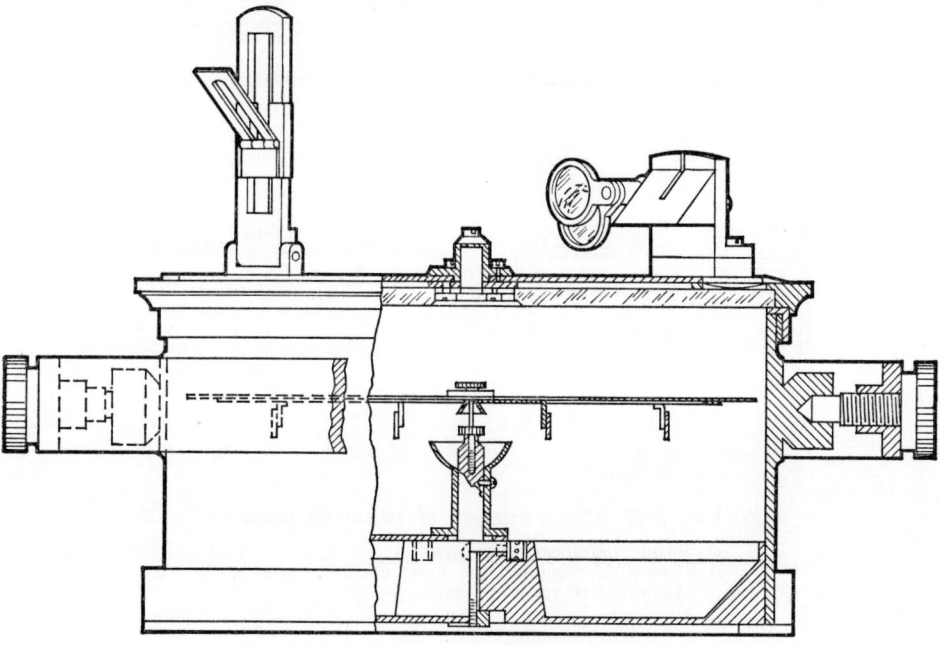

FIG. 5.20. Admiralty compass, Pattern 1

Fig. 5.21 illustrates the various forms that a prismatic compass may take.

The Admiralty compass, Pattern 1, is an exceptionally well-designed prismatic compass, illustrated in Figs. 5.20 and 5.22. The following description is taken from the catalogue of the Admiralty Compass Observatory's Museum.

This is the well-known Admiralty Standard Compass designed by the Compass Committee in 1840.

The bowl is of fairly thick copper, presumably to calm the oscillations of the card, as suggested by Snow Harris some years previously. In construction and refinements it is far ahead of any compass previously designed, also of any for many years subsequently.

A noticeable feature is the method of mounting the pivot, this being screwed into a pillar which, in turn, is mounted inside a central ring carrying four tangent screws for adjusting the pivot centrally to the periphery of the bowl. Another feature is the card lifter operated by a screw towards the aft side of the bowl.

The pivot is of iridium and the cup is of sapphire. For 'heavy weather' cards—which were much heavier than the ordinary cards—the pivot was of ruby and the cup of speculum metal.

The principle refinements lie in the azimuth circle. This consists of a verge ring with a circle of silver let into it in the fashion used for the arcs of sextants, this circle being graduated every half-degree to 360°.

The verge ring carries the verge glass which has a central hole in which is fitted a metal boss smaller in size so as to permit of lateral adjustment.

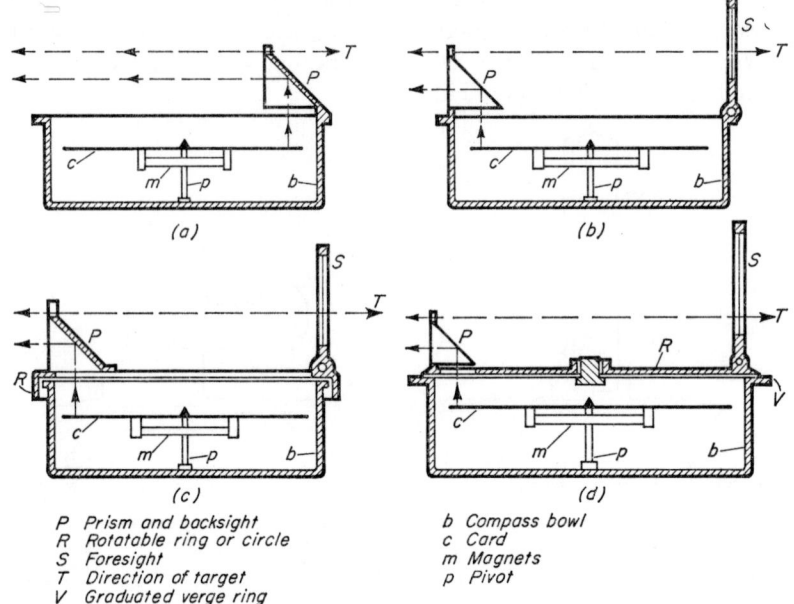

P Prism and backsight
R Rotatable ring or circle
S Foresight
T Direction of target
V Graduated verge ring

b Compass bowl
c Card
m Magnets
p Pivot

FIG. 5.21. Various forms of prismatic compass

(a) Simple handbearing compass.
(b) Surveyor's compass (the most usual arrangement).
(c) Medium landing compass.
(d) Admiralty Pattern 1.

On a pin rising from the above boss, the actual azimuth circle is shipped. This consists of a light ring with cross-pieces, the centre of these carrying a collar which fits snugly over the pin. On the ring an elaborate fitting carries the prism box with the prism and dark shades, and opposite to this is the sight vane with a folding reflector of black glass for taking celestial bearings. It is also fitted with two verniers so divided as to be capable of recording angles to one minute of arc.

The prism carrier is capable of being adjusted to a fine degree of accuracy by means of adjusting and setting screws, and the circle is secured against coming off the verge ring by means of a central screw and by a double leaf spring secured to the prism carrier.

The card is slightly over $7\frac{1}{2}$ in. in diameter and is of mica covered with paper, on which the graduations are printed. It is divided to quarter points, and on the outer edge to every 20 minutes of arc, with 0° at north and south and 90° at east and west, with the figures inverted for reading in the prism. Engraved steel plates for printing were provided by the Compass Committee so as to ensure uniformity in the cards, and one of these was still in use for the purpose in 1930.

The point of suspension is coincident with the horizontal plane of the card and of the gimbal axes in accordance with a principle laid down by the Compass Committee.

There are four edge bar needles, two about 7·3 in. and two about 5·4 in. long, the two inner ones (the longest) are about 1·9 in. apart, and the other two about 1·65 in. from their neighbouring needles respectively. This grouping of the needles gives a uniform displacement of mass about the centre of the system, and provides for an equal moment of inertia about any horizontal axis. A brass bar with sliding weights lies N. and S.

These compasses were first put into production in 1842, the appointment of Capt. Johnson dated 14th March of that year referring to 'two hundred of the compasses set forth in the Magnetic Compass Committee's report are now being manufactured by Mr. Gilbert of Fenchurch Street'.

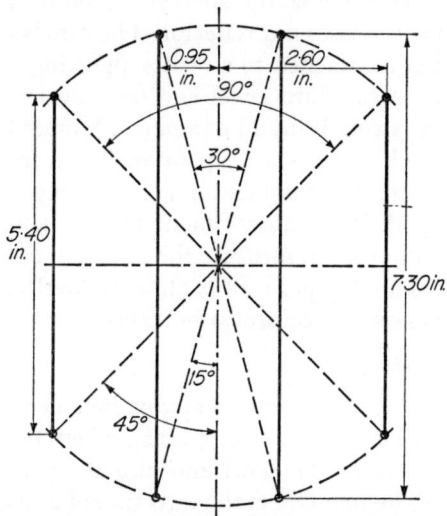

FIG. 5.22. Magnet system of Admiralty compass, Pattern 1

Originally this compass was supplied with a tripod and sprang for setting up on shore for the purpose of adjusting ships' compasses by reciprocal bearings. At one time a number were fitted with an additional gimbal ring and rubber support bands for use during gunfire or in heavy weather.

The compass described will be recognized as a dry-card compass, using air as the damping medium, assisted by eddy-current damping from the copper bowl. The card and magnet system weighs 68 grams and its period in a field of 0·18 oersted is 18 seconds. Designed over 100 years ago, this compass, though somewhat bulky and heavy, is still well in advance of most magnetic compasses for land and surveying purposes (it is no longer suitable for sea-going service). Although it is now being superseded by instruments of the type of the Hilger & Watts datum compass, the Pattern 1 compass is nevertheless capable of accurate determinations of variation, the error being as small as ±7 minutes of arc (95 per cent confidence level) after calibrating the instrument.

A typical surveyor's compass is illustrated in Plate 5.5. It is lighter and simpler than the Pattern 1 and more portable, though not capable of such a high degree of accuracy. The card is of aluminium, engraved to 360° every half-degree. It is mounted on a pivot and jewel, with a device for raising the card off the pivot when the compass is not in use. At opposite ends of a diameter of the brass bowl are a prism for viewing the card, provided with shade glasses, and a sighting vane fitted with a reflector for taking bearings of celestial objects. Flat magnets are secured to the underside of the card. The compass is mounted on a tripod having a ball-and-socket head.

A similar instrument, known as the medium landing compass, is used for aircraft compass adjustment (Plate 5.6). However, in this case the prism and sight are mounted on a movable azimuth circle; hence accurate calibration is not possible unless the circle is always secured in the same position on the bowl.

An improved version is the Admiralty Pattern 2 prismatic compass. This compass is functionally the same as the surveyor's compass and has a prism and sighting vane, but no shade glasses nor reflector. The card is mica, engraved in half-degrees through 360° and is attached to a float supporting two bar magnets. This assembly is mounted on a sapphire jewel and osmium-iridium pivot, and completely immersed in ethyl alcohol (liquid damping.) A brass bowl carries the prism and sight and a spirit level, and has an expansion chamber underneath to accommodate changes of pressure and temperature. This compass is used with a ball-and-socket head and tripod. Its primary function is the determination of variation on land, and after calibration it can measure the direction of the magnetic meridian to within ± 20 minutes of arc (95 per cent confidence level). It is also employed for adjusting aircraft compasses by reciprocal bearings.

Errors of the prismatic compass

(i) *Card error* may be due to inaccurate graduations or to an eccentric pivot or jewel. The latter case is illustrated in Fig. 5.23(a), where the sighting arrangement rotates about O (the centre of the bowl and also the pivot point) but the card centre is at the point P. The line OX is the direction of a mark, and is parallel to a diameter of the card QPR. Hence the correct magnetic bearing of the mark is $N\hat{P}Q$. The sighting device shows the bearing read off the card to be $N\hat{P}Z$, hence there is an error of $Q\hat{P}Z$. On rotating the sight about O (whether it be attached to the bowl or on a separate azimuth circle) through 180° the correct bearing is $N\hat{P}R$, but the observed bearing is $N\hat{P}S$, the error being $R\hat{P}S$, equal to $Q\hat{P}Z$ but opposite in sign. When the line of sight is at right angles to QPR, XOY passes through P and the error is nil.

The error can be written as

$$\delta = A \cos \theta$$

where $A = Q\hat{P}Z$ and θ is the angle of bearing of the mark, measured on the compass card, from QP.

Such an error may also occur when a graduated verge ring is used, if the azimuth circle rotates eccentrically about the azimuth scale. This is illustrated in Fig. 5.23(d) if the engraved scale is centred on P and the circle rotates about O.

These errors, along with graduation inaccuracies, can be calibrated out since the eccentricity is related to the scale or card markings.

(ii) *Eccentric error of the azimuth circle* may be shown in Fig. 5.23(b). The card and bowl centres and pivot are at P. The azimuth circle is eccentric about the bowl and has its centre at O, i.e. the line of sight is OX. However, the circle and its centre O rotate about the bowl and its centre P; thus the line of sight is always tangential to the small circle OP. A constant error arises, independent of bearing. The observed bearing is NP̂Z (or NP̂S) and since the correct magnetic bearing in the direction of OX (or OY) is NP̂Q (or NP̂R) the error is QP̂Z (or RP̂S), constant in magnitude and sign. This type of error can be determined by comparison with known marks, if the value of magnetic variation that must be applied is first ascertained.

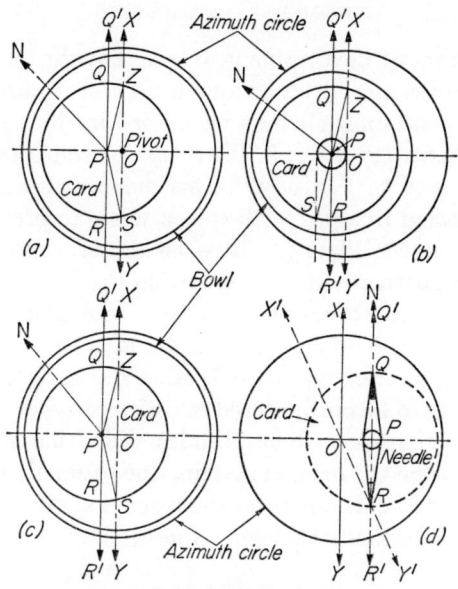

FIG. 5.23. Errors that may arise in prismatic compasses
The diagram exaggerates greatly.
P pivot point
O centre of azimuth circle
QP and XOY magnetic north
X′OY′ apparent magnetic north
XÔX′ index error

(iii) *Pivot error* is illustrated in Fig. 5.23(*c*) where the pivot P at the centre of the card is not at the centre O of the bowl and circle. The correct magnetic bearing of the mark, seen along OX, is NP̂Q. If the circle is rotated on the bowl through 180°, the bearing of a mark, seen along OY, is NP̂R, the respective errors being QP̂Z and RP̂S, equal in magnitude but opposite in sign—an eccentric error similar to that previously considered. However, since the bowl can be rotated about P and the circle about O, the eccentricity OP bears no fixed relation to the scale or card graduations and therefore cannot be calibrated out, unless the sight is fixed to the bowl and the whole compass rotated. In this case, the line XOY remains tangential to a small circle of radius OP as in Fig. 5.23(*b*) and the error is fixed and can be calibrated out.

When a graduated verge ring is used, PQ [Fig. 5.23(*d*)] is magnetic north and the needle will lie along this line. To obtain the scale reading of the circle corresponding to magnetic north, the sight must be aligned with the 'north' graduation of the card or the south end of the needle R (i.e. the image formed in the prism and the sight wire must be in line). Thus the line of sight will no longer be OX, which is parallel to PQ and therefore magnetic north, but ROX' and an index error will be introduced. Unless the compass is always set up with the north end of the needle directed to a lubber's point on the bowl, such an error cannot be calibrated out, since it varies according to the relative positions of O and P.

AIRCRAFT COMPASSES

In the early days of aircraft development, it was thought that a simple adaptation of the mariner's compass would be adequate for use in aircraft. It was, in fact, reasonably useful in small craft such as were used in the 1914–18 war, and early types are shown in Plates 5.7 and 5.8. Plate 5.7 illustrates the type of instrument that was used by Colonel S. F. Cody* in his pioneering flights, and is merely a mariner's compass packed in a box with cotton waste to prevent instability due to vibration and similar effects. Plate 5.8 shows an aircraft compass using an edgewise card, a style of presentation which is still current.

It was soon found that two types of disturbance took place in aircraft compasses, one due to vibration and the other to accelerations. The problem of vibration was overcome by designing an effective anti-vibration system after carrying out a study of the frequencies likely to be encountered in different types of aircraft at the compass position, with their respective amplitudes. Vibration at the appropriate frequency sets up a torque on the magnet system, due to small asymmetries and dynamic out-of-balance, and in extreme cases the magnet system will rotate continually whilst that particular vibration is present, the directive force of the Earth's field having lost control.

A compass adapted for mounting on an instrument panel, and usually having an edgewise card, may be mounted on suitable shock absorbers. These may be secured between the compass and the panel or, more usually, the whole instrument panel is attached to the aircraft by shock absorbers. In the Royal Air Force a 'floating' type of mounting has been found very effective for the horizontally mounted compass; in this case, the compass bowl is supported in a floating ring, the movement of which is damped by light springs, and which is itself supported in a container, having conical spiral springs attached to it to bear the weight of the compass bowl. This compass, known as the P-type, is illustrated in Plate 5.9 and Fig. 5.24.

The problem of acceleration was soon apparent, especially as aircraft became faster. It was found that large deviations of the compass occurred on changes of east-west speed and when turning through north. This northerly turning error could be so pronounced as to cause the compass to show a turn in the opposite sense from that actually made by the aircraft, or even to show no turn at all. It is worth noting that these effects are not peculiar to aircraft; they occur whenever a pivoted-needle type of magnetic compass is accelerated, as explained in Chapter 3.

The disturbances introduced into a lightly damped compass system by acceleration effects have necessitated the design of compasses especially for use in aircraft. In general, these are liquid-filled and are much more heavily damped than ships'

---

* Samuel Franklin Cody (1861–1913): British aviator born in U.S.A., becoming naturalized British subject. He was the first man to fly in Great Britain, October 1908.

compasses. The systems are very light and the magnets comparatively weak in order to give the system a long period compared with sea-going compasses. Additional damping, which increases the resistance to disturbance, is provided by filaments attached to the magnet system which increase the 'drag' on the liquid. This means that the compass cannot quickly build up an excessive error and so limits the amplitude of the disturbance as the aircraft flies through north. The compass systems are in fact aperiodic, a typical damping curve being curve (b) in Fig. 3.13.

P-type compass

The P-type compass is illustrated in Plate 5.9. Its magnet system is pressed out of nickel-silver and the use of solder is avoided in securing the various parts. There is no float, hence the great importance of keeping the system light. The pivot is in the magnet system and the jewel is mounted in a stem supported by a bridge in the bowl. This pivot and jewel arrangement is the reverse of marine practice.

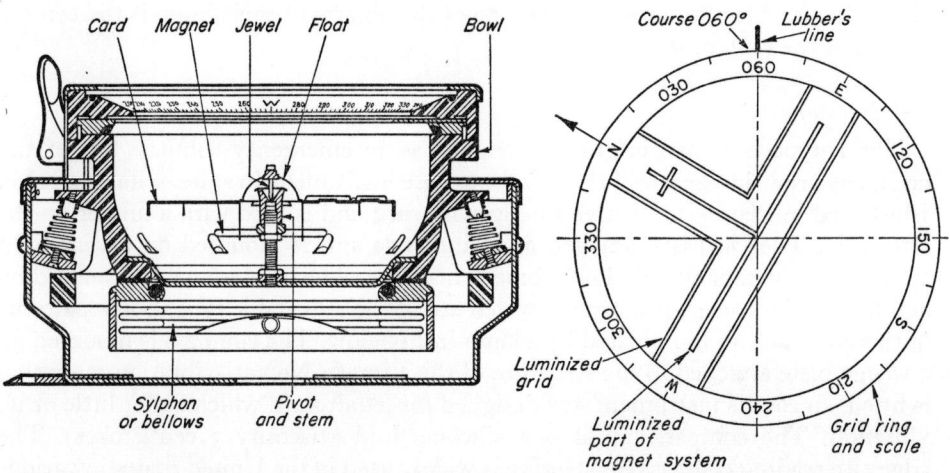

FIG. 5.24. P-type compass for aircraft   FIG. 5.25. Grid steering

The bowl, which is normally constructed of bakelite, has the usual verge glass and bezel and a bellows or expansion chamber to compensate for changes in temperature and pressure. The magnet system has no compass card, its place being taken by thin luminous tubes containing mesothorium, attached to the N., E. and W. filaments. These are read in conjunction with an external grid-ring. The bowl is mounted in the anti-vibration suspension mentioned on p. 118 and the details of the compass construction are shown in Fig. 5.23. Since the temperature range through which aircraft compasses have to work is from −40°C to +70°C, absolute alcohol (sp. gr. 0·787) is used. Consequently, stringent precautions are necessary in the manufacture of such compasses to ensure cleanliness, satisfactory quality of paint, non-corrosive properties of solder and so forth.

The P-type compass is intended for steering and uses the grid presentation illustrated in Fig. 5.25. The grid is aligned with the magnet system as shown. Around the circumference of the grid ring a scale of degrees is read against a lubber's line in the fore-and-aft axis of the aircraft, thereby indicating the latter's course, which is maintained by keeping the system and the grid aligned. To change direction, the pilot turns the grid ring so that it indicates the desired course, and steers the aircraft so that the magnet system becomes realigned with the grid once again. The grid and scale of degrees are luminized.

A variation of the P-type steering compass is the P12, in which the compass is inverted and read by means of a mirror.

O-type compass

Like the P-type, the O-type compass has a specially designed magnet system to provide extra damping and to counteract part of the effect of northerly turning error. However, this is an azimuth compass for taking bearings; hence a card is provided which is read in the prism of an azimuth circle.

Plate 5.12 illustrates the O2 compass, whose construction is generally similar in principle to the P-type except that it does not embody the grid system. The small size of the card is of interest—again, one of the objects of this design is to keep the weight of the system to a minimum.

E-type compass

The E-type compass, originally designed as an emergency compass for jet aircraft, has proved to be a useful general-purpose instrument, in spite of its small size. Illustrated in Plate 5.10, it uses an edgewise card and is read with a lubber on the after side. The card is graduated at 5° intervals and is mounted on a very light system, with comparatively low-moment magnets. The bowl is a Diakon moulding and houses the corrector magnets, which are of the adjustable type. Some patterns of this compass are illuminated by a lamp from below. The compass is mounted on a wedge plate attached to the structure of the aircraft. No anti-vibration mounting is fitted, since this instrument was designed for jet aircraft, which suffer little or no vibration. The compass liquid is a silicone fluid (viscosity 3 centistokes). The edgewise reading compass for steering is widely used in the United States in various forms and a typical compass is the B-16, illustrated in Plate 5.12. Since the magnet system of an aircraft compass is aperiodic, it is usual to express its oscillatory characteristics in terms of 'time of swing', which is the time required for the system to swing through an angle of 85° after an initial deflection of 90°. The amount of overswing is also of interest in assessing the steadiness under turning flight. The principal characteristics of various aircraft compasses are given in Table 5.4.

TABLE 5.4   *Characteristics of aircraft compasses*

Aircraft compass	Magnetic moment (c.g.s. units)	Type of magnet steel	Weight of system in air (g)	Time of swing 85° from 90° (sec)
P-type	45–55	15 per cent cobalt	3·5	4·5–6·5
O-type	80–100	15 per cent cobalt	5·0	5–8
E-type	37–46	35 per cent cobalt	2·8	2–3

For the magnetic correction of aircraft compasses, separate corrector boxes are used for P- and O-types and built-in correctors for edgewise reading types. Only permanent magnet correction is provided, no attempt being made to correct for soft-iron effects. A fixed but adjustable magnet system is used, the correctors being varieties of a 'scissors' arrangement, in which at zero setting two adjacent corrector magnets are placed so that their magnetic fields cancel, but on rotation of the adjusting spindle, the magnets may be turned through angles up to 90°, in which position the magnetic fields add together to give maximum correction. Vertical correction is not usual, but is provided in the E2 compass. Fig. 5.26 shows the principle of the adjustable corrector. The same effect may be achieved by using spiral gears instead of bevel gears. A device similar to this has been used in Germany for correcting naval compasses.

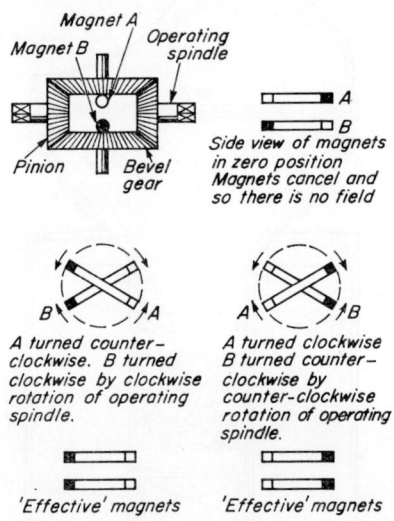

FIG. 5.26. Adjustable magnetic corrector

Continental aircraft compasses in general follow the same methods of construction as British instruments, but mention must be made of an ingenious German projector compass made by Plath of Hamburg, illustrated in Fig. 5.27. In this instrument a very small compass is used, the bowl of which is almost entirely made up of an expansion chamber or sylphon. Both ends are enclosed by a verge glass and the magnet system carries a thin metal card about $\frac{3}{4}$ in. in diameter. A scale of degrees is perforated through the card, rather in the form of a stencil. The compass is placed at the bottom of a tubular mounting and beneath the lower verge there is an electric lamp and condenser. An optical system projects an image of the card on a fine ground-glass screen of about 4 in. diameter at the upper end of the mounting. The definition of the image is very clear and gives the effect of a large compass card without recourse to a heavy compass system. The whole instrument is carried in an antivibrational suspension.

Since civil airlines tend to adopt well-tried service designs of compass, those described have necessarily been military types.

With the advent of the armoured fighting vehicle, navigation problems on land inevitably arose. The use of a magnetic compass in a tank presented almost insuperable problems owing to the effect of large masses of iron close to the compass, and although a certain degree of success has been achieved by using an aircraft type of compass with full correction similar to that used in a ship, it can hardly be said that the arrangement is conducive to accurate navigation or steering. A more effective method has been the use of a projector system on the lines of a submarine projector compass, but the fact that the projector tube extends some four feet outside the top of the tank is an evident disadvantage from the military aspect.

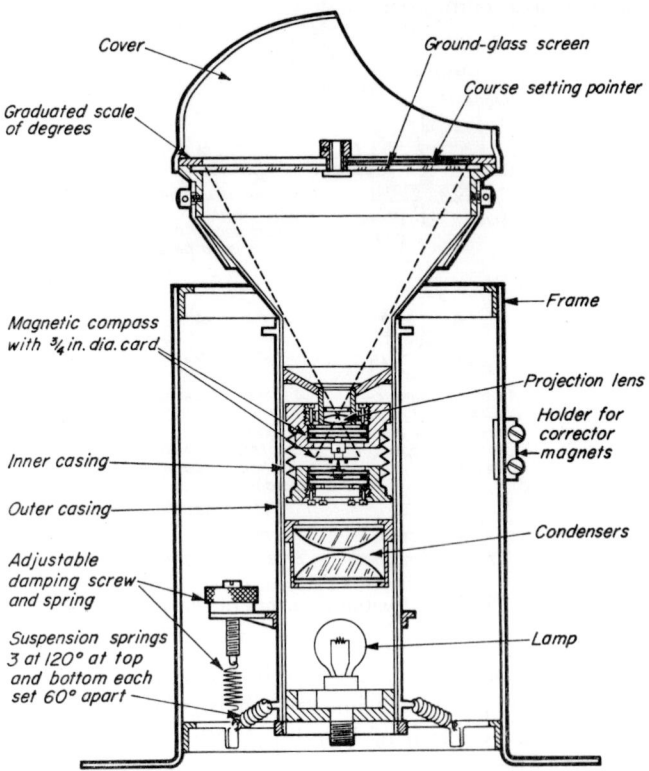

FIG. 5.27. Projector compass for aircraft
Plath type

Nowadays, several vehicles are available for use in rough country; for instance, the Landrover used in the campaign for the suppression of locusts in Africa, the 'weasels' of the North Greenland Expedition of 1952–4 and the 'snow cats' of the Trans-Antarctic Expedition of 1956–7. In tackling the problem of accurate steering of these vehicles, it has been possible to apply the lessons learnt in the construction and equipment of tanks. The E2 compass is used successfully in Landrovers and 'weasels', while Fig. 5.28 shows a simple projector compass used in the 'snow cats' in the Trans-Antarctic Expedition.

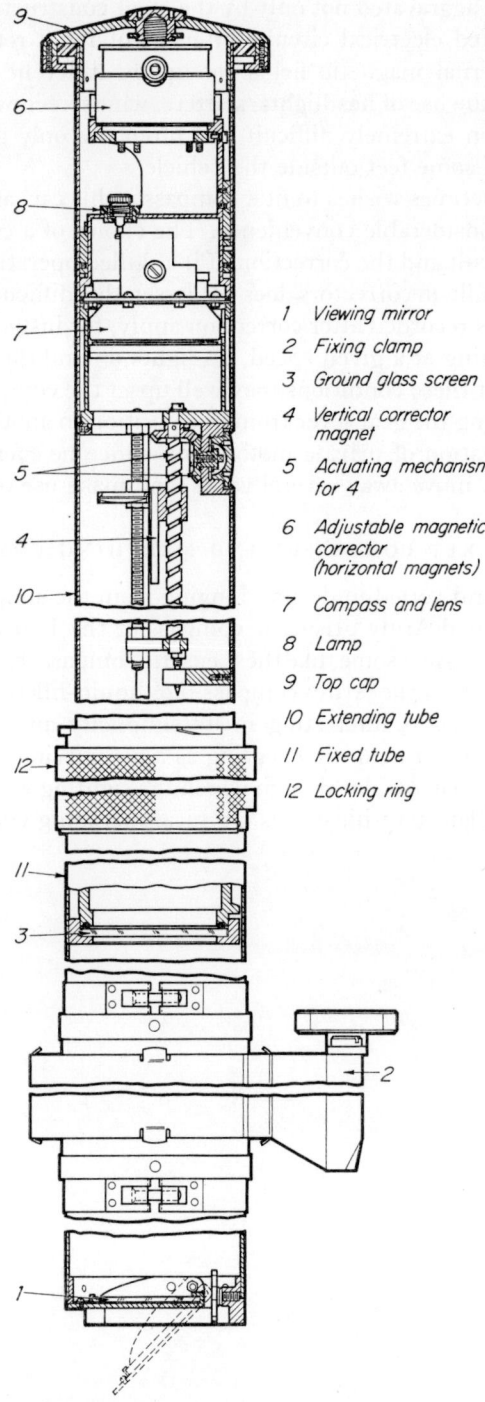

1 Viewing mirror
2 Fixing clamp
3 Ground glass screen
4 Vertical corrector magnet
5 Actuating mechanism for 4
6 Adjustable magnetic corrector (horizontal magnets)
7 Compass and lens
8 Lamp
9 Top cap
10 Extending tube
11 Fixed tube
12 Locking ring

Fig. 5.28. Simple projector compass

The problem is aggravated not only by the steel construction of the vehicle, but also by complicated electrical circuits often involving a return path through the chassis. The external magnetic fields set up by different charging rates of the vehicle's battery, the use of headlights, starters, windscreen wipers and so on render compass correction extremely difficult and often the only good compass position is at a distance of some feet outside the vehicle.

A motorist sometimes wishes to fit a compass in his car, and such an instrument can often be of considerable convenience. The choice of a compass position, however, is most difficult and the correction of it a skilled operation. The fact that some compasses have built-in correctors does not lessen this difficulty. It so often happens that the deviations recorded after correction apply, for instance, only with all doors shut, engine running at a given speed, no lights on and the screen wiper switched off. Any change of these conditions may well upset the compass, as may the simple operation of moving the gear lever from one position to another. No doubt the best practice for navigation of private motor cars—since no enemy hazards exist—is to get out of the car, move away several yards, and make use of a marching compass.

POCKET COMPASSES AND MARCHING COMPASSES

These are many and varied in design, ranging from the simple magnetic needle on a pivot to the refined Army prismatic compass or the Lensatic compass. Some of them are dry compasses; some, like the Lensatic compass, have eddy-current damping, and some, such as the Army compass, are liquid-filled. The optical system of the Lensatic compass is interesting in its simplicity and effectiveness—a single convex lens in a hinged mounting serving as a magnifying viewer of the card, the mounting having a narrow slot though which the sighting wire may be viewed when taking bearings. Plate 5.13 illustrates a typical marching compass.

Chapter 6

Magnetometers

Magnetometers are instruments for the measurement of magnetic field intensity and direction. They are used for the determination of the several components of the Earth's magnetic field, namely:

    (i) Variation
    (ii) Horizontal intensity
    (iii) Vertical intensity
    (iv) Total intensity
    (v) Angle of dip

and for magnetic measurements in the laboratory. They may be for single observations or a series of observations, or else they may be adapted to give a continuous record of the magnetic elements and their changes.

Simple magnetometer

Pivoted or suspended needles are the most usual instruments for general magnetic measurements, and the simplest form is the pivoted-needle type, such as the Cambridge, made by W. G. Pye Ltd. This is a form of compass used in conjunction with a scale of distances, as shown in Fig. 6.1, for determination of magnetic

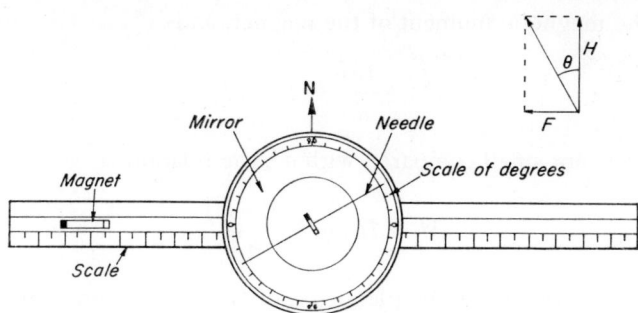

FIG. 6.1. A simple magnetometer

moments and the like. The needle is short and carries a light aluminium or glass pointer. It rests on a jewel and pivot type of suspension in a case, which has a scale of degrees graduated 0°–90°–0°–90°–0°. The pointer is read against this scale and concentric with the scale is an annular mirror in which the pointer is reflected.

By placing the eye immediately above the pointer so that the reflected image of the needle appears to be coincident with the pointer itself, errors due to parallax are avoided.

This type of instrument is subject to several kinds of error. Care must first of all be taken to see that the pointer is at right angles to the needle. This is ascertained by turning the needle system over and ensuring that the pointer takes up the same position in both cases. The point of the pivot may not be in the centre of the scales. This gives rise to eccentric errors which are eliminated by taking the mean of the readings of both ends of the pointer. Or it may not be at the centre of the scale of distances, so that when measuring the moment of a magnet by the deflection method, the magnet must be placed at equal distances on either side of the pivot and again the mean of the two readings taken. The magnet that is being measured may not be symmetrically magnetized and so it must be turned round end for end and the mean of the two readings taken. Thus eight readings are taken to ensure freedom from instrumental errors.

The magnetometer is set up, without any deflecting field being present, so that the needle lies N.–S. under the influence of the Earth's magnetic field. The pointer ends should then read 0°–0°. The deflecting magnet whose moment is required is placed at a distance so as to give a convenient deflection, and the necessary eight readings of the deflection angle are taken.

Since the needle of the magnetometer will take up a position in the resultant field of the earth and the deflecting magnet, we have from the diagram inset in Fig. 6.1 the relation

$$\frac{F}{H} = \tan \theta$$

Now if the centre of the magnet is at a distance, d, east or west from the magnetometer pivot and if $2l$ is the length of the magnet, then, provided l^2 is small compared with d^2, we can write

$$M = \tfrac{1}{2}Hd^3 \tan \theta$$

where M is the magnetic moment of the magnet, and

$$\frac{2M}{d^3} = F$$

If, however, l^2 is not small compared with d^2, the relation is

$$M = H \tan \theta \frac{(d^2 - l^2)^2}{2d}$$

The deflecting magnet may be placed N. or S. of the magnetometer if desired, but still pointing E.–W. In this case the equations for M become

$$M = H \tan \theta \times d^3$$

and

$$M = H \tan \theta (d^2 + l^2)^{3/2}$$

By using a magnet of moment M and setting up the magnetometer in different places, the relative values of H may be found.

A disadvantage of this instrument is the fact that the pointer is rather broad compared with the deflections to be measured; a longer pointer would give a larger scale and so enable more accurate readings to be taken, but this in turn means an increase in size and weight and more friction at the pivot. A long weightless pointer is required, and this may be obtained by using a reflected beam of light, as in the suspension magnetometer.

Suspension magnetometer

A suspension magnetometer, suitable for laboratory use, is illustrated in Fig. 6.2. The magnetometer itself consists of a small magnet A of low moment, carried on a filament B of quartz or tungsten about 12 in. long. The upper extremity of the

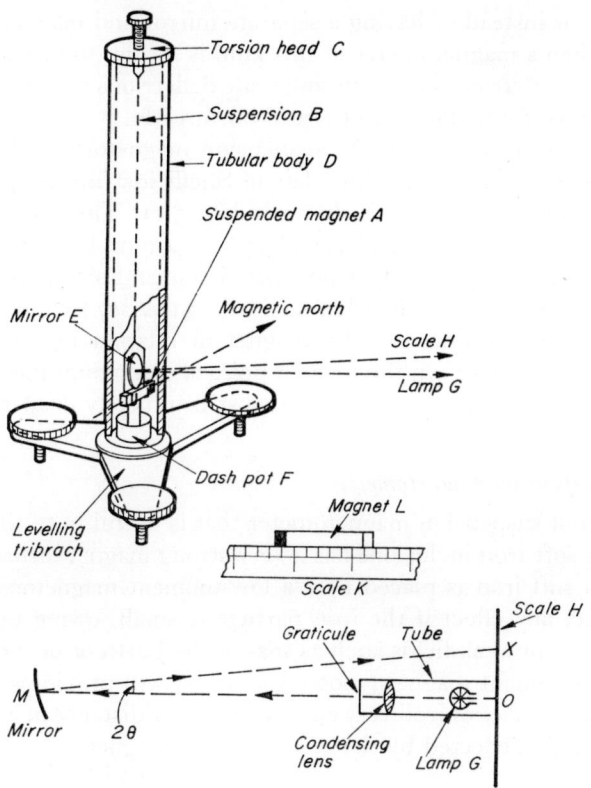

FIG. 6.2. Suspension magnetometer

filament is secured to a torsion head C mounted on the top of the enclosing tube D of the magnetometer assembly. A concave mirror E, usually having a radius of curvature of one metre, is attached to the magnet. Damping is effected by a light wire 'paddle' attached below the needle, and immersed in oil in a dashpot F. Having removed the twist from the suspension, the instrument is set up with the needle or magnet lying in the magnetic meridian, and levelled. A scale of distances K is set up at right angles to the magnetic meridian, as in the case of the simple pivoted-needle magnetometer. The lamp G and its associated scale H are placed so that an image of a graticule mounted in the lamp assembly is focused on the scale at O or some convenient graduation. The moment of a magnet L is measured as before, but the deflection angle θ is determined by the movement of the image of the lamp graticule along the scale OX. Usually OX is all that is required for comparison and similar purposes but if the actual angle of deflection is required, OM must be known. Since the reflection occurring at the mirror doubles the rotation of the beam of light,

$$2\theta = \tan^{-1}\frac{OX}{OM}$$

where θ is the deflection of the magnetometer needle.

Another form of construction is to attach a few pieces of magnetized steel to the back of the mirror instead of having a separate mirror and magnet.

It is usual, when a magnetometer of this kind is set up, to provide a conversion table so that distances from the needle and scale deflections are correlated to enable magnetic moments, for instance, to be read off directly.

An interesting development of the suspension magnetometer is an instrument designed by Professor W. Sucksmith,* late of Sheffield University. This has been referred to on p. 32 and a section is shown in Fig. 3.11. The suspension is unspun silk and carries the usual mirror and a small ($5 \times 1 \times 1$ mm) but very strong magnet, which is encircled by a thick copper pot with a conical hole. This provides eddy-current damping as described in Chapter 3 (p. 31 *et seq.*) which may be varied in degree by raising or lowering the magnet in relation to the conical cavity, hence altering the gap between the magnet poles and surrounding mass of copper. This instrument is robust, can be made almost completely dead-beat and is very suitable for laboratory use.

High-moment suspension magnetometer

Another form of suspension magnetometer that is useful in testing materials for the detection of soft-iron inclusions has a very strong magnet instead of one of low moment. When soft iron is placed near a low-moment magnetometer, it is often difficult to detect any effect if the iron particle is small, owing to the small pole strength produced in weak fields such as that of the Earth or of the magnetometer magnet. A high-moment magnet (3000 c.g.s. units) will provide sufficient field to magnetize small particles of iron very successfully at a distance of a few millimetres and is itself thereby deflected by the resulting iron magnet.

KEW MAGNETOMETER

The Kew magnetometer, a most versatile instrument, has been referred to briefly in Chapter 3.† With the aid of its various attachments, it can be used for the measurement of magnetic variation, and horizontal intensity and for performing the deflection experiment described on page 126. The arrangement of the instrument for the measurement of variation and of horizontal intensity is illustrated in Fig. 6.3. The needle used in the instrument is shown in Fig. 6.4 and consists of a magnetized

---

* Sucksmith, W., 'An improved magnetometer', *J. sci. Instrum.*, 1945, **22**, No. 7.
† A good description of the Kew magnetometer is given by C. Chree in 'Observational Methods in Terrestrial Magnetism', *Dictionary of Applied Physics*, ed. Sir Richard Glazebrook, 1922, **2**, 532–43.

steel tube with a lens at one end and a finely graduated transparent scale at the other, the scale being at the principal focus of the lens. The magnet is a collimator; when viewing through a telescope placed co-axially with the magnet and focused for infinity, an image of the scale is seen in the focal plane of the telescope. The suspension, which is illustrated in Fig. 3.6, has any twist removed from it, and the magnetometer body is then rotated until the image of the centre graduation of the scale coincides with the cross-wire of the telescope. The direction of the geometric axis of the magnetized tube is given by the reading of the azimuth scale of the instrument. The magnet is turned over (i.e. rotated about its horizontal axis, within its supporting stirrup) and the direction of the geometric axis again ascertained. The mean of the two azimuth readings gives the direction of the magnetic axis of the tubular magnet and therefore that of the magnetic meridian.

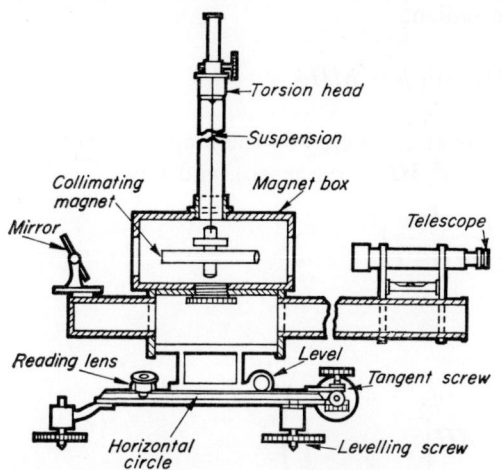

FIG. 6.3. Kew magnetometer

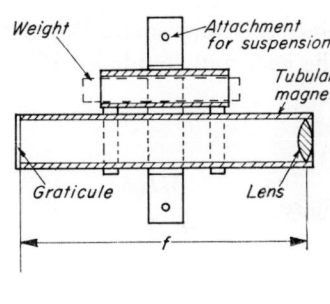

FIG. 6.4. The needle of the Kew magnetometer

Since some time may elapse before the magnet has stopped swinging after the disturbance caused by turning it, some small change in the direction of the magnetic meridian may have occurred, owing to the daily change. For a more accurate determination, therefore, four readings are taken at equal intervals of time, first with the magnet in its initial position, second and third with it rotated or turned over, and fourth with it turned over again to its initial position. The mean azimuth is the direction of the magnetic meridian at the mean time of the four observations.

If the direction of the geographic meridian is known, the magnetic variation is found from the difference between the magnetic and the geographical meridians. The geographical meridian may be found by observing the time of transit across the telescope cross-wires of the image of the sun produced by the mirror, care having been taken to see that the axis of the telescope is horizontal. Having determined this time and knowing the local latitude and longitude, astronomical tables may be used to calculate the sun's true bearing, and hence the geographical meridian may be ascertained.

On p. 30, it was shown how the period of oscillation of a suspended needle was related to the strength of the controlling field, viz.

$$T = 2\pi \left(\frac{I}{MH} \right)^{1/2}$$

Thus, if H is the strength of the Earth's horizontal component, the Kew magnetometer can be used to find its value. A small oscillation is given to the magnet and the time is observed, for, say, 100 transits of the image of the centre division of the scale across the telescope cross-wires in one direction. From this T is found. However, a correction is required owing to the fact that the fibre exerts a small controlling couple on the magnet, thereby reducing the period of oscillation.

By rotating the torsion head through 90°, an angular deviation of ϕ radians will be produced and may be measured. The twist in the fibre is then $(\pi/2) - \phi$. If c is the couple exerted for a twist of one radian,

$$c \left(\frac{\pi}{2} - \phi \right) = MH \sin \phi = MH\phi$$

since ϕ is very small. When the magnet is at an angle θ to the magnetic meridian, the restoring couple is $(MH + c)\theta$ instead of $MH\theta$ (see p. 29), and now

$$T = 2\pi \left(\frac{I}{MH + c} \right)^{1/2}$$

Since $\quad c = \dfrac{MH\phi}{(\pi/2) - \phi}$

$$T = 2\pi \left(\frac{I}{MH \left[1 + \dfrac{\phi}{\dfrac{\pi}{2} - \phi} \right]} \right)^{1/2}$$

Thus the square of the observed time of swing must be multiplied by

$$1 + \frac{\phi}{\dfrac{\pi}{2} - \phi}$$

to determine the correct value of H. Further corrections have to be applied for the effect of temperature and the Earth's field on the moment M of the magnet. The effect of temperature is corrected by reducing the moment to its value at 0°C by the correction factor $[1 + q(t - 273)]$ where t is the ambient temperature in °K and q is a constant found experimentally for the particular magnet. Therefore

$$M_0 = M_t[1 + q(t - t_0)]$$

The correction due to the effect of the Earth's field on the magnet assumes that the change in magnetic moment is proportional to the ambient field when the latter is small, and also that the change is proportional to the volume V of the magnet.

If M_0 is the moment in zero field and M is the moment in the ambient field,

$$M = M_0 + aVH$$

where a is a constant peculiar to the particular magnet and may be found experimentally.

V is a constant; hence
$$M = M_0 + bH$$
Therefore
$$MH = M_0 H \left(1 + \frac{bH}{M_0} \right)$$

Since H/M_0 is always very small, we can write
$$M_0 H = MH \left(1 - \frac{bH}{M_0} \right) .$$

By combining these three correction factors, we can correct T for torsion, temperature and magnetic field by the expression
$$T_0^2 = T^2 \left[1 + \frac{\phi}{\dfrac{\pi}{2} - \phi} - q(t - t_0) + \frac{bH}{M_0} \right]$$

The moment of inertia I of the magnet and carrier is found by adding a brass cylinder to the carrier whose moment of inertia I' is known and then finding the new period of oscillation.

Thus we have
$$T = 2\pi \left(\frac{I}{MH} \right)^{1/2}$$
$$T' = 2\pi \left(\frac{I + I'}{MH} \right)^{1/2}$$
$$T^2 = \frac{4\pi^2 I}{MH} \quad \text{and} \quad T'^2 = \frac{4\pi^2 (I + I')}{MH}$$
Therefore
$$\frac{T^2}{I} = \frac{T'^2}{I + I'}$$
Hence
$$I = \frac{T^2}{T'^2 - T^2} I'$$

M is found by performing the deflection experiment described on p. 126. The tubular magnet and its box are removed and replaced by a small magnet and mirror, seen inset in Fig. 6.5, suspended by a long fibre similar to the arrangement of Fig. 6.2. The instrument is set up as in Fig. 6.5 and the telescope is focused on the image of the scale formed by the mirror. The collimating magnet is set in the rest on the carrier at a suitable distance, d, from the suspended magnet. To avoid introducing a twist of the suspension the deflection of the suspended magnet is not observed, but the body of the instrument is rotated until the centre division of the scale coincides with the telescope cross-wires. This is done with and without the collimating magnet in the rest and so an angle θ is measured as the difference between the azimuth scale readings.

Then

$$M = \frac{H}{2} d^3 \sin \theta$$

or if l^2 ($2l$ being the length of the collimating magnet) is appreciable compared with d^2,

$$M = \frac{H}{2} \left(\frac{d^2 - l^2}{d} \right)^2 \sin \theta$$

Corrections as described on p. 130 are then applied.

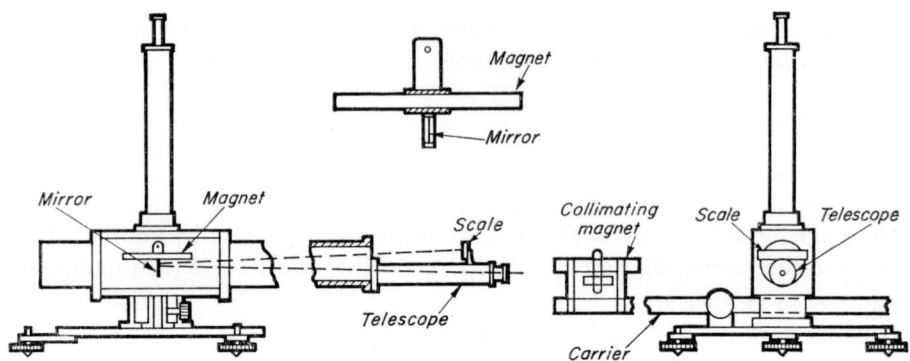

FIG. 6.5. Setting-up the Kew magnetometer

An excellent magnetometer was made by Wild,* and Fraser† described a very well-constructed instrument used for the magnetic survey of India, which gave equally accurate results at observatories and in the field. An instrument using the best features of the India Survey magnetometer combined with the design produced by the U.S. Coast and Geodetic Survey has been developed and constructed by the Department of Terrestrial Magnetism of the Carnegie Institution of Washington.‡

For merely determining the variation a simplified instrument is frequently used, such as that at the Admiralty Compass Observatory (illustrated in Fig. 6.6 and Plate 6.1). The suspended collimating magnet is mounted in a box, fitted with a suspension tube and torsion head, supported on a levelling tribrach. This is permanently installed on a concrete pillar in a magnetometer hut. The instrument is set up as described on p. 129, viewing the magnet scale through the telescope of a theodolite set up on an adjacent pillar. Having determined the direction of the magnetic axis of the magnet and hence the meridian, the azimuth of the theodolite is compared with the geographic or true meridian, and thus the magnetic variation is known. Fixed marks or observations of heavenly bodies are used to determine the geographic meridian. As an alternative to this method of

* Wild, H., 'Theodolith für magnetische Landesaufnahmen', *Vjschr. naturf. Ges. Zürich*, 1896, **41**, 1–25.

† Fraser, H. A. D., 'The unifilar magnetometer of the magnetic survey of India', *Terr. Magn. atmos. Elect.*, 1901, **6**, 65-9.

‡ Fleming, J. A., 'Two new types of magnetometers made by the Department of Terrestrial Magnetism of the Carnegie Institution of Washington.' *Terr. Magn. atmos. Elect.*, 1911, **16**, 1–12.

observation, it may be more convenient to leave the theodolite set on a fixed and approximate value of the magnetic meridian, and to read the scale of the collimating magnet against the cross-wire of the telescope. Since the collimating magnet is continually on the move, readings are taken at the extremities of ten successive oscillations. The mean value from the centre of the scale is found and applied as a correction to the set azimuth of the theodolite. It is most important that no collimating error should exist owing to the scale of the magnet not being exactly at the principal focus of the lens; otherwise scale errors will be introduced. The error is given by

$$\epsilon = \frac{\delta}{f}\left[1 - \frac{1}{\dfrac{f-D}{f} + \dfrac{xD}{f^2}} \right]$$

where δ is the deflection, f the focal length of lens, D the displacement of the graticule from its correct position, $f-D$ the distance from lens to graticule and x the distance of focal plane of telescope.

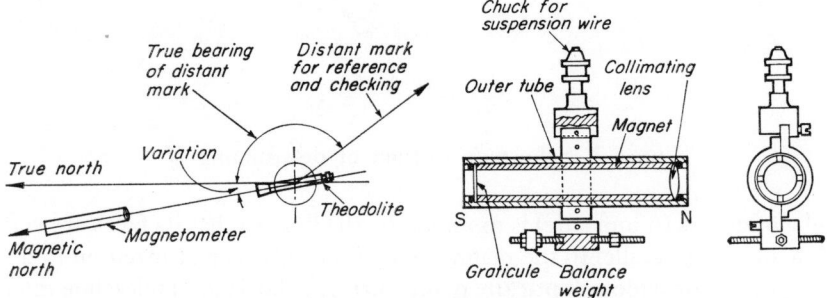

FIG. 6.6. A.C.O. magnetometer

SCHUSTER-SMITH VARIOMETER

A very accurate method of determining H at a magnetic observatory was devised by Sir Arthur Schuster.* It is based on the comparison of H with the uniform field produced by an electric current in a pair of Helmholtz coils.

A magnet M is suspended at the centre of a pair of Helmholtz coils (Fig. 6.7). With the coils unenergized, the magnet sets in the magnetic meridian. A suitable value of current in the coils produces a magnetic field Fi and the coils are then rotated through an angle θ until the suspended magnet sets E.–W. as shown. The component $Fi\cos\theta$ is then equal to H and the component $Fi\sin\theta$ sets the magnet E.–W. By calculating F from the coil dimensions and observing θ, and checking θ for a reversal of the suspended magnet at the symmetrical position of the coils, H may be measured to a few parts in 100,000.

An instrument on these lines was constructed by F. C. Smith in 1920.* This had coils 30 cm in radius wound on a hollow marble cylinder. The cylinder was carefully made and it was considered that the coil constants were known to one part in

* Smith, F. C., 'On an electro-magnetic method for the measurement of the horizontal intensity of the Earth's magnetic field', *Phil. Trans. roy. Soc.*, 1922, **223**A, 175–200; also Schuster, A., *Terr. Magn. atmos. Elect.*, 1914, **19**, 19–22.

10^5. Current being measurable to about 5 parts in 10^5, H could be determined with about the same accuracy.

An interesting description of a variometer of this type made by the Cambridge Instrument Co. Ltd is given in an Ordnance Survey report (1930).* A special feature is the liquid suspension used in the magnet assembly. Two cobalt steel magnets ($1\cdot5 \times 1\cdot5 \times 12$ mm long) are mounted horizontally side by side on the underside of an annular float, immersed in petrol contained in a gunmetal chamber. The float is located centrally by a jewel and pivot as shown in Fig. 6.8, the jewel being mounted on the spindle which projects through the centre of the float assembly.

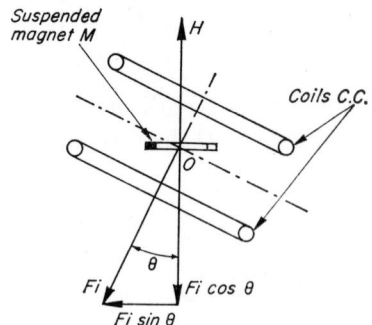

FIG. 6.7. Schuster's method of determining H

A cube of quartz (6 mm side) is mounted on the float so that its faces are vertical, two of them being parallel to the magnet axes. The cube is platinized on four sides and is worked to provide an optically plane mirror. A horizontal telescope mounted on the instrument is used to view the mirror through optically worked glass windows in the gunmetal chamber. A cross-wire illuminated by light reflected on it by a prism occupies the upper part of the telescope field. The image of the cross-wire is reflected by the mirror to a micrometer scale (20 divisions per mm) in the focal plane of the telescope and is viewed through a Ramsden eyepiece. A deflection of 20 seconds of arc can be measured. The coils and telescope are mounted so as to rotate over an azimuth circle, 25 cm in diameter, divided to 15 minutes of arc, and microscopes and verniers are provided so that readings may be taken to 20 seconds of arc. A sensitive level and levelling screws are provided. The coils, which have a mean diameter of 40 cm, are mounted 20 cm apart and each is wound with 40 turns of 24 S.W.G. copper wire. They are connected in series so as to produce axial fields in the same direction. The current usually employed is of the order of $0\cdot1$ amp and can be measured to an accuracy of 5 microamps by means of a potentiometer.

The potentiometer is designed to an accuracy of one part in 2×10^4. A standard cadmium cell is balanced against a coil which can be increased in steps of $0\cdot1$ ohm from 9 ohms up to 11 ohms. The current in the potentiometer can thus be varied in steps of about 1 per cent to the required accuracy. The current is supplied from a 6-volt accumulator, rheostats enable the value to be adjusted to the approximate level required (a 150-mA f.s.d. meter being used to indicate the approximate

* '*A portable magnetometer of the null type*', Ordnance Survey, Southampton, 1930, H.M.S.O.

current), and the final value is obtained by balancing across the standard cell. The instrument is fitted with a small reflecting galvanometer with self-contained lamp, scale and battery. For field use a tripod is provided.

A complete determination can be made in about 10 minutes and the measurements are not likely to be more than 1 gamma in error.

QHM magnetometer

The QHM, quartz-fibre horizontal intensity magnetometer, designed by D. La Cour in 1933,* is a portable instrument for measuring horizontal intensity by comparison with calibrating measurements made at a base-station. It consists of a vertical brass tube, adapted to be mounted on a theodolite type of base, and containing a quartz suspension to which a mirror and magnet are attached. The mirror is viewed through a small telescope.

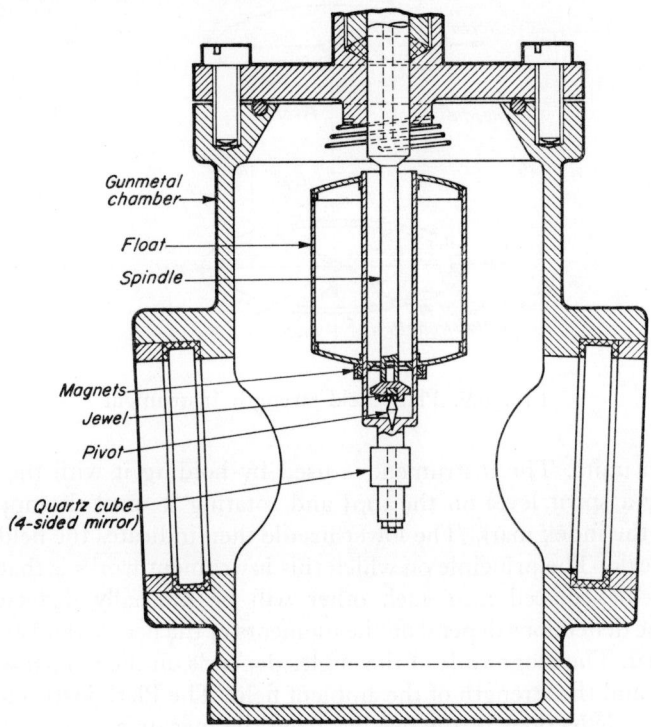

Gunmetal chamber

Float

Spindle

Magnets

Jewel

Pivot

Quartz cube (4-sided mirror)

FIG. 6.8. Float chamber and magnet system of Schuster–Smith variometer

To use the instrument, after being calibrated, the suspension is released so that it will swing freely in the magnetic meridian. By rotating the base by a given amount, first clockwise and then counter-clockwise, a known amount of torsion is applied and the deflection of the magnet noted. This is compared with the standard calibration and hence the local value of H may be determined.

* La Cour, D., 'Le Quartz magnétomètre QHM'(quartz horizontal force magnetometer) Copenhagen (1936) *Met. Inst. Comm. Magn.* No. 15; Fanselau, G., *Z. Geophys.*, 1936, **12**, 192 (report on use of this instrument).

Since the instrument is small and light, it may be sent by post after calibration at the base-station for use at any desired site, and it will give results accurate to a few gamma.

Plath field-strength instrument

Among portable magnetic measuring instruments, mention must be made of an interesting field-strength-measuring instrument made by C. Plath of Hamburg (Fig. 6.9). It consists of a cylindrical body with a verge glass at its upper end. Two magnetic compass needles are placed one above the other, the pivots being on the vertical axis of the cylinder. The upper needle is associated with a single index mark on the casing and the lower indicates readings on an annular scale marked in

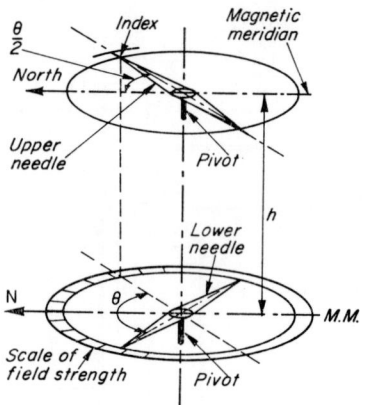

FIG. 6.9. Plath field-strength instrument

field-strength units. The instrument is used by holding it with the axis vertical (indicated by a spirit level on the top) and rotating it until the upper needle is aligned with the index mark. The lower needle then indicates the field strength on the annular scale. The principle on which this instrument works is that two pivoted magnetic needles placed near each other will be mutually deflected from the meridian. The deflections depend on the moments of the needles and are equal when these are equal. The magnitude of the angles depends on the moments of the magnetic needles and the strength of the ambient field. The Plath instrument measures the sum of the deflections of the needles from the meridian.

The relationship between field strength H and angular displacement for equal needles is of the form

$$H = \frac{M}{h^3} \cos \theta (B + C \cos 2\theta + D \cos 4\theta + \ldots)$$

where θ is the deflection of either needle from the meridian and $B, C, D, \ldots$, are constants depending on powers of l^2/h^2, l being the half-length of the needle and h the vertical separation of the needles. As h becomes large compared with l, so H becomes more nearly equal to $(BM \cos \theta)/h^3$, M being the magnetic moment of either needle.

The compass variometer is another instrument of this type, used for rapid measurements of magnetic field strength. In this instrument two equal magnets are used and the distance between them is adjusted to give a fixed angle of deflection at any point on the Earth's surface. It is estimated that an accuracy of one gamma is obtainable with such an instrument suitably calibrated.

THE DIP CIRCLE

The dip circle, illustrated in Fig. 6.10, is used to measure the inclination or angle of dip of the Earth's total field. The instrument can be adapted to measure the vertical intensity.

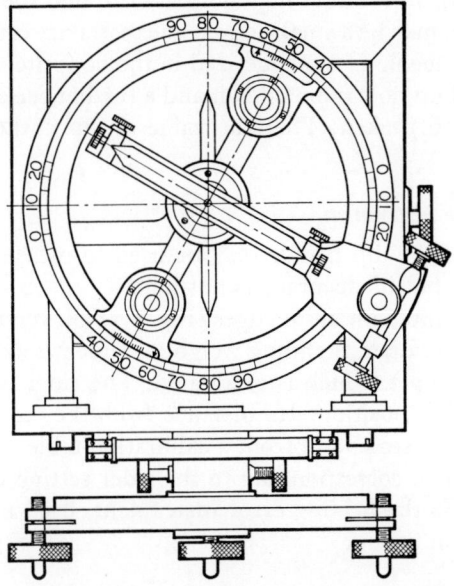

FIG. 6.10. Dip circle

The dip needle is a thin steel magnet with a fine axle which rests on agate knife edges. It is carried in vee-shaped supports so that it may be lifted off the knife edges when not in use. The needle, both ends of which can be read, is at the centre of a vertical circle marked with a scale of degrees, and to ensure that the axle of the needle is at the centre of this scale the needle needs to be raised and lowered repeatedly when making a reading. The body of the instrument may be rotated about a vertical axis and its azimuth read upon a horizontal circle with a scale of degrees. Microscopes and verniers enable the circles to be read accurately, and tangent screws permit fine adjustments to be made. Levelling screws and a sensitive level are fitted.

When the body of the instrument is levelled and orientated so that the dip needle sets vertically, the plane in which the needle rotates is at right angles to the magnetic meridian. The azimuth scale now reads 90°–90°. On rotating the instrument through 90° about its vertical axis, the plane of rotation of the needle is brought into the meridian and the needle will set in the direction of the total field. The angle at which it sets is measured on the vertical circle and is the angle of dip.

To ensure accurate determination of dip, certain errors need to be eliminated.

   (i) Both ends of the needle are read to correct for eccentricity (i.e. the fact that the centre of rotation of the needle may not be at the centre of the scale).

  (ii) Since the 0°– 0° line of the vertical scale may not be truly horizontal, the instrument is rotated about its vertical axis through 180° and the two readings in (i) repeated.

 (iii) Since the magnetic axis of the needle may not coincide with its geometric axis (see Fig. 3.1), the needle is turned over on its bearings and four more readings taken as above.

 (iv) Since the axis of rotation may not pass through the centre of gravity of the needle, there may be a small couple due to gravity causing a wrong indication of dip. The needle is remagnetized in the opposite direction so that the end that pointed up now points down and a further series of eight readings as in (i), (ii) and (iii) taken. The true value of dip is the mean of these sixteen readings.

Kelvin vertical-force instrument

A modification of the dip needle may be used to measure the vertical intensity of the Earth's field, but its accuracy is not high.

A bar magnet mounted on knife edges (Fig. 6.11) lies so that its plane of rotation is in the magnetic meridian. A sliding weight or 'rider' is arranged so that it may be moved along the magnet, which is graduated. The rider is set so that the magnet rests in the horizontal position. Its distance from the centre of the needle is read off along the engraved scale, and from a calibration table supplied with the instrument the vertical force corresponding to the rider setting can be ascertained.

A similar device is the heeling error instrument, used in the adjustment of the mariner's compass in a ship.

Schmidt vertical and horizontal magnetometers

The Schmidt vertical and horizontal magnetometers, based on the dip needle, are used for measuring anomalies in the vertical and horizontal intensities of the Earth's magnetic field. They are set up at a base-station and calibrated and the base-reading is determined. On the assumption that the value of magnetic intensity at the station is normal, measurements at other sites disclose any anomalies.

In the vertical instrument (Fig. 6.12), a magnet system is balanced on knife edges at right angles to the magnetic meridian. The deflections of the system are measured by an autocollimating telescope. The centre of gravity is arranged to be on the south side below the axis. The system is surrounded by a case containing eddy-current copper dampers and thermometers and surrounded by a cork-lined jacket to insulate the system from variations in temperature. The magnet system rests on a bridge with quartz pieces which act as bearings for the knife edge, and a clamping device is provided. In use, the instrument rests on a tripod provided with a graduated scale so that it may be orientated as required. Before the instrument is mounted, a magnetic compass is used on the tripod to determine the direction of the magnetic meridian. A scale of 60 divisions is provided in the optical system and the position of a pointer reflected from the mirror of the magnet system is read on this scale. A temperature-compensating construction is used in the magnet system.

The operation of the Schmidt magnetometer is expressed by equation (3.9), viz.

$$\tan \eta = \frac{MH \sin \alpha \cos \beta - MZ \sin \beta - wga \sin \beta}{MH \cos \alpha + wgb \sin \beta}$$

If S_0 is the centre reading corresponding to zero deflection, S the reading corresponding to deflection η, and f the focal length of the objective,

$$\tan 2\eta = \frac{S - S_0}{f}$$

or, since η is a small angle, $2 \tan \eta$ may be written for $\tan 2\eta$.

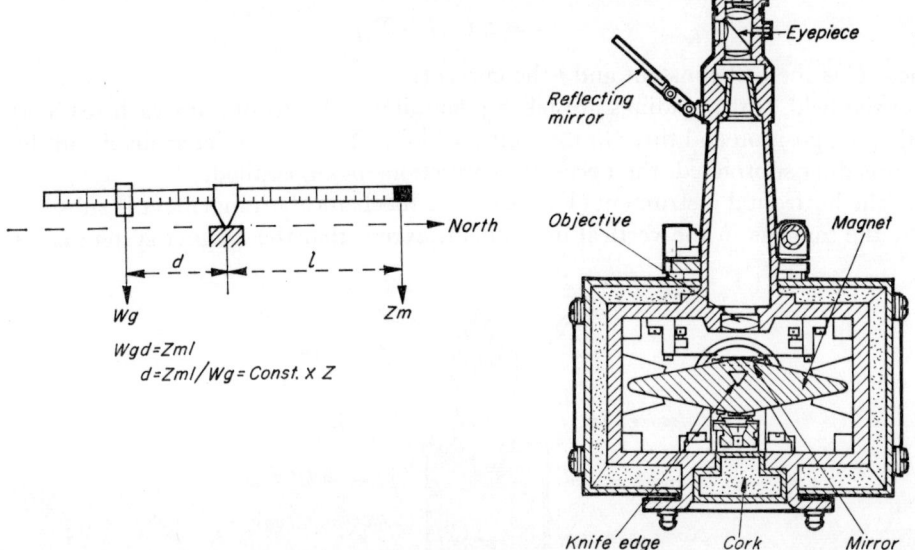

FIG. 6.11. Kelvin vertical-force instrument   FIG. 6.12. Schmidt vertical magnetometer

Since, in use, $\alpha = 90°$ or $270°$ and $\beta = 90°$, we can now write

$$S - S_0 = 2f(MZ - wga)/wgb$$

remembering that both a and b are negative.

Substituting the scale value $\epsilon = wgb/2fM$, we now have

$$\epsilon(S - S_0) = Z - \frac{wga}{M}$$

Thus when the deflection in scale divisions is multiplied by the scale value, the difference between vertical intensity and gravity is obtained. (The term wga/M is the vertical intensity Z_0 for which the system has been adjusted.) Thus multiplication of the scale value by the scale reading from o determines the anomaly in vertical intensity.

Errors will arise in the use of the Schmidt magnetometer if certain precautions are not observed. The instrument must be correctly orientated in the magnetic east–west or west–east positions; it must be most carefully levelled in the direction of the magnetic meridian, and most carefully placed in a level position, in the east–west direction, upon the tripod. Variations in temperature are normally compensated, but small changes of the position of the system's centre of gravity can cause appreciable errors.

The scale-value is determined by the use of auxiliary magnets or Helmholtz coils. For magnets,

$$\epsilon = 4Mk(S_2 - S_1)r^3$$

where M is the moment of the magnet, k a deflection constant, r the distance between centres of magnet and magnet systems, and S_1 and S_2 are readings with the north pole of the magnet up and down respectively. For the Helmholtz coil,

$$\epsilon = 2iC(S_2 - S_1)$$

where C is the coil constant and i the current.

In the field, three readings are taken (clamping and releasing for each reading) in the east position and three in the west position. The mean is determined and the base-reading subtracted, the necessary corrections being applied.

In the horizontal instrument (Fig. 6.13) the mechanical arrangement is substantially the same as in the vertical instrument, except that the magnet system is set

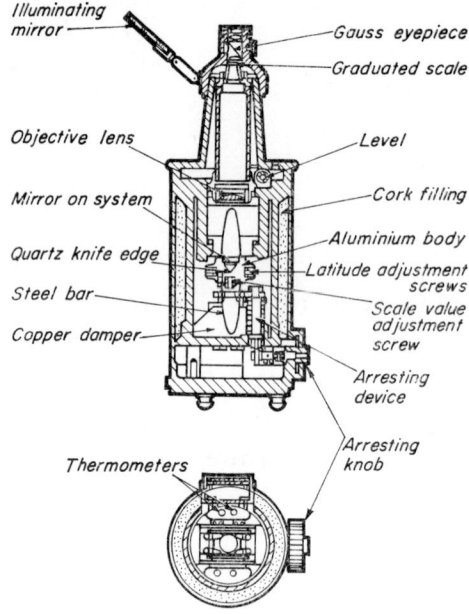

Illuminating mirror
Gauss eyepiece
Graduated scale
Objective lens
Level
Mirror on system
Cork filling
Aluminium body
Quartz knife edge
Latitude adjustment screws
Steel bar
Scale value adjustment screw
Copper damper
Arresting device
Arresting knob
Thermometers

FIG. 6.13. Schmidt horizontal magnetometer

up with its magnetic axis vertical and readings are taken in the magnetic meridian. This is possible in one position only. If the instrument were rotated 180°, gravity would be added to instead of subtracted from the magnetic couple and would put the system off balance completely. If ϕ is the deflection from the vertical of the magnet system, from equation (3.9), by writing $\eta = (90° - \phi)$ and by interchanging a and b, we have

$$\tan \phi = \frac{MH \cos \alpha - wga \sin \beta}{MH \sin \alpha \cos \beta - MZ \sin \beta + wgb \sin \beta}$$

Now

$$\alpha = 0° \quad \text{and} \quad i = 90°$$

so

$$\tan \phi = -\frac{MH - wga}{MZ - wgb}$$

or in terms of scale readings

$$S - S_0 = -\frac{2fM}{MZ - wgb}\left(H - \frac{wga}{M}\right)$$

where $2fM/(MZ - wgb)$ is the reciprocal scale value of the magnetometer and wga/M the horizontal intensity H_0 for which the system is adjusted.

The same precautions against errors when setting up must be observed as in the case of the vertical magnetometer, though the effect of mislevelling is less serious. Misorientation, however, is far more important in the horizontal instrument. A correction must be applied for changes in Z (see equation for scale readings above). Other corrections and the method of calibration follow the same pattern as for the vertical magnetometer. In the field, a number of readings, say six, are taken (clamping and releasing each time); these are averaged, the base-reading subtracted and corrections made as necessary.

EARTH INDUCTOR

The Earth inductor, illustrated in Fig. 6.14, is a more precise instrument than the dip circle. Its operation depends on the induction of an electric current in the windings of a closed coil of wire when rotating in a magnetic field in such a manner as to change the number of lines of force threading the coil. This has already been discussed in Chapter 4, p. 47. The inductor rotates about an axis in the plane of its coil and to determine the angle of dip, it is necessary to find the position in the plane of the magnetic meridian of the inductor's axis of rotation when no current flows in the coil.

The axis of rotation is first directed into the meridian by means of a magnetic compass. The inclination of the coil is then changed until, with the coil rotating, no current flows in the coil windings, a sensitive galvanometer being used for verification. The angle of dip is read off a vertical circle attached to the body of the instrument. Readings are taken with the instrument rotated 180° and the mean of the observations is the correct angle of dip. The instrument is provided with a level and with micrometers and verniers to facilitate reading; it is mounted on a levelling tribrach. In earlier models, rotation of the coil was carried out by hand, but nowadays a constant-speed motor and flexible drive are frequently used.

Induction variometer

The induction variometer provides another method of measuring the vertical intensity of the Earth's magnetic field. This is based on the property of very soft iron rods to become magnets when placed in a magnetic field, their pole strength depending on the field's magnitude. Although the idea of making such instruments was put forward many years ago, it has only recently been practicable to make a sufficiently sensitive variometer, owing to the invention of perminvar, an alloy of nickel, cobalt and iron, which has a constant permeability for all practical purposes and negligible hysteresis at low field strengths. Fig. 6.15 illustrates the principle

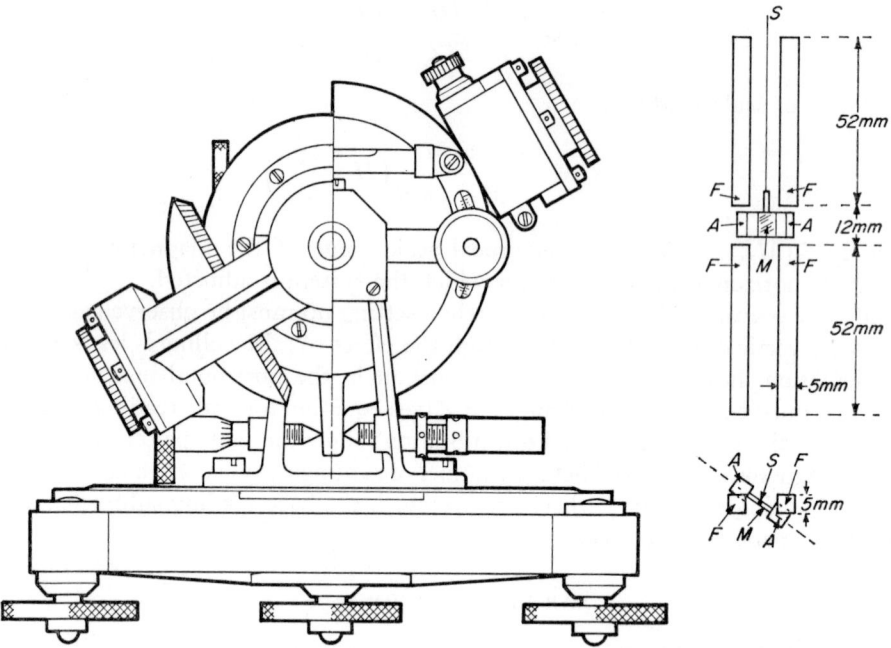

FIG. 6.14. Earth inductor

FIG. 6.15. Induction variometer

of such an instrument designed by A. G. McNish.* Four vertical pieces of perminvar, the 'field pieces' $5 \times 5 \times 52$ mm long, are arranged vertically in pairs as at F F F F, and supported on Pyrex glass rods. Between the upper and lower field pieces, there is air gap of about 12 mm, across which there is a non-homogeneous field. In this space two smaller pieces of perminvar—A A, the 'armature'—are suspended by a quartz fibre S. These are mounted with a mirror, M. By induction, F F F F and A A all become magnetized by the Earth's vertical field, and the armature tries to rotate in the air gap so that the flux through field pieces and armature is a maximum. Torsion in the quartz fibre restrains this rotation, and the angular position of the armature, which is usually recorded photographically using a light beam reflected by the mirror M, provides a measure of the Earth's vertical

* McNish, A. G., 'A new type of vertical intensity induction variometer', *Terr. Magn. atmos. Elect.*, 1936, **41**, 161–72; also *Rev. sci. Instrum.*, **7**, 336–8.

field intensity. Alternatively, a scale could be used. The sensitivity may be changed by using different quartz fibres or by altering the distances between the field pieces. A convenient temperature-compensating property of this instrument is that the magnetic torque on the armature varies with its vertical position in the air gap. Thus as the armature rises and falls with alterations in temperature, the change in torque can be made equal and opposite to the effect of the change in the permeability of perminvar, which also varies with temperature.

Induction variometers on this principle can be constructed to measure the east-component and north-component of the Earth's magnetic field, with the advantage that the measurement of one component is independent of the changes in the direction of the magnetic field.

Saturable inductors as magnetometers

Saturable inductors can be very usefully employed as magnetometers. Their form and construction have been described and some account of their operation given in Chapter 4.

In use, it must be remembered that the output from the inductor is proportional to the strength of the axial field and that the direction of any field, whether that of the Earth or the resultant of two or more fields, is most easily determined by the orientation of the inductor to indicate zero output, when it is at right angles to the ambient field. Consequently, provided certain precautions are taken, an inductor can be used to make the usual magnetic measurements made with a pivoted or suspended needle. The necessary precautions include those taken when using the simple pivoted-needle instrument, namely reversing the direction of the inductor to obtain a reciprocal reading to avoid eccentricity and asymmetry and, if the instrument is fitted with a pointer and scale, reading the two ends of the pointer. In addition, the inductor must be accurately levelled. Its use can be extended to multiple axis systems (Fig. 6.16) and to field strength difference measurements or gradiometers (Fig. 6.17).

However, it must be realized that the saturable inductor depends on electrical supplies and that the output, and hence final field measurement, must be free from distortion and accurately determined. Hence, although it may appear more attractive, it can, for example, determine a field direction only by inference, whereas a pivoted needle lies in the direction of the field and can be observed directly. There is also the question of signal-to-noise ratio in the measurement of small fields, and thus in the accurate determination of zero field for finding the direction of a magnetic field. It can be said, therefore, only that the inductor type of magnetometer is complementary to the classical instruments; although at the same time, given suitable conditions, it is very much more convenient, and can be readily adapted to a continuous recording device.

A saturable inductor is generally used in two ways as a magnetometer:

(i) as a direct meter of the ambient field;

(ii) in conjunction with a 'backing off' circuit to reduce the output to zero.

The direct method involves a detector head consisting, for example, of a twin-core saturable inductor. An oscillator or any convenient and suitable source of alternating current may supply excitation. A filter will probably be required to cut off all but the second harmonic signal fed to an amplifier, which delivers its output to a voltmeter, cathode-ray oscilloscope or other suitable measuring device—which could be continuously recording (Fig. 6.18). Constancy of calibration, an accurate and linear amplifier and meter and a linear inductor characteristic are essential features of this system.

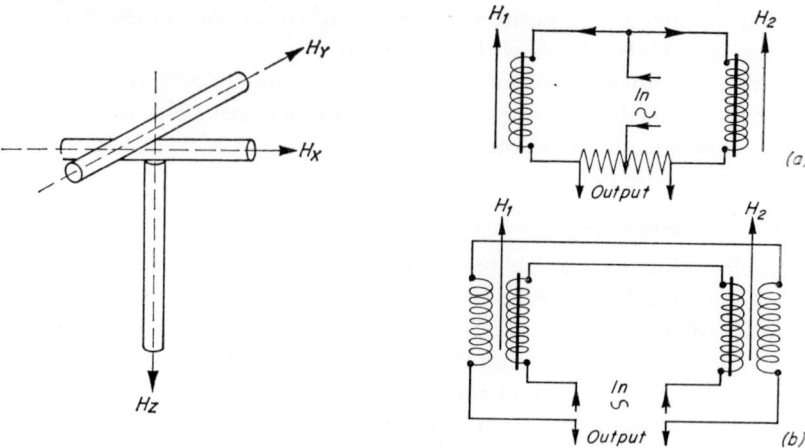

FIG. 6.16. Multiple-axis inductor system   FIG. 6.17. Simplified circuits for gradiometry

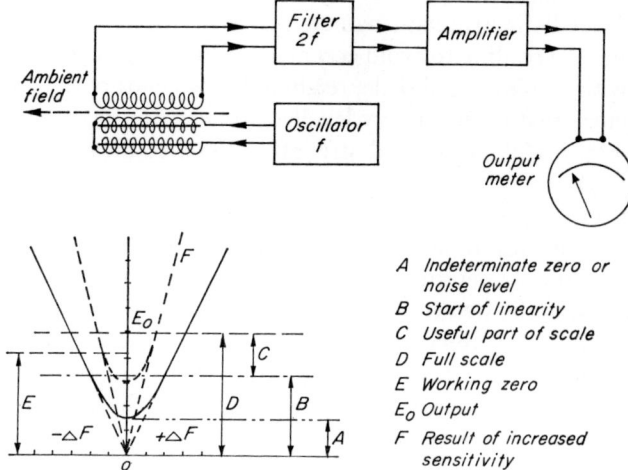

A Indeterminate zero or noise level
B Start of linearity
C Useful part of scale
D Full scale
E Working zero
E_0 Output
F Result of increased sensitivity

FIG. 6.18. The saturable inductor as a magnetometer

A simple magnetometer of the kind described has certain limitations. The output signal, when rectified, will show changes in the ambient field in amplitude only. The system is not directionally sensitive, i.e. equal fields in opposite directions will give similar readings at the meter. Moreover, the zero is seldom sharply defined and only when a fairly large signal exists does the output bear a linear relation to the input or field strength. The limitation imposed on the output meter is a great disadvantage, for instance, when measuring small changes in the Earth's magnetic

144

field; for with a sensitive instrument, half the scale can well be taken up by the ill-defined zero, leaving only one quarter of the scale on either side of a mean reading for recording positive and negative changes. A semblance of directional conscious-ness is achieved by biasing the detector using a small magnetic field or a small rotation in the Earth's field, so that the apparent zero for measuring purposes is in the centre of the meter scale. Only at the expense of sensitivity can the zero

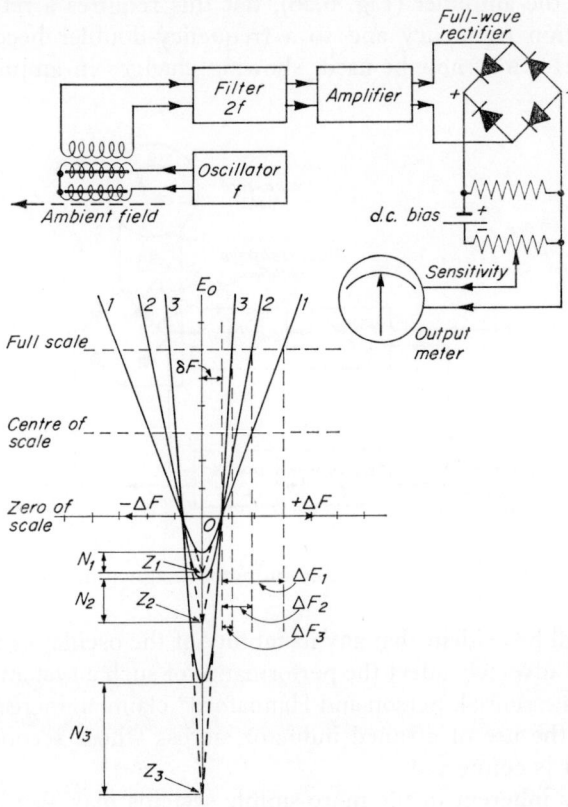

FIG. 6.19. Effect of biasing the meter of a saturable-inductor magnetometer

δF is the 'bias' field and is constant for all ranges. Curves 1, 2 and 3 show the relation between E_0 and ΔF for three values of sensitivity.
Z_1, Z_2 and Z_3 are the virtual zeros for the curves 1, 2 and 3. N_1, N_2 and N_3 are proportional to sensitivity.
ΔF_1, ΔF_2 and ΔF_3 are the linear full-scale ranges for ΔF for the three curves.

signal be reduced to a small part of the instrument scale. Excursions of the output greater than the usable part of the scale result in ambiguity, since an indication increasing towards zero will change its apparent sense on reaching and exceeding zero.

By biasing the meter with a small d.c. supply the effect of a 'suppressed' zero is obtained (Fig. 6.19). Hence in order that the meter may start to show indications of the output, the input signal must be large enough for its output to cancel the bias.

This enables a working level of input to be used, well removed from the indeterminate zero of the system and random noise. The range of the output displayed on the meter can be in the linear region of operation, with linear and symmetrical excursions to either side of a working zero in the centre of the scale. The sensitivity may be increased to almost any desired value as the virtual zero of the scale moves correspondingly.

True directional sensitivity can be achieved by using a phase-sensitive rectifier at the output of the amplifier (Fig. 6.20), but this requires a reference signal at twice the excitation frequency and so a frequency-doubler becomes necessary. A centre-zero meter may now be used, showing changes in amplitude and direction.

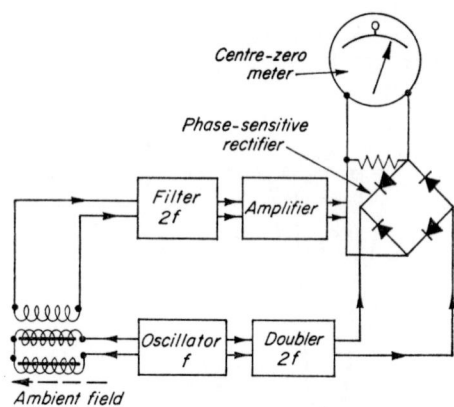

FIG. 6.20. A saturable-inductor magnetometer

However, it will be evident that any instability of the oscillator, and particularly phase-shifts, will adversely affect the performance of such a system and the careful design of filters is essential. Serson and Hannaford* claim an increase in sensitivity and stability by the use of a tuned inductor, across whose secondary winding a suitable capacitor is connected.

The difficulties inherent in the more simple systems may also be overcome by applying to a recording magnetometer (Figs. 6.21 and 6.22) the modulated high-frequency method of operating a second-harmonic inductor, described in Chapter 4.

The use of feedback or backing off has, in some circumstances, advantages over the methods just described. The inductor may be provided with an extra winding, or placed within a suitable axially arranged coil. This coil is supplied with direct current from a battery through a potentiometer and an ammeter. Thus a field of a desired value may be applied along the axis of the inductor at will. The inductor is placed in the ambient field and by adjusting the potentiometer an equal and opposite field may also be applied. A zero reading of the output meter will indicate the equality, and the magnitude of the backing off or opposing field (and hence the ambient field) will be seen from the current indicated on the ammeter, provided the dimensions and number of turns of the coil are known.

* Serson, P. H., and Hannaford, W. L., 'A portable electrical magnetometer.' *Canadian Journal of Technology*, 1956, **34**, 232–43.

Therefore every measurement of field is reduced to a zero determination and the inductor is required only to provide a low noise level in zero field. The certainty with which the size and turns of a coil are known and the ease with which direct current can be read make this method accurate and reliable [Fig. 6.23(a)].

An extension of this application is to rectify the output from the amplifier and to feed it back into a tertiary winding on the inductor, the magnitude of the direct current therein being a measure of the field. Complete 'backing off' is never achieved, a small signal being required to drive the amplifier, but a very high value of negative

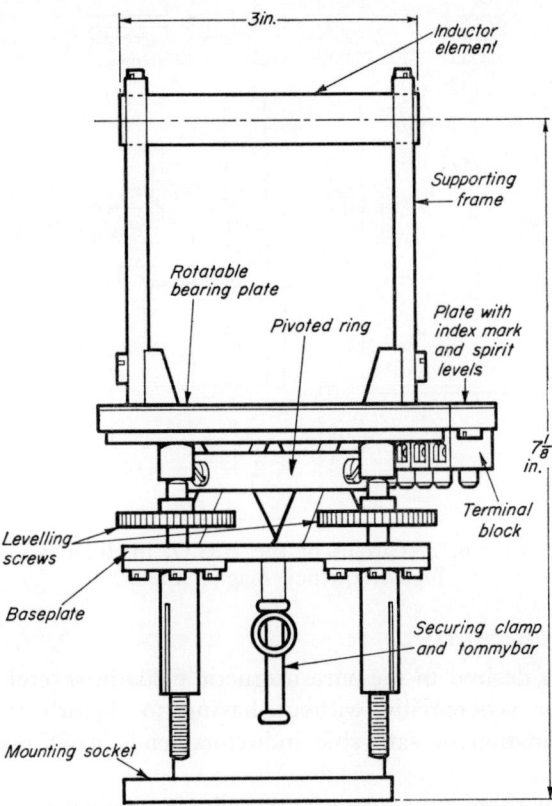

FIG. 6.21. The A.C.O. modulated high-frequency magnetometer

feedback is reached and a consequent independence of inductor linearity and stability, as the inductor is operating very close to zero field at all times [Fig. 6.23(b)]. Continuous recording is possible with this system.

Measurement of field differences. If the circuit of the saturable inductor is re-arranged, it can be used for measuring the difference in field strengths. The rearrangement, the gradiometer, consists in placing one of the cores of the system in the region affected by each of the fields under consideration.

In Fig. 6.17(a), the cores are excited so that in zero field or equal fields, the bridge system illustrated is balanced. If $H_1 \neq H_2$, the bridge becomes unbalanced and an e.m.f. appears across the output terminals. In Fig. 6.17(b), the inductor of

Fig. 4.16 and 4.17 is separated, having two primaries and two secondaries. The primaries are excited in series and the secondaries are connected so that when $H_1 = H_2$ or when both are zero, the secondary voltages cancel out. When $H_1 \neq H_2$ an output voltage appears, proportional to the field difference. The operation of the field difference detector can be illustrated in the same way as that of the saturable inductor used for direct field strength measurement, given on pp. 143 and 144. An interesting application of this detector is to use the output to control a feedback circuit whereby the two fields can be maintained at equal magnitudes.

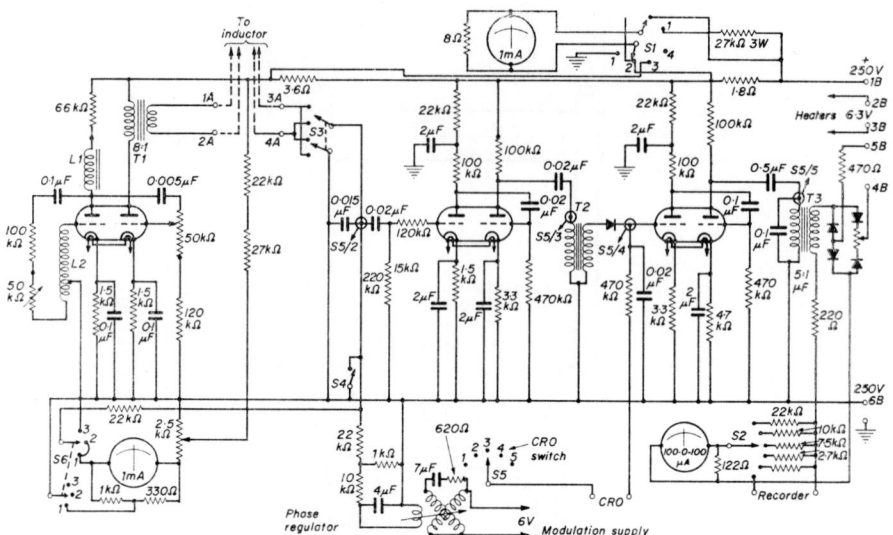

FIG. 6.22. Circuit of the A.C.O. modulated
high-frequency magnetometer

Multiple axis magnetometers

It is sometimes desired to measure magnetic fields in several directions, either simultaneously or sequentially, without having to disturb the magnetometer system. A combination of saturable inductors lends itself readily to such an operation.

In Fig. 6.16, the three inductors are mutually perpendicular, enabling the X, Y and Z components of the Earth's field [see Fig. 2.3(a)] to be measured and Z, H, T, D and V to be determined. The three fields H_x, H_y and H_z could be the athwart-ship, fore-and-aft and vertical components of the total ambient field in a ship, whose values are required, shall we say, for a magnetic survey. By having three indicators (a common excitation is suitable), the three components of the field can be measured simultaneously or, if such a refinement is unnecessary, a simple switching arrangement can be used, employing only one indicator. Feedback may be employed, as already described. (See also Pioneer 3-axis magnetometer, p. 151.)

With a fixed system such as this, and with the aid of a suitable resolving mechanism, the three magnetic components in a ship, aircraft or vehicle can be used to determine the field in the horizontal plane and its direction with respect to the ship. Thus the device becomes a compass.

Given suitable dimensions, the saturable inductor performs very well as a probe for exploring the magnetic fields in confined and restricted spaces, notably inside electrical apparatus; an interesting example of an application of this kind is found in a precision permeameter, described by Mee and Street.* No typical form of saturable inductor magnetometer needs to be illustrated, as the form, dimensions and construction of such detectors depend on the particular application.

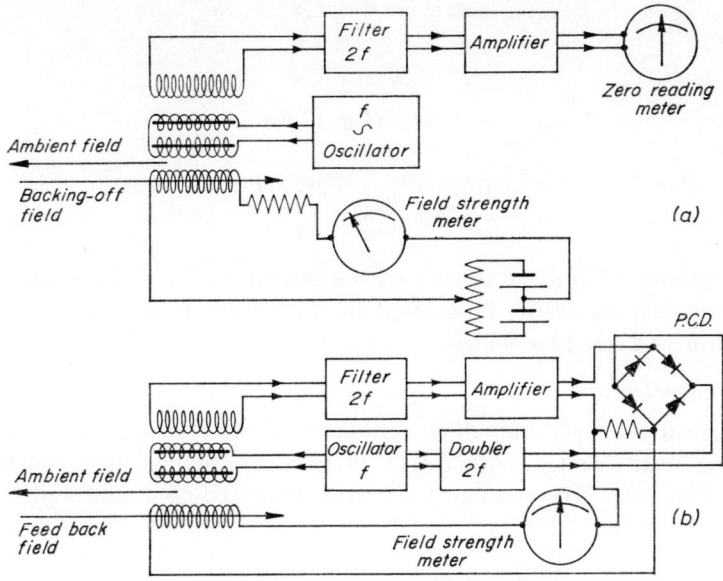

Fig. 6.23. Backing-off and feedback arrangements

RECORDING INSTRUMENTS

Where continuous records of the components of the Earth's magnetic field are required, magnetographs are used. These are usually combinations of magnetometer or variometer and a clockwork-driven recorder, on which a sensitized paper chart is mounted.

Variation magnetograph

The variation magnetograph is, in fact, an adaptation of the magnetometer shown in Fig. 6.2, in which the scale is replaced by a sheet of sensitive photographic paper wound upon a drum driven at a constant speed. Time is measured in the direction of rotation of the drum and variation at right angles to it.

Horizontal force magnetograph

The horizontal force magnetograph, or variometer, uses a similar suspended magnet system and mirror, but the suspension is twisted until the needle lies at right angles to the meridian. This was the method employed by Eschenhagen† in

* Mee, C. D., and Street, R., 'An improved precision permeameter', *Proc. Instn. elect. Engrs.*, 1954, **101**(2), 639.
† Eschenhagen, M., 'Über erdmagnetische Intensitätvariometer', *Verh. dtsch. phys. Ges.*, 1899, **I**, 147–52.

his variometer. Compensating magnets may be used instead of torsion in the suspension to bring the needle at right angles to the meridian. The operation of this system depends on the equilibrium of the magnetic and torsional couples.

If M is the magnetic moment, H the horizontal intensity of the Earth's field, θ the angle between the axis of the needle and the magnetic meridian and α the angular twist in the suspension,

$$MH \sin \theta = C\alpha$$

If $\theta = 90°$,

$$MH = C\alpha$$

$$M.\delta H = C.\delta\alpha$$

Therefore

$$\delta\alpha = \frac{M.\delta H}{C}$$

Thus the greater M and the smaller C, the greater $\delta\alpha$ for a given value of δH.

The recording apparatus is as used in the variation magnetograph, namely, lamp, mirror and sensitive paper.

Watson's magnetograph

The vertical intensity induction variometer described on p. 138 is primarily a recording instrument. Another type of magnetograph is that constructed by Watson,* Fig. 6.24. Here a magnet system is mounted on a quartz plate to which

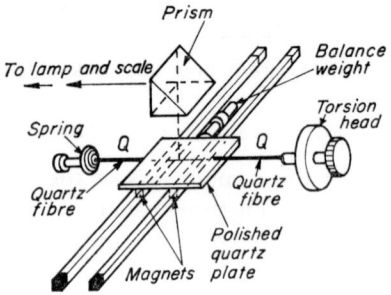

Fig. 6.24. Levelling arrangement

horizontal suspension fibres QQ are fused. One fibre is attached to a support in the form of a spring and the other is attached to a torsion head. The instrument is set up, as in the case of a dip circle, with the plane of rotation in the magnetic meridian, and the adjustment of the torsion head sets the magnet system horizontally. The small balance weight is placed so that the ends of the needles which usually point upwards now tend to point downwards, the rotation of the torsion head bringing the magnets horizontal. The quartz plate is polished to serve as a mirror and a prism enables the movement of the mirror to be recorded, using a horizontal beam of light from a lamp and a photographic recorder. By virtue of its

* Watson, W., 'A quartz thread vertical force magnetograph', *Phil. Mag.*, 1904, **7**, 393–9; *Terr. Magn. atmos. Elect.*, **9**, 62–8.

construction, this magnetograph is easily compensated for alterations in temperature, by setting off the change in magnetic moment of the magnets against that in the rigidity of the quartz fibres as the temperature varies.

La Cour's magnetograph

La Cour designed a very satisfactory and widely used magnetograph of the quartz suspension type, with several novel constructional features including an optical method of temperature compensation. This magnetograph was first used in the International Polar Year of 1932–3. Instruments of this type are made for the measurement of changes of variation, horizontal force and vertical force, and the three together may be set up at a magnetic observatory, in conjunction with a single source of light directed through an array of prisms to each separate instrument, to provide a photographic record of the changes in the three magnetic elements simultaneously on one chart. By using a further system of prisms to deflect the light beam at each revolution of the chart drum (say once in two hours), a 'quick run' magnetogram may be produced where, for 24 hours, each element is displayed in twelve lanes or channels, the light beam being 'switched' to the next channel as the drum completes each full revolution.

The Pioneer 3-axis magnetometer is a conveniently portable saturable inductor system for making magnetic surveys. Field strength readings are possible in three directions mutually at right angles, a switch selecting the required axis and direct indication of field strength being shown on a meter. The whole equipment, including oscillator and amplifier, is self-contained and battery-driven.

Saturable inductor magnetographs

Saturable inductors are particularly adaptable as the sensitive elements in recording magnetometers, since the detector may be remotely situated from the recording gear. The range and sensitivity may be determined by choosing suitable types of inductor of appropriate dimensions. Stable amplifiers and oscillators preventing drift and changes of sensitivity are readily constructed. A recording milliammeter or millivoltmeter provides a robust output instrument and a choice of clock speeds gives an additional facility for normal and quick runs. The arrangement of such a magnetometer is shown in Figs. 6.18, 6.19, 6.20 and 6.23(b), the output meter being a recorder of the type already mentioned.

A.C.O. modulated high-frequency magnetometer. A recent development of this type of magnetometer, carried out at the Admiralty Compass Observatory, using the modulated high-frequency mode of operating the inductor described in Chapter 4 is illustrated in Figs. 6.21 and 6.22. The advantages of this mode are apparent from the results obtained. The equipment indicates the direction of the applied field unambiguously, permitting a centre-zero recording instrument to be used. A sensitivity of about 4γ per mm of chart is a convenient value, enabling large disturbances to be fully recorded without losing the day-to-day irregularities of a minor nature.

However, the sensitivity may be arranged to suit any particular application. With the inductor placed magnetic east–west, the instrument becomes a variation variometer or declinometer. By placing the inductor in line with any component of the Earth's field, the intensity of that component may be measured. Through a calibrated resistance network a proportion of the d.c. voltage from the stabilized H.T.

source is supplied to the secondary winding of the inductor, so that a magnetic field of known strength may be applied at will. When this is equal and opposite to the field being measured, the output signal is reduced to zero. The magnitude of the secondary or backing-off current is now a measure of the applied, and therefore of the ambient, field. This is conveniently ascertained by reference to the setting of the calibrated potentiometer. In the equipment illustrated, two ranges are available: 0–5000γ (5γ per division) and 0–$0\cdot5$ oersted (50γ per division) giving accuracies of $0\cdot3$ per cent and 1 per cent respectively. For accurate results the inductor needs to be levelled correctly with its magnetic axis horizontal. This is best done by finding a zero field position approximately east–west and then again in the reciprocal direction. The angular change will be $180°$ only if the inductor is levelled correctly. To correct for level, the mean of the two zero positions is determined and the inductor orientated accordingly. The level is then adjusted to give zero output, when $180°$ rotation should also give zero. By rotating the inductor a further $90°$ its magnetic axis is aligned with the field to be measured. A potentiometer reading is taken in this position and again when turned through $180°$, using the field direction selector switch in the amplifier. The mean of these readings multiplied by the scale factor for the potentiometer gives the field intensity. By the same means the equipment may be used as a horizontal or vertical force variometer, the potentiometer applying the field required for cancellation and subsequent changes in field strength being displayed on the recording output meter. Measurement of variation is, to some extent, indirect, a change of east–west field being interpreted as an angular change, since

$$\delta F/H = \sin \delta V$$

where δF is the change in the east–west field, H is the horizontal intensity and δV is the change in field direction. For small angles, the expression may be rewritten $\delta V = \delta F/H$. The chart may be read directly in terms of arc. The results agree very closely with those obtained from a quartz suspension variometer, the correspondence being evident for both large and small changes in form and amplitude. A correlation of the order of $0\cdot8$ has been found between an inductor magnetograph of this kind at Slough and a La Cour variometer at Hartland.

Instead of a single inductor, a three-axis orthogonal system as in Fig. 6.16 may be used, and three field components, which may be the vertical, true northerly and true easterly components of the Earth's field, may be measured in turn.

Butterworth's magnetograph

Butterworth* has devised an electrically recording magnetograph worthy of special notice. This makes use of the change in impedance of a length of Mu-metal wire when situated in a varying magnetic field.

Harrison, Turney, Rowe and Gollop† showed that the change in impedance in a wire of high permeability could be used for studying a magnetic field. The effect

---

* Butterworth, A., 'A sensitive recording magnetometer', *J. Instn. elect. Engrs.*, 1947, **94** (2), 325.

† Harrison, E. P., Turney, G. L., and Rowe, H., *Nature*, 1935, **135**, 961.

Harrison, E. P., Turney, G. L., Rowe, H., and Gollop, H., *Proc. roy. Soc.*, 1936, **157**A, 451.

Harrison, E. P., and Rowe, H., *Proc. phys. Soc. Lond.*, 1938, **50**, 176.

British Pat. 562755, 1943.

Harrison, E. P., and Smith (Miss), E. H., *Proc. phys. Soc. Lond.*, 1944, **56**, 31.

is sometimes known as the 'skin effect' and is due to the current density at the surface of the wire being greater than at the centre, on account of the greater back e.m.f. induced at the centre by the field caused by the flow of current. This effect is proportional to $(\mu f)^{1/2}$ and thus, whilst it was usually associated with radio frequencies, if μ is high enough, it is significant at audio frequencies. Since permeability depends on the magnetic field in which a specimen is situated, the impedance of the wire will depend on the ambient magnetic field and on the field due to the current flow.

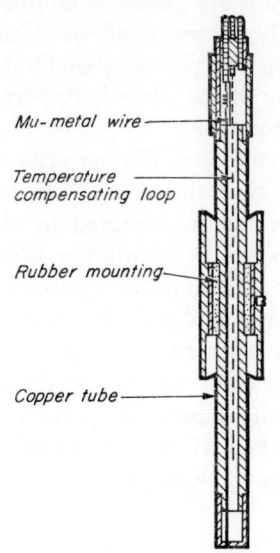

FIG. 6.25. The Mu-metal wire in Butterworth's magnetometer

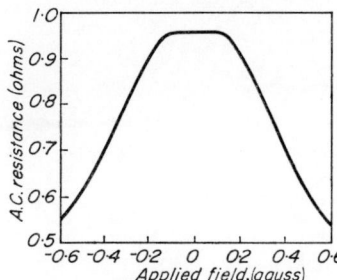

FIG. 6.26. Variation with ambient field of Mu-metal resistance

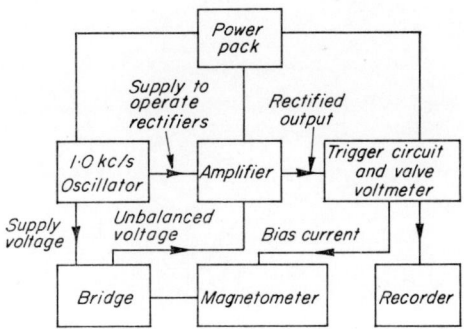

FIG. 6.27. Butterworth's magnetometer

The method was thought to be unreliable for a recording instrument owing to possible variations of frequency, current, temperature and tension on the wire, but effective compensating methods have been employed in Butterworth's instrument and changes of a magnetic field can be recorded to an accuracy of 1 gamma.

A length of Mu-metal wire is connected in a bridge circuit supplied with alternating current. Once the bridge is balanced, any subsequent output signal will be due to a change of resistance of the wire which may be produced by a change in temperature or field. The wire is mounted as shown in Fig. 6.25—one end being rigidly attached to the mounting tube and the other to a light tension spring. A length of copper wire housed adjacent to the Mu-metal wire is connected in the bridge circuit so that any temperature variations of the unit produce the same change of resistance in both. The unit is placed in an oil jacket and lagged.

The variation of resistance of a Mu-metal wire with ambient field is shown in Fig. 6.26, from which it will be seen that a biasing field of about 0·4 oersted is required. To produce a system free from supply and similar variations, the power is obtained from a stabilized source, the output from the bridge is amplified in a negative feedback amplifier and two units or sensitive elements are used, mounted parallel and arranged as a bridge circuit. The biasing fields are arranged to be equal and opposite. The heat treatment and straightening of the wires is important and is similar to that for the saturable inductor magnetometer.

Fig. 6.27 shows a schematic diagram of the magnetometer. The sensitive element of the magnetometer is usually placed E.–W. on a rigid mounting. The bridge is operated at low gain and the orientation of the element corrected to give zero reading on the output meter. The gain is increased and orientation adjusted as required until full sensitivity is reached. The magnetometer now records changes in the east–west component of the Earth's magnetic field and may therefore be used as a variation variometer (or declinometer). This instrument shares the advantages of the saturable inductor magnetometers referred to on p. 143 in that the detector or sensitive element may be placed at a considerable distance from the recording instrument, where magnetic conditions are undisturbed. If the run of cable is long, suitable precautions should be taken in matching the cable impedance to the magnetometer.

Chapter 7

Surveying Compasses

This chapter deals primarily with portable instruments for field use. Although many of the observatory instruments already described can well be adapted to this end, there are a number of simpler and more robust devices commonly used for surveying. These are principally compasses of various kinds, either as complete instruments or as part of a more elaborate instrument such as a theodolite. Their purpose is to establish values of magnetic variation, to align the telescope of a theodolite or to obtain magnetic bearings, as in the process of calibrating or adjusting aircraft compasses. The usual forms of compass are prismatic or tubular. Prismatic compasses are described in Chapter 5.

DATUM COMPASSES

A more recent development in land compasses for surveying, determining variation, and adjusting aircraft compasses by reciprocal bearings is the datum compass. This class of instrument uses optical means for reading the bearing scale and is fitted with telescopic sights. In spite of modern technical refinements, its standard of accuracy is little better than that of the Pattern 1, usually about 4 to 6 minutes of arc (95 per cent). Two examples are described here, the Admiralty Compass Observatory datum instrument and the Hilger & Watts datum compass.

A.C.O. datum compass

The A.C.O. datum compass (Plate 7.1) uses a buoyant magnet system similar to the system in the Schuster-Smith variometer described on page 133. The card, which is made of glass 0·020 in. thick, is 3⅝ in. in diameter and is graduated in degrees and fifths of degrees through 360°. It is mounted on a float, in the centre of the upper surface of which the pivot is situated. The float assembly is enclosed in an aluminium alloy bowl and is restrained from floating by a claw-like mechanism operated from outside through a sylphon tube or packless gland. The ring magnet, with a moment of 3000 c.g.s. units, is attached to the float. Surrounding the magnet is a thick copper ring, secured to the outer chamber, to provide additional damping. The system is immersed in silicone fluid.

The outer chamber or bowl stands on a tribrach with levelling screws, provided with spirit-levels. The top of the chamber is closed by a cast cover, in whose centre is housed the sapphire jewel in which the osmium-iridium pivot in the magnet system centres, and rotates, when the claw-like mechanism restraining the float is released. The jewel centre is the reference point for the machining of the cover and for the attachment of the sighting and reading devices, and is thus the one point to which all parts of the compass are related.

The card is read by a simple microscope, light being admitted to the bowl through a small window and reflected by a 45° prism through the glass card. A fixed-focus telescope, with its axis in the vertical plane through the jewel centre, provides a sighting device and can be elevated over an arc of about 20°. The float release mechanism is operated by a camera type of cable release.

The compass bowl rotates on a bearing at the centre of the tribrach and the instrument is used on a specially constructed tripod having considerable lateral adjustment at its head.

Hilger & Watts datum compass

The Hilger & Watts datum compass, illustrated in Plate 7.2, is an adaptation or modification of the firm's Microptic mining transit, with the vertical circle omitted. It is virtually a small theodolite, fitted with a magnetic compass specially designed for aligning the horizontal circle of the instrument in the meridian. The horizontal circle is mounted in a casting supported by a levelling tribrach. The circle may be rotated and clamped in any position required, but usually in that where 0°–180° is the magnetic meridian as defined by the compass. An upper casting, carrying the magnetic compass, telescopic sight and levels, rotates about the horizontal circle or can be clamped to it. Tangent screws enable fine adjustments to be made in the positions of both the upper casting and the horizontal circle. The latter, graduated in degrees and tenths of degrees, is viewed through a microscope eyepiece in the casting, in which there is also a small window for illumination.

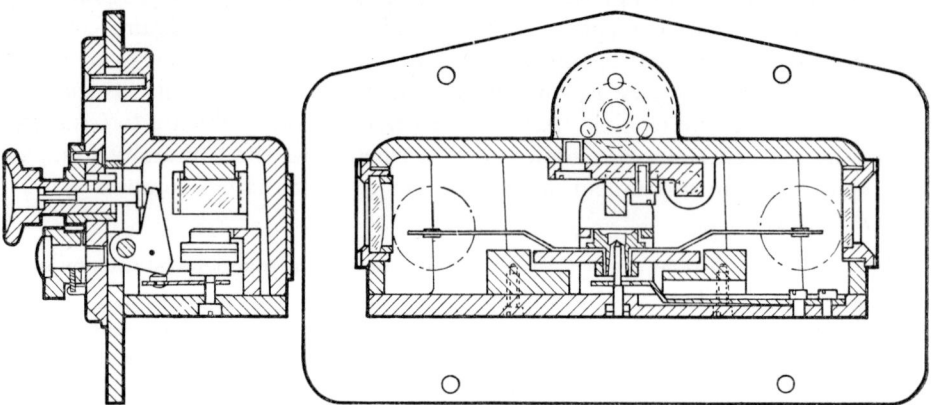

Fig. 7.1. The magnetic assembly of the Hilger & Watts datum compass

The magnetic compass, Fig. 7.1, has several novel features. It consists of a long, flat magnetic needle supported at its centre on a fine pivot and jewel. There are thick copper blocks adjacent to the ends of the needle to provide eddy-current damping and a lifting device allows the needle to be raised off the pivot when not in use. When required, the needle is placed on the pivot by a cable release. An interesting optical system enables both ends of the needle to be viewed simultaneously, ensuring that the instrument is correctly aligned in the magnetic meridian. The needle has upstanding filaments at either end, viewed through a convex lens. The near filament is so close to the lens that no visible image is formed, but a reflection of it is seen in a vertical mirror placed halfway along the needle. The image thus formed appears to be at half the length of the needle behind the mirror and therefore in the same plane as the further filament which is viewed directly through the lens. Correct alignment is indicated by coincidence of the images of the filaments. The compass is attached to the side of the telescope casting and can be adjusted to suit changes of dip angle.

In use, the instrument is first levelled, the lower or circle clamp tightened, and the upper or telescope clamping screw released. The telescope casting is then rotated until the scale reading seen through the microscope is 0°. The upper clamp is tightened and the lower clamp released. The instrument, including the circle, can then be rotated until the magnetic compass indicates that it is aligned in the magnetic meridian. The lower clamp is tightened and the upper clamp released. Any bearings now taken through the telescope will be read off the horizontal circle as correct magnetic bearings. Calibration is necessary to establish the index error of the magnetic compass and any misalignment of the telescope axis.

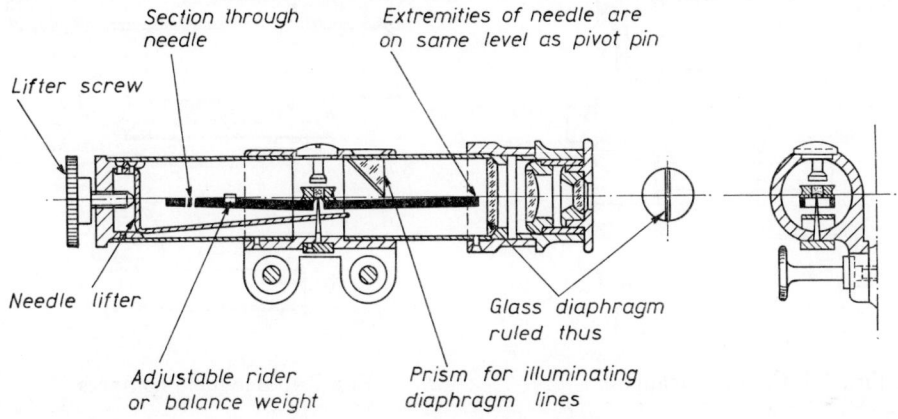

FIG. 7.2. Theodolite compass

The instrument is used on a tripod similar to that used with the A.C.O. datum compass.

A theodolite compass, though not, strictly speaking, a separate instrument, is illustrated in Fig. 7.2. It consists of a long, thin pivoted needle which may be raised from its pivot by the lifter when not in use. Coincidence of the south end of the needle with the lines of a graticule in the eyepiece indicates that the optical axis of the theodolite's telescope, to which the compass is attached, is aligned in the magnetic meridian.

MASTER COMPASSES

For the determination of the magnetic meridian in the field, there are a number of types of master or absolute compasses which, in theory, need no calibration and have no index error. The Kew magnetometer, although primarily a laboratory instrument, can also be used in the field, and attempts have been made to adapt its principle to portable instruments.

157

Connelly compass

The Connelly compass, shown in Fig. 7.3, uses a tubular magnet suspended in a pivoted stirrup, and provided with an eyepiece so that a line of sight may be taken through the magnet and the meridional direction observed. Reversing the magnet about its axis and obtaining two readings of the meridian eliminates collimation error. The small size of the optical system is a drawback of this instrument and it is possible to take only approximately level sights.

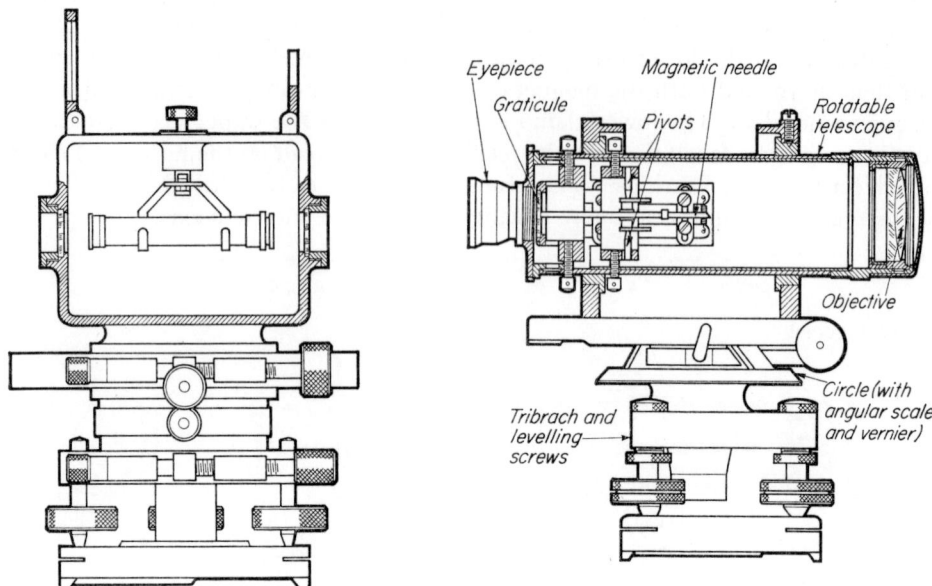

FIG. 7.3. Connelly compass   FIG. 7.4. Wingfield compass

Wingfield compass

The Wingfield compass, Fig. 7.4, has a magnet system pivoted within and along the optical axis of a telescope, the south end of the needle being read against a graticule. A double-sided jewel and two pivots are used, so that when the telescope is rotated about its axis through 180° the needle rests on the second pivot, thus being reversed and enabling asymmetry of needle and collimating errors to be corrected by taking the mean of two observations. It is essential, however, that the jewel be symmetrical and that the pivots be in the same vertical plane. Some impairment of the telescope optical system is inevitable owing to the presence of the magnet system.

A modification by Watts provided the compass with its own reading microscope and mounted it on the side of the telescope.

Hilger & Watts absolute compass

The Hilger & Watts absolute compass, illustrated in Fig. 7.5 and Plate 7.3, is an attempt to avoid the difficulties experienced in these earlier compasses. The compass consists of a needle with a double-sided jewel, as in the Wingfield compass, so that it may rest on either of the two pivots by inverting the compass housing. The needle carries a small mirror which reflects a collimated light beam, so that an image of a target graticule mounted in the compass body is observed on an engraved scale in the reading eyepiece. As the needle swings, the image moves across the eyepiece scale, which is graduated in 5-minute divisions. The needle is damped by eddy currents in a copper block set close to the magnet system.

Observations are taken with the compass in two positions and the mean of these enables the magnetic orientations of an azimuth circle to be determined. Hence the bearing of a given object from magnetic north may be found, or magnetic variation ascertained by taking celestial observations. A small theodolite provides a suitable azimuth device, and the compass is particularly designed for attachment to the Watts Microptic transit.

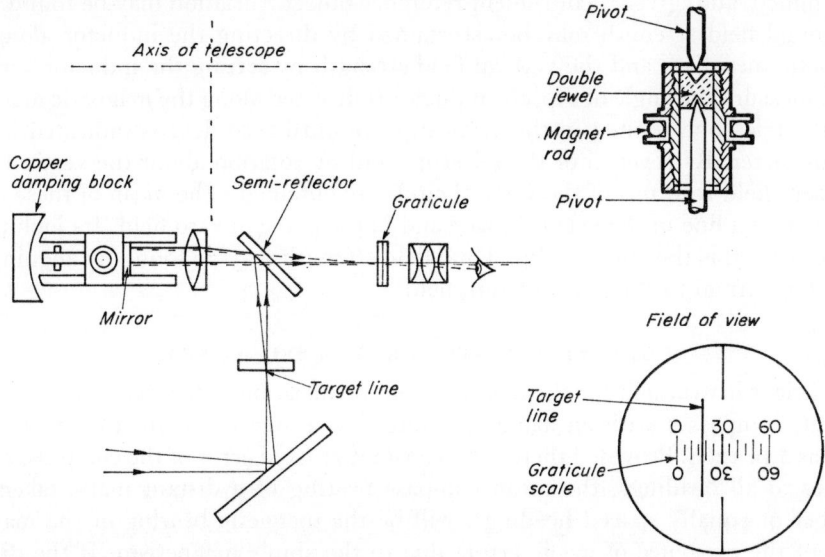

FIG. 7.5. Hilger & Watts absolute compass

The use of collimated light makes the alignment of the jewel and pivots unimportant. The principal limitation is the use of jewel and pivot bearings, which inevitably introduce a small amount of friction.

Inductor theodolite

The inductor theodolite, appearing in Plate 7.4, is an instrument in which a saturable inductor is mounted on the telescope of a theodolite with the magnetic axis of the inductor parallel to the optical axis of the telescope. Not only does this instrument provide an absolute or datum compass for determining the magnetic meridian, and hence variation, but it is also a versatile surveying instrument with which angle of dip and total, vertical and horizontal field strengths may be rapidly measured. It is provided with an electrical unit which includes power supply, oscillator amplifier and output meter. Field strength is measured by cancellation with a backing-off field, the value of which is controlled and indicated by a calibrated potentiometer. Owing to the use of transistors, the electrical unit is compact

159

and easily portable, the whole equipment being little more bulky or heavy than a conventional theodolite. Needless to say, the normal operation of the theodolite is unimpaired. Care needs to be taken to see that the proximity of the electrical unit has no significant magnetic effect on the inductor, and a suitable 'safe distance' must be allowed.

To determine the magnetic meridian, the theodolite is levelled and the vertical circle set to zero. The instrument is rotated in azimuth until the output meter also reads zero. The horizontal circle reading is noted. A second reading is taken with the theodolite rotated to a reciprocal zero position, which may not necessarily be exactly 180° from the first zero if there is any small misalignment between inductor and telescope. The telescope is then reversed and two more readings taken. The mean of the four is the magnetic east–west direction from which the meridian is determined, and, given a convenient reference object, variation may be found. The horizontal field strength may be ascertained by directing the inductor along the magnetic meridian, and the vertical field strength by setting the inductor vertical.

To measure the angle of dip, the inductor is directed along the magnetic meridian and the telescope rotated in the vertical plane until zero field is indicated by the output meter. By reversal of the telescope and by rotation about the vertical axis, four zero field readings of the vertical circle are obtained. The mean of these is the direction of a line in the vertical plane and in the plane of zero field. Its inclination to the vertical is the angle of dip. Total field strength may be found by aligning the inductor normal to the plane of zero field.

SURVEYING INSTRUMENTS USED AT SEA

An efficient instrument for the measurement of variation is the mariner's compass. In fact, compasses with an azimuth-taking device were used for this purpose as early as A.D. 1269. Provided there is no constant or fixed error of the compass, which applies to all headings, the mean compass bearing of a distant mark, taken at a number of equally spaced headings, will be the magnetic bearing of the mark in spite of the existence of cyclic errors due to the ship's magnetism. If the distant mark is the sun, preferably at low altitude, its true bearing is known, given the time of observation and the ship's position. Thus the ship is 'swung', i.e., placed on sixteen equally spaced headings, starting from north or any cardinal or inter-cardinal heading, and a compass bearing is taken of the sun. This is repeated with the ship swinging in the opposite sense of rotation. The differences between each compass bearing and the corresponding true bearing of the sun are averaged, and give the value of magnetic variation (see also Chapter 12).

Collimating compass

The collimating compass, illustrated in Fig. 7.6, enables more accurate determinations of variation to be made than are possible with an ordinary azimuth circle and mariner's compass, since movements, however small, of the compass card in the horizontal plane, and the azimuth circle in the vertical plane, detract from accurate observations.

In 1908, W. J. Peters designed the Carnegie Institution collimating compass. It consists of a normal mariner's liquid-filled compass with the card removed. Four 10° scales are placed at the north, south, east and west points of the instrument, and each is at the principal focus of a concave mirror. The four mirrors are mounted on the float. The bowl, which is gimballed in the usual way, has four windows through which the scales may be seen, openings being provided in the gimbal system so that these windows are visible (Fig. 7.6).

The angle between the central graduations of the scale and a suitable celestial body is measured with a sextant, enabling variation to be determined. The constants of the collimating compass (scale errors and the like) are determined by calibration. Given a calm sea and a body at not more than 10° altitude, an accuracy of 0°·1 is claimed in measuring variation.

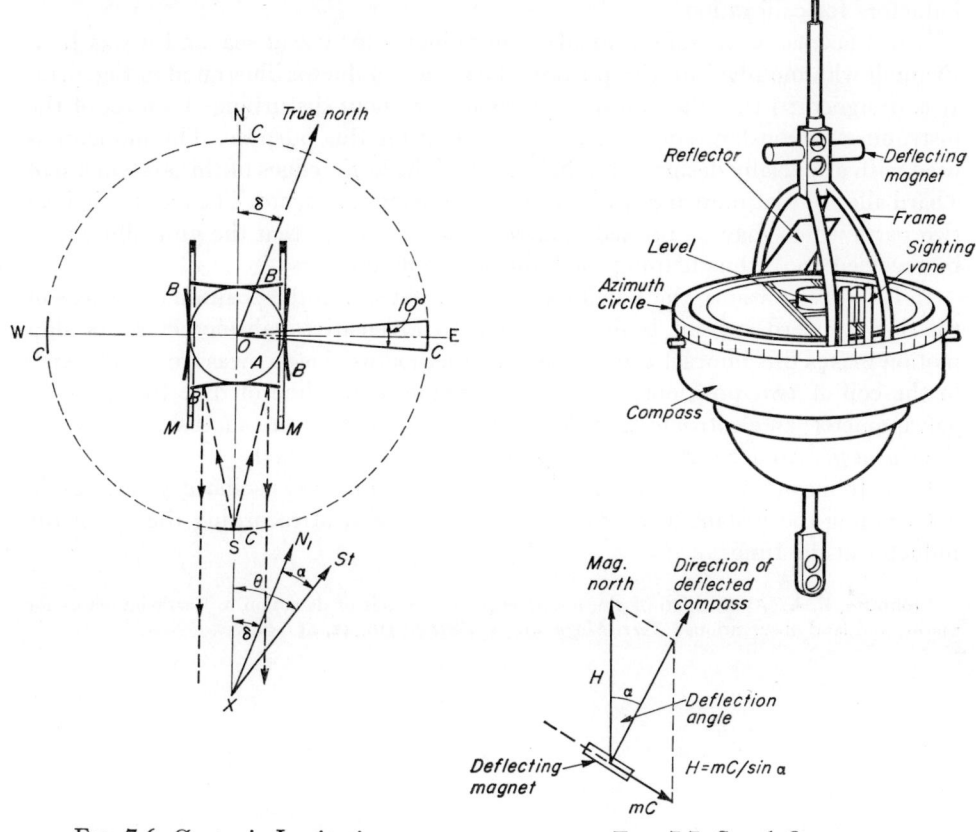

FIG. 7.6. Carnegie Institution collimating compass

FIG. 7.7. Sea deflector

Sea deflector

The sea deflector, illustrated in Fig. 7.7, is used for measuring horizontal intensity at sea. The method was developed by Bauer and Fleming;[*] a deflecting magnet of known moment is mounted vertically above a mariner's compass, the angle of deflection being determined by comparison with an undeflected compass. If α is

---

[*] Bauer, L. A., and Fleming, J. A., 'The C.I.W. deflector in use on the *Carnegie* for determining the magnetic horizontal intensity and magnetic declination at sea', *Terr. Magn. atmos. Elect.*, 1913, **18**, 57–62.

the deflection angle, M the moment of the deflecting magnet, C a constant depending on the distance of the magnet, its temperature coefficient and the induction and distribution coefficients, and H the horizontal intensity of the Earth's field, then $H = MC/\sin \alpha$. C is determined from observations ashore. An accuracy of one part in a thousand is claimed for this method.

Dip circle

The dip circle may be adapted for use at sea and mounted on a gimbal-stand, but owing to irregularities in the needle-axles, corrections that have to be applied vary with so-called 'magnetic latitude'. This disadvantage called for the use of Earth inductors for calibration.

It was necessary, therefore, to adapt an inductor for use at sea, and it was J. A. Fleming who modified for this purpose the form of inductor illustrated in Fig. 6.14. It is designed so that the coil may be rotated without disturbing the level of the instrument, a constant-speed motor being used for this purpose. The inductor is used with a specially designed gimbal system. The knife edges of this system are of a hard alloy of platinum and iridium and the bearings are agate. The outer ring is in two parts which may be rotated relative to each other so that the gimballing may be reversed 180°, thus helping to eliminate levelling errors.

In land observations, the Earth inductor is rotated and the position of the axis of rotation for zero current is determined with a sensitive galvanometer. As ship motion makes this impracticable at sea, the method used is to measure the currents in the coil at two positions, one on either side of the line of dip. Instead of a galvanometer, an electronic amplifier and microammeter were successfully used in the *Carnegie*. An accuracy of 0°·1 is claimed with this method.

E. A. Johnson* describes a method of using a cathode-ray oscilloscope to assist in determining the instant when the output is zero, and of recording the tilt of the inductor at the time.

* Johnson, E. A. 'Application of alternating current methods of detection to Earth inductors for marine and land observations', *Terr. Magn. atmos. Elect.*, 1936, **41**, 251–60.

Chapter 8

Transmitting Compasses

The usefulness of a magnetic compass is much enhanced if it can be made remote-indicating or transmitting and therefore capable of operating repeaters or slave compasses in different parts of a ship or aircraft. Not only are simultaneous and identical compass headings available for different members of a crew or for feeding into automatic navigational instruments, but the magnetic compass proper can be placed in the best available position having regard to the proximity of magnetic material and electrical equipment and not necessarily to the convenience and accessibility of the instrument. Attempts to develop a compass of this kind have been made for about a century. Some of the earliest of these, though somewhat crude, were very interesting; other more recent ones only achieved the distinction of being course indicating, rather than fully transmitting.

A transmitting magnetic compass can never be more accurate than the master magnetic compass or detector. Its performance can sometimes be improved in certain directions by careful design of the follow-up system, but the precision with which the meridian is defined is determined by the master compass. It should therefore be the aim of the compass-designer to reproduce as closely as possible at the repeater instruments the behaviour of the master compass. The information on the ship's heading must be derived from the magnet system without imposing any restraint on its freedom of movement, the system must respond instantaneously to the movement of the ship (there must be the minimum of 'lag' or 'dead sector'), no error must be introduced by the means of obtaining information from the master compass, the transmission should be smooth and free from 'hunt' and should respond fast enough to record the fastest rate of turn without lagging or falling out of step.

Many proposed systems only partly achieve these desiderata, yet they are interesting in that they demonstrate the possibility of various methods of deriving transmission from a magnetically sensitive device.

Both pivoted needles and inductor systems provide satisfactory detector elements, but it must be remembered that the torque available from a pivoted needle and the electrical power available from an inductor are both so small that any form of mechanical or electrical loading is inadmissible.

A transmitting compass consists of a master magnetic element with a means of sensing compass heading and communicating the information to repeater instruments, either directly or with the aid of a follow-up system or torque amplifier. The sensing means or pick-up device is usually electrical in nature though there have been one or two isolated examples of pneumatic devices, notably the Askania compass in the VI flying bomb. Two forms of pick-up may be used: one that converts the angular position of the ship with respect to the magnetic compass into an electrical signal proportional to angular displacement, and one that detects the change of position from a null-balance, so driving a follow-up system from which angular information backed up by a power drive can be obtained.

The functions of the detector unit of a transmitting compass can conveniently be carried out by the saturable inductor. It can be used with equal facility both as a proportional means of conveying angular information and as a null system; but it has disadvantages when compared with a pivoted-needle system in that (a) electrical failure renders the system useless, whereas a needle can be read directly, (b) electrical signals have to be combined and converted into a directional indication, thus introducing an extra step in the process, and (c) its dynamic characteristics render it liable to certain instabilities.

It is for this last reason that inductor systems have not been used in simple transmitting compasses. It is normal practice, however, to incorporate them in a more advanced form of transmitting compass, namely the gyro-magnetic compass to be described in Chapter 9.

Here a common method is to use two inductors, perpendicular to each other (fore-and-aft and athwartships), maintained by a gimbal system or a gyroscope in the horizontal plane. Fig. 8.1(a) illustrates this arrangement in which it is apparent that one inductor measures $H \cos \zeta$ and the other measures $H \sin \zeta$, where H is the Earth's magnetic field and ζ the magnetic heading of the ship. Alternatively, four inductors as shown in Fig. 8.1(b) may be used. Other arrangements using three inductors are illustrated in Figs. 8.1(c) and (d).

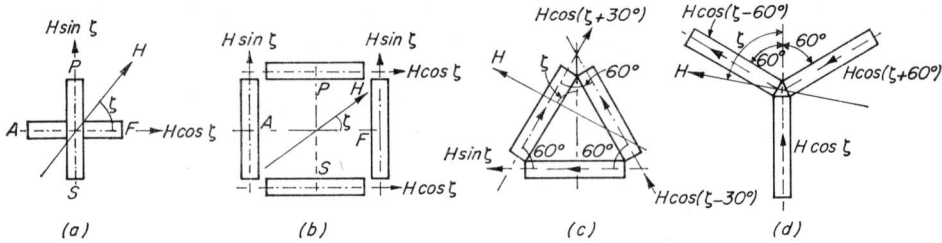

FIG. 8.1. Arrangements of inductors

It has been shown that deviations due to tilt can arise in a double-pivoted compass system (p. 26) and with an inductor (p. 72). Owing to accelerations, deviations can also occur with a single-pivoted compass (p. 41 and Figs. 3.16–3.18). Similarly an inductor element, though mounted in gimbals or suspended by a universal joint, exhibits the same deviations as a needle compass when under the influence of accelerations that tend to tilt it out of the horizontal.

The dynamic characteristics of a pivoted-needle compass with liquid damping are more readily adjusted to counter the acceleration errors discussed in Chapter 3 than are those of a simple inductor system, since alterations to the periods of the needle system (in tilt and in azimuth) can be made and the damping modified. When a pivoted-needle compass is tilted, rotation of the magnetic system must first be set up before deviation occurs, whereas with an inductor system, the deviation is instantly apparent in the electrical output.

Damping could, of course, be applied to an inductor arrangement, but experience and practice have shown that, as far as unstabilized transmitting magnetic compasses are concerned, the pivoted-needle system is generally to be preferred.

METHODS OF OBTAINING A SIGNAL
FROM THE PIVOTED-NEEDLE COMPASS:
'PICK-UP' DEVICES

Given a well-designed magnetic compass, there now remains the problem of obtaining a signal from it. There are many methods and elaborations, which fall into the following main groups:

(1) 'Slave' system, in which a signal generator is driven through magnetic coupling between the compass magnet and a slave magnet on the generator.
(2) Capacity system in which the compass either is part of a capacitative bridge or drives a variable condenser with a given azimuth–capacity law.
(3) Resistance system in which the compass is part of a liquid resistance bridge or drives a suitable potentiometer.
(4) Photoelectric system in which the compass determines the balance of a light-sensitive bridge arrangement.
(5) Inductive system, usually employing some form of saturable inductor operated by the field from the compass magnet.

System (1) includes varieties of systems (2) and (3) where the slave magnet operates signal generators of the capacitative or resistive type.

Some of these systems will give a distant reproduction of the master compass heading without the use of any intermediate follow-up system or amplification, but, in general, the torque produced at the repeater is very small and not conducive to high accuracy. These may be described as remote-indicating magnetic compasses. Where torque and accuracy are required, a follow-up system and a power-driven data transmission system are essential.

REMOTE-INDICATING MAGNETIC COMPASSES

Distant-reading compasses that do not use a follow-up system have the great advantage of simplicity and lightness though their accuracy may suffer. They are used mainly as aircraft instruments.

Kollsmann Telegon compass

The Kollsman Telegon compass (Fig. 8.2 and Plate 8.1) is possibly the simplest form of distant-reading compass. It consists of three units—the master compass, the repeater, and the a.c. motor generator, driven from a 28-volt d.c. supply. The generator supplies 26 volts at 400 c/s to the master compass and repeater.

The master unit is a liquid-filled compass of high moment (2000 c.g.s units). This drives, through magnetic coupling, a Telegon transmitter mounted immediately above the compass. The spindle of the transmitter carries a small slave magnet; thus the rotor of the transmitter is aligned with the axis of the compass needle while the stator which is attached to the ship turns around it.

The master unit is totally enclosed in a plastic bowl with an adjustable magnetic corrector fitted to the underside. The Telegon transmitter is a very light 2-phase synchro type of unit (Fig. 8.3). The stator has a primary winding and a 2-phase arrangement of secondary windings, the primary being coupled magnetically to the secondary windings by a Z-shaped soft-iron rotor mounted on the spindle which carries the slave magnet. This spindle runs in jewelled bearings.

The repeater is similar to the transmitter but, instead of the slave magnet, a light pointer is fitted to the spindle, moving over a dial marked with compass directions.

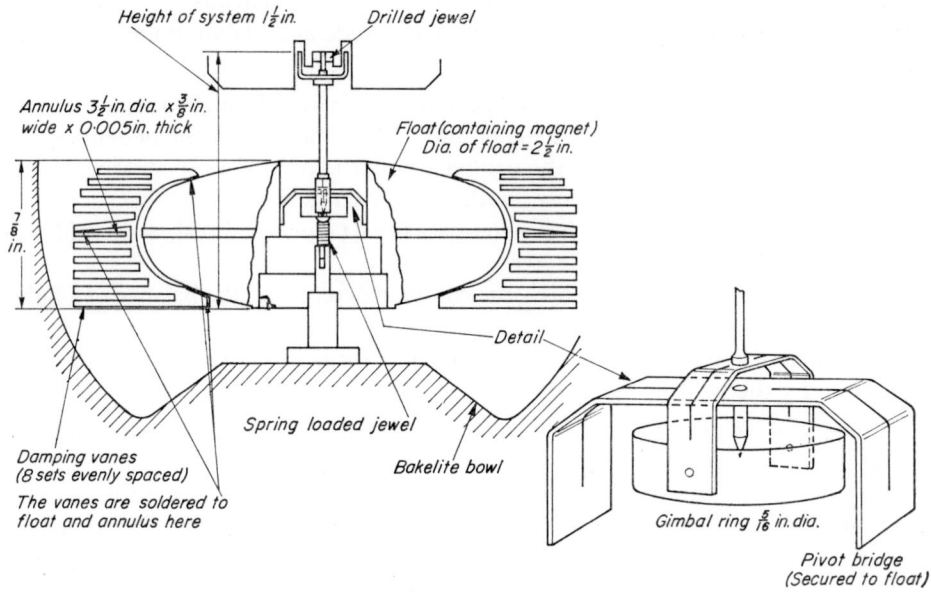

FIG. 8.2. Kollsmann Telegon compass
The magnet system

The primary windings of each instrument are energized at 400 c/s and are connected in parallel, as also are the secondary windings. The position of the soft-iron rotor of the transmitter determines the relative coupling between the primary and the two secondary coils. Corresponding fields are set up in the repeater windings, whereupon the repeater rotor sets itself in a given angular relation to the transmitter rotor and reproduces any change of its angular position.

Thus in this compass any change of ship's head is transmitted through the magnetic coupling to the Telegon transmitter and hence to the repeater. The torque, however, is small and the accuracy is of the order of $\pm 4°$ (95 per cent confidence level).

The repeater is a standard type of aircraft instrument with graduations every $2°$ and figures every $30°$, the final 'o' being suppressed. Rotating grid lines enable a predetermined course to be set and flown by keeping the pointer within the lines. The system is self-aligning.

Patin compass

The Patin compass, which was used extensively by the German Air Force, has a liquid-filled compass with a high-moment magnet directly coupled to a very finely wound toroidal potentiometer. The system works on a direct-current supply and is illustrated in Fig. 8.4.

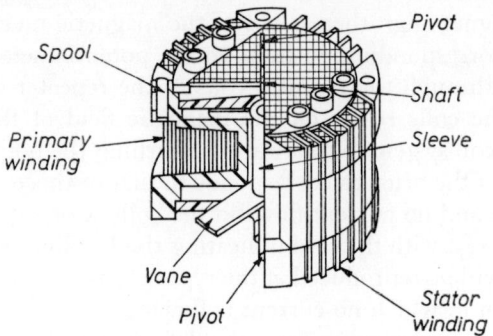

FIG. 8.3. Kollsmann Telegon unit
A sectional view

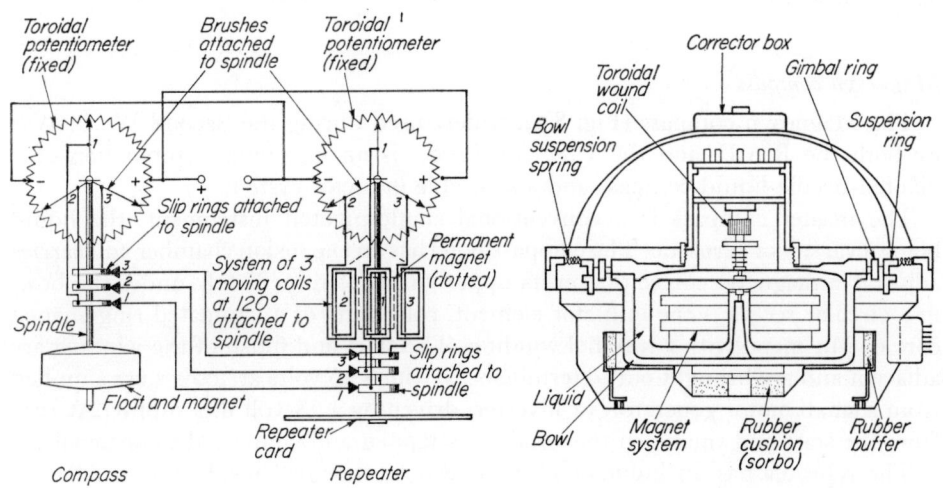

FIG. 8.4. Patin compass

FIG. 8.5. Patin compass
Section through the master unit

The magnet has a moment of 12 000 c.g.s. units and is enclosed in the float (Fig. 8.5). The system is double-pivoted and thus the compass bowl has to be gimballed to avoid deviations owing to tilt. The gimbal system is mounted in a sprung suspension consisting of a number of radial springs connecting concentric rings of the gimbal and mounting (Fig. 8.5 and Plates 8.2 and 8.3).

The spindle of the magnet system carries three slip-rings, a compass card marked in degrees and three very light gold-plated brushes which wipe the periphery of the stationary potentiometer. This potentiometer is a remarkable example of toroidal winding, having a diameter of only about $\frac{5}{8}$ in. An e.m.f. of 24 volt d.c. is applied to its diametrically opposite points. The repeater (Plate 8.4) is similar to the compass in having a toroidal potentiometer, but in series with the moving brushes and the slip-rings are three coils of about 450 ohms each, suspended, as in a moving coil galvanometer, about a strong permanent magnet. These coils, with the brushes and slip-rings, are mounted on a spindle which carries the compass repeater card.

The float and magnet align themselves to the magnetic meridian and the three brushes take up a corresponding position on the potentiometer. Current flows in the three lines and through the repeater coils to the repeater potentiometer. The magnetic field of the coils reacts on the magnetic field of the fixed permanent magnet causing the coil system to rotate and so turning the potentiometer brushes. When the position of the brushes corresponds to that of those on the compass, the system is in balance and no current flows in the coils. Consequently, the repeater movement comes to rest with the card indicating the heading of the compass. Any turn of the aircraft will be reproduced at the repeater, as the coils always endeavour to attain the position in which no current is flowing.

Not only does this system suffer from the defects usually associated with high-moment magnet systems, but it is inclined to be frictional owing to the pressure of the brushes on the potentiometer and slip-rings. It possesses the advantage of being self-aligning.

Magnesyn compass

The Magnesyn compass (Fig. 8.6), widely used during the Second World War by both the R.A.F. and the U.S. Air Force, is an ingenious combination of a pivoted-needle liquid compass and a saturable inductor system.

The master compass is a conventional single-pivoted instrument, the liquid being a variety of kerosene. The compass bowl has an expansion chamber and carries adjustable magnetic correctors on its upper surface. Immediately under the bowl in a circular recess is the inductor element, consisting of a laminated ring-shaped core of Mu-metal with a toroidal winding. The start and finish of the winding are adjacent and are brought out to terminals, to which 26 volts at 400 c/s are supplied from a small motor generator, or inverter, driven by a 28-volt d.c. supply. At 120° from the start and the finish the winding is tapped and connected to terminals.

The repeater has an inductor element which, though considerably smaller, is made in the same way as that on the compass. The repeater element is mounted in a Mu-metal screening can to avoid interference from external fields. Concentric with this inductor is a small permanent magnet mounted on a spindle with jewelled bearings. The spindle carries a light pointer moving over a dial marked in degrees. In external appearance, the Magnesyn repeater is the same as the Kollsman repeater, illustrated in Plate 8.1.

The exact theory of the Magnesyn is somewhat obscure, the normal analysis used for a conventional saturable inductor being difficult to apply. The simple explanation is that the compass magnet sets up a diametral field across the inductor ring and produces a flux in the ring, as shown in Fig. 8.7. The a.c. field of the toroid drives the core material into saturation during each half-cycle, producing a second-harmonic component of the exciting voltage. The distribution of this component depends on the angular position of the diametral field, i.e. upon the compass heading. Since the windings of the compass and repeater are connected similarly, the combination of second-harmonic and fundamental frequencies and their phase distribution are reproduced in the repeater, resulting in the generation of a diametral d.c. field along which the repeater magnet sets. Thus the angular position of the repeater pointer corresponds to a given position of the compass magnet with respect to the compass inductor, and compass heading is reproduced by the repeater.

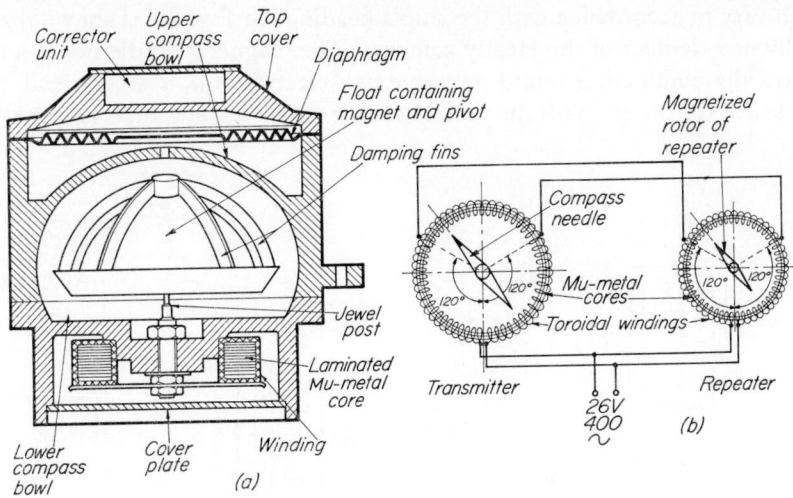

FIG. 8.6. Magnesyn compass

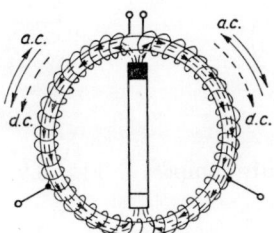

FIG. 8.7. Theory of the Magnesyn repeater

The system is robust, convenient, light in weight and self-aligning. It has an accuracy of about ± 3° (95 per cent zone) but very little torque is developed at the repeater and so frictional troubles are liable to occur.

Heatly compass (sine-cosine system)

This instrument demonstrates an interesting method of obtaining a compass direction by using two pivoted needles. It is not, strictly speaking, in the same category as the examples first quoted, yet it is not exactly a follow-up system in the accepted sense. It is a null-system employing magnetic feedback.

Consider a magnetic needle freely pivoted as a compass and mounted in a ship. When the ship is heading north, the compass needle will lie along its fore-and-aft line. If the ship be turned, for example, 45° to starboard, i.e. towards the east, the apparent or relative movement of the compass needle will be 45° in an anticlockwise direction. Now if an additional magnetic field $H \sin 45°$ be applied to the needle at right angles to the fore-and-aft line, the needle can be made to point along the fore-and-aft line once again. To go a stage further, if the magnetic field is produced by a coil, the current flowing in the coil is a measure of the sine of the angle through which the compass needle has been moved, or through which the ship has turned.

To obtain a continuous indication of heading, the current in such a coil must be made to vary in accordance with the ship's heading. In Fig. 8.8 is shown diagrammatically one element of the Heatly compass. The magnetic needle lies in a trough of electrically conducting liquid between two electrodes a, a at one end and an electrode b at the other. With the vessel heading north, the system is as shown, with

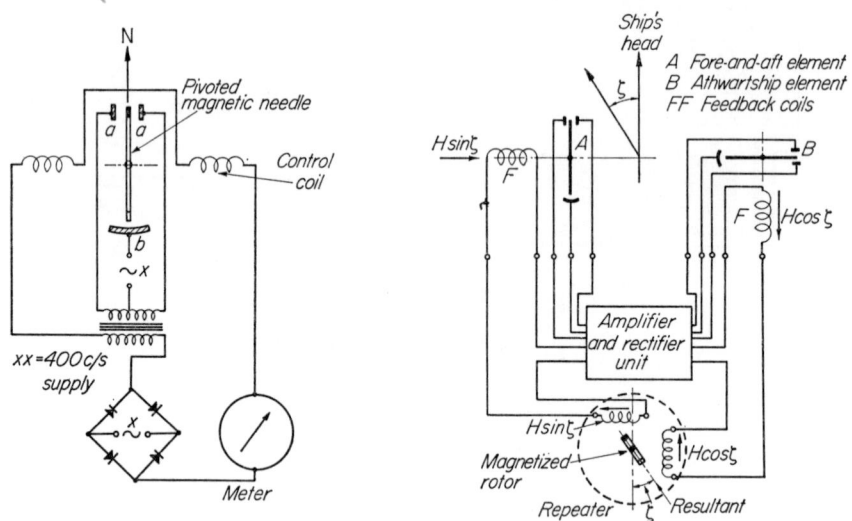

FIG. 8.8. An element of the Heatly compass   FIG. 8.9. Simplified diagram of the Heatly compass

the needle resting midway between electrodes a, a. The resistance paths between the needle and either of these are equal, and the a.c. bridge system, consisting of the needle and electrodes and the centre-tapped transformer, is in balance. No voltage is applied to the rectifier system and no current flows in the control coil.

On turning the system away from the magnetic meridian, the needle approaches one or other of the electrodes, thus unbalancing the bridge system and causing an alternating voltage to be applied to the rectifiers. A direct current now flows in the control coil, which produces a magnetic field at right angles to the centre line of the electrode system, that is to say, in opposition to the athwartship component of the Earth's field. This tends to pull the needle into the fore-and-aft line where balance is restored. The current in the control coil will thus be a measure of the sine of the angle of heading from magnetic north. In fact, this ideal condition is not completely realized, as will be explained later (p. 171).

To avoid ambiguity, two magnetic systems are employed, disposed fore-and-aft and athwartships. The currents required to keep the two needles respectively fore-and-aft and athwartships are combined and used to operate a repeater instrument instead of merely giving indications on meters. The repeater instrument consists of two coils at right angles through which the control currents flow, the resultant magnetic field orientating the rotor of the instrument, which may in its simplest form be a permanent magnet, in accordance with the direction of the fore-and-aft line of the ship. A pointer and dial complete the instrument, a simple version of which is shown in Fig. 8.9.

The two currents are proportional to $H \sin \zeta$ and $H \cos \zeta$ respectively; when these are correctly combined in a repeater, ζ is determined, since the tangent of the resultant angle of orientation of the repeater magnet is $(H \sin \zeta)/(H \cos \zeta)$, namely $\tan \zeta$.

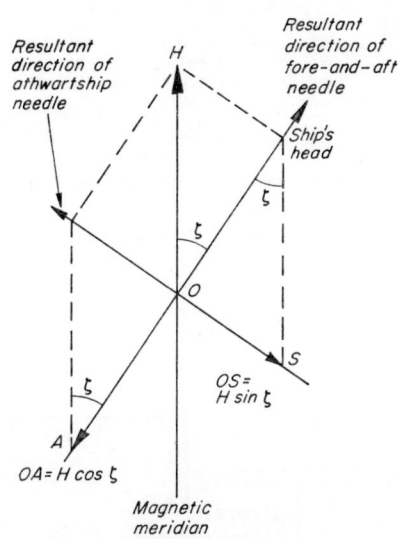

FIG. 8.10. Vector diagram for the Heatly compass

The vector diagram showing the relationship of the control fields and ζ is given in Fig. 8.10. In practice, the relationship is not as simple as it appears, since in order to establish the flow of the control current, the needles must of necessity be displaced from the fore-and-aft and athwartship direction by a small angle. As a result, each needle is acted on by a component of the field which would normally be parallel to the mid-position of the needle. To obtain equilibrium, the control fields must now balance components of both fore-and-aft and athwartship fields, with the result that the control currents will no longer be accurate measures of $H \sin \zeta$ and $H \cos \zeta$ (see Fig. 8.11).

This error has been avoided by using 'reflex' coils in series with the control coils, the sin ζ reflex coil being placed at right angles to the cos ζ control coil and the cos ζ reflex coil at right angles to the sin ζ control coil, so that the unwanted component is eliminated by cancellation of the field parallel to the mid-position of the needle. The arrangement is shown in Fig. 8.12.

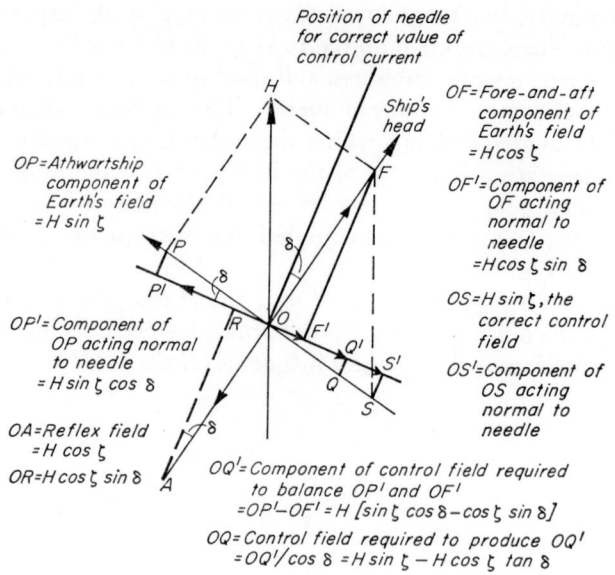

FIG. 8.11. Errors in control currents of the Heatly compass

Error $= OS - OQ' = H \cos \zeta \tan \delta$.
This is corrected by a reflex field $= H \cos \zeta$ which provides component OA equal and opposite to OF.

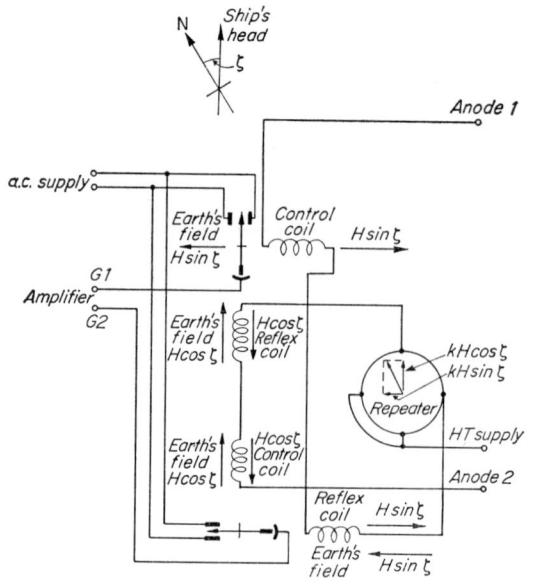

FIG. 8.12. The use of reflex coils with the Heatly compass

The amplifier circuit, shown in Fig. 8.13, is interesting. The diagram shows half of a double triode connected to one needle system and its associated coils and circuit; the other half is similar, its anodes being excited by an a.c. supply derived from a vibrator and transformer. When the positive half-cycle is being applied to the anode, and the potentiometer P formed by the needle and electrolyte system is in a position where no voltage difference exists between P and Q, there is enough negative grid bias to stop the flow of anode current. The battery therefore causes current to flow in the external circuit in the direction of the full arrow. As the potentiometer P moves towards the positive end, the grid bias is reduced and anode current begins to flow. The circuit constants are such that with half the signal

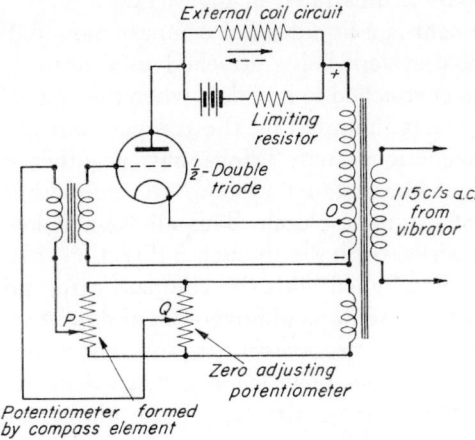

FIG. 8.13. Amplifier for the Heatly compass

value, the anode current balances the battery current. As the potentiometer P moves to a still more positive position, the anode current increases to such an extent that the current in the external circuit is reversed and flows in the direction of the dotted arrow. Eventually, for a full movement of P, the anode current in the external circuit is equal and opposite to the original battery current.

TRANSMITTING COMPASSES THAT USE A FOLLOW-UP SYSTEM

Transmitting compasses using a follow-up system are by far the most refined class of transmitting instrument giving high precision and adequate power for driving numbers of distant repeaters. The operation of the system resembles that of a simple closed-loop servo.

A pick-up system detects any movement about a magnetically sensitive element in terms of an error signal. This is passed to an amplifier, which usually drives a follow-up motor. The motor is coupled to the pick-up system and re-sets it in the zero or null position. Thus, as the ship or aircraft turns, the pick-up system is continually kept in alignment with the magnetically sensitive element by being rotated in a contrary sense to the movement of the vessel; in fact, the following element becomes north-seeking. Since this rotation is effected by the follow-up motor, which draws its power from an external source, it can impart the same rotation to a data transmission system. Hence, within limits, any number of repeaters may be driven without imposing a reactive load on the magnetically sensitive element.

Chevalier von Peichl's compass

A primitive arrangement, and one not suitable for modern requirements, is the simple 'on-off' or 'bang-bang' control, in which the compass magnet operates between two electrical contacts and controls a follow-up system. A historically interesting application of this is found in Chevalier von Peichl's electric patent compass, described in 1892.* This is one of the earliest complete designs in which an attempt was made to meet all the essential requirements for a transmitting magnetic compass. The compass is illustrated in Fig. 8.14.

The magnetic element is a liquid-filled compass mounted in gimbals, the outer member of the gimbal system being attached to a vertical spindle. The magnet has an electrical contact attached to it which, when the system is at rest and aligned, lies between two contacts mounted on the compass bowl. Any movement of the ship around the magnetic system brings one or other of the bowl contacts against the contact of the magnet system, so completing an electrical circuit controlling a form of magnetic clutch. This allows a follow-up device to engage with and rotate the vertical spindle in such a direction that the bowl and gimbal system are once more realigned with the compass card and the electrical circuit is broken. Doubtless the system would over-run and hunt violently, but it demonstrates the basic principle of a successful transmitting system. (This principle of a follow-up system and rotating bowl is, in fact, used in present-day compasses.)

A compass card mounted on the top of the gimbal system indicates ship's head. The follow-up 'motor' is ingenious, being simply a falling weight, wound up before use as in a grandfather clock. A simple form of electrical step or impulse transmission is used to control distant repeaters, and even a course recorder is included in the binnacle.

Peichl's description claims for his compass advantages that are found in the transmitting compasses in use today.

Capacitative system

Proposals for capacitative systems have been made by the Sperry Gyroscope Co. Ltd. One is virtually a slave system and is illustrated in principle in Fig. 8.15. A high-moment magnetic compass (liquid-filled and pivoted in the usual way) drives a slave magnet attached to the spindle of a light differential condenser. The single 'moving' vane is on the spindle and remains effectively stationary in space as the ship turns about it owing to the coupling through the slave magnet with the compass needle. The two 'fixed' vanes turn with the ship, but are geared to a follow-up motor. The condenser is supplied with high-frequency alternating current (a high frequency is necessary, of the order of 1 Mc/s, to permit sufficient transfer of energy across the small capacity of the condenser). When the moving vane is symmetrically placed with respect to the two fixed vanes, no voltage is applied to the amplifier or to the motor. A small change of direction brings one of the fixed vanes nearer to the moving vane than the other. A signal is therefore passed to the amplifier and the follow-up motor is energized so that the fixed vanes (or stator) are rotated in such a direction that electrical balance is once more restored, by bringing the moving vane into the symmetrical position with respect to the fixed

* J. von Peichl, Brit. Pat. 1734, 28th Jan. 1892: Electric compasses and course recorders.

vanes. Continuous rotation of the ship causes a continuous counter-rotation of the condenser; thus the ship's heading can be transmitted to repeaters by a suitable data transmitter.

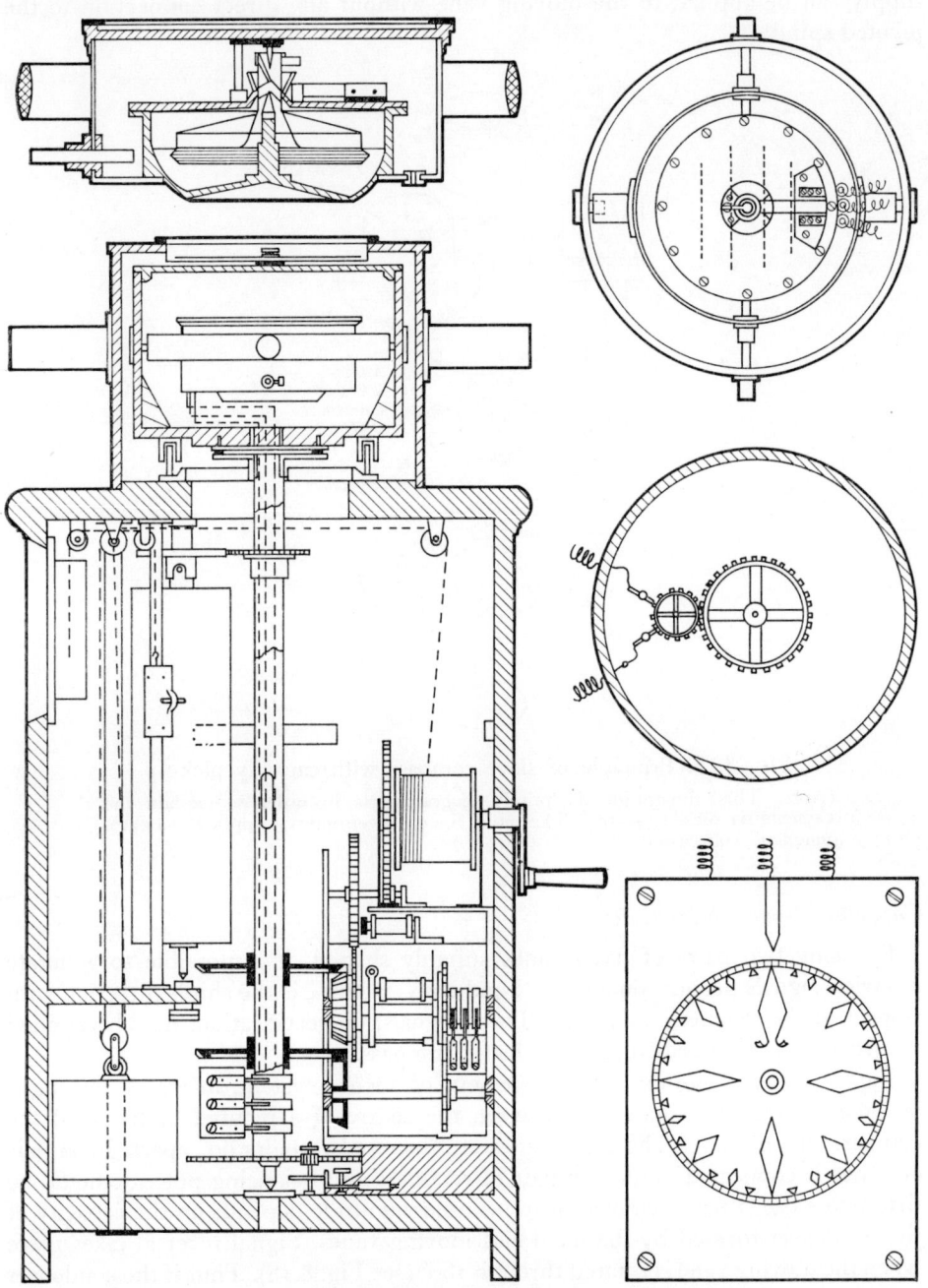

FIG. 8.14. Chevalier von Peichl's compass

Fig. 8.16 suggests a method whereby the moving vane is directly attached to the compass needle and the other vanes are in the form of copper plates which also provide eddy-current damping. The compass needle is double-pivoted; hence the whole assembly needs to be gimballed. By using an auxiliary plate, the electrical supply can be applied to the moving vane without any direct connection to the pivoted spindle.

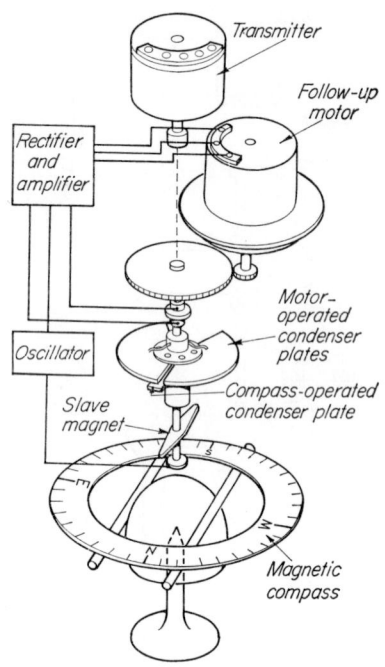

FIG. 8.15. Principle of slave compass with capacity pick-up

(*Note.* This illustration is purely diagrammatic in order to explain the capacitative pick-up system. The actual Sperry invention was applied to a gyro-magnetic compass.)

Capacitative sine–cosine system

By using two pairs of fixed vanes suitably shaped, it is possible to generate electrical signals proportional to sin ζ and cos ζ, where ζ is the ship's direction with respect to the magnetic meridian. These signals, on rectification, may be applied to a repeater, as described for the Heatly compass.

Fig. 8.17 shows the circuit arrangement of such a system, in which the high-frequency supply is connected between the centre of a rectifier system and the compass-driven vane. The fixed vanes are disposed in pairs 90° apart. Each pair of vanes is connected to its own pair of rectifiers and balancing potentiometer, so that across the latter a signal appears whose magnitude depends on the capacity of the condenser formed by the fixed and moving vanes. Signal reversal takes place when the moving vane is rotated through 180° (see Fig. 8.18). Thus if the condenser vanes are shaped so that the signals across the potentiometers are proportional to sin ζ and cos ζ respectively, the direct currents flowing in the repeater coils will produce corresponding magnetic fields, along the resultant of which the magnetized rotor will set itself, reproducing the direction of the ship's head. This arrangement suffers from the lack of a follow-up system and its attendant advantages. By modulating the high-frequency supply, however, a.c. instead of d.c. signals at the

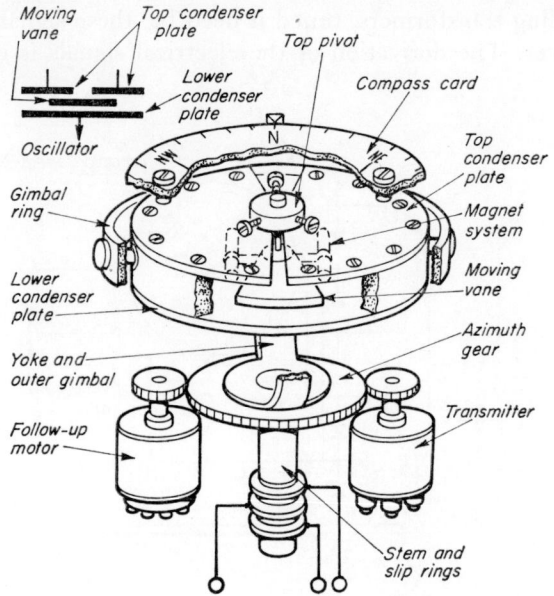

FIG. 8.16. A capacitative system

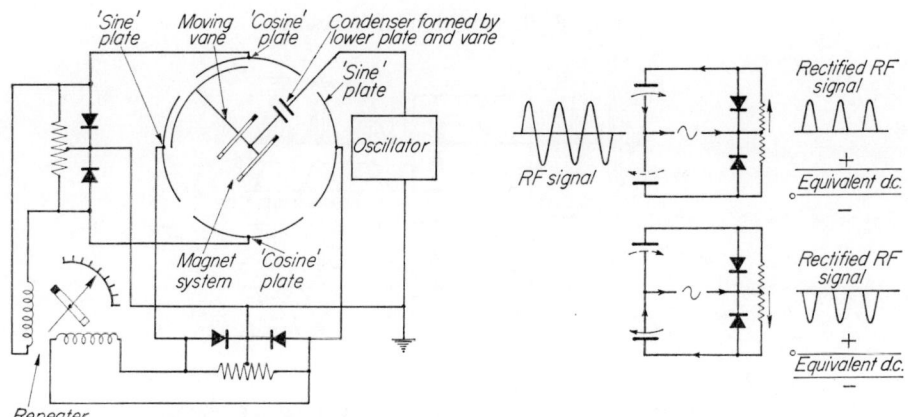

FIG. 8.17. Circuit for a capacitative sine–cosine system

FIG. 8.18. Signal reversal in the system of Fig. 8.17

modulation frequency may be obtained across the potentiometers. These may be combined in a resolver (p. 75 *et seq.*) and a follow-up system with data transmission added.

Fig. 8.19 shows a suggested circuit arrangement. The rectified signals, instead of being constant, would vary in amplitude in accordance with the modulation.

177

When smoothed, only the envelope at the modulation frequency would remain, changing phase by 180° as the moving vane turned in azimuth through 180°. By the proper choice of the circuit constants and the correct form of the fixed and moving vanes, separate electrical signals of the form $k \sin \zeta \sin pt$ and $k \cos \zeta \sin pt$ would be available ($p = 2\pi f$, where f is the modulation frequency). Through the medium of coupling transformers, tuned if need be, these signals would be combined in a resolver. The derivation of the electrical signals is explained in Fig. 8.20.

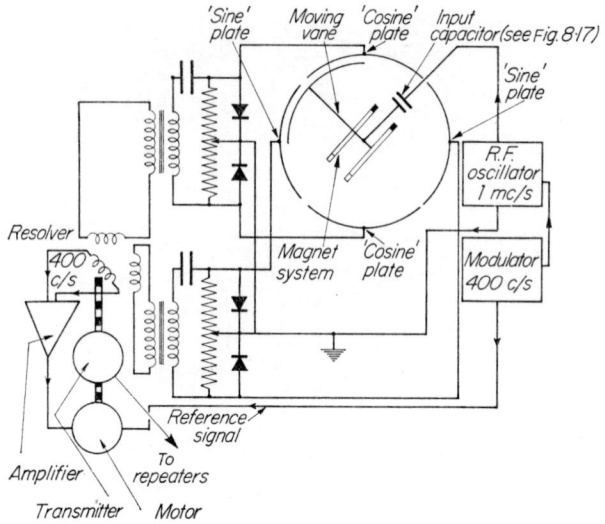

FIG. 8.19. Addition of a resolver

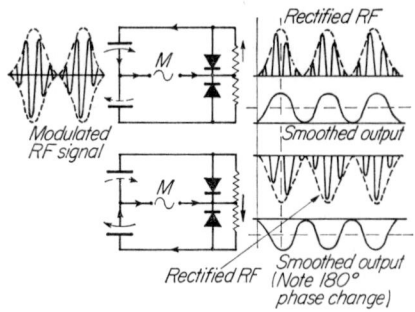

FIG. 8.20. Derivation of phase-changing a.c. signal

The output from the resolver would be a phase-changing signal as shown in the vector diagram, Fig. 8.21. This, applied to an amplifier, would provide power to drive a follow-up motor geared to the rotor of the resolver, which it would set in a direction corresponding to that of the ship's head, and to a suitable transmitter.

Convenient frequencies for a system of this kind would be a high frequency of 1 to 2 Mc/s, modulated at 400 c/s, the latter being a usual frequency for small servo-systems.

The shape of the condenser vanes is critical. A form of moving vane which can be adapted for sine–cosine applications is a disk pivoted at the circumference. The fixed vanes are, theoretically, extremely narrow strips (Fig. 8.22). The capacity of such a combination is proportional to the length of the strip intercepted by the disk, which interception is proportional to the sine or cosine of the angle between the strip and the tangent to the disk (see Fig. 8.23).

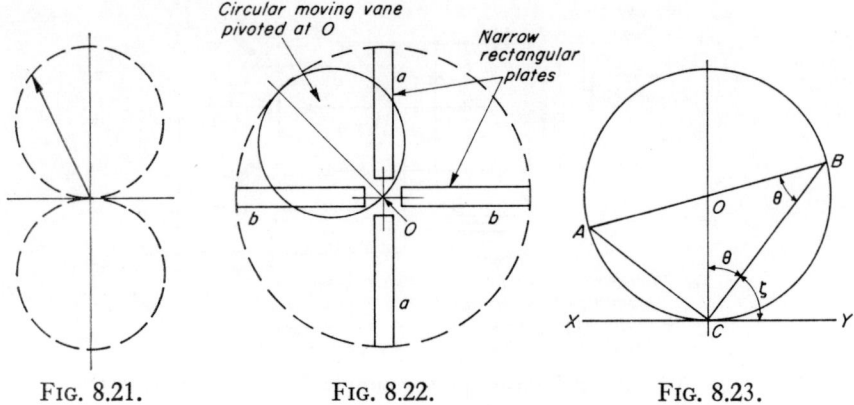

FIG. 8.21. FIG. 8.22. FIG. 8.23.

FIG. 8.21. Vector diagram of phase-changing signal

FIG. 8.22. Vane arrangement to give sine–cosine law

FIG. 8.23. Capacity of the system in Fig. 8.22

XCY is a tangent to the circle ACB, centre O.
CB is the intercepting vane.
AOB is a diameter of the circle.
ABC is therefore a right-angled triangle.

Thus $\quad \dfrac{AC}{AB} = \sin \theta = \cos \zeta$

and $\quad \dfrac{BC}{AB} = \cos \theta = \sin \zeta$

The theoretical results are rarely possible with the simple system indicated, and the introduction of trimming condensers which have to be adjusted and calibrated is almost inevitable. Owing to irregularities in the vanes, 'fringing' and other effects, the theoretical sine–cosine law is not obeyed. It is possible that the use of earthed 'guard vanes' could improve the accuracy.

Fig. 8.24 shows how a modulated high-frequency supply may be used with the compass systems illustrated in Figs. 8.15 and 8.16. The arrangement is equivalent to one pair of condenser vanes and an associated rectifier-potentiometer circuit, and provides a phase-changing signal as in Fig. 8.20, which is applied to a follow-up motor through the medium of an amplifier.

Some errors of vector or sine–cosine systems

Two forms of sine–cosine system have been described and others could be devised using inductors. This, in fact, is common practice in gyro-magnetic compasses, but for reasons already stated the unstabilized inductor does not lend itself to a transmitting compass application.

Consider the case illustrated in Fig. 8.25 where OH is a vector representing the magnitude of the horizontal component of the Earth's magnetic field. OA and OB are two pick-up systems perpendicular to each other in the horizontal plane and rotated about the vertical so that OB makes an angle θ with the magnetic meridian. The component of OH along OB is OH cos θ and along OA is OH sin θ.

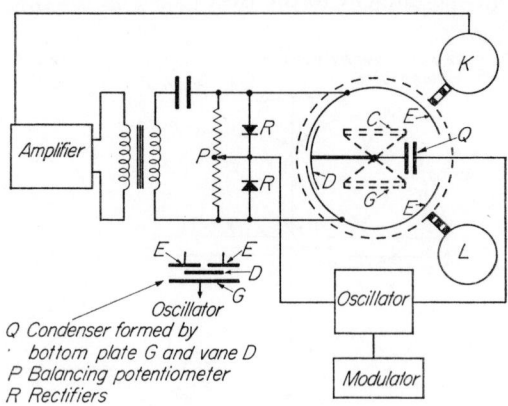

Q Condenser formed by
   bottom plate G and vane D
P Balancing potentiometer
R Rectifiers

FIG. 8.24. Modulated high-frequency supply for capacitative systems

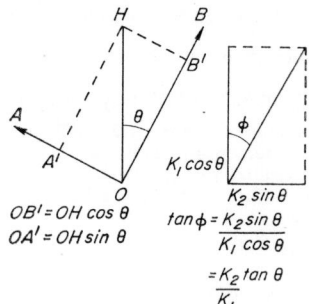

$$OB' = OH \cos \theta$$
$$OA' = OH \sin \theta$$

$$\tan \phi = \frac{K_2 \sin \theta}{K_1 \cos \theta}$$
$$= \frac{K_2 \tan \theta}{K_1}$$

FIG. 8.25. Errors of sine–cosine systems

The output at 45° ($\sin^{-1} 0.707$) equals the output at 45° ($\cos^{-1} 0.707$), i.e. O'A = OB. Amplification of O'A to O'B', or attenuation of OB to OA', makes the two outputs equal at 45°.

Suppose electrical signals proportional to OH cos θ and OH sin θ are generated by the devices represented by OA and OB. We now have

$$E_1 = K_1 \cos \theta \quad \text{for} \quad OB$$
$$E_2 = K_2 \sin \theta \quad \text{for} \quad OA$$

If these are combined in a resolver, the resultant voltage vector will be orientated with respect to one or other of the stator coils at an angle ϕ.

$$\tan \phi = \frac{K_2 \sin \theta}{K_1 \cos \theta} = \frac{K_2}{K} \tan \theta$$

It will be seen that unless $K_2 = K_1$, ϕ can be equal to θ only when $\theta = 0$, $\pi/2$, π or $3\pi/2$, i.e. when one vector = 0.

It is therefore important that the output signals from the compass pick-up elements be carefully matched before application to the resolver.

If $K_2 \neq K_1$ and we write $(\theta - \epsilon)$ for ϕ, where ϵ is the error in indication,

$$\tan(\theta - \epsilon) = \frac{K_2}{K_1} \tan \theta$$

Now $\tan(\theta - \epsilon)$, where ϵ is small, can be written

$$\tan \theta \left(1 - \frac{2\epsilon}{\sin 2\theta}\right)$$

Thus

$$\frac{K_2}{K_1} \tan \theta = \left(1 - \frac{2\epsilon}{\sin 2\theta}\right) \tan \theta$$

or

$$\frac{K_2}{K_1} = 1 - \frac{2\epsilon}{\sin 2\theta}$$

therefore

$$\epsilon = \frac{\sin \theta}{2}\left(1 - \frac{K_2}{K_1}\right)$$

This is the type of error which is zero at $0°$, $90°$, $180°$, $270°$ and at a maximum, alternatively positive and negative, at $45°$, $135°$, $225°$ and $315°$, so that it is sometimes called inter-cardinal or quadrantal error.

An error of this kind can readily be corrected by making $K_1 = K_2$, usually by alternating the larger signal; hence, providing the relation between E and $\cos \theta$ and $\sin \theta$ is linear, accuracy is maintained.

More intractable errors occur when the relationship between E and $\sin \theta$ and $\cos \theta$ is not linear (Fig. 8.26). Almost any type of error could arise according to the form of non-linearity. A common form is when the curve relating E and $\sin \theta$ or $\cos \theta$ starts as a straight line but curves as it approaches higher values of θ, as shown.

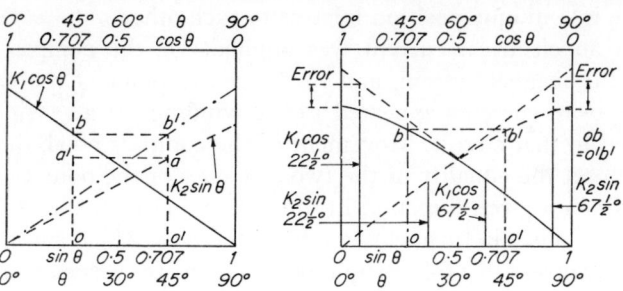

FIG. 8.26. Errors of sine–cosine systems

This condition gives rise to an error related to 4θ—known as an octantal error. It is best explained graphically. The two vectors, $K_2 \sin \theta$ and $K_1 \cos \theta$, are shown, K_1 and K_2 having been adjusted so that their amplitudes at $45°$ are equal, thus producing no error. At $0°$ and $90°$ there is also no error, since at $0°$ one vector, i.e. $\sin \theta$, is 0, so $\tan \theta = 0$ ($= \tan 0°$), and at $90°$ one vector, i.e. $\cos \theta$ is 0, so tan $\theta = \infty$ ($= \tan 90°$). But, say, at $22\frac{1}{2}°$, $K_2 \sin \theta$ is the expected value from a linear relation, but $K_1 \cos \theta$ is smaller than it should be; thus $\tan \phi$ is greater than tan $22\frac{1}{2}°$, giving a positive angular error. At $67\frac{1}{2}°$, $K_2 \sin \theta$ is smaller than it should be, whereas $K_1 \cos \theta$ is the expected value; hence $\tan \phi$ is less than tan $67\frac{1}{2}°$, giving a negative angular error. This pattern recurs for every quadrant, giving four positive peaks and four negative peaks, and the error curve is of the form $\epsilon = h \sin 4\theta$, where h is a constant.

Other forms of non-linearity would give error curves of different harmonic content. An error of the form $\epsilon = a + e \cos 2\theta$, where a and e are constants, can arise if the pick-up systems are not mutually perpendicular in the horizontal plane. If they are at $(90° - \psi°)$ instead of $90°$, $a = e = \psi°/2$.

Resistive systems

The practical difficulties with capacitative compass systems are considerable, but the use of a resistive system has resulted in the development of a very satisfactory transmitting compass of the follow-up type. A resistance bridge system using the compass fluid as a resistive medium was suggested by E. L. Holmes.* The idea was further developed at the Admiralty Compass Observatory in 1943 and an instrument produced known as the Admiralty Transmitting Magnetic Compass.

Holmes compass

The Holmes compass appeared in a number of forms. The first one was primarily a ship's standard compass, and demonstrated the basic principle of this type of instrument.

By its angular relation to electrodes in the compass bowl, an electrode attached to the magnet system divided the electrical path between the bowl electrodes, provided by a conducting compass liquid, so as to form a potentiometer system. In conjunction with an external circuit, the potentiometer system formed a resistance bridge circuit, which was balanced when the electrode on the magnet system divided the liquid potentiometer equally; that is to say, when the compass card and the bowl were in a unique position relative to each other.

Now when an electrical current was applied, via the compass pivot, to the electrode on the magnet system, and to the centre point of the external circuit, equal and opposite currents appeared in the windings of a sensitive, differential relay, included in that circuit. Movement of the compass bowl, due to change of ship's head, upset the equality of the two branches of the potentiometer, causing current to flow in a given direction in the external circuit, thereby operating the differential relay. This, in turn, closed a further circuit which controlled the direction of the field current in a follow-up motor. The motor was mechanically coupled to the bowl and rotated it within the binnacle in such a direction as to restore the balance of the potentiometer system, and to realign the compass bowl in its initial position with respect to the compass card. The bowl was virtually 'locked' to the compass system; whichever way the ship turned, the potentiometer would be unbalanced one way or the other, the relays would operate and cause the follow-up

---

* E. L. Holmes, Brit. Pat. 345 362: Improvements in course recording or like apparatus for use in ships or the like.

motor to drive the bowl back to restore the balance position. There was thus a north-seeking bowl driven by a powerful motor, ensuring a large driving torque for data transmission without imposing any reactive load on the magnet system—in other words, fulfilling the requirement of a magnetic compass capable of operating a number of distant repeaters and yet retaining the properties of a normal magnetic compass. Since the card and bowl were always in the same relative position, the former could not indicate ship's head unless used as a non-transmitting compass. An external card mounted on the bowl provided the indication required. In other respects the compass met all the normal navigational requirements. In the event of a power failure, the compass could be used in the ordinary way as a magnetic compass without transmission by manually rotating the bowl until the lubber's line was in the required position. The circuit arrangement is seen in Fig. 8.27. An azimuth circle could be used, as on any other compass, for taking bearings.

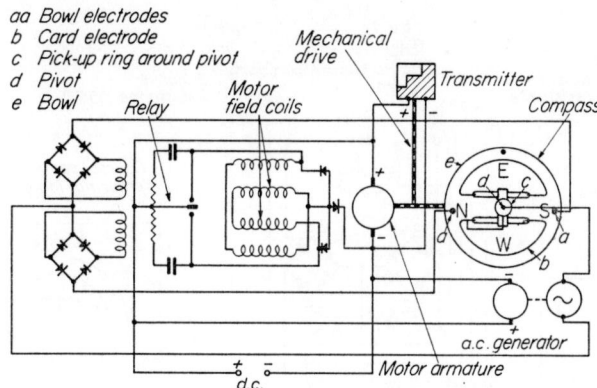

FIG. 8.27. Holmes compass
Circuit arrangement

In the early Holmes compass, it was the construction rather than the principle which gave rise to a number of disadvantages. The use of hydrochloric acid as an electrolyte or bowl liquid meant a glass bowl and platinum electrodes—a very costly construction. The design of the glass bowl, with a bubble trap instead of an expansion disk, caused aeration troubles, particularly when subject to vibration, and the glass magnet system was heavy and had dynamic characteristics not calculated to give the best performance. Perhaps the weakest feature was the use of a relay and a direct-current follow-up motor. The relay was unsuitable for conditions involving shock and vibration and the large external magnetic field of the motor meant that it had to be well removed from the compass bowl, the follow-up being communicated by a system of Hooke's jointed rods.

Holmes Tele-Compass

The Tele-Compass, illustrated in Fig. 8.28 and Plate 8.5, was a later form of the Holmes compass. Although little more than a course indicator, the follow-up being effected manually, it showed an advance in the construction of the bowl and electrode system. A glass bowl was still used, but with three electrodes symmetrically spaced in the liquid, thus providing an equally divided resistive path to an a.c.

supply connected to the centre one. A normal type of magnet system, but without a card, carried conducting strips which by their angular relation to the bowl electrodes altered the proportion by which the potentiometer was divided. The liquid was isopropyl alcohol with a trace of hydrochloric acid added, allowing a current of about 14 mA to pass. To avoid ambiguity, the movement of the magnet system was limited by a stop. The centre electrode was connected to one terminal of an a.c. supply (400 c/s), the outer electrodes to a phase-sensitive bridge rectifier system whose centre was taken to the other a.c. terminal. The symmetrical or aligned position of the bowl was indicated by zero current flowing in the external circuit of the rectifier bridge, by means of a centre zero milliammeter. Change of course upset the resistance balance in the bowl, which was restored by manually rotating the bowl through a flexible shaft and worm gear. The amount of rotation corresponded to the change of course and was indicated on a convenient dial.

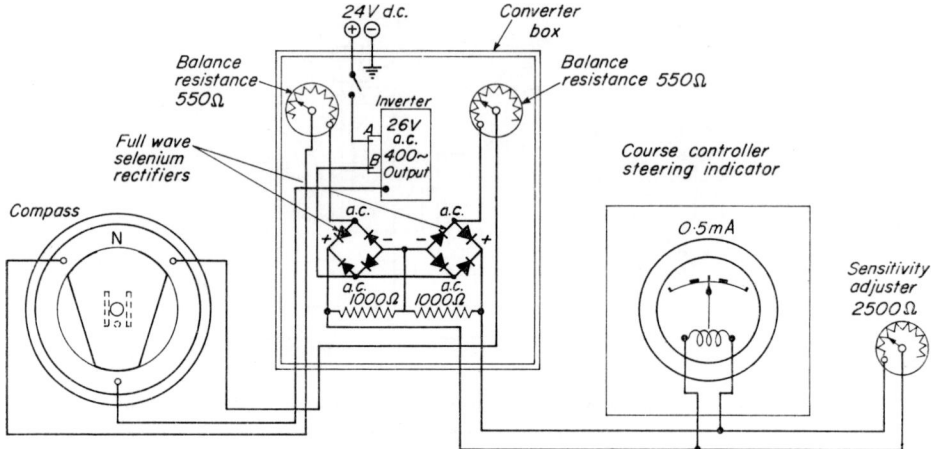

FIG. 8.28. Holmes Tele-Compass system
Wiring diagram

To maintain a given course, the dial was set accordingly and the craft's head directed so that the angular relation of the bowl to the magnet system gave zero indication at the meter. Deviations from the course were indicated by a right or left movement of the meter pointer.

While this was hardly a useful transmitting compass, the improvement in the bowl potentiometer system and external rectifier bridge was marked and provided the basis of a neat follow-up compass, giving synchro transmission and taking advantage of electronic methods to operate an a.c. follow-up motor, thus eliminating many of the defects of the original Holmes compass.

Holmes repeater compass

The Holmes repeater compass (Plate 8.6 and Fig. 8.29) was entirely an a.c. system. The rectifiers were no longer required and the bowl liquid potentiometer in conjunction with a bowl transformer resembled a Wheatstone bridge system. This bridge system was in balance when the magnet system electrodes were symmetrically disposed about the bowl electrodes in the position shown in Fig. 8.29. An a.c. supply was fed from the bowl transformer to the outer electrodes. The centre tap of the bowl transformer was connected to the input transformer of the amplifier, the other terminal of which was connected to the third electrode in the compass bowl. The bowl and the magnet system are shown in Fig. 8.30.

Any movement of the compass bowl with respect to the magnet system altered the balance of the bridge system, causing a voltage to appear between the centre electrode of the bowl and the centre tap of the bowl transformer—and therefore across the primary terminals of the input transformer. The signal from the secondary winding of this transformer was amplified by the valve and, through the output transformer, power was applied to the variable phase 2, 3 of the chaser motor.

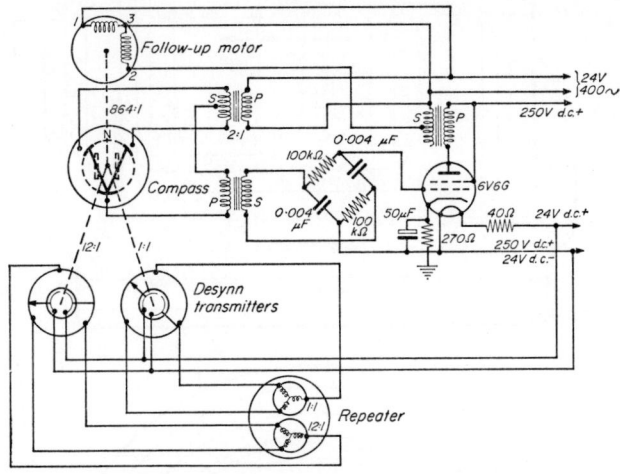

FIG. 8.29. Holmes repeater compass

The fixed phase 1, 3 was supplied in parallel with the primary of the bowl transformer. The motor armature rotated according to the phase relation between the two windings of the motor. A quadrature relation was required and this was arranged by a suitable phase-shifting circuit in the amplifier. A change from a quadrature lead to a quadrature lag in the variable phase with respect to the fixed phase brought about a change in direction of rotation of the motor, caused by the bowl electrodes moving from one side to the other of the balance position. Thus the direction of displacement of the bowl electrode system determined the direction of rotation of the motor shaft, and, for small angles, the amount of displacement, and also determined the magnitude of the signal.

The motor was geared to the bowl so as to restore the balance of the bridge if it became disturbed. As in the earlier Holmes compass, the bowl became 'locked' electrically to the magnet system and was therefore north-seeking. A card indicated ship's head and the two Desynn transmitters provided coarse (1 : 1) and fine (12 : 1) transmission to a repeater. This equipment was small and light and was primarily designed for aircraft. Its characteristics were similar to those of a normal aircraft liquid compass.

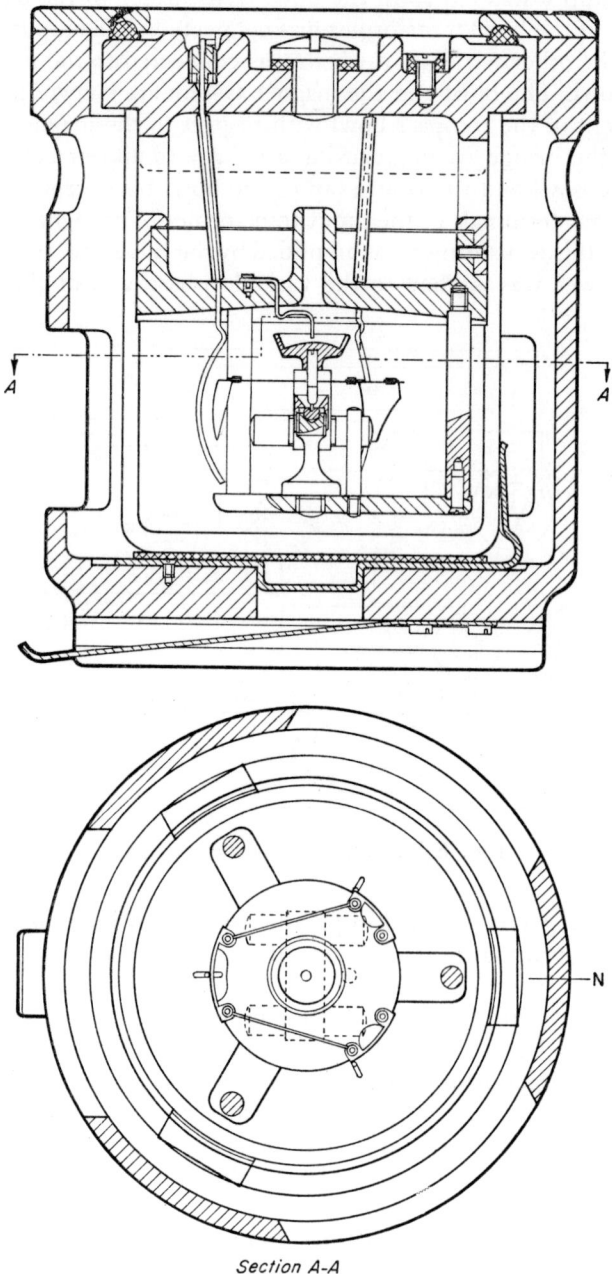

Section A-A

FIG. 8.30. Holmes repeater compass
Bowl and magnet system

Admiralty Transmitting Magnetic Compass

The Admiralty Transmitting Magnetic Compass* uses the same liquid resistance principle as the Holmes compass. Designed in 1943 to meet the requirements of small naval craft, its use has extended to larger units of the Fleet. Like the Holmes compass, it is primarily a marine instrument, but a small version of it was used in aircraft before the appearance of the Holmes repeater compass just described, which it closely resembled.

Several versions of the A.T.M.C. were designed: Type 1, the original model, using a magnetic compass with a $3\frac{1}{2}$-in. card and designed for coastal craft; Type 2, with a magnetic compass having the same dimensions and characteristics as the Admiralty liquid compass Pattern 195, and intended for use in ships up to the size of destroyers and minesweepers; Type 3, in which the compass was basically a P-type aircraft compass; and Type 4, designed for armoured fighting vehicles.

In all A.T.M.C.'s except Type 4, an existing compass having the desired characteristics was adapted, being fitted with electrodes to utilize the Holmes principle. It was found that by using a small trace of lithium chloride in the compass liquid, a painted metal bowl could be used, a special glass bowl being unnecessary. This meant that any standard type of compass could be readily adapted by the addition of platinum bowl electrodes inserted through insulating bushes of bakelite or glass, and by the addition to the card of a light platinum strip.

The electrode system was rearranged to give complete freedom of the card without introducing any ambiguity, and the pivot was not used as a central electrode. The system is illustrated in Fig. 8.31. The arrangement is seen to correspond

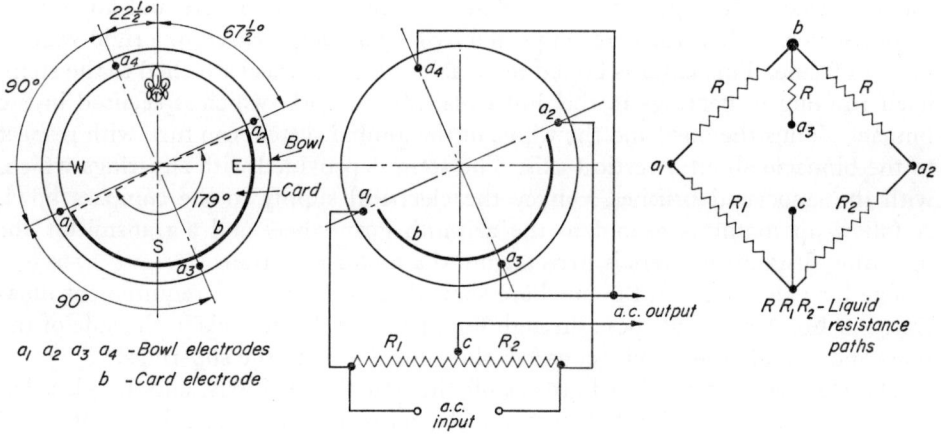

FIG. 8.31. Admiralty Transmitting Magnetic Compass
Electrode system

more nearly to a true resistance bridge, though the resistive network represented by the liquid and its associated electrodes is extremely complex. The fourth electrode ensures correct operation if the card is 180° from the balance position. The card is

---

* *Handbook of the Admiralty Transmitting Magnetic Compass*, Admiralty Compass Observatory, 1948, B.R. 1795; 1952, B.R. 1795 (1).

completely free, and the card and magnet system are in accordance with normal compass practice, with the addition of a thin platinum strip along the card's south-south-easterly semi-circumference, *b*.

The shape of the electrodes is important. An angle of 179° subtended by the card electrode results in a higher sensitivity than an angle of 180° or angles of less than 179°. The bowl electrodes need to be curved so that as the bowl tilts about the magnet system, the balance of the bridge does not alter, due correction to the curvature being made to allow for the 'screening' effect of the compass card (see Fig. 8.32).

The use of a completely resistive system rather than a system incorporating a transformer winding is significant. A bridge system that is partly resistive and partly inductive suffers from the introduction of phase changes in the signal as the temperature of the bowl is altered. This is owing to the increase or decrease of the resistive part of the bridge, with no change in the inductive part. Although the balance may be unchanged, the L : R ratio is altered and the quadrature relation between the ultimate amplified signal and the motor reference phase is affected, resulting, under certain conditions, in an insensitive system.

Admiralty Transmitting Magnetic Compass, Type 2

The Admiralty Transmitting Magnetic Compass, Type 2, whose master compass is illustrated in Plate 8.7, is a typical example of the kind of equipment described above.

The master compass is an adaptation of the Admiralty Pattern 195 magnetic compass. It is mounted in gimbals which are in the form of an inner ring carrying the athwartship (compass bowl) gimbal axis and the outer fore-and-aft gimbal pivots, which in their turn are supported by a stout ring secured to a rigid yoke or support frame. This latter is bolted to an azimuth gear wheel attached to the stem, itself running in bearings in the unit's main framework, which is secured in the binnacle. Thus the bowl and the whole of the gimbal system can turn with respect to the binnacle about a vertical axis. The stem is provided with slip-rings which, with the associated brushes, convey the electrical supply to the compass bowl. A follow-up motor is geared to the azimuth gear wheel and a transmitter for operating distant repeaters is driven from the motor gear train.

The four electrodes in the bowl are shaped according to the requirement illustrated in Fig. 8.32. They pass through liquid-tight bakelite seals in the side of the bowl and flexible leads connect across the gimbal axes to the slip-rings.

The electrodes in the bowl are set off the outer gimbal axis, and are placed at $67\frac{1}{2}°$, $157\frac{1}{2}°$, $247\frac{1}{2}°$, and $337\frac{1}{2}°$, respectively, measured clockwise. This is to avoid interference with the internal lubber's line which is along the outer gimbal axis; for the coincidence of the lubber's line and the north point of the compass card represents the balanced state of the resistance bridge system. When the gimbal system is set with its outer axis fore-and-aft, and the azimuth gear locked in that position, the instrument can be used, in the event of a power failure, as an ordinary non-transmitting magnetic compass.

The resistance bridge system is completed by a matched pair of resistors with a balancing potentiometer between them, as shown in Fig. 8.33. The output from the bridge is taken to a follow-up amplifier with three stages of amplification and terminating in the output transformer. The circuit is a normal arrangement with no attempt at phase-advance or derivative control.

The voltage output from the bridge system is proportional to the displacement from the balance position up to $\pm 2°$ and the 'dead sector' is $\pm \frac{1}{4}°$. There is normally sufficient frictional damping in the system to prevent hunting, but if it should occur, it may be stopped by adjusting the sensitivity of the amplifier to the point where hunting just ceases, without impairing the sensitivity.

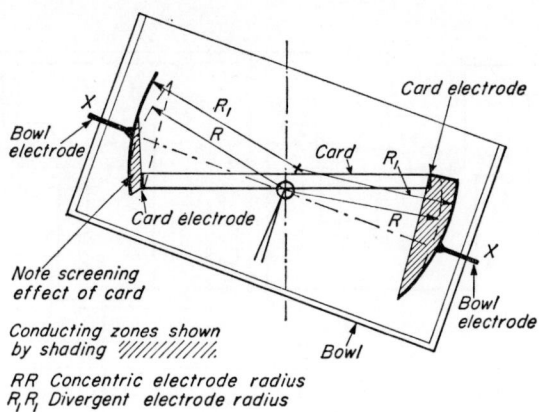

FIG. 8.32. Admiralty Transmitting Magnetic Compass

Since the conducting zone on the left is less effective than that on the right, owing to the screening of the card, the electrodes are wider apart at the top than at the bottom. If this were not so, an error in balance would occur when the compass was tilted, the sign of the error changing with reversal of tilt. The error is at a maximum when the tilt is about a diameter at right angles to XX, i.e. about O, and zero when the tilt is about XX. It takes the form $B \sin \theta$ where θ is the angle between the tilt axis and XX. B depends on the angle of tilt and electrode dimensions.

The output from the amplifier is taken to the two-phase follow-up motor geared to the gimbal and bowl system and is so arranged that the balance of the bridge is always maintained, making the rotating gimbal-and-bowl system north-seeking. A scale encircling the outer gimbal enables ship's head to be read against a lubber's line. A rate of follow-up of 360° in three minutes is obtainable, with a lag of less than $\frac{1}{4}°$. Transmission is derived from an M-type step-by-step commutator transmitter which will drive, if necessary, up to fifteen repeaters.

The master compass in Plate 8.7 is mounted in a brass binnacle provided with the usual facilities for permanent magnetic and soft-iron correction. The electrical equipment—amplifier, meter panel, distribution fuse panel for repeater circuits and a 24-volt d.c. driven generator—is fitted in a control unit. The generator provides 300 volts d.c. high tension, 6·3 volts a.c. for heaters, and 60 volts a.c. for motor reference phase and bowl supply. The a.c. supply is at 400 c/s. Typical compass repeaters are shown in Plates 8.8 and 8.9—a tape-type steering repeater, and a card-type azimuth or bearing repeater with azimuth circle.

A useful accessory is the total-error corrector, described in Chapter 10 and illustrated in Fig. 10.16 and Plate 10.1. This allows variation and deviation to be applied to the transmission system so that the repeaters read true headings. The outgoing transmission from the master compass drives a pair of M-type repeater motors in tandem, coupled to a further transmitter through a pair of differential gears. Variation and deviation are fed into the differentials by manually operated knobs and the outgoing transmission, which is fed into the repeater distribution panel, is advanced or retarded by the appropriate correction. The deviation is applied in accordance with the deviation table, showing the correction appropriate to any given course.

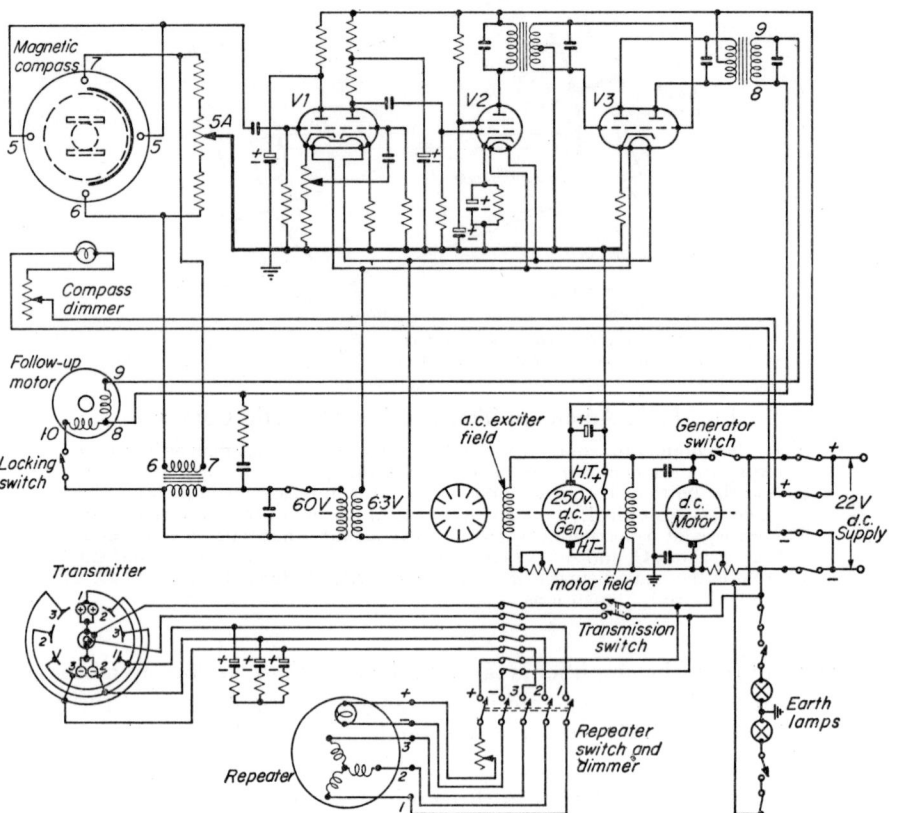

FIG. 8.33. Circuit for the Admiralty Transmitting Magnetic Compass, Type 2

The earlier A.T.M.C. Type 1 was a similar but less refined instrument. Types 3 and 4 were small outfits with limited repeater capacity, but resembled Type 2 in principle and general circuit arrangements.

A noteworthy feature of the A.T.M.C. is that, since the compass bowl is north-seeking, an azimuth circle mounted on it is stabilized in azimuth, and when trained on a distant mark will maintain that direction as the ship turns and manoeuvres. The performance of the repeaters is indistinguishable from that of a Pattern 195 magnetic compass.

The Siemens-Halske MKFE Anlage transmitting compass equipment, illustrated in Plate 8.10, is an interesting German instrument developed during the War and following the principle used in the Holmes and A.T.M. compasses.

The application of the principle is somewhat different in that a fixed bowl and rotating electrode system are used. The bowl follows conventional German design and has a hemispherical glass cover, which magnifies the card. There is no gimbal system to support the bowl, which is secured directly to the binnacle. The card and magnet system pivot about the vertical axis of an electrode system shaped like a pair of concentric U's, as shown in Fig. 8.34.

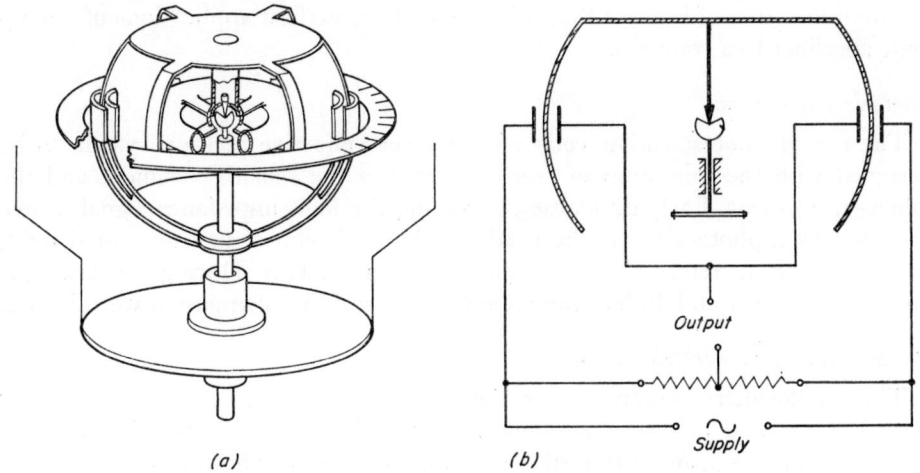

(a)　　　　　　　　(b)

FIG. 8.34. Siemens-Halske MKFE Anlage transmitting compass

The outer U is divided into two limbs, insulated from each other, and the inner U is continuous. These correspond to the bowl electrodes in the Holmes and A.T.M. compasses. They are attached to as pindle and connected to three slip-rings mounted on it, the whole assembly passing through a liquid-tight gland in the bottom of the bowl.

The liquid in the bowl is alcohol, to which phenoxyacetic acid is added to make it conducting. If the electrodes are connected to an external centre-tapped resistance as shown, and an a.c. supply is connected to the outer electrodes, a balanced Wheatstone bridge is formed, two of the arms being the resistive path formed by the liquid between the limbs of the U.

The compass card and magnet system, which is carried by the pivot, is fitted with a hemispherical shutter, with its open side vertical, which passes between the electrodes. When the shutter is so disposed that the resistance path between the limbs of the electrode system is interrupted equally on both sides, the bridge is in balance.

An alteration of course moves the bowl and electrode system with respect to the card and shutter; hence one path is further interrupted and the other is opened, with a consequent unbalancing of the bridge. The resulting signal from the bridge is amplified by a magnetic amplifier and used to energize a two-phase follow-up motor geared to the stem of the electrode system. The motor drives the electrodes to a new balance position, and stops; the electrode system thus becomes north-seeking. A synchronous transmitter is also driven by the follow-up motor for operating distant repeaters.

The use of a liquid-tight gland is a weakness of this design and the system of rotating electrodes is not as effective as that of the rotating bowl of the A.T.M.C., the effects of swirl in the latter being largely avoided. Moreover, the card and magnet system, with the hemispherical shutter, is unduly heavy on the pivot.

In the early models of this compass the magnetic amplifier had a propensity to hunt and so long a time-constant that a tachogenerator was geared to the follow-up motor to provide a velocity feed-back component in the input signal to the amplifier. It is of interest that this is one of the earliest applications of a magnetic amplifier to a compass.

Photoelectric systems

The use of photosensitive cells allows a very effective pick-up system to be designed with the minimum of reaction between the sensitive element and the follow-up element. Early difficulties, involving the high-impedance signal source provided by a photocell and the need for the exclusion of any unwanted light, have appeared to retard the development of this system. During the last war, however, German and Italian photoelectric transmitting compasses were in use.

Plath-MKF: photoelectric compass

This photoelectric compass was designed for small vessels, such as motor torpedo boats and minesweepers, and was used during the Second World War. The instrument is again of typical German design with a spherical bowl, enclosed by a transparent upper half. The bowl, which is not gimballed, is mounted in a binnacle with permanent magnetic, soft-iron and anti-degaussing correctors. A slot in the compass card, subtending 180°, provides an opening through which light from a lamp, mounted on a rotatable platform below the compass, is reflected upwards and then downwards again on two photosensitive cells, also mounted on the platform. Fig. 8.35 shows the optical system on the platform whereby the light from the lamp travels through the slot by two different paths and is reflected back again from a mirror in the top of the compass bowl so as to illuminate the photocells equally. This is the balanced position of the system since the outputs from the cells are equal and, when applied to the grids of the amplifier valves (also mounted on the rotatable platform), produce no output signal from the 'control coupling' or transformer.

Any change of course of the ship results in a relative movement between the binnacle, which carries the photocell and valve assembly, and the compass card with its slot, thus causing one photocell to be illuminated more than the other.

A signal now goes to the valve amplifier and an output voltage appears at the transformer windings, energizing the follow-up motor. This two-phase motor is coupled to a synchronous transmitter, connected electrically to a repeater motor geared to the rotatable platform. Operation of the motor causes the repeater motor to turn the platform until the photocells are once again equally illuminated, whereupon the motor stops. The rotatable platform always assumes the same position relative to the compass card and therefore to magnetic north and is effectively a power-driven north-seeking element. Further distant repeaters can be operated from the follow-up transmitter. In many respects this compass is similar to and shares the advantages of the A.T.M. compass.

Fig. 8.36 shows the circuit employed. The complete equipment includes the binnacle, in which the master compass is mounted. The photocell assembly, lamp and amplifier on the rotatable platform, and the follow-up repeater motor are part of the master compass assembly. The follow-up motor and transmitter are mounted in a compartment on the side of the binnacle.

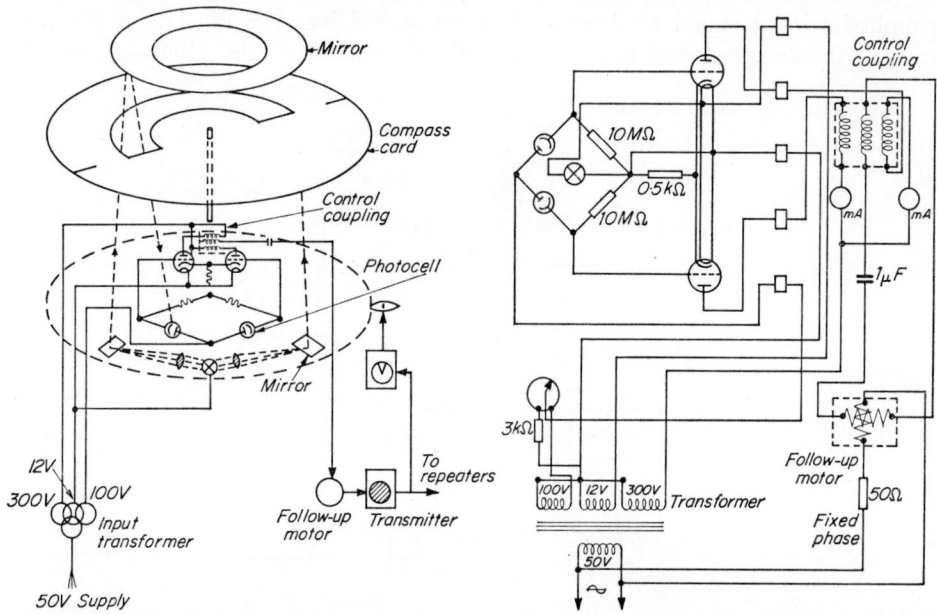

FIG. 8.35. Plath photoelectric compass
Schematic diagram of the optical system

FIG. 8.36. Plath photoelectric compass
Schematic diagram of the master unit

External equipment consists of a meter panel in which the milliammeters, indicating the anode currents to the amplifier valves, are mounted. This panel includes a potentiometer so that the sensitivity of the system may be adjusted. Sometimes, rather than being included in the binnacle, the follow-up motor and transmitter are fitted in a bulkhead-mounted box. Terminal boxes, repeaters and dimmers and a 50-volt 50 c/s d.c.-driven alternator complete the equipment, whose current consumption is 22 amps. The repeaters have coarse and fine dials, the former marked in degrees to 360°. The latter is marked to 10 degrees and rotates thirty-six times for each revolution of the coarse dial. In the event of a power failure, the compass can be used as a simple magnetic compass.

Bussola Magnetica Asservita O.G.

The Bussola Magnetica Asservita O.G., whose operation is illustrated in Fig. 8.37, is an Italian photoelectric transmitting compass. This again was designed to fill a need in small fast craft. As with the A.T.M.C., the compass bowl is free to rotate with respect to the compass card and the binnacle. The bowl carries an optical system and an illuminating lamp, which shines through a slot in the card on a photocell, also mounted on the bowl assembly. The beam of light is interrupted rapidly by a rotating perforated disk or shutter, so that the output from the cell is a form of alternating voltage (or chopped d.c.).

If the card is in a position where the cell is fully illuminated, the consequent interrupted signal from the photocell is amplified by a valve amplifier which in turn supplies power to a follow-up motor, which is part of a unit in the base of the binnacle. The motor drives a synchronous transmitter through gearing and, by means of a vertical shaft and further gearing, rotates the compass bowl. When supplied with a signal as described, the motor drives the bowl in a direction causing the compass card to cut off the illumination from the photocell, and the

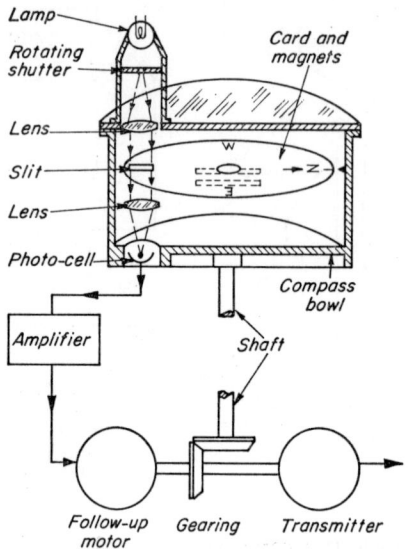

FIG. 8.37. Bussola Magnetica Asservita O.G.

signal from the cell ceases. The amplifier is designed so that cessation of the input signal results in a reversal of the phase of the output signal. Thus the bowl is hunted back towards the fully illuminated state of the photocell. The bowl oscillates or hunts about the mean position of the slot in the card when the ship is on a steady course.

An alteration of course changes the relative position of the card slot with respect to the bowl (and lamp and cell). The bowl is therefore driven by the follow-up motor to seek the new mean position of the slot and, as in the A.T.M.C., becomes north-seeking. The instrument can be used as a simple magnetic compass, with the power switched off.

Kelvin-Hughes transmitting magnetic compass

The Kelvin-Hughes transmitting magnetic compass, seen in Fig. 8.38, is a second-harmonic inductor system, driven by a normal liquid-filled magnetic compass. The system has been designed to be attached to any conventional type of ship's

compass (see *Compass-aided inductor systems*, Chapter 4). The inductor system is
an east–west sensing device with a follow-up as illustrated in Fig. 4.37. East–west
is defined in this instance as the direction at right angles to the compass magnets.
The inductor, which is of unusual design, is energized at 200 c/s from an oscillator.
The 400 c/s output from the inductor is amplified and used to drive the follow-up
motor, which is geared through an ingenious epicyclic gear to the inductor and to
the M-type transmitter which supplies the remote repeating instruments. The
compass can be used either with transmission or as a normal standard or steering
instrument. Externally the binnacle appears the same as those usually fitted and the
compass card is always visible and available for use, whether or not the compass is
energized.

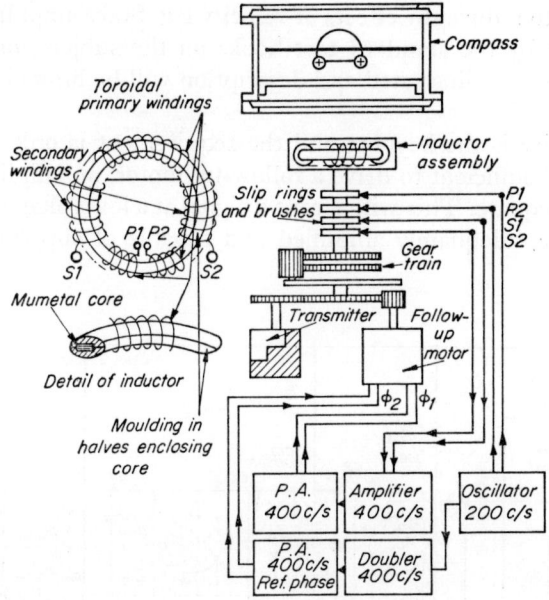

FIG. 8.38. Kelvin-Hughes transmitting magnetic compass

FOLLOW-UP AMPLIFIERS

The follow-up system of a transmitting compass is a simple form of servo-
mechanism having a pick-up element at the compass proper, providing an input
signal (sometimes called an error signal when using systems that operate on
misalignment from a given datum). Various forms of these devices have already
been described with their specific compass systems. In order to use the information,
amplification of this signal is often necessary to enable the follow-up or resetting
mechanism in the compass (usually a small electric motor) to be operated; hence
an important part of a transmitting compass is the amplifier.

The fundamental design of compass servo- or follow-up amplifiers follows
conventional practice. The frequencies encountered in this field range from about
150 c/s to 800 c/s, though higher frequencies may sometimes be found in saturable-
inductor excitation circuits.

The complexity or otherwise of a compass servo-amplifier depends on the sensitivity, accuracy and response characteristics to be provided by the whole equipment. In general the amplifier will have an input stage consisting of one or more voltage amplifying valves feeding a power or output stage. The input stage is supplied either from the pick-up system of the compass or from the winding of a synchro transmission element, and the output stage normally feeds a follow-up or driving motor, which may be either a.c. or d.c.

Again depending on the design characteristics, there may be filter circuits preceding and following the input stage, and between the latter and the output stage, other intermediate stages may be inserted to modify the original signal characteristics by introducing phase-shifting, demodulation, integration, differentiation and similar effects in order to provide damping or non-hunting characteristics to the output signal, or to reduce the effects of velocity lag. Since amplifier design as such is amply covered by the standard text-books on the subject, only a few typical compass amplifiers are illustrated and description will be limited to certain salient features.

Where simplicity is the keynote and the requirement is only for amplification of an input signal sufficient to drive a follow-up motor, an amplifier as illustrated in Fig. 8.39 is adequate. This will accept a signal at a low voltage level and deliver an output voltage, adequately amplified and capable of supplying power to the motor windings.

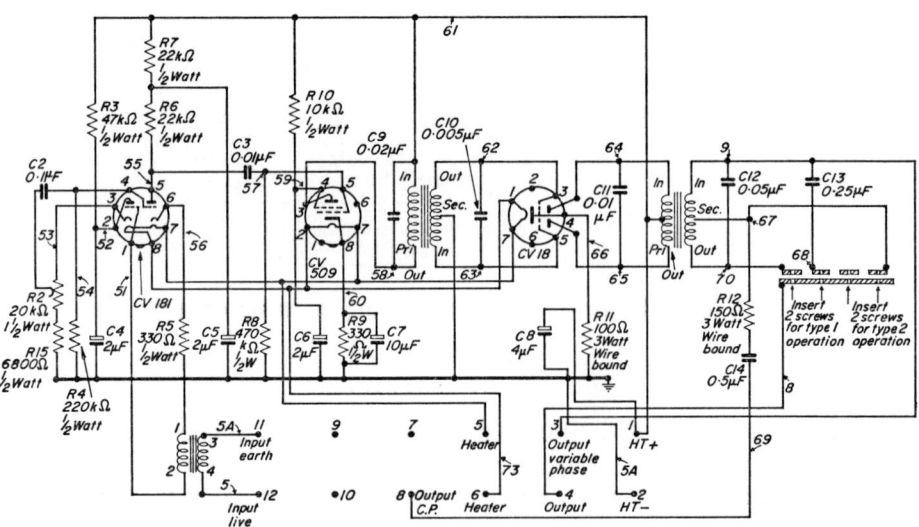

Fig. 8.39. A simple amplifier

A similar amplifier having two channels, as would be required in a gyro-magnetic compass, is illustrated in Fig. 8.40. Double triode valves are employed for the voltage amplifying stages, one double valve handling a separate channel in each of its sections. The output stage has two pentodes in push-pull. Otherwise the circuit is conventional.

Fig. 8.41 shows a more refined type of follow-up amplifier in which a phase-advancing circuit is introduced between the input and output stages. Its purpose is to avoid hunting in the motor without reducing the sensitivity and without relying on the frictional losses in the system to prevent the motor from over-running.

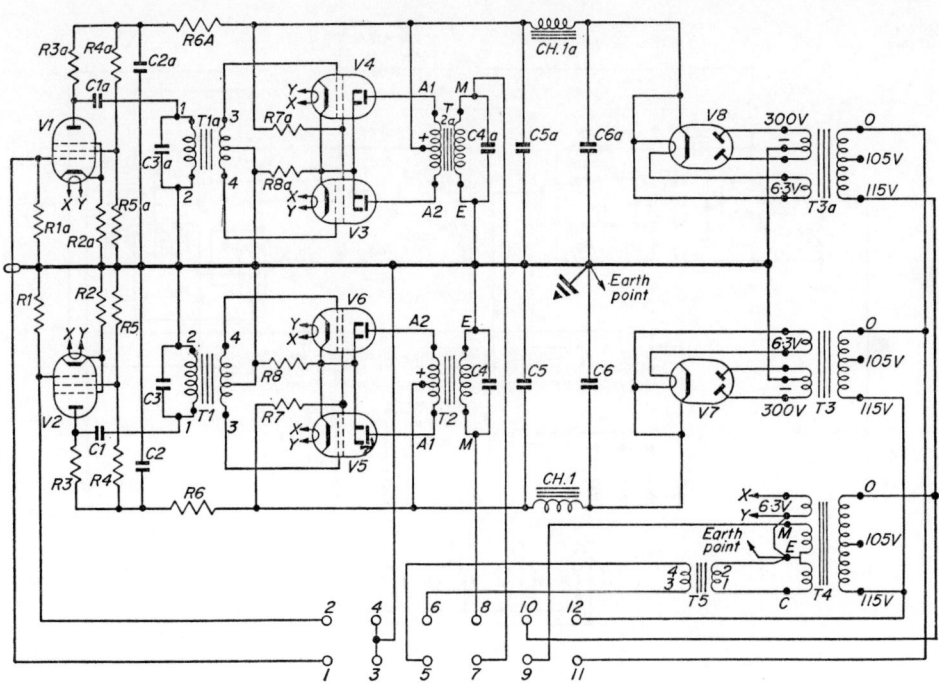

R1	1MΩ	R1a	1MΩ	R2	150Ω	R2a	150Ω	R3	47kΩ	R3a	47 kΩ	R4	68 kΩ	R4a	68 kΩ
R5	33kΩ	R5a	33 kΩ	R6	10 kΩ	R6a	10 kΩ	R7	4.7kΩ	R7a	4.7kΩ	R8	330Ω	R8a	330Ω

C1	0.05 μF 350 V	C1a	0.05 μF 350 V	C2	1 μF 250 V	C2a	1 μF 250 V
C3	0.005 μF 500 V	C3a	0.005 μF 500 V	C4	1 μF 250 V	C4a	1 μF 250 V
C5	2 μF 400 V	C5a	2 μF 400 V	C6	0.5 μF 350 V	C6a	0.5 μF 350 V

V1	CV 138	V2	CV 138	V3	CV 2136	V4	CV 2136	V5	CV 2136	V6	CV 2136	V7	CV 493	V8	CV 493

Fig. 8.40. Two-channel amplifier for gyro-magnetic compass

The tendency to hunt, or the possibility of hunting, is present in most servo-systems since inertial forces cannot be eliminated. If the input is displaced from the aligned state with the output shaft locked, a large voltage is present at the input terminals to the amplifier. On releasing the output shaft, the motor will run 'home' at full speed. Although zero voltage at the output is reached, the inertia of the system causes the motor to over-run, so misaligning the system in the reverse sense. A new impulse is given to the system and an over-run occurs again on the opposite side. If the frictional losses are low and the sensitivity high, an oscillatory or hunting condition occurs. In these circumstances, the signal displays the excitation frequency (usually several hundreds of cycles per second), modulated at the hunt frequency.

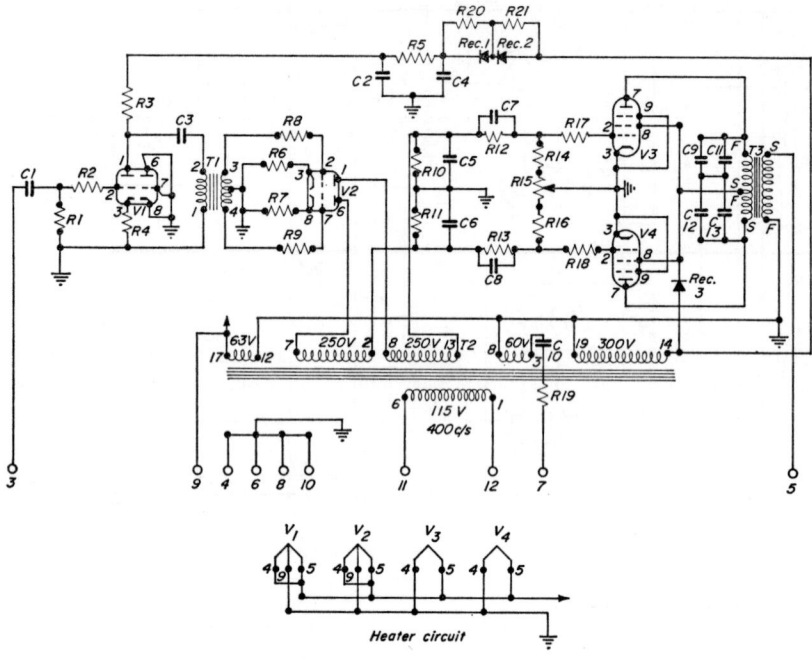

Capacitors							
C1	1 μF 200 V	C5	0·25 μF 200 V	C9	0·05 μF 500 V	C13	0·05 μF 500 V
C2	0·5 μF 350V	C6	0·25 μF 200V	C10	1·0 μF 200 V		
C3	0·1 μF 350V	C7	0·1 μF 200V	C11	0·05 μF 500V		
C4	0·5 μF 350 V	C8	0·1 μF 200V	C12	0·05 μF 500 V		

Resistors							
R1	470KΩ ¼W	R7	470 Ω ¼W	R13	470 KΩ ¼W	R19	100 Ω 6W W/W
R2	150 KΩ ¼W	R8	470 KΩ ¼W	R14	47 KΩ ¼W	R20	470 KΩ ¼W
R3	220 KΩ ¼W	R9	470 KΩ ¼W	R15	0-25 KΩW/W POT	R21	470 KΩ ¼W
R4	470 Ω ¼W	R10	47KΩ ¼W	R16	47 KΩ ¼W		
R5	47KΩ ¼W	R11	47KΩ ¼W	R17	22 KΩ ¼W		
R6	470 Ω ¼W	R12	470 KΩ ¼W	R18	22 KΩ ¼W		

Valves							
V1	12AX7	CV4004	V2	12AT7	CV4024	V3	6CH6	CV4055	V4	6CH6	CV4055

FIG. 8.41. Follow-up amplifier

After demodulation, this hunt signal may be advanced in phase, so that a braking or retarding voltage component is applied to the motor. If the constants of the circuit are correct, the motor operates as though it were heavily damped or 'dead-beat'. In the example illustrated, the second stage, a double triode with a.c. applied to the anodes, acts as a phase-sensitive rectifier, or demodulator. The resulting unidirectional signal may be constant or sinusoidal. If the latter, it is the hunt signal and is phase-advanced by the RC network which follows. The modified, unidirectional voltage is applied to the grids of the output valves, the anodes of which are fed in phase at an appropriate frequency to drive the motor. Thus with zero input, the outputs of the two valves cancel. When a signal is applied to the grids of these valves, one or other is 'gated' and a driving voltage appears across the output terminals of the amplifier, whose phase corresponds to that of the signal applied to the grids.

An output stage fed with an alternating signal can be used to drive a d.c. motor by using a.c. excitation of the anodes. An amplified input signal is applied in antiphase to the grids of a pair of output valves whose anodes are fed at the signal frequency, but in phase with each other and in phase or 180° out of phase with the incoming signal. For a given phase sense of the input, one grid is driven more positive and the other grid more negative; thus one valve passes more current on the positive half-cycle of anode excitations than the other. On the negative half-cycle, neither valve passes current. This phase-conscious rectification permits a flow of unidirectional current in the output circuit, the direction of which depends on the phase of the grid signal. Fig. 8.42 illustrates this circuit.

	Forward signal		Reverse signal	
G_1	+	−	−	+
G_2	−	+	+	−
A_1	+	−	+	−
A_2	+	−	+	−
Greatest current in	F_1	X	F_2	X

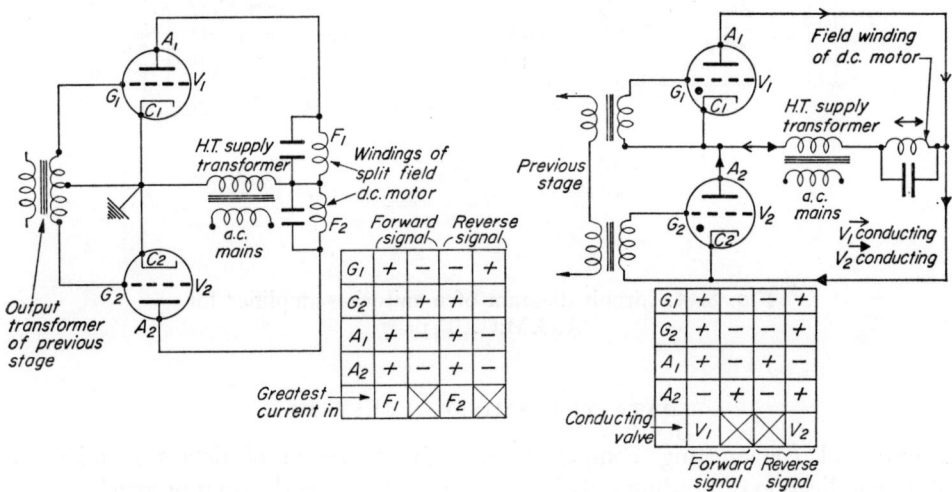

FIG. 8.42. Phase-conscious rectification

Anodes supplied in phase; grids in antiphase, but in phase or antiphase with anode supply.

	Forward signal		Reverse signal	
G_1	+	−	−	+
G_2	+	−	−	+
A_1	+	−	+	−
A_2	−	+	−	+
Conducting valve	V_1	X	X	V_2

FIG. 8.43. Follow-up amplifier with thyratrons

Anodes supplied in antiphase; grids in phase, and in phase or antiphase with corresponding anode supply.

An alternative d.c. output stage may be contrived by using thyratrons or gas-filled valves. These are fed with a.c. at the anodes and are non-conducting until a threshold grid potential is provided. The valve suddenly conducts, or 'fires', and passes current regardless of any further increase in the grid voltage, until the grid becomes negative or the anode potential drops to zero. Thyratrons normally handle more anode current than a 'hard' valve, and a follow-up amplifier with a thyratron output is shown in Fig. 8.43.

Fig. 8.44 illustrates an amplifier similar to that shown in Fig. 8.41, but designed to operate from a saturable inductor. Since the input signal is double the excitation supply, the amplified signal is also at double that frequency. A frequency-doubling circuit is therefore needed to provide the reference signal to the demodulating circuit preceding the phase-advancing network.

If a separate oscillator is used to excite the inductor, part of its output may be used to energize a frequency-doubling amplifier, capable of handling enough power to provide the reference phase of the follow-up motor, which would then operate at the signal frequency of the inductor.

Transistors and magnetic amplifiers provide a useful alternative to thermionic valves and result in economy of space and power requirements.

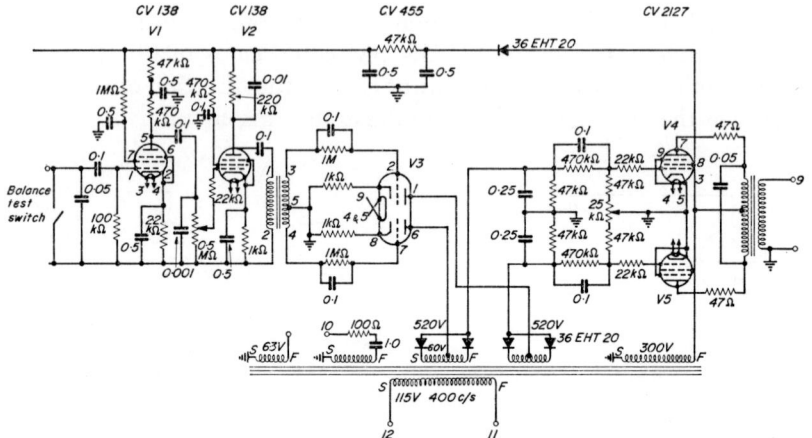

FIG. 8.44. Circuit diagram of monitoring amplifier for A.G.M.C. Type 7

DATA-TRANSMISSION SYSTEMS

Almost all transmitting compasses embody a system of data transmission. Where indication of heading only is necessary, either step-by-step or synchronous transmission may be used. Where power is required to be transmitted as well as angular position, step-by-step transmission is suitable, but where auto-synchronization is needed, synchronous transmission with some form of torque amplifier (usually electronic) is used.

Step-by-step transmission is usually either the 4-wire system used by the Sperry Gyroscope Company Ltd and by S. G. Brown Ltd or the 3-wire or M-type system used in naval applications and invented by the British Admiralty. As its name implies, this form of transmission communicates angular position by a series of discrete steps—a common step value is 10 minutes of arc, though a value of 30 minutes is used in aircraft and 2 minutes for certain naval precision requirements. The system operates on a d.c. supply.

The transmitter, which initiates the data, is virtually a 3-pole single-throw or double-throw switch, according to the type of transmission used, and when driven by some rotary mechanism energizes the external circuit sequentially in steps of a given duration.

The external circuit is a motor having a permanent magnet or soft-iron armature and field windings consisting of three sets of coils. These coils are energized in succession both singly and in combination, producing a magnetic field across the stator which alters its angular position in a series of steps and in which the armature aligns itself. The size of the step as an output angle is determined by the gear ratios between the driving mechanism and the transmitter and the repeater motor and a dial or other appliance.

Four-wire system

The 4-wire system is shown in Fig. 8.45. The transmitter is a cam-operated 3-pole switch arranged to connect the positive line in turn to lines 1, 2 and 3, either

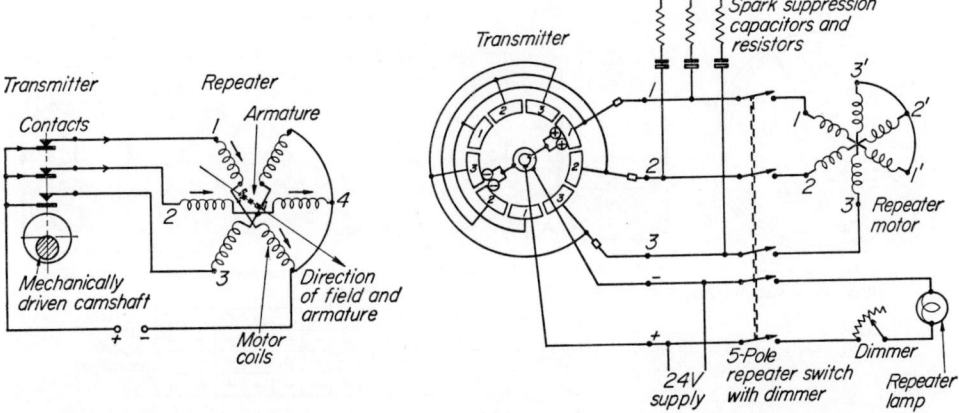

FIG. 8.45. Four-wire system   FIG. 8.46. M-type or three-wire system

singly or in pairs. The sequence is therefore: 1, 12, 2, 23, 3, 31—six steps in all for one revolution of the transmitter shaft. In each step either one or two of the motor coils are joined to the positive supply, the return path to negative being through the common junction, 4, of the coils. The successive positions of the rotor are in line with coil 1, in between coils 1 and 2, in line with coil 2, and so on, making the six steps in 180°, or half a revolution. A repetition of the transmitter sequence giving another six steps will cause the repeater rotor to complete 360° of rotation, since the rotor is not polarized, being of soft iron and becoming magnetized only because of the field due to the coils. The transmitter therefore has to turn twice as fast as the repeater armature to provide the twelve steps per revolution that are required.

If the transmitter is geared 360 : 1 to a driving shaft and the repeater armature 180 : 1 to a dial or pointer, every revolution of the transmitter is one degree, in which there are six steps; thus each step is 10 minutes of arc. Similarly, the repeater armature makes a revolution for every 2° of the movement of the dial or pointer, corresponding to twelve steps or again 10 minutes of arc per step. Other step values may be obtained with different gear ratios.

This system operates on 22, 50 or 70 volts as a general rule. It develops maximum torque in the stationary position and is capable of driving not only repeater dials, but other transmitters, plotting tables and similar apparatus.

M-type or 3-wire system

The M-type or 3-wire system is an improvement on the 4-wire system in that it develops more torque and it may be used with either a polarized or non-polarized armature in the repeater motor. It is illustrated in Fig. 8.46. The transmitter is a 3-pole double-throw switch which connects each line in turn to the positive or negative supply line, or leaves them connected to neither. The coils are wound similarly to the 4-wire arrangement, but the common or 'star' point of the coils is not connected to any external circuit. The sequence makes, for instance, 1 and 2 positive and 3 negative, followed by 1 disconnected, 2 positive and 3 negative and so on, as illustrated in Fig. 8.47. There are therefore, as previously, 12 steps per

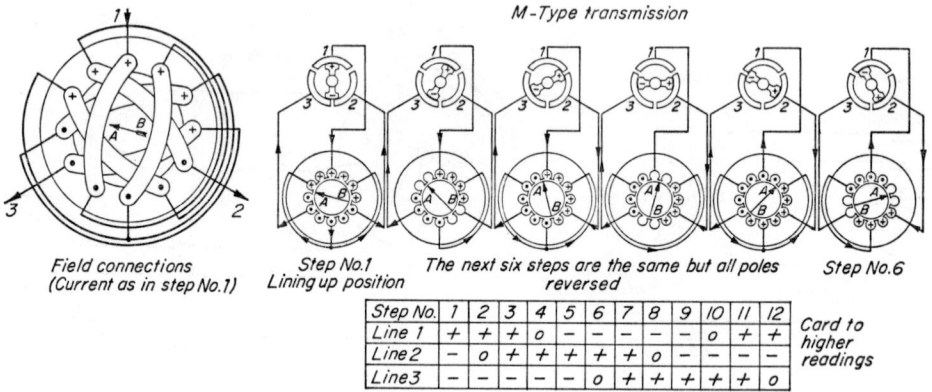

Step No.	1	2	3	4	5	6	7	8	9	10	11	12	
Line 1	+	+	+	o	−	−	−	−	−	o	+	+	Card to
Line 2	−	o	+	+	+	+	+	o	−	−	−	−	higher
Line 3	−	−	−	−	−	o	+	+	+	+	+	o	readings

FIG. 8.47. M-type transmission
Operation of repeater motor and sequence of steps

revolution of the repeater armature, but also twelve steps per revolution of the transmitter shaft. Gearing both units to driving and driven elements through 180 : 1 ratio produces 10-minute steps. Fig. 8.48 shows torque curves of M-type repeater motors. The same result can be achieved by using, instead of the cam-operated transmitters as described above, rotary or commutator transmitters, and one example is illustrated in Fig. 8.49. The gear ratio depends on the number of segments in the commutator. The M-type transmitter illustrated has nine segments, corresponding to three cycles of the transmission sequence; thus to obtain 10-minute steps, the gear ratio is 60 : 1.

It will be realized that in the examples quoted, the sequence of twelve steps repeats itself with every revolution of the repeater armature. In 10-minute steps this is every 2°, and hence the system cannot distinguish between any 2° sector. It is therefore neither self-aligning nor synchronous and needs to be lined up as required. Once it is aligned, it remains in step unless a speed of 144 steps per second (or in the above examples, 4 r.p.m. of a 10-minute repeater dial) is exceeded. This corresponds to 720 r.p.m. of the repeater motor.

Attempts have been made to design automatically aligning step-by-step transmission but this adds complications to what is otherwise a very simple and reliable system.

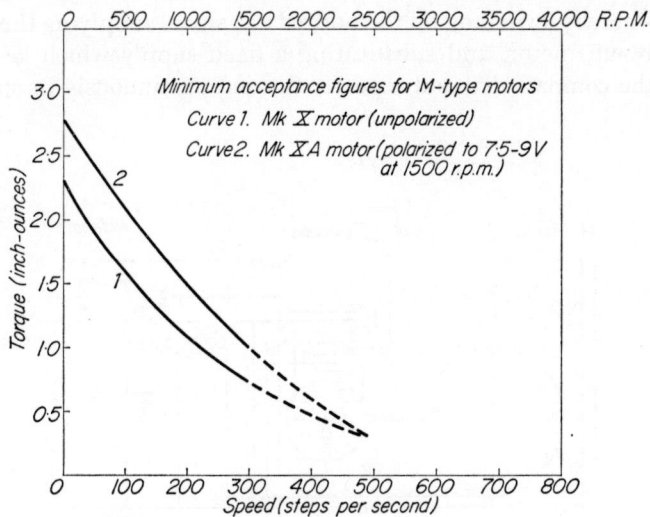

FIG. 8.48. Torque–speed curves for M-type motors

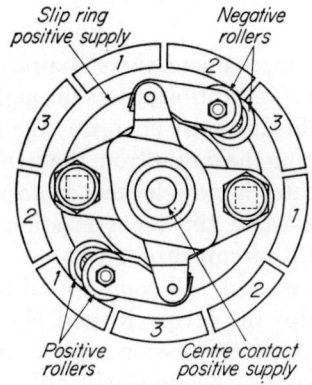

FIG. 8.49. Commutator of M-type transmitter

One such arrangement, applied to an Admiralty type of transmitting compass, is illustrated in Fig. 8.50. C is the repeater dial driven by the repeater motor M; T is the transmitter, geared to the follow-up motor F of a master compass or other source of transmission, which is provided with a dial D. If the transmitter/repeater system is in step and aligned before switching on, the card C will continue to reproduce exactly the angular indications of the dial D until some breakdown of the circuit, such as an interruption of the supply, occurs. When transmission is ultimately resumed, it may be out of step.

Assume that the system is misaligned. To realign, the push button P is pressed. The d.c. supply through the push button energizes the 'holding-on' relay R_1, and the contacts K close. The circuit through R_1 and K includes a contact X at the compass, closed unless the compass reading is exactly north, and parallel contacts Y which are also normally closed, associated with the relay R_2. The relay now holds on the contacts K when the push button is released, and also changes over the contacts L, thereby switching off the normal a.c. signal supplying the driving phase of the follow-up motor, and substituting a fixed supply which keeps the motor and hence the compass dial and repeater turning continuously in one direction.

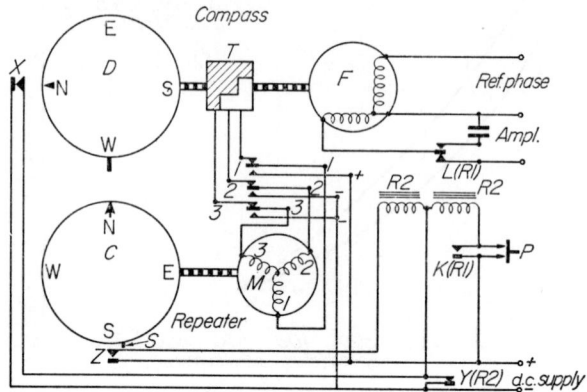

FIG. 8.50. Automatic aligning in step-by-step transmission

Adjacent to the repeater mechanism there is a pair of contacts, Z, which are normally open. As the repeater indication passes through north, a striker is operated which closes the contacts. This instantly energizes the relay R_2 and stops the rotation of the repeater by switching the motor from the transmitter circuit to a standard step energization. The repeater is now locked on to north but the compass dial continues to rotate. At the same time, energization of R_2 opens contacts Y, leaving X to maintain the d.c. circuit through R_1.

In due course, the compass card will come round to north, and at this instant the contact X opens and trips the relays R_1 and R_2, restoring the circuit to its initial state and restarting the transmission with compass and repeater rotating together and in step. The correct heading will quickly be sought as the normal a.c. driving signal has been restored to the follow-up motor.

A warning lamp in the relay circuit shows that the synchronizing cycle is operating. Should the push button be pressed with the system in step, the contacts X and Y both open together as compass and repeater indicate north; thus the relays are tripped and the system continues to run in step without interruption.

The system is not limited to one repeater. Where several are used, the contacts Z are in parallel. Should each repeater be misaligned to a different extent, the repeaters will stop, in turn, on north. When the last one is locked on north, the relays are tripped the next time the compass indicates north, and all the repeaters start off together and in step. Any compass using a follow-up system can be adapted in the manner described to provide automatic alignment of the step-by-step transmission.

Step alignment in step-by-step transmission systems. It has already been stated that the sequence of 10-minute steps repeats every 2°. This is strictly true only if the repeater motor is fitted with a polarized or magnetized armature. With a soft-iron armature, the induced magnetism reverses every 180° of rotation or every six steps, so that this type of armature can be synchronized at each degree.

Thus, it is not possible to align the transmitter with the repeater more closely than 2° for a polarized armature or 1° for a soft-iron armature in a 10-minute system unless precautions are taken to ensure that both instruments are interconnected to give corresponding step positions at either end.

The choice of a 'lining-up' step is arbitrary, but Admiralty practice, for instance, is to make line 1 negative and lines 2 and 3 positive (or vice versa) for exact degree indications where a soft-iron armature is used, and 1 negative, 2 and 3 positive where a polarized rotor or armature is used.

With different step values, the alignment sector varies; for example, with $\frac{1}{2}°$ transmission steps it is not possible to align more closely than 6° for a polarized armature or 3° for a soft-iron armature unless initial lining-up precautions have been taken.

By altering the sequence of connections, synchronization can be adjusted by two steps, i.e. by $\frac{1}{3}°$ in the case of 10-minute transmission or 1° in the case of $\frac{1}{2}°$ transmission; that is to say, instead of connecting line 1 of transmitter to line 1 of repeater, line 2 to line 2 and line 3 to line 3, they are connected 2 to 1, 3 to 2, 1 to 3. Conversely, an error in connection will result in a corresponding error in indication.

Reversal of rotation may be effected by changing over lines 2 and 3 without altering the lining-up position.

Suppression. A problem that arises with step-by-step transmission is that an inductive circuit is constantly being interrupted. The back e.m.f. causes sparking at the transmitter contacts, with consequent burning and deterioration of the contact points. A circuit containing resistance and capacity, or capacity only, connected from line to earth, or to negative in the case of four-line transmission, or to the star point in the case of M-type transmission, is an effective and commonly used method of spark suppression (Fig. 8.51).

The value of the capacitors and resistors depends on the number of repeaters connected to the transmitter and on the speed of transmission.

It frequently happens that, in spite of adequate spark suppression, considerable interference with radio equipment may be caused by step-by-step transmission. Effective suppression of this interference can usually be arranged by inserting radio-frequency chokes in each transmitting line and supply line and connecting the 'line' terminal of each choke to earth through a small capacitor; $0.1\ \mu\text{F}$ is a convenient value.

Repeaters

Repeaters take a variety of forms. Plates 8.8 and 8.9 illustrate the usual dial type of repeater, in which the motor, through gearing, drives a compass card from which bearings may be taken with an azimuth circle or by which a ship may be steered. There are various types of open-scale repeaters of which one made by S. G. Brown Ltd, has a dial which rotates at four times the speed of the compass (see Fig. 8.52). The dial is engraved with a peripheral scale with 90 divisions, each one therefore representing one degree. The dial is pierced by nine equally spaced windows through each of which a disk carrying groups of four numbers can be

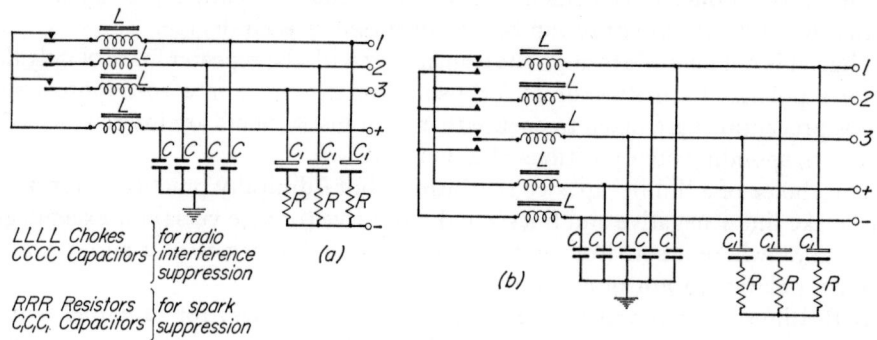

LLLL Chokes | for radio
CCCC Capacitors | interference
| suppression

RRR Resistors | for spark
$C_iC_iC_i$ Capacitors | suppression

(a)

(b)

FIG. 8.51. Spark suppression

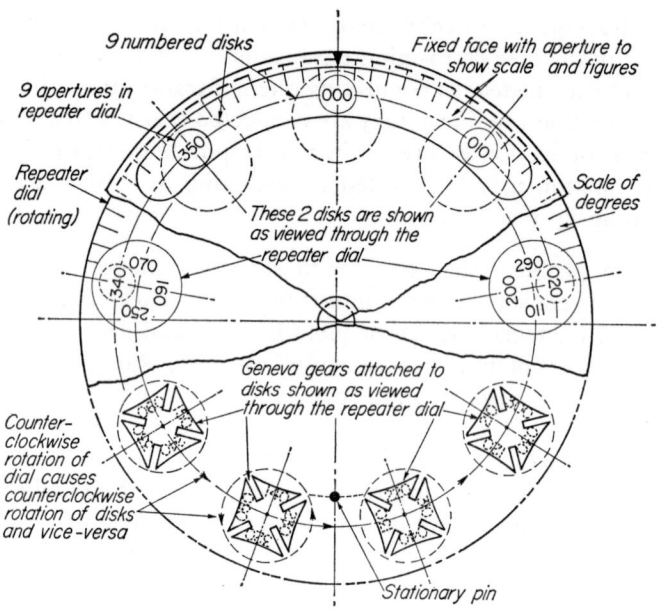

9 numbered disks

Fixed face with aperture to
show scale and figures

9 apertures in
repeater dial

Repeater
dial
(rotating)

These 2 disks are shown
as viewed through the
repeater dial

Scale of
degrees

Counter-
clockwise
rotation of
dial causes
counterclockwise
rotation of disks
and vice-versa

Geneva gears attached to
disks shown as viewed
through the repeater dial

Stationary pin

FIG. 8.52. Dial of open-scale repeater

viewed, only one number at a time being visible at each window. The numbers are in groups 90° apart thus: 0, 90, 180, 270; 10, 100, 190, 280; 20, 110, 200, 290; up to 80, 170, 260, 350. By means of a Geneva gear and a fixed pin, each disk is moved round one quarter of a revolution in turn for each revolution of the main dial, so that at the windows, three only of which are visible at a time, three consecutive numbers are shown, e.g. 170, 180, 190, the intervening arcs being covered by the

206

degree scale indications. As the dial continues to turn, 190 disappears, and 160, 170, 180 are seen; and so on until in due course 190 is changed to 280 and the three original windows after one revolution of the dial show 260, 270, 280.

Another open-scale repeater, used in ships of the Royal Navy, is the tape repeater, illustrated in Plate 8.11. An endless tape, 90 in. long, is driven over rollers by a repeater motor and is divided into 360 degrees; one degree takes up $\frac{1}{4}$ in. The tape, which is 35-mm cinema film, is illuminated from behind. This instrument is equivalent to a repeater with a card some 29 in. in diameter. Needless to say, these open-scale repeaters cannot be used for taking bearings; they are essentially steering repeaters.

Synchronous d.c. systems

No form of step-by-step transmission is self-aligning; that is to say, it may be a given number of degrees out of line when switched on according to the gear ratio driving the transmitter or the step-value. On the other hand, a synchronous system always lines up when power is switched on, however far apart angularly the transmitter and repeater indications may be. The Patin compass, described on p. 167, has a synchronous transmitting system using d.c. energization. A similar system is the Desynn transmission, illustrated in Figs. 8.53 and 8.54.

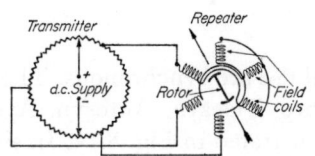

FIG. 8.53. Desynn transmission

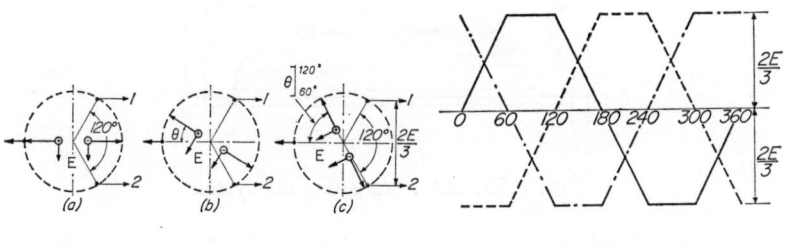

FIG. 8.54. Trapezoidal signal from Patin and Desynn systems

The repeater motor is rather like an M-type motor, but the transmitter is a toroidally-wound resistance, with tappings for the three transmission lines spaced at 120°. Two brushes, driven by the input shaft or gears, move round the toroid and are connected to the d.c. supply. In effect, it may be likened to a 30° step (1:1 gear ratio) M-type system but with the steps 'smoothed off' to give a continuous instead of an intermittent rotation. There is only one repeater position corresponding to any angular position of the transmitter brushes.

Ideally, the voltage across a pair of transmission lines should be sinusoidal for accurate angular data to be transmitted. The Patin and the Desynn systems do not provide sinusoidal outputs, but rather a trapezoidal form of signal and thus the accuracy of the system is somewhat limited unless a calibration chart or some means of correction is used.

Referring now to Fig. 8.54.... At (a), let $\theta = 0°$. The voltage across 1 and 2 lines is zero. At (b), the potential at 1 is $-E[(120° - \theta)/180°]$ and at 2 is $-E[(120° + \theta)/180°]$ compared with positive.

Thus the voltage across 1 and 2 is $E\{[(120° + \theta)/180°] - [(120° - \theta)/180°]\} = E\theta/90$. This obtains only between $\theta = \pm 60°$, for once θ exceeds this value, the potential across 1 and 2 is $E \times 120/180 = 2E/3$ and remains so until $\theta = 120°$. From $\theta = 120°$ to $\theta = 180°$ the voltage is $E(180 - \theta)/90$ and thus is zero when $\theta = 180°$. From $\theta = 180°$ to $\theta = 240°$, the potential is $-E(\theta - 180)/90$ and from $\theta = 240°$ to $\theta = 300°$ the voltage across 1 and 2 is $-2E/3$. From $\theta = 300°$ to $0°(360°)$, the voltage across 1 and 2 is $-E(360 - \theta)/90$. Thus the line voltage is

$$E\left\{\left[\frac{\theta}{90}\right]_{\theta=0°}^{\theta=60°} ; \left[\frac{2}{3}\right]_{\theta=60°}^{\theta=120°} ; \left[\frac{180-\theta}{90}\right]_{\theta=120°}^{\theta=180°} ; \left[\frac{\theta-180}{90}\right]_{\theta=180°}^{\theta=240°} ; \left[-\frac{2}{3}\right]_{\theta=240°}^{\theta=300°} ; \left[\frac{\theta-360}{90}\right]_{\theta=300°}^{\theta=0 \; (360°)}\right\}$$

which can be seen to consist of two sectors in which the voltage is constant at $\pm 2E/3$ connected by straight lines where the voltage is proportional to θ. The other pairs of lines (2, 3; 3,1) show a similar effect, the voltage–angle graphs being displaced by 120° and 240°. Little torque is transmitted by this system.

A.C. transmission systems

All a.c. systems in general use are synchronous. There are many varieties of the basic system, such as Synchro, Telesyn, Autosyn, Aysynn, but the fundamental circuit of these systems is illustrated in Fig. 8.55. In simple terms, the transmitter

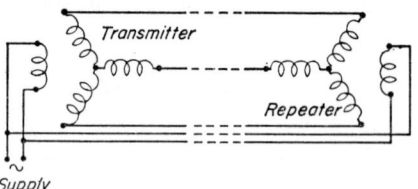

Fig. 8.55. A.C. transmission system

and repeater are identical instruments consisting of a laminated soft-iron stator with a distributed three-phase winding, and a laminated soft-iron wound rotor. When connected to a suitable supply as in Fig. 8.55, both rotors are fed in parallel. Each rotor induces voltages in the three-phase windings according to its position, and the rotors will come to rest in such a position that no current flows in the transmission lines. If the angular position of one of them is altered, current flows in the lines and the other rotor seeks a new position to restore the 'no current' state. Thus the repeater rotor moves through the same angle as the transmitter rotor. The voltage across the lines is sinusoidal and, given careful manufacture, a high degree of accuracy is obtainable though the torque transmitted is small. This system is suitable only for indications of angular position, such as those to be made by a compass repeater. Increased accuracy can be obtained by using coarse and fine sector transmission.

One pair of elements is geared 1:1 to the input and output ends of the system and the other pair is geared, for example, 36:1. Thus the latter makes 36 revolutions whilst the former makes one, and one revolution of the latter represents only 10° of angular change of the system. This by itself is ambiguous, but the coarse sector system identifies the correct 10° sector. The ultimate errors of such a system are therefore $\frac{1}{36}$ of those of a 'coarse only' link.

Additional information can be added to the circuit by using the synchro-elements as electrical differential units, i.e. the stator can be mounted so that it can be rotated as well as the rotor. Thus the sum or difference of two angles can be transmitted to the repeater.

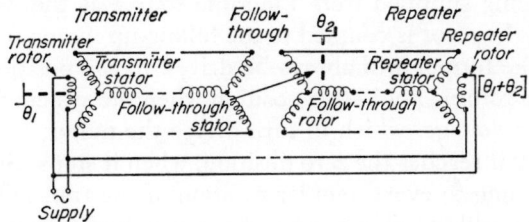

FIG. 8.56. A 'follow-through' synchro-element

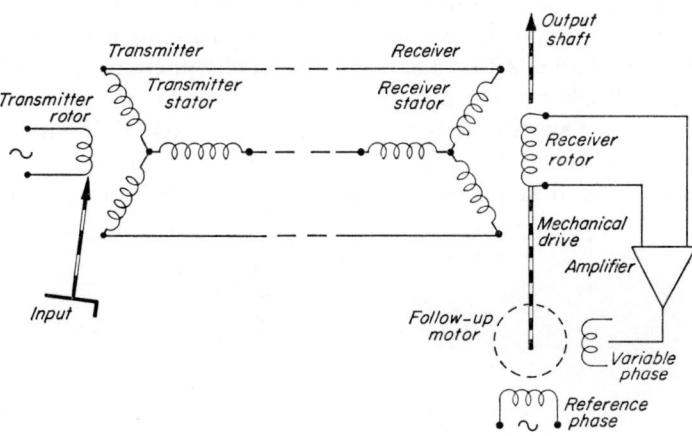

FIG. 8.57. Power synchro-system

The same result can be achieved by using a 'follow-through' synchro-element. This has a three-phase stator as in the normal element, but in addition its rotor has a three-phase winding. When connected in the circuit of Fig. 8.56, additional angular information can be fed in by the rotation of the follow-through rotor.

Power synchro-systems

If synchronous transmission is required, capable of developing sufficient torque for driving instruments other than very light indicators, a power synchro-system using an amplifier and follow-up arrangement (see Fig. 8.57) is necessary.

The transmitter rotor is connected to the supply (as in Fig. 8.55) and the stator connected to that of the receiver, but the rotor of the receiver is connected to the input of a suitable amplifier.

The coupling between the rotor and the stator windings of the transmitter causes voltages to be induced in the latter which appear also in the stator of the receiver. The rotor of the receiver will also pick up a voltage owing to its coupling with the stator windings, this voltage depending on the angular position of both transmitter and receiver rotor. It is evident that the relative positions of transmitter rotor and stator determine the voltages in each stator winding and that the voltage in the receiver rotor depends also on its position relative to its stator; there will therefore be a position of maximum and minimum voltage for the receiver rotor for every position of the transmitter rotor. The receiver voltage is applied to the amplifier, whose output supplies the variable phase of a follow-up motor, the reference phase being supplied from the same source as the input supply to the transmitter. Since the rotor is coupled to the follow-up motor by gearing or other mechanical link, the former will always be driven to the position of minimum or zero voltage. If the rotor is not in this position when the system is switched on, the amplifier receives a voltage, which in turn drives the motor, so rotating the rotor of the receiver until it reaches the zero position, when it stops. Since there is a zero position corresponding to every angular position of the transmitter, the system is synchronous and capable of delivering power up to the capacity of the follow-up motor. Such a system may be elaborated by the use of a follow-through element as in Fig. 8.57, between the transmitter and receiver, or by the use of coarse and fine links, in which case the two circuits are combined by a mixing network or by a relay so that when the system is within a narrow angle of alignment the fine link operates, but the coarse link takes over when this angle is exceeded.

The whole range of synchronous power transmission systems is outside the scope of this work and reference should be made to standard publications on the subject. Much information is to be gained from the manufacturers' own literature.

Chapter 9

The Gyro-Magnetic Compass

The problem of accelerations acting on pivoted-needle compasses has been discussed in Chapter 3, and it can be shown that the same difficulties arise with a pendulous inductor compass. Apart from the simple expedient of using a heavily damped magnet system with a low magnetic moment, as described under *Aircraft compasses* in Chapter 5, a more elegant solution to the problem is gyro-stabilization of the magnetic compass. Instruments employing this method are known as gyromagnetic compasses. The gyroscope has the property of being inherently stable; in a perfect instrument, the axis of spin remains in a fixed direction in space. Its effective inertia is high and the forces and accelerations required to disturb it are comparatively large.

THE VERTICAL GYROSCOPE

In its simplest form a vertical gyroscope consists of a heavy flywheel rotating at several thousands of revolutions a minute. The axle is carried in bearings in an inner gimbal, which is pivoted within an outer gimbal, which, in turn, is pivoted within the framework of the instrument attached to the craft or vehicle. The three axes so defined—spin, inner gimbal and outer gimbal—are orthogonal, with the spin axis vertical.

The gyroscope, when run up to speed, will resist any effort to alter the direction of the spin axis, which acts as a reference or datum direction as the attitude of its vehicle alters. The direction of the spin axis can, however, be changed by the application of torques about the gimbal axes, and it is the existence of such torques, due to frictional forces, that destroy the perfection of the gyroscope as a directional instrument and cause it to wander. The provision of bearings of high quality and low friction is therefore essential. Lack of balance of the wheel or gimbal structure can cause torques which will cause wandering of the gyroscope. The rotation of the Earth will cause an apparent movement of the gyro-axis and, except at the north or south poles, a vertical gyroscope will appear to the observer to tilt.

Precession

The movement of the spin axis in space, whether caused deliberately or accidentally, is called precession, and the way it occurs is explained in Fig. 9.1(*a*) and (*b*). If the spin axis is to be directed along a line other than that in which it rests, there will be a change of angular momentum of the system, and this must be provided by a torque as shown in Fig. 9.1(*a*). The torque vector is at right angles to the precession vector, rotations being vectorally represented by the 'right-hand screw' rule.

In the diagram, the spin vector is upwards. A clockwise torque on the horizontal axis OT is a vector at right angles to the spin axis and directed to the right. The consequent precession is represented by a third vector OP directed away from the observer and producing a clockwise movement of the spin axis. Reversal of either the spin or torque will reverse the precession. Again in Fig. 9.1(*b*), the wheel is rotating anticlockwise looking from the top, so the spin vector points upwards. Any attempt to overturn the gyroscope so that the top of the spin axis moves away from the observer into the page will result in the spin axis actually tilting to the left in the plane of the page, while any attempt to tilt the spin axis in the plane of the page will result in an actual tilt into the page.

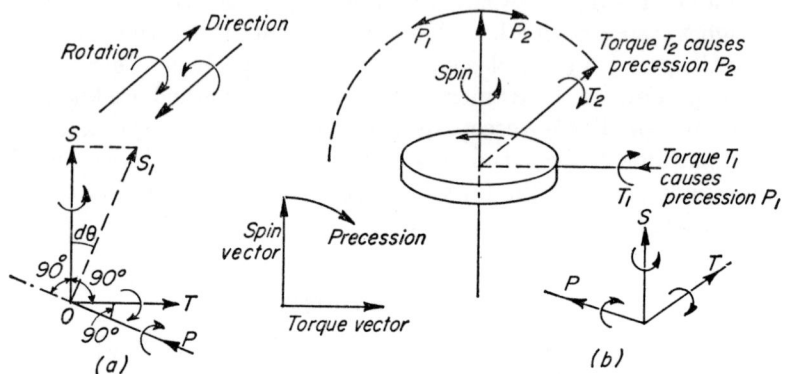

FIG. 9.1. Vertical gyroscope

If OS is the spin axis and it is desired to tilt it to the position OS₁ through an angle $d\theta$, a torque vector SS₁ is necessary to alter the angular momentum of the system. Thus, to rotate OS clockwise, represented by the precession vector PO, torque SS₁, represented by OT, and therefore clockwise looking along OT, needs to be applied.

The rate of precession depends inversely on the angular momentum of the wheel and directly on the magnitude of the applied torque. Thus the rate of precession, $\dot{\theta}$, is given by

$$\dot{\theta} = \frac{\tau}{I\omega} \tag{9.1}$$

where the torque applied is τ and the angular momentum is $I\omega$, I being the moment of inertia and ω the angular velocity. A heavy, fast-running wheel therefore needs more torque to disturb it than a light, slow-running wheel, and thus has greater gyroscopic stability.

Given a reasonably precise form of construction, a vertical gyroscope may be used for maintaining a magnetic compass or detector unit horizontal. The magnetic element may be mounted directly on the inner gimbal or a set of similar gimbals carrying the magnetic element may be coupled by servo-systems to a remote gyro, thus providing a stable platform for the magnetic element.

A magnetic element mounted on such a stable platform must not be pendulous. Acceleration errors are caused by forcing a pendulous detector element into the false vertical, and this will occur even if the platform upon which the element is mounted is perfectly stable. The types of magnetic element, therefore, that are used with this form of stabilization are inductor units or double-pivoted magnet systems mounted directly on the platform.

Tilt errors

Departure of the spin axis from the vertical will tilt the magnetic element, thus causing appreciable deviations in azimuth.

A fixed inductor or double-pivoted needle tilted about its north-south axis by an angle β will have an azimuthal deviation δ given by

$$\tan \delta = \tan D \sin \beta$$

In the United Kingdom, a tilt of one degree will produce a deviation of $2°.32$. To ensure that deviations are not greater than $\frac{1}{2}°$, the gyro axle must never be more than 13 minutes of arc from the true vertical, a requirement not always easy to achieve in a small instrument. Though an average accuracy of that order may be possible, many small gyros execute a conical precession of a greater magnitude about the true vertical. Although it is true that a gyroscope will not readily depart from its attitude even when mounted on a moving vehicle, imperfections and extraneous forces will act upon it and so it is necessary to provide means of keeping and restoring the vertical direction of the axle.

Erection control

Gyroscopes may be held vertical by the use of gravity-controlled switches attached to the gimbal axes. These may be mercury switches, on-off contacts operated by a pendulum, or pick-up devices (capacitative, inductive or resistive) also operated by a pendulum. Departure of the spin axis from the vertical as indicated by the pendulum will cause a displacement of one or both of the pick-up or switch devices. These are arranged to switch an electrical supply (either directly or with an amplifier) to torque motors, mounted about the appropriate gimbal axis, so that a torque is applied in the correct sense to restore the gyro axle to its vertical position; in other words, any disturbing force is sensed by a resultant precession. A counter-torque is then applied and the axle precessed back to the desired position. The applied torque may be proportional to tilt, 'proportional control', or of a constant value, 'on-off' control (see Fig. 9.2). The gravity switches are P_1 and P_2 and the respective torque motors are M_1 and M_2.

If there is a tendency for the gyroscope to wander off continuously, whether this be owing to Earth rotation or extraneous torques, a corresponding restoring torque is needed. To this end, a small displacement of the gyro is necessary to provide a signal, requiring the gyroscope axle to depart from the true vertical by a small angle θ_{ST} called the 'stick-off' error.

The azimuthal deviation of the magnetic compass is given [see equation (3.8)] by

$$\tan \delta = \frac{\theta_{ST} \tan D \cos \alpha}{1 - \theta_{ST} \tan D \sin \alpha}$$

Proportional control of erection. Suppose the drift is due to a disturbing torque τ and the gyro is drifting at a rate $\dot\theta$. Let the correcting torque be $\kappa\theta$. For equilibrium,

$$\tau - \kappa\theta = H\dot\theta \qquad (9.2)$$

where H is the angular momentum of the gyro.

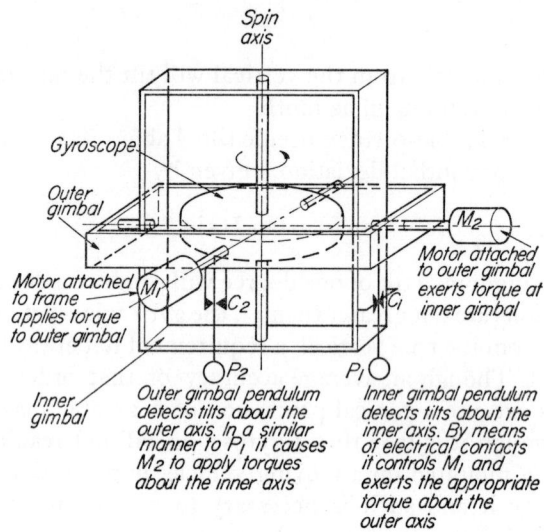

Spin axis

Gyroscope

Outer gimbal

Motor attached to frame applies torque to outer gimbal

M_1

C_2

C_i

M_2

Motor attached to outer gimbal exerts torque at inner gimbal

P_2 P_1

Inner gimbal

Outer gimbal pendulum detects tilts about the outer axis. In a similar manner to P_1 it causes M_2 to apply torques about the inner axis

Inner gimbal pendulum detects tilts about the inner axis. By means of electrical contacts it controls M_1 and exerts the appropriate torque about the outer axis

FIG. 9.2. Erection control

For $(\tau/\kappa) - \theta$ write $-\phi$; then by differentiation, $\dot\theta = \dot\phi$. Thus

$$-\phi = \frac{H}{\kappa}\dot\phi \quad \text{or} \quad \frac{d\phi}{\phi} = \frac{-\kappa \, dt}{H}$$

Integrating,

$$\log_e \phi = \frac{-\kappa t}{H} + c$$

or

$$\phi = c'e^{-\kappa t/H} \qquad (9.3)$$

Since $\phi = \theta - \tau/\kappa$,

$$\theta = c'e^{-\kappa t/H} + \tau/\kappa \qquad (9.4)$$

When t is very large, θ is the ultimate stick-off error. Thus we can write

$$\theta_{ST} = \tau/\kappa \qquad (9.5)$$

where κ is the control constant, in torque per radian of error.

Alternatively, in terms of a control constant, κ', in rad/min/rad ($\kappa' = \kappa/H$)

$$\theta_{ST} = Ce^{-\kappa't} + \frac{\dot\theta}{\kappa'}$$

For θ_{ST} to be small, τ must be small or κ large. If κ is large, the gyro tends to follow small perturbations and is no longer a stable reference.

'*On-off*' or '*bang-bang*' *control.** Again considering a gyroscope with a steady drift, $\dot{\theta}$, and a constant control rate, $\dot{\phi}$, let the 'dead sector' of the switching device be an angle $\pm\,\Theta$. Until a tilt of Θ is reached, the gyro is tilting at $\dot{\theta}$. Then the full control is switched on and the gyro is restored to its normal position at a rate of $(\dot{\phi}-\dot{\theta})$.

When the tilt again is zero, the cycle recommences. There is thus no equilibrium state as with proportional control. The oscillation of the gyro can be represented, therefore, by a triangular wave of amplitude Θ and sides sloping at $\dot{\theta}$ and $(\dot{\phi}-\dot{\theta})$.

The mean stick-off error is

$$\Theta/2 = \theta_{ST} \tag{9.6}$$

and the period of the oscillation (see Fig. 9.3) is

$$T = \frac{\Theta}{\dot{\theta}[\mathrm{I}-(\dot{\theta}/\dot{\phi})]} \tag{9.7}$$

For smooth control, $\dot{\phi}$ must not be too large. The stick-off is independent of $\dot{\phi}$ or $\dot{\theta}$, τ_1 or τ.

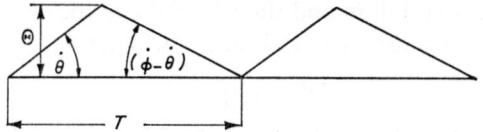

FIG. 9.3. The period of oscillation

In terms of torque, we have a disturbing torque $\tau = H\dot{\theta}$ and a control torque $\tau_1 = H\dot{\phi}$. Thus

$$\dot{\theta} = \tau/H$$
$$\dot{\phi} = \tau_1/H$$

So

$$T = \frac{\Theta}{(\tau/H)[\mathrm{I}-(\tau/\tau_1)]} \tag{9.8}$$

Bendix erection control. An interesting erecting device for a vertical gyroscope is used in the Bendix Gyro Flux-Gate compass, in which the stable platform is the gyroscope casing itself. In a track around the periphery of the casing, a steel ball is propelled by eddy currents generated by the spinning gyroscope. If the gyroscope tilts, the ball slows on the uphill part of its track and hastens on the downhill part. The delay and the speed of the ball's progress are such that the weight of the ball produces the requisite torque in the correct relation to the tilt to bring the spin axis into the vertical. So no electrical contacts or motors are required in this mechanism.

Acceleration errors

Although a magnetic element mounted on or stabilized by a controlled gyroscope may itself be unaffected by accelerations, the gyroscope will be influenced by them,

---

* With large displacements, proportional control degenerates into an 'on-off' system.

as the control pendulums themselves will seek the false vertical. A false controlling signal will be supplied to the torque motors and an otherwise correctly directed gyroscope will be deflected towards the false vertical, which a sufficiently prolonged acceleration will cause it to assume, the controlling torques determining the rate of precession. The amount of tilt depends on the precession rate and time, the first of which can be a function of the tilt error and governs the recovery time of the gyroscope after acceleration.

The control torques need be sufficient only to counter the random torques in the system, and in a good gyroscope these may be small. It will therefore take a considerable time for the gyroscope to re-erect after being tilted.

It is evident that, owing to the imperfections in the gyroscope and the consequent necessity for some control system, accelerations can adversely affect a gyro-stabilized magnetic element. As the gyroscope tilts towards the false vertical, the magnetic element also is tilted and consequently azimuth deviations will arise. It must, however, be made clear that, given careful design, the deviations will be far less than those in an unstabilized system.

Proportional control. The control rate is proportional to the difference between the inclination of the pendulum and the tilt of the gyro.

Let κ' be the control constant in rad/min/rad. Then

$$\dot{\theta} = \kappa'(\beta - \theta) \tag{9.9}$$

where β is the inclination of the false vertical to the true vertical.

Let $\beta - \theta = -\gamma$. Then

$$\dot{\theta} = \dot{\gamma}$$

So

$$\dot{\gamma} = -\kappa'\gamma$$

or

$$\frac{d\gamma}{\gamma} = -\kappa'\,dt$$

$$\log_e \gamma = -\kappa't + C$$

and

$$\gamma = Ce^{-\kappa't}$$

But $-\gamma = \beta - \theta$ and therefore

$$\theta = \beta + \gamma$$

Thus

$$\theta = \beta + Ce^{-\kappa't}$$

When $t = 0$, $\theta = 0$, and therefore

$$C = -\beta$$

Therefore

$$\theta = \beta(1 - e^{-\kappa't}) \tag{9.10}$$

and

$$\tan \delta = \frac{\beta(1 - e^{-\kappa't}) \tan D \cos \alpha}{1 - \beta(1 - e^{-\kappa't}) \tan D \sin \alpha} \tag{9.11}$$

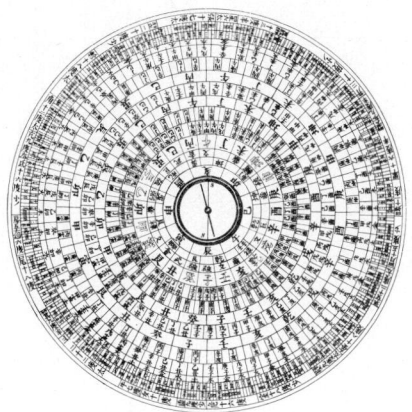

Plate 1.1. Examples of early Chinese compasses

Plate 1.2. Chinese junk's compass

Plate 1.3. Chinese compass and
sundial

Plate 1.4. The *Rosa Ventorum*, A.D. 1521,
from *The Rose of the Winds; the Origin and
Development of the Compass Card*, by
Sylvanus P. Thompson, D.Sc., F.R.S. (read
at the International History Congress, 1913)

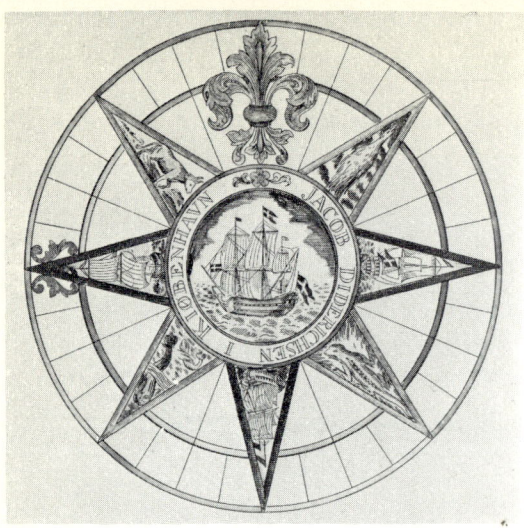

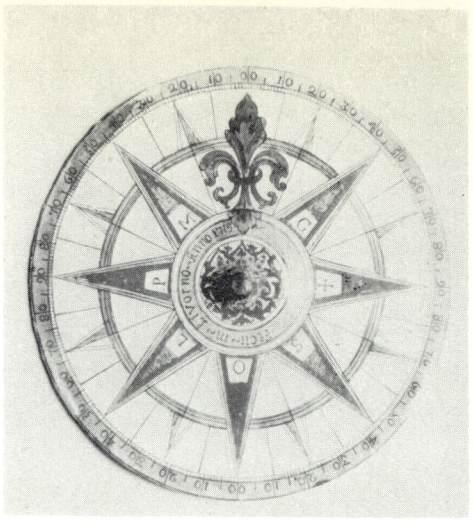

Plate 1.5. Danish compass card

Plate 1.6. Italian compass card
(Leghorn, 1719)

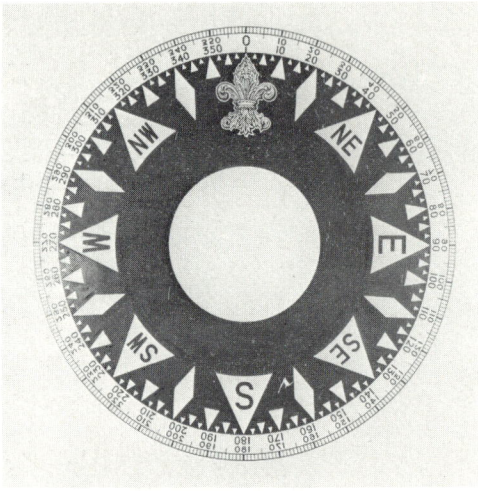

Plate 1.7. British compass card
(Merchant Navy)

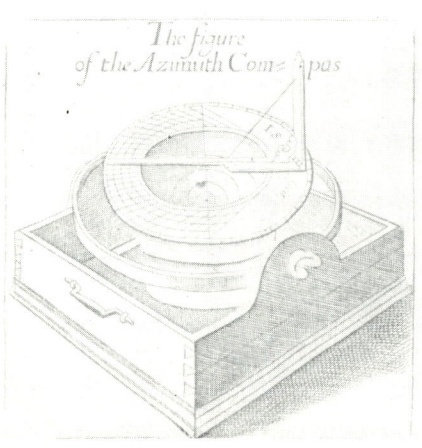

Plate 1.8. Azimuth compass of A.D. 1669
(*Practical Navigation*, John Seller, 1672)

Plate 1.9. Gavin Knight's compass of A.D. 1750
(*Phil. Trans. roy. Soc.* **XLVI,** 1750, Plate 1)

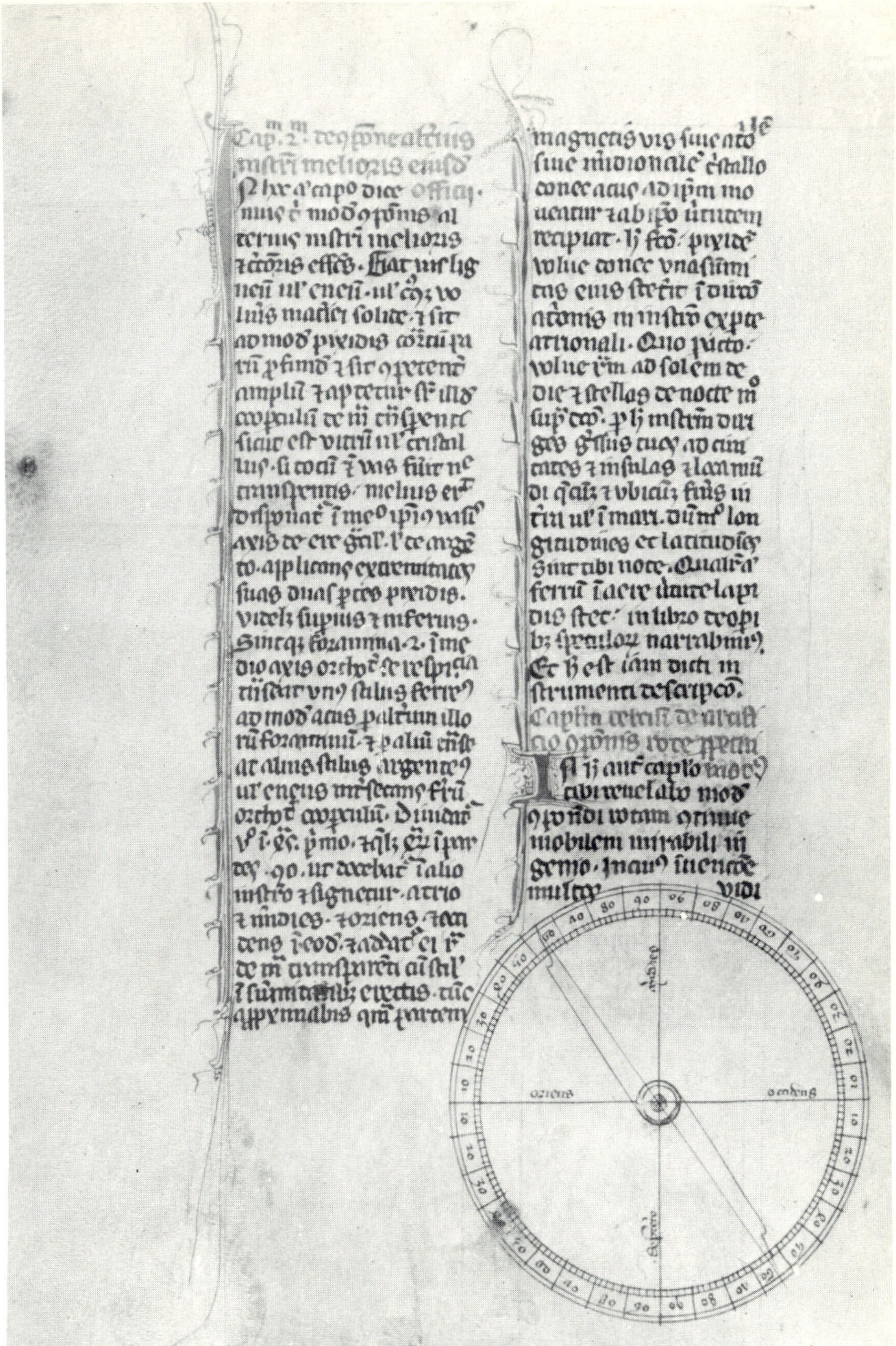

Plate 1.10. Part of the *Epistola de Magnete* of Petrus Peregrinus de Maricourt, A.D. 1269

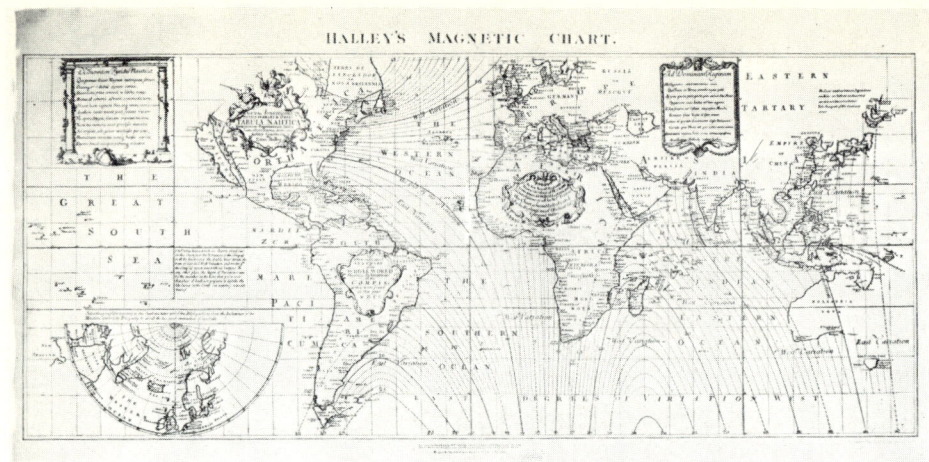

Plate 2.1. Halley's magnetic chart, A.D. 1701

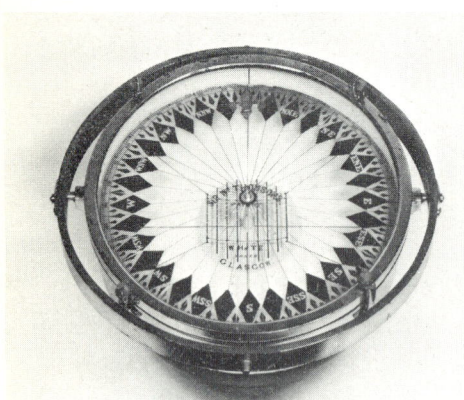

Plate 5.1. Kelvin compass

Plate 5.2. Trawler's compass, Admiralty Patt. 992

Plate 5.3. Sestrel-Moore yacht compass with
shadow pin and alternative mounting

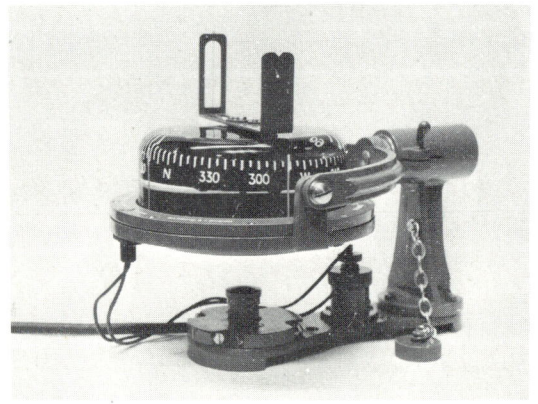

Plate 5.4. Sestrel-Moore yacht compass with
alidade

Plate 5.5. Surveyor's prismatic compasses

Plate 5.6. Medium landing compass

Plate 5.8. Early type of aircraft compass
Admiralty Patt. 259

Plate 5.7. Compass used
in Col. S. F. Cody's flying machine

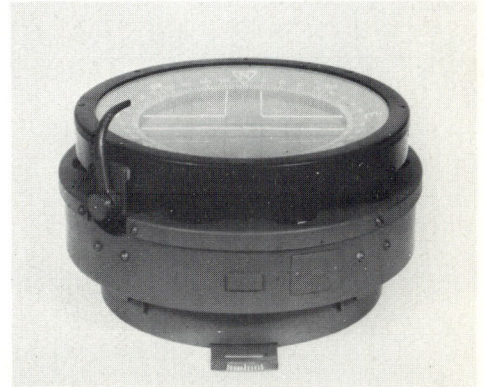

Plate 5.9. P-type compass for aircraft

Plate 5.10. E-type compass for aircraft

Plate 5.11. Type B-16 compass for aircraft (USA)

Plate 5.12. O-type compass for aircraft
with azimuth circle

Plate 5.13. Army prismatic compass

Plate 6.1. Variation magnetometer
(*Admiralty Compass Observatory*)

Plate 7.1. Datum compass
(*Admiralty Compass Observatory*)

Plate 7.2. Datum compass
(*Hilger & Watts Ltd*)

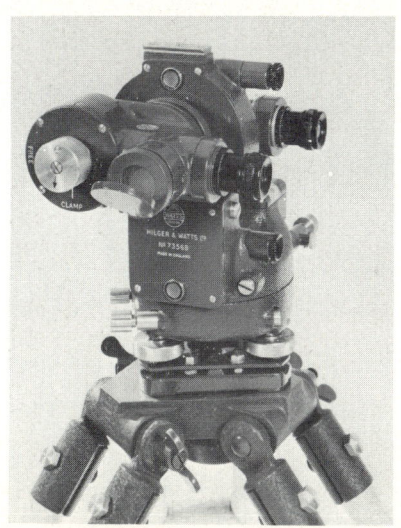

Plate 7.3. Absolute or 'Master' compass
(*Hilger & Watts Ltd*)

Plate 7.4. Inductor theodolite
(Crown copyright)

Plate 8.1. Telegon compass repeater
(Magnesyn compass repeater looks similar)

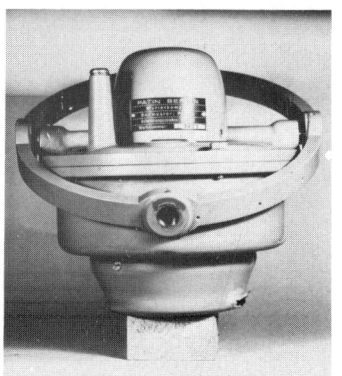

Plate 8.2. Patin compass
master unit—side view

Plate 8.3. Patin compass
master unit—view from above

Plate 8.4. Patin compass—repeater

Plate 8.5. Holmes tele-compass—
power unit, course indicator and compass unit

Plate 8.6. Holmes repeater compass—repeater, compass and amplifier

Plate 8.7. Admiralty transmitting magnetic compass,
Type 2—master unit

Plate 8.8. Azimuth repeater, Admiralty Patt. 1900
with azimuth circle Patt. 6703 and bracket Patt. 4783

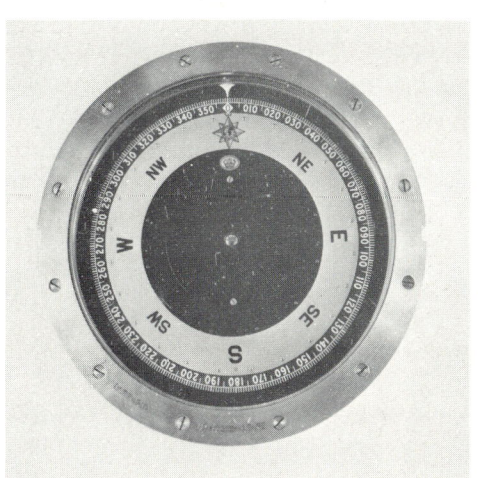

Plate 8.9. Steering repeater, Admiralty Patt. 1910.

Plate 8.10. Siemens-Halske transmitting magnetic
compass—compass unit

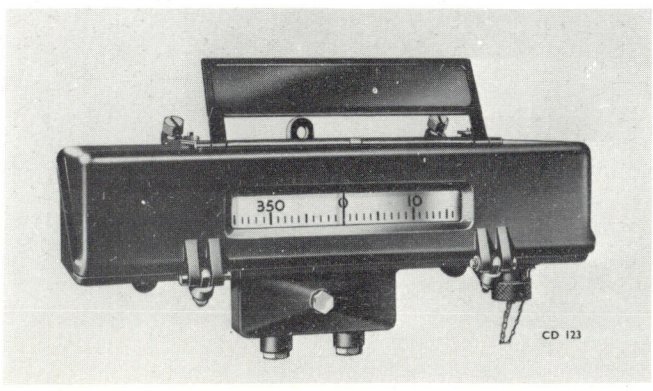

Plate 8.11. Steering repeater (tape type), Admiralty Patt. 3012

Plate 9.1. Air Ministry D.R. compass, Mk.I

Plate 9.2. Sperry Flux Valve detector unit

Plate 9.3. Navigator's repeater or master
indicator for gyro-magnetic compass
equipment

Plate 9.4. Gyroscope unit (pilot's repeater)
for gyro-magnetic compass equipment

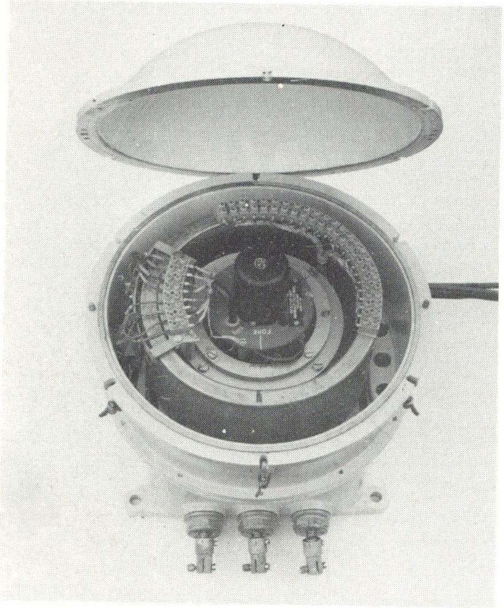

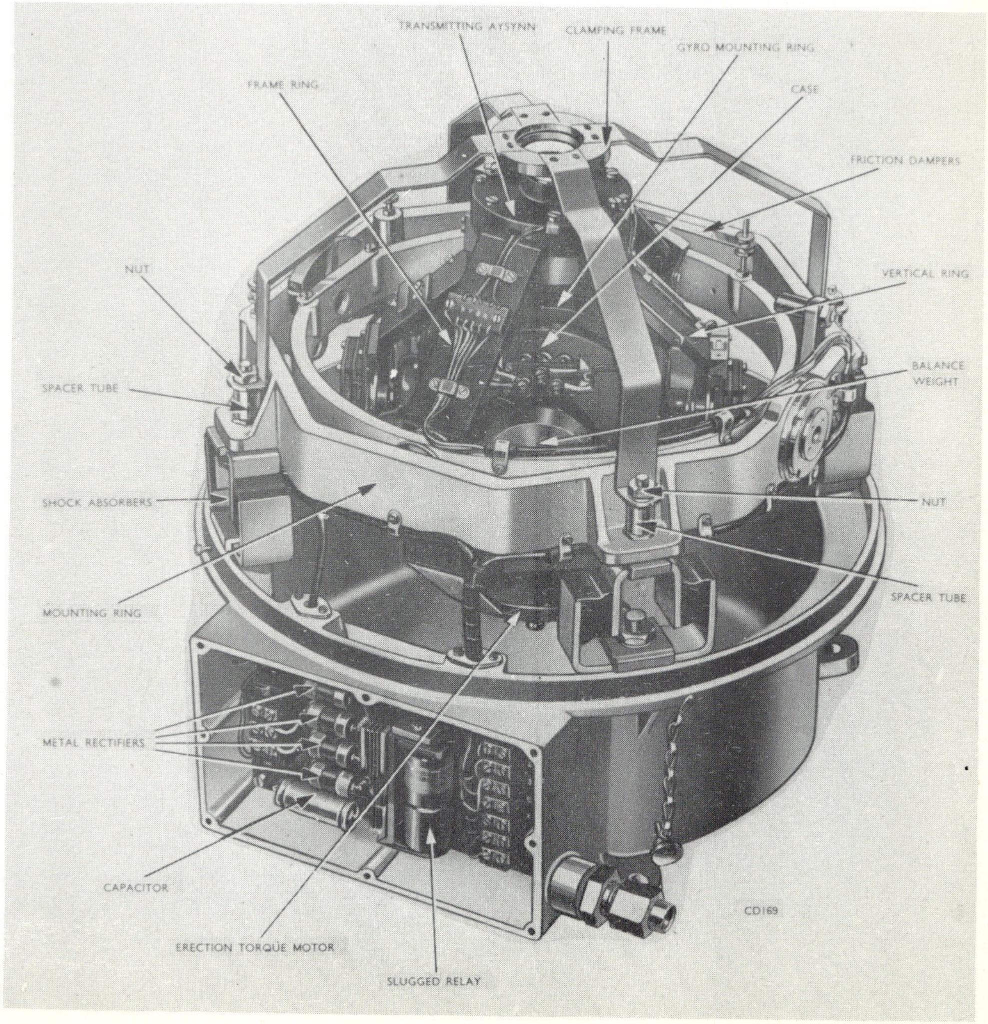

Plate 9.6. Admiralty gyro-magnetic compasses, Types 6 and 7—gyroscope unit

Plate 9.5. Admiralty gyro-magnetic compass, Type 7—detector unit with corrector coils and binnacle

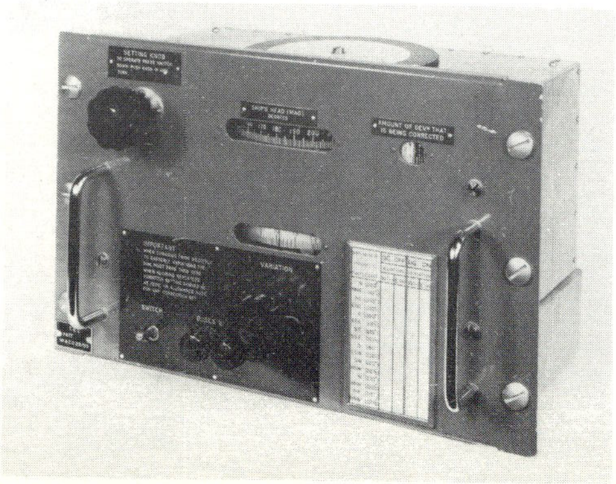

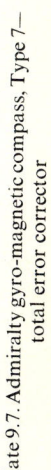

Plate 9.7. Admiralty gyro-magnetic compass, Type 7—total error corrector

Plate 9.8. Admiralty gyro-magnetic
compass, Type 6—console

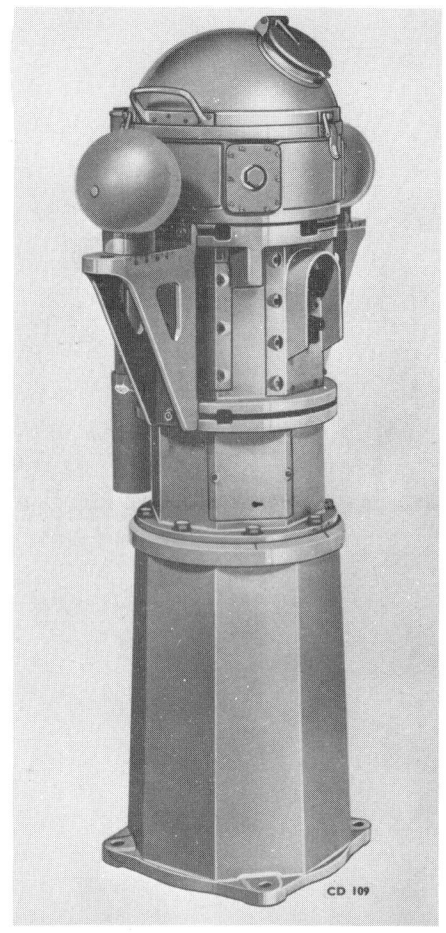

Plate 9.9. Admiralty gyro-magnetic compass,
Type 6—binnacle and master unit
(Type 5 is similar in external appearance)

Plate 10.1. Total-error corrector,
Admiralty transmitting magnetic compass

Plate 12.1. Turntable to test magnetic compasses

A compromise is obvious, in that $\dot\phi$ and κ' must be large enough to cope with random precessions, the effects of the Earth's rotation (see p. 240) and the results of gimbal friction, but not too great to permit the system to approach the false vertical rapidly. This, of course, means that a rapid return from a displaced attitude is not possible.

Acceleration 'cut-outs' are sometimes proposed to leave the gyroscope free during accelerations, but these have a questionable value.

On-off control. Let $\dot\phi$ be a fixed precessional control rate; then the tilt of the gyroscope's axle, after time t, is

$$\theta = \dot\phi t$$

The azimuthal deviation, δ, is given (see also p. 213) by

$$\tan\delta = \frac{\theta \tan D \cos \alpha}{1 - \theta \tan D \sin \alpha} \qquad (9.12a)$$

where α is the angle between north and axis of tilt, or

$$\tan\delta = \frac{\dot\phi t \tan D \cos \alpha}{1 - \dot\phi t \tan D \sin \alpha} \qquad (9.12b)$$

The false vertical is β from the true vertical where $\tan\beta$ is a/g and a is the acceleration on the system. The time taken for the gyroscope to be tilted into the false vertical, i.e. for θ to be equal to β, is given by

$$t = \frac{\tan^{-1}(a/g)}{\dot\phi} \qquad (9.13)$$

Oscillatory errors

If an oscillatory acceleration acts on the erection control system, e.g. the gravity control switches, and provided the precession rates in each direction are equal, the axis of the gyroscope will oscillate symmetrically about the true vertical, with a very much smaller amplitude than that of the disturbing oscillation. In Fig. 9.4(a) the oscillation of the gravity switch is represented by $\theta = A \sin pt$.

Assume the cycle of events to begin at o. The gyroscope will be tilted towards the 'false vertical' at a rate equal to $+\dot\phi$. If $\dot\phi \ll pA$, the oscillation will have passed its maximum displacement and will be returning towards zero before the gyroscope's tilt has caught up with it. At a, the gyroscope's tilt will equal the instantaneous oscillatory tilt and from there the precession will be reversed as the oscillation proceeds. The gyroscope will now tilt at a rate equal to $-\dot\phi$ until at b the tilt is again reversed and will proceed at $+\dot\phi$.

Now $t_a < T/2$ and so if $+\dot\phi = -\dot\phi$ in magnitude, $-\dot\phi$ cuts the zero line after time $2t_a$ and $2t_a < T$. Hence point b must be on the negative side of the zero line. Thus point c is less positive than point a. Stability is reached ultimately when successive points such as c and e are the same distance from the zero line and form similar starting points for successive excursions of $\dot\phi$. By symmetry, the point d must be on the curve $A \sin pt$ and equidistant from c and e, and so the gyroscope's axle settles down to an alternating precession represented by c, d, e, and nearly 90° out of phase with $A \sin pt$.

217

H

The greatest error is that in the first half-cycle, at a. If this is θ_a

$$\theta_a = \dot{\phi} t_a$$
$$\theta_a = A \sin p[(T/2) - t_a]$$

Provided the period T and the control rate $\dot{\phi}$ are such that $(T/2) - t_a$ is small, we may write

$$\theta_a = Ap\left(\frac{T}{2} - t_a\right)$$

but $t_a = \theta_a/\dot{\phi}$.

Therefore

$$\theta_a = Ap\left(\frac{T}{2} - \frac{\theta_a}{\dot{\phi}}\right)$$
$$= A\pi\dot{\phi}/(\dot{\phi} + Ap)$$
$$= A\pi T\dot{\phi}/(T\dot{\phi} + 2\pi A)$$

If $(T/2) - t_a$ is not small, reference should be made to the graphical solution on p. 234.

At the stable condition, the alternating precession is symmetrical about zero with amplitude of $df/2$.

$$df = T\dot{\phi}/2$$

therefore

$$\theta_a = -\theta_f = T\dot{\phi}/4$$

If, however, the precession rates are not equal, owing for instance to a small drift of the gyroscope, a cumulative error in tilt will arise and the axis will stick off from the vertical, as demonstrated in Fig. 9.4(b), causing a corresponding compass deviation in azimuth.

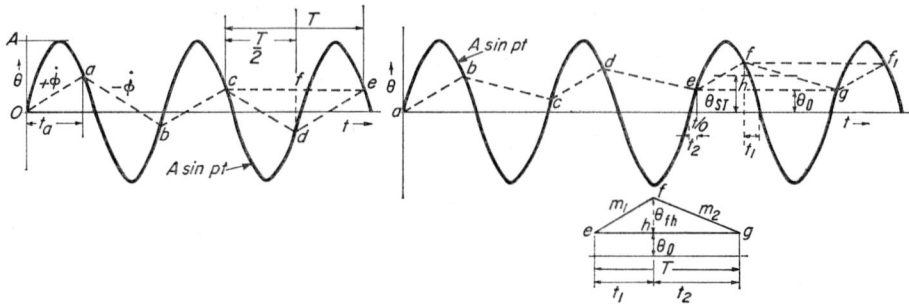

FIG. 9.4. Stick-off errors

Again, the oscillation is represented by $A \sin pt$. The control system receives a false control signal and precession takes place for instance, along ab, which can be represented by $y = m_1 x + c$ $(m_1 = \dot{\phi}_1)$. At b, the tilt of the gyroscope and the deflection of the gravity switch are equal and, as the oscillation proceeds, the control precession reverses and may now be represented by bc, or $y = -m_2 x + d$ $(m_2 = \dot{\phi}_2)$.

Now if m_2 is less than m_1, c may be displaced positively from the zero line, $y = 0$, and e may be even further displaced. This progression continues until two successive points such as e and g are equally displaced from zero. From then on, the cycle repeats itself. The parameters of the sine wave and the values of m_1 and m_2 determine when this equilibrium state will occur.

It will be seen that not only does the gyroscope axle stick off by an amount θ_{ST}, but in addition it undergoes a triangular oscillation efg whose period is T, that of the sine curve $A \sin pt$. Now $m_1 = \dot{\phi} + \dot{\theta}$ and $m_2 = \dot{\phi} - \dot{\theta}$, where θ is the drift rate of the gyroscope and $\dot{\phi}$ is the control rate. The magnitude of θ_b in the first half-cycle is determined in the same manner as in the previous example, but m_1 replaces $\dot{\phi}$ in the expression for θ_b.

Now in the stable condition, ehg is parallel to the line $y = 0$ and e, f and g all lie on the curve $A \sin pt$. Since $eh + hg = T$

$$eh = \frac{m_2 T}{m_1 + m_2}$$

and the error superimposed on eg is

$$\theta_{fh} = \frac{m_1 m_2}{m_1 + m_2} \cdot T$$

$$= T(\dot{\phi} + \dot{\theta})(\dot{\phi} - \dot{\theta})/2\dot{\phi}$$

$$= \frac{T}{2}\dot{\phi}\left(1 - \frac{1}{k^2}\right) \quad \text{where} \quad \frac{1}{k} = \frac{\dot{\theta}}{\dot{\phi}}$$

The amplitude of the error about its mean level is

$$\frac{T}{4}\dot{\phi}\left(1 - \frac{1}{k^2}\right)$$

and when $\dot{\theta} = 0$, this becomes $T\dot{\phi}/4$ which corresponds to the result in the previous example.

The stick-off error is given by

$$\theta_{ST} = \theta_h + \tfrac{1}{2}\theta_{fh} = \theta_e + \tfrac{1}{2}\theta_{fh} \quad (\theta_h = \theta_e = \theta_g)$$

$$\theta_f = A \sin pt_1$$

$$\theta_e = A \sin pt_2$$

$$eh = \frac{T}{2} - (t_1 + t_2)$$

If, as before, T and $\dot{\phi}$ are such that t_1 and t_2 are small compared with T, write

$$t_1 = \theta_f / Ap$$

$$t_2 = \theta_e / Ap$$

Then, since $eh = \dfrac{m_2 T}{m_1 + m_2}$, we have

$$\frac{m_2 T}{m_1 + m_2} = \frac{T}{2} - \frac{\text{I}}{Ap}(\theta_f + \theta_e) = \frac{T}{2} - \frac{\text{I}}{Ap}(2\theta_f - \theta_{fh})$$

$$\theta_{fh} = \frac{T}{2}\dot{\phi}\left(\text{I} - \frac{\text{I}}{k^2}\right)$$

Therefore

$$\frac{m_2 T}{m_1 + m_2} - \frac{T}{2} - \frac{T^2}{4\pi A}\dot{\phi}\left(\text{I} - \frac{\text{I}}{k^2}\right) = -\frac{T}{2\pi A}.2\theta_f$$

and

$$\frac{\theta_f}{\pi A} = \frac{\text{I}}{2k} + \frac{T}{4\pi A}\dot{\phi}\left(\text{I} - \frac{\text{I}}{k^2}\right)$$

i.e.

$$\theta_f = \frac{A\pi}{2k} + \frac{T}{4}\dot{\phi}\left(\text{I} - \frac{\text{I}}{k^2}\right)$$

and

$$\theta_e = \frac{A\pi}{2k} - \frac{T}{4}\dot{\phi}\left(\text{I} - \frac{\text{I}}{k^2}\right)$$

Therefore the mean level of the error or stick-off is given by $\theta_{ST} = A\pi/2k$. When $\dot{\theta} = 0$, $\theta_{ST} = 0$, the stable condition in Fig. 9.4(a). To summarize:

Even if the gyroscope has frictionless bearings and is perfectly balanced, it will depart from the vertical owing to the rotation of the Earth. Small imperfections in its construction also cause random departures from the vertical and sometimes a continuous slow drift. To correct these errors, a gravity control, such as a pendulum-operated switch, is applied to the gyroscope. Any tilt of the gyroscope brings the switch into play and correcting torques are applied to restore the gyroscope's axis to the true vertical. The control may be proportional to tilt or on-off.

Although the attitude of the gyroscope itself is not seriously affected by accelerations, the pendulum is very responsive and, when subject to accelerations, applies unwanted control torques which disturb the gyro. Thus a device necessary to correct small discrepancies and errors becomes a source of error during accelerations, whether sustained or oscillatory.

The errors that arise are as follows:

(A) Error due to persistent wander, which is a small residual tilt or stick-off.

    (i) Proportional control $\qquad\qquad\qquad \theta_{ST} = \tau/\kappa = \dot{\theta}/\kappa' \qquad\qquad$ (9.14)

    (ii) On-off control $\qquad\qquad\qquad\qquad \theta_{ST} = \Theta/2 \qquad\qquad\qquad$ (9.15)

where τ is the disturbing torque, κ the control constant (torque/rad), κ' the control constant (rad/min/rad) and Θ the dead sector.

(B) Error that increases with time until the false vertical is reached, owing to a sustained acceleration.

(i) Proportional control $\qquad \theta = \beta(1 - e^{-\kappa' t})$ (9.16)

(ii) On-off control $\qquad \theta = \dot{\phi} t$ (9.17)

where θ is the angle of tilt, κ' the control or monitoring constant (rad/min/rad), t the time, $\dot{\phi}$ the control rate in rad/min and β the angle of false vertical.

(c) Stick-off and oscillatory errors due to an oscillation of the pendulum, such as might be caused by pitch or roll. In the interests of simplicity, an on-off control only is considered.

(a) No gyro drift, equal control rates in either direction:

(i) Cumulative or stick-off error $\qquad \theta_{ST} = 0$ (9.18)

(ii) Oscillatory error $\qquad \theta = +\dfrac{T\dot{\phi}}{4}$ (9.19)

(b) Gyro drifting, or control rates unequal:

(i) Cumulative or stick-off error $\qquad \theta_{ST} = \dfrac{A\pi}{2k}$ (9.20)

(ii) Oscillatory error $\qquad \theta = \pm\dfrac{T\dot{\phi}}{4}\left(1 - \dfrac{1}{k^2}\right)$ (9.21)

where $\dot{\phi}$ is the control rate in rad/min and $k = \dot{\phi}/\dot{\theta}$.

The magnitude of the compass deviation δ owing to acceleration and stick-off errors is given by the following expressions.

If the gyro tilt is θ (or θ_{ST} as the case may be)

$$\tan \delta = \frac{\frac{1}{2}\sin 2\alpha(1 - \cos \theta) + \tan D \sin \theta \cos \alpha}{1 - \frac{1}{2}(1 - \cos 2\alpha)(1 - \cos \theta) - \tan D \sin \theta \sin \alpha} \qquad (9.22)$$

or, for small angles of tilt [equations (3.5) and 3.8)],

$$\tan \delta = \frac{\theta \tan D \cos \alpha}{1 - \theta \tan D \sin \alpha}$$

It may be wondered why, with so many errors introduced by gyro-stabilization, its use is considered to be worth while. It is true to say, however, that for short-term and oscillatory accelerations, the deviations are all very much less than would arise with a simple magnetic compass or inductor system.

The greatest objection to the vertical gyroscope as a stabilizing means is its bulk and weight. It is not easy to accommodate a vertical gyroscope in the wing tip of a modern aircraft, for instance, and the use of remote stabilization necessitates a fairly elaborate servo-system.

A good example of this type of gyro-magnetic compass is the Bendix Pioneer Gyro Flux-Gate Compass.

Another method of stabilization is the use of a horizontal or azimuth gyroscope as a directional instrument, providing transmission through synchro or other means as may be required.

The horizontal gyroscope as a directional indicator

The simplest form of horizontal gyroscope is shown in Fig. 9.5. It will be recognized as similar to Fig. 9.2 turned through 90°. The spin axis is horizontal and the gimbal axes are horizontal and vertical, all being orthogonally disposed.

As the ideal vertical gyroscope tends always to keep its axle pointing vertically, so the horizontal gyroscope ideally keeps its axle in a fixed direction in the horizontal plane, provided the effect of the Earth's rotation is corrected.

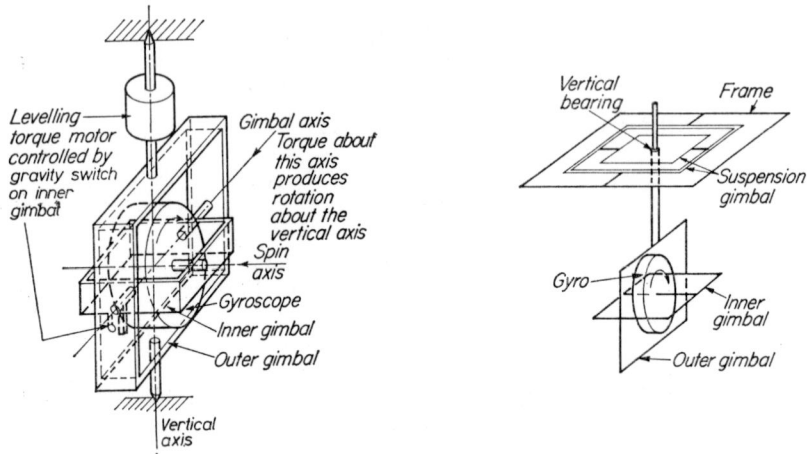

FIG. 9.5. Horizontal gyroscope   FIG. 9.6. Addition of a second gimbal system

Owing to the same imperfections that spoil a vertical gyroscope, a horizontal or directional gyroscope tends to drift in azimuth and is therefore useful as a directional instrument only for a short time.

As with the vertical gyroscope, a gravity-controlled switch may be used to control a torque motor acting about the vertical axis; and thus to restore a tilted axle to the horizontal in accordance with the rules for precession already stated.

A steady rate of tilt, or oscillations of the gravity-controlled device, or a sustained acceleration will tend to deflect the axle from its correct horizontal attitude. These errors are similar to those derived in equations (9.14) to (9.21), but the effect of tilt in this case is not as detrimental as in the vertically stabilized compasses since a tilt of the gyroscope axle could introduce only small errors in azimuth. These so-called geometric errors are given by the expression

$$\tan \epsilon = \frac{\tan^2(\nu/2) \sin 2\psi}{1 + \tan^2(\nu/2) \cos 2\psi} \tag{9.23}$$

where ϵ is the geometric error, ν is the tilt of the axle, and ψ is the angle between the gyroscope axle and the axis of tilt. In the usual form of gimballing, $\psi = 90°$, so $\epsilon = 0$.

The only significant error introduced by tilting of the axle is that due to friction about the inner gimbal axis, which, in turn, will cause precession about the vertical axis (i.e. in azimuth). It must be remembered that a large tilt will reduce the angular momentum of the gyroscope measured as a horizontal vector.

The levelling control can be operated relative to the outer gimbal ring instead of to gravity, and it is probably immaterial which method is used when angles of tilt are small. Gimbal-ring levelling may introduce more frictional torques than gravity-levelling when the attitude of the craft or vehicle is frequently changed. During accelerations, particularly those of an oscillatory nature, gravity-levelling control may well be a disadvantage.

In Fig. 9.5 the reference system is effectively the inner gimbal and the reference direction may be defined equally as the inner gimbal axis or the gyroscope's spin axis. As long as the gimbal axis is horizontal, rotation about the vertical axis represents true azimuth changes. The vertical axis, being supported in the craft's framework, gives what is commonly known as a 'deck plane' transmission, since the inner gimbal axis is restrained in a plane parallel to the deck. When the craft or vehicle tilts, rolls or pitches, the deck plane is no longer horizontal, and neither is the gimbal axis.

Geometric errors now arise between the apparent rotation of the craft in the deck plane, i.e. about the gyroscope's 'vertical' axis, and in azimuth, i.e. about the true vertical (equation 9.23). It is therefore preferred in some instances to suspend the gyroscope and its gimbals in a pendulous system, and a Hooke's joint or second gimbal system may be added (Fig. 9.6). The transmitter on the vertical axis now indicates true azimuth or 'horizontal' angles. When the pendulous system is accelerated, geometric errors appear, as the system swings towards the false vertical. In this kind of construction the gimbal ring levelling control is, in fact, gravity control.

Any form of transmitter attached to the vertical axis imposes a small frictional torque about that axis and a consequent tilt of the spin axis.

Azimuth control of the gyroscope

The levelled gyroscope will tend to drift in azimuth owing to Earth rotation and to random frictional torques and out-of-balance forces. A magnetic compass may be used in a number of ways for imposing a control on the gyroscope, which may then become a smoothing device to even out the transient perturbations of the magnetic compass.

In the simplest arrangement, the bowl or container of the compass may be mounted directly on the outer gimbal ring of the gyroscope, as illustrated in Fig. 9.7. A pick-up device on the compass bowl—various kinds of which are described in Chapter 3—detects misalignment between a fiducial or index mark on the bowl (which is in the plane of the gimbal ring) and the north point of the compass. The pick-up device will provide a signal, amplified if necessary, to impose a torque about the inner gimbal (horizontal) axis. Precession about the vertical axis then takes place in the correct sense so as to effect (or restore) alignment between the index mark and the north point. The outer gimbal ring is thus aligned correctly with respect to magnetic north, and any subsequent drift of the gyroscope's axle, in azimuth, will be discerned by the pick-up device and corrected. By the same means, perturbations of the magnetic compass are applied as corrections, though erroneous, to the gyroscope. However, the rate of correction should be large compared with any drift rate the gyroscope may exhibit, but small compared with the rate of the perturbations of the compass.

Having ensured the correct orientation of the gyroscope (and transmission, if any) with relation to compass north, it now remains to show how the acceleration and transient errors in the magnetic compass may be smoothed or averaged. In the usual form of magnetic compass, the transient and quasi-periodic disturbances are usually associated with a comparatively short period of oscillation.

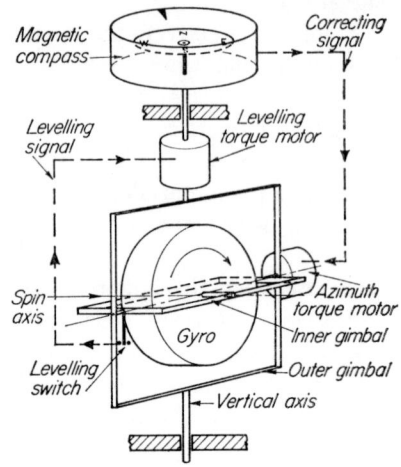

Fig. 9.7. Azimuth control by magnetic compass

For instance, a pendulous inductor system may have a period of from $\frac{1}{2}$ to 2 seconds, and a well-designed pivoted compass may have a period of from 20 to 30 seconds. It is during these times of short-lived perturbations causing unsteadiness in a magnetic compass, that the inherent stability of a gyroscope is exploited.

A properly designed gyroscope will not drift to any great extent during the short time that the compass is disturbed and, if the monitoring rate is correctly chosen, the effect of false monitoring is unlikely to be excessive, so that reliable azimuth information may still be provided. Any small error that may have accrued is removed by the magnetic compass when it settles down. The average direction of magnetic north is, in fact, continually communicated to the gyroscope during oscillatory disturbances of the compass.

The action of monitoring is illustrated in Fig. 9.8, where the compass is subjected to a transient or damped disturbance. The effect of oscillatory disturbances of the magnetic compass on a gyro-magnetic system, and the behaviour of such a system in a moving craft are discussed on pp. 229–40.

During turns, when northerly turning error (p. 237 *et seq.*) occurs, the stability of the gyroscope is relied on to maintain an accurate heading reference until the turn is completed.

Thus a levelled horizontal gyroscope can, by its long-term stability, provide a heading reference while the magnetic compass is seriously disturbed and, in the event of the gyroscope drifting or wandering, the north-pointing property of the system is maintained (or restored) during the undisturbed condition of the magnetic compass.

It is important that the gyroscope and its gimbal system should have no magnetic effect on the compass. A further refinement, and one that removes any magnetic interference due to the construction of the gyroscope, is to transmit the heading information electrically from the gyroscope to the magnetic compass, thus providing the appropriate means of comparison between the directions of the two instruments, and then to signal electrically the appropriate correction to the gyroscope. Alternatively, the information from the magnetic compass may be

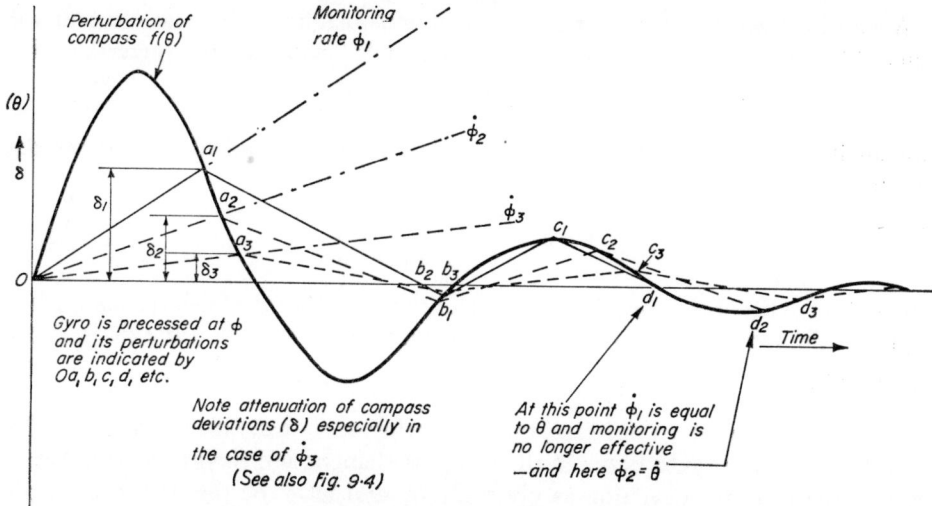

FIG. 9.8. Monitoring action

transmitted to the gyroscope. For example, a two-axis saturable inductor system may be arranged to supply signals to a resolver, mounted directly on the gyroscope's vertical axis, and constituting a means of comparison. The output signal from the resolver rotor provides the correction to the gyroscope.

A combination of these two systems is one in which transmissions from the magnetic compass and from the gyroscope are compared in a separate unit, such as a resolver coupled to a servo-system or to a coincidence synchro-link.

Stick-off or drift errors

In a state of rest or constant velocity of the craft, a stick-off error can arise owing to the Earth's rotation and to random forces causing azimuthal precession. This error is given by expressions similar to equations (9.5) and (9.6):

$$\theta_{ST} = \tau/\kappa = \theta/\kappa'$$

$$\theta_{ST} = \frac{\Theta}{2}$$

and as, instead of tilt of the gyroscope's axis, we are now considering deflection, for θ we write δ'.

For the azimuthal deviation of the compass, the two expressions above for θ_{ST} need, in the case of vertical stabilization, to be used in conjunction with equation (9.22), whereas for the horizontal or azimuth gyroscope, the values of θ_{ST} given by the same equations are themselves azimuthal deviations. It will be realized that, owing to the comparatively large value of magnetic dip in the United Kingdom, a vertical gyroscope and a horizontal gyroscope having comparable drift rates will produce very different compass deviations, the vertical instrument showing a deviation about $2\frac{1}{2}$ times that of the horizontal instrument, an argument in favour of the azimuth gyroscope form of stabilization.

Deviations due to acceleration

A sustained acceleration acting on a pendulous magnetic compass element will cause it to tilt into the false vertical making an angle β with the true vertical, so that

$$\tan \beta = a/g$$

For small angles of tilt, the azimuthal deviation that will be applied to the gyroscope will be given by

$$\tan \delta = \frac{\beta \tan D \cos \alpha}{1 - \beta \tan D \sin \alpha}$$

where β is the tilt of the compass element (or inclination of the false vertical), D the angle of dip and α the angle between magnetic north and the tilt axis [see also equation (3.8)].

Proportional control. In this case, the monitoring or control rate is proportional to the difference in azimuth between the compass element and the gyroscope. Again, let δ be the compass deviation as given above, and let δ' be the deviation of the gyroscope (i.e. the deviation of the system). Then

$$\delta' = \kappa'(\delta - \delta') \tag{9.24}$$

This corresponds to equation (9.9), from which follows

$$\delta' = \delta(1 - e^{-\kappa't}) \tag{9.25}$$

so

$$\delta' = \left(\tan^{-1} \frac{\beta \tan D \cos \alpha}{1 - \beta \tan D \sin \alpha}\right)(1 - e^{-\kappa't}) \tag{9.26}$$

Compare this result with equation (9.11).

On-off control. The deviation of the system, i.e. the actual deflection of the gyroscope axis, will depend on the control or monitoring signal. If this is ϕ then the deviation is given by

$$\delta' = \phi t \tag{9.27}$$

where t is the duration of the acceleration. Compare this with equation (9.12b).

The whole of the compass deviation δ is introduced after a time t_1:

$$t_1 = \tan^{-1}\left(\frac{\beta \tan D \cos \alpha}{1 - \beta \tan D \sin \alpha}\right)\Big/\dot{\phi} \qquad (9.28)$$

Here again, for small angles the horizontal gyroscope leads to a more accurate result.

Oscillatory errors or deviations

If the compass element is oscillating and hence transmitting an incorrect control signal of the form

$$\delta = A' \sin pt$$

the arguments of p. 217 *et seq.* may be applied to show that the directional error or deviation of the gyroscope will be composed of stick-off and oscillatory components. These will be:

$$\delta'_{ST} = A'\pi/2k \quad \text{(stick-off)} \qquad (9.29)$$

and

$$\delta' = \pm\frac{T}{4}\dot{\phi}\left(1 - \frac{1}{k^2}\right) \quad \text{(oscillatory)} \qquad (9.30)$$

where k, $\dot{\phi}$ and T have the same significance as previously.

A', however, is the amplitude of the compass deviation and depends largely on the dynamics of the compass system. For example, a heavily damped liquid compass will show a very different value of A' from that shown by an undamped pendulous inductor. For the best results, small values of A' and T are desirable. To avoid instability due to the use of a large value of $\dot{\phi}$, k needs to be large, hence δ' must be small, i.e. a gyroscope with very low random rates.

To summarize:

(A) Deviation due to persistent wander or stick-off:

    (i) Proportional control $\qquad\qquad \delta'_{ST} = \dot{\delta}'/\kappa' \qquad (9.31)$

    (ii) On-off control $\qquad\qquad\quad \delta'_{ST} = \Theta/2 \qquad (9.32)$

(B) Deviation that increases with time owing to persistent acceleration:

    (i) Proportional control $\qquad \delta' = \delta(1 - e^{-\kappa' t}) \qquad (9.33)$

        where δ is the ultimate deviation of the compass due to acceleration.

    (ii) On-off control $\qquad\qquad\qquad \delta' = \dot{\phi}t \qquad (9.34)$

(c) Stick-off and oscillatory deviations due to oscillations of the compass system, such as might be caused by pitch and roll. For the sake of simplicity on-off control only is considered.

    (*a*) No gyro drift, equal monitoring in both directions:

        (i) Cumulative or stick-off deviation $\quad \delta_{ST} = 0 \qquad (9.35)$

        (ii) Oscillatory deviation $\qquad\qquad \delta' = \pm T\dot{\phi}/4 \qquad (9.36)$

(b) Gyro drifting, or monitoring rates unequal:

(i) Cumulative or stick-off deviation $\qquad \delta'_{ST} = A'\pi/2k$ $\qquad$ (9.37)

(ii) Oscillatory deviation $\qquad \delta' = \pm\dfrac{T}{4}\dot{\phi}\left(1-\dfrac{1}{k^2}\right)$ $\qquad$ (9.38)

These are all azimuthal deviations, whereas the corresponding equations (9.14) to (9.21) are errors in tilt and need multiplication by the correct expression in terms of $\tan D$ and α to convert them to azimuthal deviations.

COMPARISON OF VERTICAL AND HORIZONTAL GYROSCOPES AS STABILIZERS OF A MAGNETIC COMPASS

In the simple case of a slowly wandering gyroscope, the advantage is with the horizontal instrument since, given the same drift and control rates, the error in the direction of the axis of a vertical gyroscope has a greater effect on the compass deviation than in the horizontal instrument, owing to the effect of magnetic dip.

If in equation (9.14) and (9.31) $\delta' = \theta$, the azimuthal deviations are respectively

$$\tan^{-1}\frac{\dfrac{\theta}{\kappa'}\tan D \cos \alpha}{1-\dfrac{\theta}{\kappa'}\tan D \sin \alpha} \quad \text{and} \quad \frac{\theta}{\kappa'}$$

or, when the deviations are small and when $\alpha = 0$, $(\theta/\kappa')\tan D$ and θ/κ'. Similarly for equations (9.15) and (9.32).

The following expressions are for sustained accelerations. In the one instance, the vertical gyroscope is being forced towards the false vertical; and in the other instance, the pendulous magnetic element controlling the horizontal gyroscope is in that attitude.

With proportional control, equation (9.16) gives the azimuthal deviation as

$$\tan^{-1}\frac{\beta(1-e^{-\kappa't})\tan D \cos \alpha}{1-\beta(1-e^{-\kappa't})\tan D \sin \alpha}$$

and equation (9.33) gives the azimuthal deviation as $\delta(1-e^{-\kappa't})$. But

$$\delta = \tan^{-1}\frac{\beta \tan D \cos \alpha}{1-\beta \tan D \sin \alpha}$$

Thus when β is small, there is little to choose between the systems.

For example, if $\alpha = 0$, the azimuthal deviation in both cases is

$$\beta(1-e^{-\kappa't})\tan D$$

With on-off control, equation (9.17) for the vertical gyroscope gives an azimuthal deviation of

$$\frac{\dot{\phi}t \times \tan D \cos \alpha}{1-\dot{\phi}t \times \tan D \sin \alpha}$$

which reaches a maximum value of $\beta \tan D$.

Equation (9.34) for the horizontal gyroscope gives an azimuthal deviation of $\dot{\phi}t$ which also reaches a maximum of $\beta \tan D$. (Again a small value of β and $\alpha = 0$ are assumed.)

The stick-off or steady deviation due to oscillatory accelerations is given by equations (9.20) and (9.37) for the vertical and horizontal gyroscopes respectively.

In the case of equation (9.20), if A is the tilt amplitude of the pendulum, the azimuthal deviation of the system is

$$\delta' = \tan^{-1}\left[\frac{(A\pi/2k) \tan D \cos \alpha}{1 - (A\pi/2k) \tan D \sin \alpha}\right]$$

In equation (9.37), A' is the amplitude of the deviation, and if the magnetic detector is oscillating in tilt with amplitude A, as equation (9.20),

$$A' = \tan^{-1}\left(\frac{A \tan D \cos \alpha}{1 - A \tan D \sin \alpha}\right)$$

Thus:

$$\delta' = \tan^{-1}\left(\frac{A \tan D \cos \alpha}{1 - A \tan D \sin \alpha}\right) \times \pi/2k$$

Again, for small values of A and making $\alpha = 0$,

$$\delta' = \frac{A\pi}{2k} \tan D \quad \text{(vertical gyroscope)}$$

$$\delta' = \frac{A\pi}{2k} \tan D \quad \text{(horizontal gyroscope)}$$

Ideally k should be large and A small; thus, provided the tilt angle involved is small, there is practically no difference in the two systems under conditions of acceleration.

For static conditions and when large tilt angles are involved, the horizontal gyroscope has advantages over the vertical.

THE GYRO-MAGNETIC COMPASS IN A MOVING CRAFT

Deviations due to acceleration

The deviations of an unstabilized magnetic compass when acted upon by accelerations have been considered in equations (3.18) to (3.38). These deviations, when present in a gyro-magnetic compass in which an azimuth gyroscope is monitored by a magnetic compass, give rise to deviations in the output of the system owing to false monitoring, i.e. the attempt of the monitoring system to catch up with a disturbed or oscillating compass magnet system.

The acceleration producing the compass deviation may be due to change of speed of the craft, rolling, pitching or steaming in a circular path; and the magnitude of the deviation due to false monitoring depends, among other things, on the duration of the acceleration, its magnitude and the monitoring rate. Some aspects of this have been examined under *Oscillatory errors* on page 217.

Equation (9.27) states that

$$\delta' = \dot{\phi}t$$

where t is the duration of the acceleration (or magnetic compass deviation), $\dot{\phi}$ is the monitoring rate and δ' the ultimate deviation of the system. When the acceleration ceases, δ' will disappear at the monitoring rate. It is assumed that the gyroscope is not wandering.

Once again, in the interests of simplicity, only on-off control is discussed, which includes systems in which the full monitoring rate is achieved with a very small error signal.

When a vessel is rolling or pitching, the magnetic compass deviation is given (equation 3.24) by

$$\delta = -\kappa . \frac{\dfrac{4\pi^2}{gT^2}.hA \tan D \cos \alpha \sin \dfrac{2\pi t}{T}}{1 + \dfrac{4\pi^2}{gT^2}.hA \tan D \sin \alpha \sin \dfrac{2\pi t}{T}}$$

where $\alpha = \zeta$ for a rolling vessel and $\alpha = (\zeta - \pi/2)$ for a pitching vessel, and κ is the appropriate attenuating term.

This equation may conveniently be written in the form

$$\delta = \frac{-\kappa . a \cos \alpha \sin \dfrac{2\pi t}{T}}{1 + a \sin \alpha \sin \dfrac{2\pi t}{T}}$$

where $a = (4\pi^2/gT^2)\, hA \tan D$.

In order to demonstrate the operation of monitoring, a much simplified and hypothetical set of initial conditions is assumed, which in practice would be regarded as unrealistic, namely, that an oscillation of the form $\delta = A \sin pt$ is initiated suddenly at $t = 0$. The deviation due to false monitoring is thus likely to be greater than that normally encountered.

During the first half-cycle of the oscillatory disturbance, from $t = 0$ to $t = T/2$, the magnitude of δ' is given by

$$\left.\begin{aligned}\delta' &= -\kappa . \frac{a \cos \alpha \sin (2\pi t/T)}{1 + a \sin \alpha \sin (2\pi t/T)}\\[2mm]\delta' &= -\dot{\phi}t\end{aligned}\right\} \tag{9.39}$$

As the oscillations continue after the first half-cycle the sequence of events follows the argument on p. 217, *Oscillatory errors*, and equation (9.36) on page 227. After a lapse of time depending upon the parameters of the oscillation and the monitoring rate, the system settles down with an oscillatory deviation of amplitude $T\dot{\phi}/4$ and period T.

There is a limiting condition to equation (9.9) when

$$\delta' = \frac{T\dot{\phi}}{4} \tag{9.40}$$

The deviation due to false monitoring is now already equal to the peak value of the oscillatory deviation in the first half-cycle. At this point and for smaller amplitudes, the monitoring ceases to be effective and the output of the system follows the perturbations of the magnetic compass.

For example, let $\alpha = 0$, $\tan D = 1\cdot73$ $(D = 60°)$, $h = 25$ ft, $T = 15$ sec, $T_1 = 30$ sec, $\dot\phi = 0\cdot4°$ per min ($0°\cdot0067$ per sec). From equations (3.27) and (9.39)

$$\delta' = -\frac{1}{225-900} \times \frac{4\pi^2 \times 25 \times A \times 1\cdot73}{32\cdot2}$$

$$= 0\cdot08 A$$

Also

$$\delta' = \frac{0\cdot0067 \times 15}{4} = 0°\cdot025$$

Therefore

$$A = 0°\cdot31$$

In this particular instance, the roll or pitch and resultant acceleration have almost negligible values at the point when monitoring ceases to be effective. The deviation of the system is also insignificant.

Even if $\dot\phi$ is $4°$ per minute, δ' increases to only $-0°\cdot25$ and A to $3°\cdot3$, so that monitoring is effective except for the smallest amplitudes of roll or pitch.

An extreme case occurs when A is very large and the monitoring persists for nearly half a cycle. Here, then, δ' is very nearly equal to $T\dot\phi/2$ and in the numerical example above, δ' is nearly $-0°\cdot05$ for $\dot\phi = 0°\cdot4$ per min and nearly $-0°\cdot5$ for $\dot\phi = 4°\cdot0$ per min.

Figs. 9.9 and 9.10 show in a graphical form the behaviour of a monitored system in a number of typical cases. In these illustrations, the *sign* of the deviation is disregarded.

The condition $\alpha = 0$ has been chosen and $\kappa = T^2/(T^2 - T_1{}^2)$ [see equation (3.27)]. Equation (9.39) can now be written

$$\delta' = \frac{T^2}{T^2 - T_1{}^2} \cdot a \sin\frac{2\pi t}{T} \tag{9.41}$$

$$\delta' = \theta t$$

Fig. 9.9(a) is a general case and curves (a), (b), (c) and (d) represent oscillations of various amplitudes, all having the same period. The line OX represents the monitoring effect where $\delta' = \dot\phi t$.

Line OX cuts the curves at points 1, 2, 3 and 4 respectively, from which δ' is determined for each instance. As the oscillation continues, these several deviations will gradually decrease to the value $T\dot\phi/4$ except for curve (d), where δ' is already $T\dot\phi/4$.

Fig. 9.9(b) shows a series of oscillations of constant amplitude but with different periods, as with a pivoted-needle compass where $T_1 > T$ [see equation (3.28)]. It is assumed that $\tan D$ and amplitude of roll or pitch remain constant and that only T varies.

It will be observed that δ' increases to δ'_{max} for curve (c) since the monitoring line OX intersects the curve at the point $t = T_3/4$. The limiting condition has therefore been reached and $\delta' = T_3\dot\phi/4$. If the oscillation represented by (c) continues, the deviation of the gyro-magnetic compass remains unchanged. With oscillations of a longer period, the same peak value of deviation occurs, since the slope of OX is greater than $4\delta_{max}/T$. Thus monitoring is no longer effective and the output of the system follows the perturbations of the magnetic compass. As oscillations corresponding to (a) and (b) continue, the ultimate deviation will be $T\phi/4$ where T is the appropriate period.

Fig. 9.9(b) may be redrawn as Fig. 9.9(c) where a common deviation curve $\delta_{max} \sin pt$ is shown with several monitoring lines OX, each of which represents the same value of ϕ but whose slopes are proportional to the respective periods illustrated in Fig. 9.9(b), namely, T_1, T_2, and T_3.

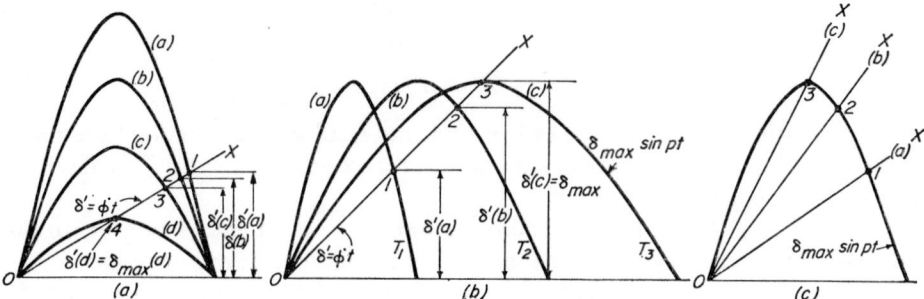

FIG. 9.9. Behaviour of a monitored system

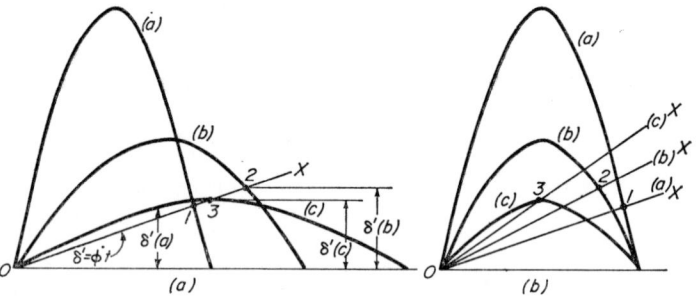

FIG. 9.10. Behaviour of a monitored system

Fig. 9.10(a) also shows a series of oscillations of differing periods, but in this instance, the amplitudes diminish with increase of period, as would occur with a saturable inductor magnetic element ($T_1 = 0$). Tan D and A are assumed to be constant.

Now in this case, the maximum value of the deviation δ' does not occur in the circumstances represented by curve (c) as it did in Fig. 9.9(b), since among the points such as 2 on curve (b) will be found one that indicates a maximum value of δ' for the particular set of conditions and determined by a specific value of T.

Fig. 9.10(b) is drawn in a similar manner to Fig. 9.9(c), showing a set of deviation curves for the various values of $\delta_{max} \sin pt$. Monitoring lines OX have slopes proportional to the several values of T but nevertheless all represent $\delta' = \phi t$. The deviations at points 1, 2 and 3 correspond with those in Fig. 9.10(a). As before, sustained oscillations will result in an ultimate deviation of $T\phi/4$ but the greatest value will be displayed by curve (c).

232

When α is not equal to o, π, $\pi/2$ or $3\pi/2$, the picture becomes rather more complex, as will be gathered from equation (9.39). The characteristics of this expression have been discussed on p. 38, Chapter 3. The small constant deviation gives rise to a small stick-off deviation which, in practice, is negligible. For the parameters encountered in rolling and pitching ships, the departure of the deviation curve from a sine curve is unimportant as far as the monitoring system is concerned, and equation (9.41) may be regarded as adequate for graphical illustration for any value of α.

A similar illustration may be given of a gyro-magnetic compass in a ship that is steaming in a circle. Here it is usually the situation during the first half-cycle or 180° of the turn which is of practical interest (whereas in a rolling or pitching ship it is probably of academic interest only), as several complete turns would have to be made before the ultimate deviation of $T\dot{\phi}/4$ were reached, unless δ_{max} were small.

From equations (3.36) and (3.37),

$$\left. \begin{aligned} \delta' &= \frac{a\cos\zeta}{1 - a\sin\zeta} \\ \delta' &= \dot{\phi}t \end{aligned} \right\} \tag{9.42}$$

where $a = \pm(2\pi V/gT)\tan D$ according to whether the turn is clockwise or counterclockwise. For equation (9.42) to be in the same form as equation (9.39), write

$$\zeta = \frac{2\pi}{T}\left(t - \frac{T}{4}\right) \qquad (\delta' = 0, \text{ when } t = 0).$$

Then

$$\delta' = \frac{a\sin\dfrac{2\pi t}{T}}{1 + a\cos\dfrac{2\pi t}{T}}$$

$$\delta' = \dot{\phi}t \tag{9.43}$$

Figs. 9.10(a) and 9.10(b) are appropriate to this case, since increasing period leads to a decreasing amplitude of deviation, and a point such as 2 on curve (b) indicates the maximum deviation that may occur with a given set of conditions. The equation

$$\delta' = \frac{a\sin\dfrac{2\pi t}{T}}{1 + a\cos\dfrac{2\pi t}{T}}$$

may be expressed as a series of the form

$$\delta' = a\sin\frac{2\pi t}{T} + a^2\sin\frac{2\pi t}{T}\cos\frac{2\pi t}{T} + a^3\sin\frac{2\pi t}{T}\cos^2\frac{2\pi t}{T}\ldots$$

Since in practice a is small compared with unity, the use of sine curves in this illustration is considered to be justified for first-order accuracy.

As the use of sinusoidal deviation curves is considered permissible to illustrate to a first order of accuracy the ultimate deviations of a gyro-magnetic compass in the general cases of rolling, pitching and turning in a circle, a simple and instructive chart, as shown in Fig. 9.11, may be drawn from which the deviations are readily and approximately determinable.

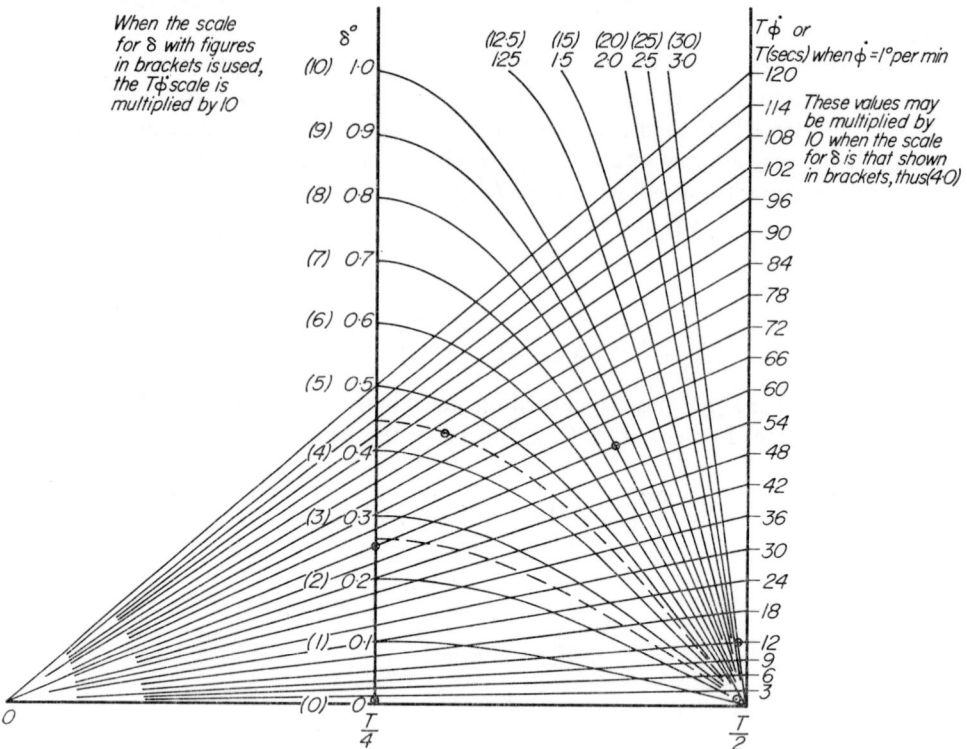

FIG. 9.11. Chart for approximate determination of deviations

The chart illustrates a family of sine curves representing $\delta_{max} \sin pt$ from $t = T/4$ to $t = T/2$. From the point o $(t = 0)$, monitoring lines are drawn, representing a rate of one degree per minute, their slopes being proportional to the various values of T covered by the chart. For example, for a period T of 15 seconds, the appropriate monitoring line will indicate a value of $\delta = 0°\cdot062$ at $t = T/4$ and $\delta = 0°\cdot125$ at $t = T/2$. Similarly, for a period of 120 seconds, $\delta = 0°\cdot5$ at $t = T/4$ and $\delta = 1°\cdot0$ at $t = T/2$.

The right-hand scale shows the period, T, up to 120 seconds for a monitoring rate of $1°$ per minute. For other values of the monitoring rate, this scale represents the quantity $T\dot{\phi}$; thus when the period is 15 seconds, for example, with a monitoring rate of $4°$ per minute, the line marked 60 is used. It should be noted that T is in seconds and $\dot{\phi}$ is in degrees per minute. The chart has scales of deviations from o to $1°$ and from o to $10°$.

To use the chart, the value of δ_{max} is ascertained for the prevailing conditions.

234

For a rolling ship,

$$\delta^{\circ}_{\max} = \frac{T^2}{T^2 - T_1^2} a \cos \zeta \times 57 \cdot 3$$

For a pitching ship,

$$\delta^{\circ}_{\max} = \frac{T^2}{T^2 - T_1^2} a \sin \zeta \times 57 \cdot 3$$

where $a = 1 \cdot 23 \dfrac{hA}{T^2} \tan D$ (h in feet, A in radians)

or $a = 0 \cdot 022 \dfrac{hA}{T^2} \tan D$ (h in feet, A in degrees)

For a turning ship,

$$\delta^{\circ}_{\max} = 11 \cdot 2 \frac{v}{T} \tan D \qquad (v \text{ in feet per second})$$

or

$$\delta^{\circ}_{\max} = 18 \cdot 9 \frac{v}{T} \tan D \qquad (v \text{ in knots})$$

The appropriate curve for $\delta_{\max}$ (in degrees) is selected and the intersection with this curve of the monitoring line for the period of roll or pitch, or for the time to complete a full steaming circle, gives the value of the deviation δ' which will arise in the first half-cycle. The intersection of the monitoring line with $t = T/4$ gives the ultimate deviation with which the system settles down, viz. $T\dot{\phi}/4$.

Example 1

A ship is rolling with a period T of 15 seconds and an amplitude A of 10°. The compass is 25 feet (h) above the centre of roll and its period T_1 is 30 seconds. The monitoring rate is 4° per minute and $\tan D$ is 1·73. Find the greatest error during the first half-cycle and the deviation when settled down.

$$a = \frac{0 \cdot 022 \times 25 \times 10 \times 1 \cdot 73}{225} = 0 \cdot 42$$

$$\delta^{\circ}_{\max} = \frac{T^2}{T^2 - T_1^2} \cdot a \times 57 \cdot 3 \qquad (\cos \zeta = 1)$$

$$= -\frac{225}{675} \times 0 \cdot 042 \times 57 \cdot 3$$

$$= -0^{\circ} \cdot 8$$

From the chart,

$$\delta' = 0^{\circ} \cdot 41 \text{ for the first half-cycle}$$
$$\delta' = 0^{\circ} \cdot 25 \text{ when settled down}$$

Example 2

A ship, heading 030° (ζ), is rolling at a period of 12 seconds with an amplitude A of 15°. The compass, which is of the inductor type, is 10 feet (h) above the centre of roll. The monitoring rate is $1°$ per minute and tan D is 2.33. Find the deviation in the first half-cycle and the deviation when settled down.

$$a = \frac{0{\cdot}022 \times 10 \times 15 \times 2{\cdot}33}{144}$$

$$= 0{\cdot}053$$

$$\delta°_{max} = a \cos \zeta \times 57{\cdot}3$$

$$= 0{\cdot}053 \times 0{\cdot}866 \times 57{\cdot}3$$

$$= 2°{\cdot}62$$

From the chart,

$$\delta' = 0°{\cdot}1 \text{ for the first half-cycle}$$
$$\delta' = 0°{\cdot}05 \text{ when settled down}$$

Example 3

A ship is steaming at 30 knots (v) on a circular course and in a clockwise direction. The time T for a complete circle is 3 min 40 sec (*i.e.* 220 sec). The monitoring rate is $4°$ per minute and tan D is $1{\cdot}73$. Find the maximum deviation during the turn.

$$\delta°_{max} = 18{\cdot}9 \frac{v}{T} \tan D$$

$$\delta°_{max} = \frac{18{\cdot}9 \times 30 \times 1{\cdot}73}{220}$$

$$= 4°{\cdot}45$$

From the chart

$$\delta' = 4°{\cdot}3$$

Fig. 9.12 illustrates how Fig. 9.11 may be used to determine the deviation at the conclusion of successive complete turns when on a circular course.

In this diagram, $\delta \sin pt$ is the deviation curve and OY is the monitoring line whose slope is ϕ, as in Fig. 9.11. OY intersects $\delta \sin pt$ at P_1 and so determines the deviation during the first half-cycle. As the oscillatory deviation continues, the monitoring action is represented by P_1A, with a slope of $-\phi$ which cuts the base line at A. Continuing, it cuts the sine curve during the second half-cycle at Z, giving a negative value of deviation.

Now if OA_1 is made equal to XA and A_1P_2 drawn with a slope of ϕ to cut the deviation curve at P_2, the deviation at P_2 equals in magnitude that at Z. A further line with a slope of $-\phi$ is now drawn to cut the base line at B. OB_1 is made equal to XB and B_1P_3 drawn with a slope ϕ, so P_3 determines the deviation (now positive) for the next half-cycle. Successive lines such as P_3C, C_1P_4 etc. determine the alternately positive and negative values of deviation for successive half-cycles, points such as $P_1, P_3 \ldots$ being positive and $P_2, P_4 \ldots$ being negative. Eventually the condition is reached when a line M_1m is drawn, at which stage the alternate positive and negative deviations are equal and the steady state is reached, where the deviation is $T\phi/4$, and where $OM_1 = XM$ and $M_1M = T_2$.

Figs. 9.13 and 9.14 are further illustrations of the action of monitoring.

236

When the rate of turn becomes great, as is common in aircraft, these approximations become invalid, since the unsymmetrical conditions and eventually the discontinuities evident in Fig. 3.18 are encountered.

When conditions are extreme, the monitoring control may become random or even totally ineffective, exercising no control over the gyroscope and causing unpredictable errors. In these circumstance an acceleration cut-out, which interrupts the monitoring at a given magnitude of acceleration, may be beneficial; a slow monitoring rate is sometimes reasonably satisfactory and a reliable gyroscope with a low drift rate is essential, not only for high-speed flight but in all other circumstances.

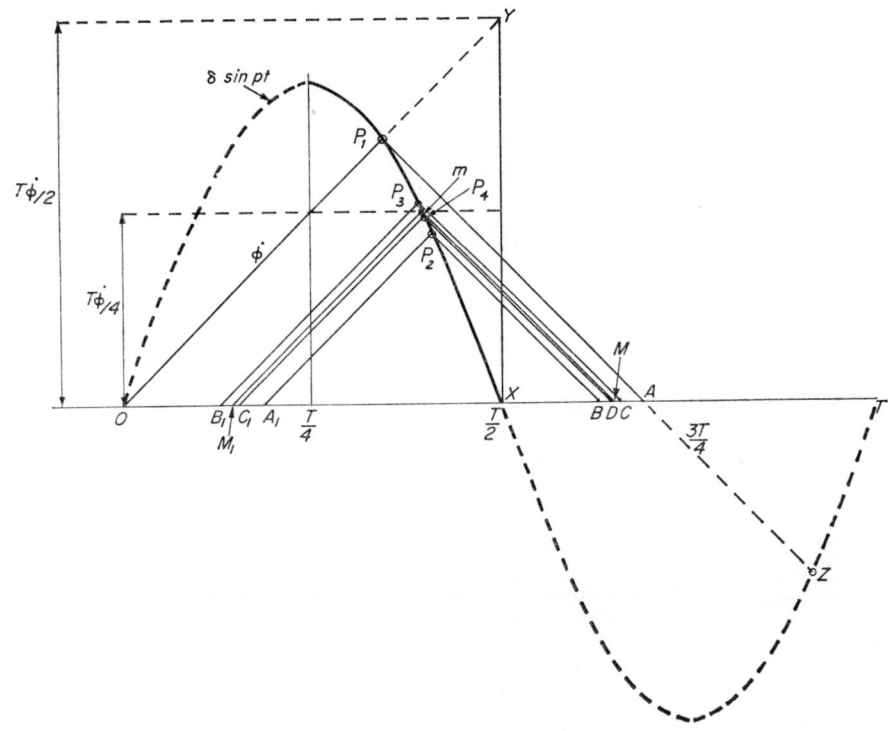

FIG. 9.12. Use of Fig. 9.11

Northerly instability

Although the gyro-magnetic compass is claimed, and rightly so, to be a means of alleviating the northerly turning errors of an unstabilized compass in aircraft, with very fast-moving craft a special case of northerly turning error may arise, even when using a gyro-magnetic compass.

Consider a north-flying aircraft, which owing to some instability in the automatic pilot or to deficiencies in the control system, develops a weaving motion or yaw. The aircraft is flying along a curving track alternately in a clockwise and a counterclockwise direction. During these alternating periods, the pendulous magnetic compass element experiences a centrifugal force to the west and to the east in turn, and tilts accordingly. Deviations then occur owing to the effect of the vertical

237

component of the Earth's magnetic field, thereby setting up a monitoring action at the gyroscope.

When the aircraft is at A, Fig. 9.15, it may be regarded as flying at that instant along a curved path of radius R. Its angular velocity or rate of change of heading is V/R towards the east. The gyroscope is deviated to the east at the monitoring rate. If the angular velocity of the aircraft on its curved track is equal to the monitoring rate, the compass repeater will indicate a constant heading, viz. north.

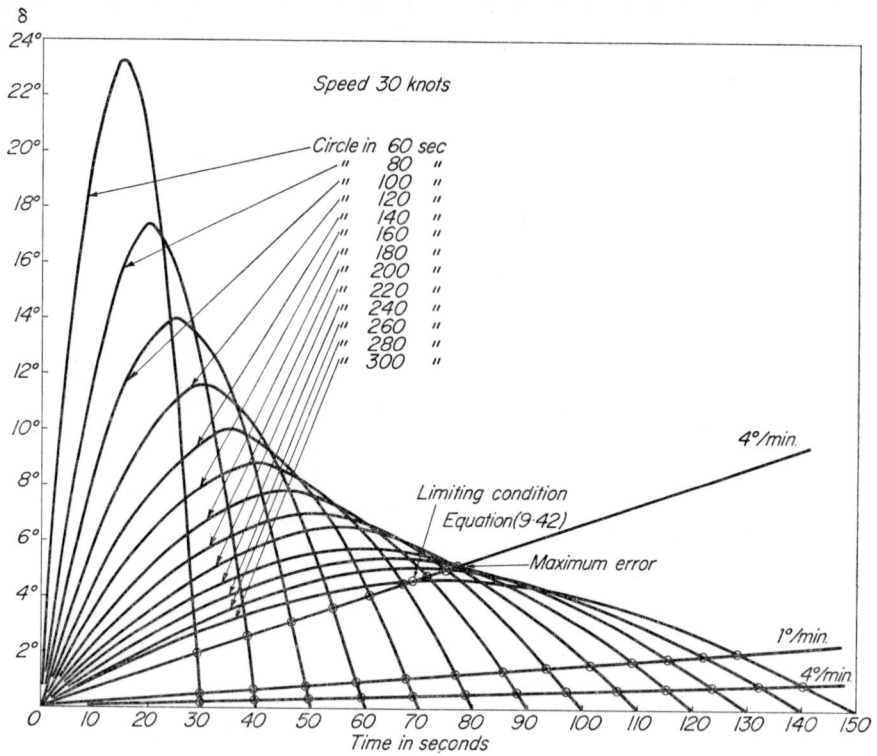

FIG. 9.13. Monitoring action

This diagram shows the deviation curves of a magnetic compass in a ship on a circular course for different rates of turn. Three different monitoring rates are also shown. The errors of a gyro-magnetic compass for the various conditions are shown where the lines for each monitoring rate cut the deviation curves. Dip = $67\frac{1}{4}°$.

At B the angular velocity of the aircraft changes its sense and becomes counter-clockwise instead of clockwise. The acceleration and the deviation are reversed and the gyroscope is again monitored at approximately the same rate at which the aircraft is altering course. The compass repeater therefore still indicates a northerly course. At C we have similar but opposite conditions to those at A, when the course by compass repeater and the direction of the aircraft are both north.

Consequently, in spite of the sinuous track followed by the aircraft, a condition has arisen where the northerly heading of the compass is steadily maintained and the instability of the control system could continue undetected for a considerable period of time.

238

Meridian convergence

In high latitudes, the convergence of the meridians can cause errors in compasses, since the apparent direction of north is changing continuously and rapidly with respect to the track of the ship or aircraft. The same is true when the magnetic variation pattern of the isogonals is changing rapidly in the same way. In fact, the direction of magnetic north may be altering faster than the monitoring control can precess the gyroscope into alignment, an instance where a fast monitoring rate is to be preferred, in spite of its other disadvantages.

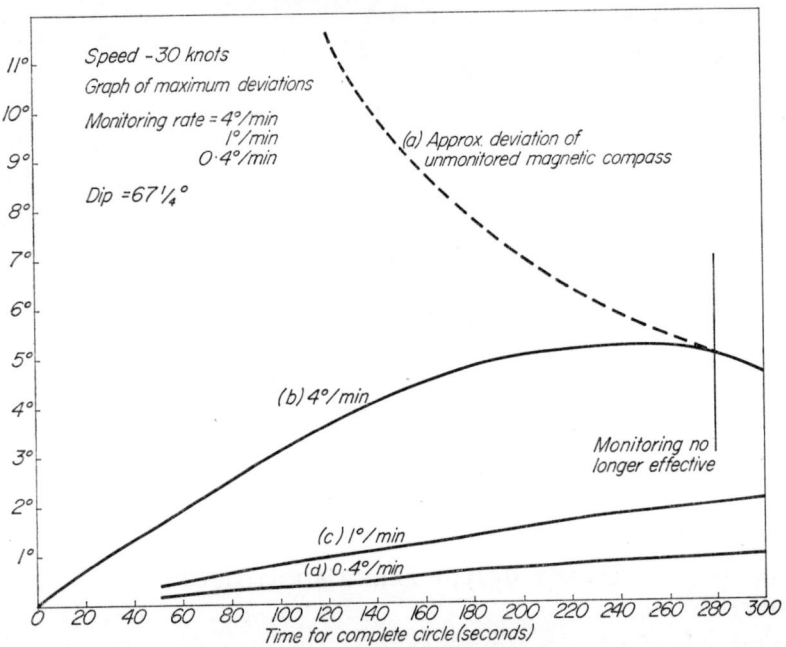

FIG. 9.14. Monitoring action

This diagram shows the deviations that occur with
(a) a magnetic compass in a ship on a circular course at 30 knots, for different rates of turn expressed in time for a complete circle;
(b) a gyro-magnetic compass with a monitoring rate of 4° per min under similar conditions;
(c) as (b) but with a monitoring rate of 1° per min;
(d) as (b) but with a monitoring rate of 0·4° per min.

In terms of true course, ζ'',

$$\text{Meridian convergence} = \frac{V}{R} \sin \zeta'' \tan L$$

where V is the speed, L the latitude, and R the Earth's radius.

These Earth rates appear as gyroscope drift in a gyro-magnetic compass system and are capable of correction by the monitoring control at some sacrifice of accuracy; since a perfect gyroscope would appear to have a drift rate at all times unless maintained level at the equator (in the case of an azimuth or horizontal gyro) or vertical at the pole (in the case of a vertical gyro).

However, azimuth rates can be corrected by means of an adjustable balance weight or latitude rider on the inner gimbal ring which exerts a torque causing the necessary precession in azimuth to cancel the Earth rate exactly. Other methods involve feeding the necessary correction into the control system of the gyroscope.

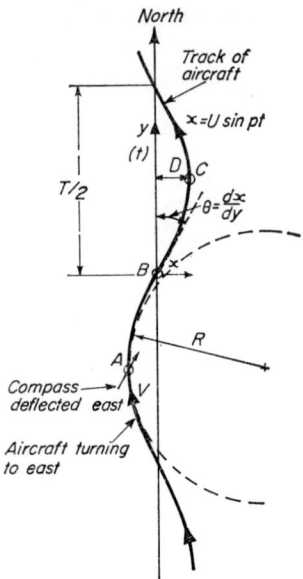

FIG. 9.15. Effect of yaw

EFFECT OF THE EARTH'S ROTATION

Earth rates

Since the axle of a free gyroscope tends to point in a fixed direction in space, when the gyroscope is mounted on the Earth it will appear to drift in azimuth and tilt with respect to the observer. This is owing to the motion of the Earth relative to the direction of the gyroscope's axle. At the equator, a gyroscope will appear to turn completely over in the course of twenty-four hours as shown in Fig. 9.16(*a*); that is to say, it will tilt at the rate of 15° per hour. At the pole, the Earth will appear to rotate underneath the gyroscope, which will have a relative (turntable) motion of 15° per hour [Fig. 9.16(*b*)].

In fact, the only position of the gyroscope axle which gives rise to neither tilt nor azimuthal movement is when it is parallel to the Earth's axis of rotation. In Fig. 9.16(*c*), the gyroscope at A is horizontal at the equator and pointing north, at B it is vertical and therefore pointing north, at C it is tilted at the angle of latitude and is in the plane of the meridian. In each case the axle is parallel to the axis of rotation NS and there is neither tilt nor drift.

In any other position, the gyroscope axle will appear to move and in fact its motion will be that of the change in altitude and azimuth of a star, since the axle of the gyroscope would tend to remain pointing to a star if initially set in that direction. The apparent motion at the equator is, as already stated, a tilting motion, and that at the pole a turntable motion. In between it is a combination of both, depending on the latitude. Let β be the angle of tilt, L the latitude, and α the azimuth measured from north; then the two expressions for the apparent motion of a free gyroscope are

$$\frac{d\beta}{dt} = 15\cdot04 \cos L \sin \alpha \ (\text{deg/hr})$$

$$\frac{d\alpha}{dt} = 15\cdot04 \sin L - 15\cdot04 \cos L \cos \alpha \tan \beta \ (\text{deg/hr})$$

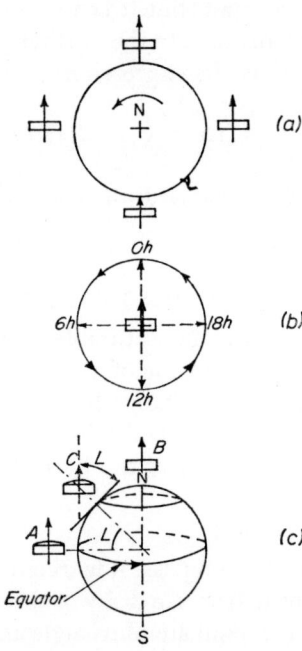

FIG. 9.16. Effect of the Earth's rotation

When $L = 0$,

$$\frac{d\beta}{dt} = 15\cdot04 \sin \alpha \ (\text{deg/hr}) \ (= 0 \ \text{when} \ \alpha = 0)$$

$$\frac{d\alpha}{dt} = -15\cdot04 \cos \alpha \tan \beta \ (\text{deg/hr}) \ (= 0 \ \text{when} \ \alpha = \pi/2 \ \text{or when} \ \beta = 0)$$

When $L = 90°$,

$$\frac{d\beta}{dt} = 0$$

$$\frac{d\alpha}{dt} = 15\cdot04 \ \text{deg/hr}$$

Coriolis acceleration

The effect of the Earth's turning counter-clockwise as viewed from above its north geographic pole is to cause a westward drift of a body moving over the Earth's surface from the pole towards the equator. The acceleration experienced by the body is known as the Coriolis acceleration. A pendulous magnetic compass system in a ship or aircraft will thus appear to be acted upon by a centrifugal acceleration due to the curved path taken by the craft as it moves from northerly latitudes to the equator. In effect, the compass element will be tilted and will suffer from an easterly deviation. When the course is to the north, the deviation is westerly, and in the southern hemisphere, the effect is reversed. The magnitude of the error is extremely small in ships and is of no account, but it can become significant in fast-moving aircraft since the error is proportional to the northerly (or southerly) component of the speed. The error is given by the expression

$$\delta = 4 \cdot 37 \cdot 10^{-4} \cdot V \sin L \tan D \cos \zeta$$

where V is the speed in knots, L the latitude, and ζ the magnetic course.

GYRO-MAGNETIC COMPASSES USING A VERTICAL GYROSCOPE

The compass is usually one of the saturable inductor systems described in Chapter 4. This is mounted on the casing of a vertical gyroscope, itself mounted in a gimbal system and levelled, for instance by a gravitational control. Since the compass element is now always held in the horizontal plane, it will indicate the direction of the horizontal component of the Earth's magnetic field.

Pioneer Gyro Flux Gate compass

The Pioneer Gyro Flux Gate Compass, illustrated in Fig. 9.17, was widely used in aircraft in the Second World War. An array of three saturable inductors, or flux gates, is set up in the form of an equilateral triangle and mounted on the casing of a vertical gyroscope, driven by a 400 c/s 115 volt a.c. supply. The gyroscope is gimballed in the conventional way, and the various electrical leads are taken by ligaments in the form of hairsprings across the gimbal axes. A Bendix ball-erection device, described on p. 215, keeps the gyroscope axle vertical, and the array of inductors horizontal.

The operation of saturable inductors is fully described in Chapter 4 and Fig. 4.38(a) shows the form of inductor array. The inductors are excited by a supply at $487\frac{1}{2}$ c/s obtained from an oscillator, and the consequent 975 c/s outputs from the three inductors are combined by a three-phase synchro-element as in Fig. 4.38(a). The amplifier, which is fed with the resultant signal from the synchro-element, is sharply tuned, with rejector and acceptor circuits, to 975 c/s, and the output drives a two-phase follow-up motor, which is geared to the rotor of the synchro-element.

A compass card, or pointer, is attached to the rotor so that stabilized indications of the compass course of the aircraft are obtained.

The equipment comprises:

   (i) *The gyro unit* containing the inductor or compass assembly and gyroscope;

(ii) *The master indicator*, containing the synchro-resolver, follow-up motor and a secondary Magnesyn transmitter (see Chapter 8) for operating a remote Magnesyn repeater. In this instance, the transmitter is controlled by a mechanically driven magnet instead of a pivoted magnetic needle system as in the Magnesyn compass;

(iii) *The amplifier* and oscillator.

A fourth unit, the inverter, or 400 c/s 115-volt a.c. generator driven from the 28-volt d.c. supply, is needed if no suitable source of main a.c. power is available.

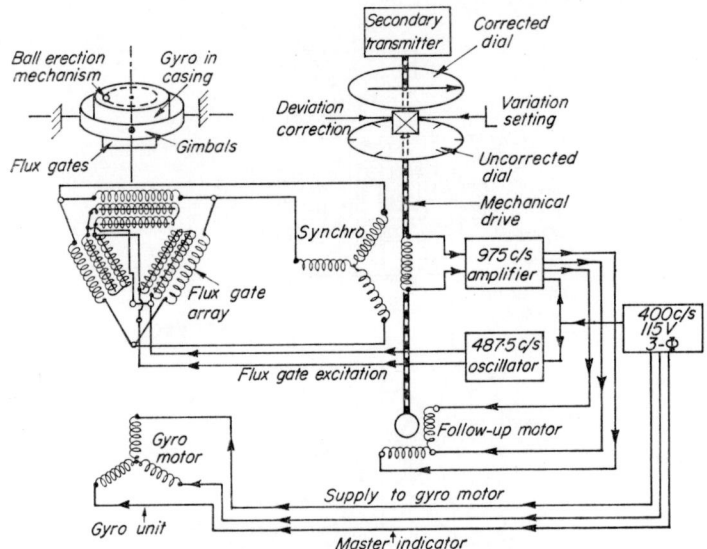

FIG. 9.17. Pioneer Gyro Flux Gate compass

The master indicator has an ingenious corrector system, similar to that shown in Fig. 10.17, to avoid having magnetic correctors at the inductor position. Such a system is permissible only where very small corrections are required since changes of deviations occur which are due to permanent magnetic components. These are proportional to $1/H$, when mechanically corrected, and complete correction can be effected only for one place.

It is carried out by turning a number of equally-spaced screws (every 15°) which are arranged round the circumference of the master indicator, and which distort a flexible metal strip, the contour of which determines the correction applied.

The master indicator has an uncorrected dial which shows the heading as indicated by the detector, and a corrected dial giving the final correct heading. Magnetic variation can also be set into the corrected heading.

Askania D-compass (experimental)

The Askania D-compass is a German instrument similar to the Gyro Flux Gate compass in that it includes a vertical gyroscope which is used to stabilize a saturable inductor. In this instance, however, a single inductor is used, combined with a follow-up system, after the manner of Fig. 4.36, so that the inductor seeks east-west.

243

In the Askania system, illustrated in Fig. 9.18, the follow-up motor is in the master repeater, where it is geared to the indicating pointer, which with a suitable dial indicates compass heading, and also to a pair of transmitters, one of which completes the follow-up loop by driving a repeater element in the gyro unit geared to the inductor. Thus the east–west orientation of the inductor, relative to the vessel in which it is mounted, determines the position of the indicating pointer and hence the magnetic heading is shown on the master repeater dial. The second transmitter is available for extra repeaters. A compact amplifier and oscillator complete the system, which operates at a frequency of 500 c/s.

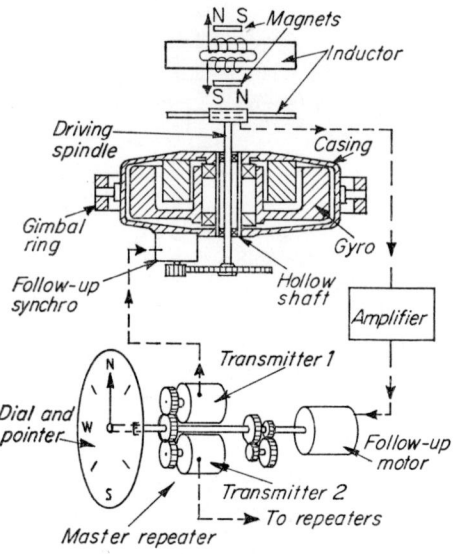

FIG. 9.18. Askania D-compass

The gyro-unit consists of an outer casing mounted on shock absorbers. Within the casing and suspended by a gimbal system is the gyroscope, which has a hollow shaft with a spindle passing through it, on whose upper end the inductor element is mounted, the synchro-repeater element being geared to the lower end. The inductor system used is the 'bridge' form described in Chapter 4, p. 52 and Fig. 4.8, and operating at fundamental frequency. Two small permanent magnets provide the polarizing field.

The gyroscope is levelled by torque motors about the gimbal axes and controlled by a levelling switch in the form of a capsule containing five electrodes and partly filled with an electrolyte.

Hudson's 3-axis inductor compass

A disadvantage of stabilizing a magnetic compass by mounting it on a vertical gyroscope is that the sensitive element is inevitably somewhat bulky. An interesting system has been described by C. S. Hudson,* in which a fixed array of three orthogonally-mounted inductors is used in conjunction with a plane conversion system stabilized by a vertical gyroscope. The inductor array is therefore separated

* Brit. Pat. 591019, 1945.

244

from the bulky elements of the stabilizer and can be small and compact, without any moving parts.

The three inductors are arranged in the fore-and-aft, athwartship and normal-to-the-deck axes. Three components of the Earth's magnetic field with respect to the craft are thus obtained, namely, x'', y'' and z''.

If x and y are the components in the horizontal plane, fore-and-aft and athwartships respectively, and if P and R are pitch and roll angles, then

$$x = x'' \cos P + (y'' \sin R - z'' \cos R) \sin P$$

$$y = y'' \cos R + z'' \sin R$$

The values of pitch and roll are obtained from the vertical gyroscope and applied to a sine-cosine potentiometer or any other convenient resolving mechanism so as to obtain the appropriate functions of P and R. Each inductor is connected electrically to a potentiometer or resolver.

The system is shown diagrammatically in Fig. 9.19, from which it is apparent that voltages proportional to x and y may be derived from amplifiers A_1 and A_2. These

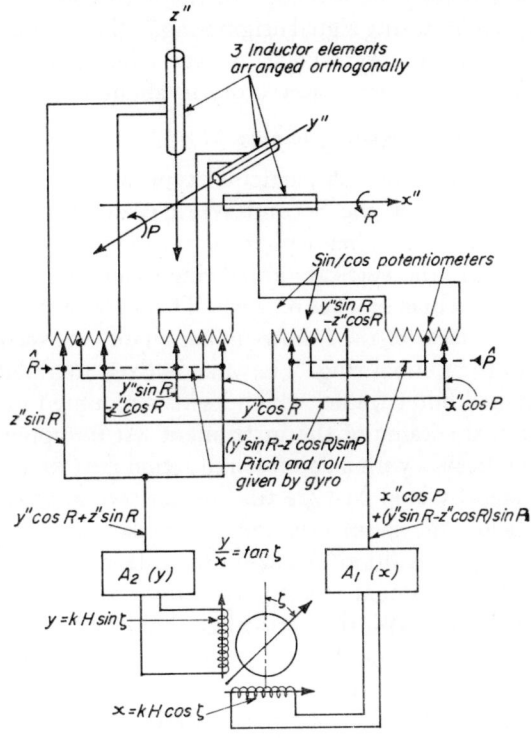

FIG. 9.19. Hudson's 3-axis inductor compass

voltages are applied to a quadrature type of indicator, from which the heading of the craft ζ' may be determined, since

$$y = kH \cos \zeta'$$

$$x = kH \sin \zeta'$$

245

More versatile and flexible equipment may be constructed using azimuth stabilization than by the use of a vertical stabilizer. Heading information is obtained from the gyroscope, usually through the medium of a data transmission system, and the gyroscope's orientation with respect to magnetic or compass north is ensured by the control system derived from the magnetic compass. At the same time, the perturbations of the compass due to acceleration and similar causes are effectively smoothed by the stability of the gyroscope. The control signal from the magnetic compass may be provided by any of the pick-up devices described in Chapter 8 and the control may be applied by direct mechanical coupling or by the aid of electrical transmission.

The earlier forms of gyro-magnetic compass combined all the component members of the system in a single instrument, the magnetic compass being actually mounted on the gyroscope's inner gimbal or mechanically connected to it. This system is that shown in Fig. 9.7, the combination of gyroscope and compass achieving the comparison between the direction of the gyroscope's axle and that of the compass needle. The monitoring signal originating in the compass is used to operate the precession control on the gyroscope. An early instrument on these lines was made at the Admiralty Compass Observatory in about 1925 by C. Chaffer.

RAF distant-reading gyro-magnetic compass

The best-known compass of this particular type was the D.R. compass Mk I, extensively used in aircraft in the Second World War. The instrument is shown in Fig. 9.20 and in Plate 9.1. A hemispherical gyroscope encased in a copper shell (Fig. 9.21) is mounted in the vertical gimbal ring, which itself is pivoted about its vertical axis in a semicircular bracket or yoke. This carries the precession magnets and one of the two contacts of the follow-up contactor, the second one being on a member attached to the gimbal ring. The yoke is fixed to a platform at the lower end of a stirrup-shaped outer frame, which itself is mounted in bearings about its vertical axis in the main frame of the instrument. At its upper end, it carries an azimuth gear which meshes with a suitable reduction gear driven by the follow-up motor and with a step-by-step M-type transmitter (see p. 202). At the lower end of the frame and below the gyroscope and its bracket, a special form of double-pivoted magnetic compass is fitted. An azimuth scale is attached to the base of the frame and the whole instrument is suspended by a form of universal joint. The outer frame is kept aligned with the vertical gimbal ring, which is maintained in a fixed direction by the gyroscope, by the action of the follow-up contactor. If the circuit is completed, the relay switches the supply to the azimuth motor so that the latter drives the outer frame in the appropriate direction. This movement breaks the circuit made by the contactor, hence the relay changes over and reverses the azimuth motor. The outer frame thus hunts about a mean position indicated by the gyroscope and will continue to do so as the craft in which it is mounted turns, the azimuth scale indicating the angle of turn. If the gyroscope drifts in azimuth, the outer frame will follow accordingly.

The magnetic compass, which is attached to the outer frame, is unusual in design. It is double-pivoted and fitted with contacts and a clamping mechanism, so that every few seconds the needle is pressed down on one of two contacts, completing an electrical circuit which energizes the windings of the precession magnets. Thus if the gyroscope, by its random drift, has moved the outer frame, and therefore the compass contacts, to the east of the magnetic meridian, the precession magnets will be so energized on the next and successive clamping cycles that the gyroscope will be precessed in a westerly direction until the clamping of the magnet takes place with the outer frame to the west of the magnetic meridian. The mean position of the frame is thus the magnetic meridian. The magnetic compass and its contacts provide the comparison means and the precession magnets the correcting means.

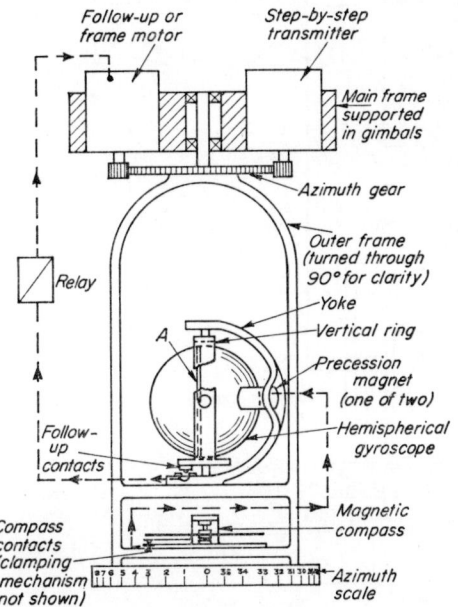

FIG. 9.20. RAF distant-reading gyro-magnetic compass

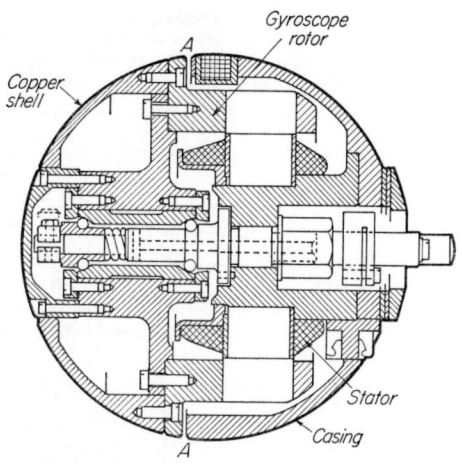

FIG. 9.21. RAF distant-reading gyro-magnetic compass

The gap at A and the edges of rotor and casing show as two parallel lines in Fig. 9.20.

The clamping mechanism is driven by a gear train from a small a.c. motor generator (which supplies the 230 c/s 3-phase current for the gyroscope) and consists of an eccentric cam which raises and lowers an operating rod connected to a clamping frame on the compass.

The precession magnets apply the requisite azimuthal and erection torques to the gyroscope by the eddy currents they induce in the copper shell of the gyrorotor.

A 28-volt d.c. supply is required for this compass.

Holmes Mag-Gyro compass

This instrument, illustrated in Fig. 9.22, is similar in general construction to the D.R. compass in that its gyroscope has a follow-up system which drives an outer frame and keeps it aligned with the gyroscope frame or outer gimbal. A liquid-damped magnetic compass with electrodes as in the Holmes tele-compass (see p. 183) is mounted on the outer gimbal. In this case also, the compass bowl with its electrodes and the compass card and its magnet system provide the comparison means.

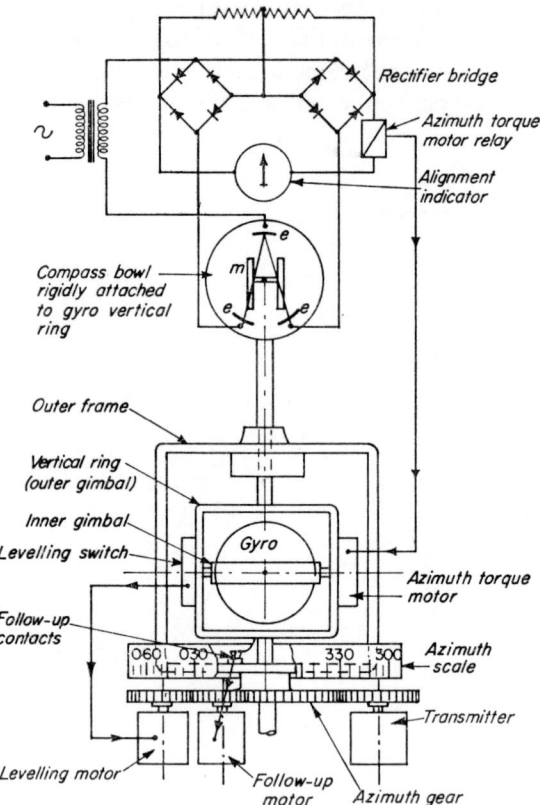

FIG. 9.22. Holmes Mag-Gyro compass

When the gyroscope axle is aligned with the meridian, the compass bowl is so disposed that its electrodes are symmetrical with respect to the magnet system and its electrode. The resistance bridge system formed by the liquid, the electrodes and the external circuit is balanced and there is no output signal. Drift of the gyroscope, however, upsets the balance and an output signal appears. This energizes a sensitive relay which in turn controls the electrical supply to a torque motor mounted about the horizontal gimbal axis of the gyroscope. Precession in azimuth takes place until the compass is once more in balance and the gyroscope directed in the magnetic meridian. The apparatus is provided with a levelling switch and torque motor, and a step-by-step transmitter.

Current practice in the design of gyro-magnetic compasses is to divide the equipment into separate components—magnetic compass, gyroscope, data transmission and resolving mechanisms and amplifiers—thereby exploiting the flexibility of the system and taking advantage of the facilities offered by electrical and electronic

248

circuits instead of using mechanical couplings, relays and d.c. motors. The magnetically sensitive element may be sited in a good magnetic position, remote from electrical interference, and being, as a rule, small and compact, may occupy a site of small dimensions without particular regard to accessibility since it is divorced from the moving mechanical gear of the equipment.

Fig. 9.23 illustrates diagrammatically a usual arrangement of a gyro-magnetic compass and is a logical development of the arrangement of Fig. 9.7. The gyroscope, being provided with a dial, is the primary direction or steering indicator and transmission from it operates a master indicator or navigational heading instrument and other remote repeaters. Since the control signal is applied direct to the gyroscope through the medium of precession coils, this system may be described as being *directly monitored*.

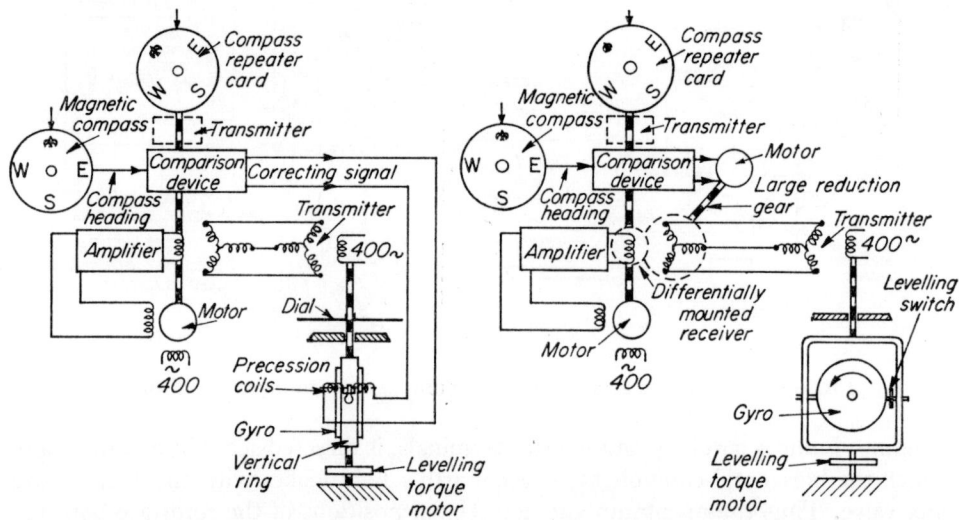

FIG. 9.23. Directly monitored system for a gyro-magnetic compass

FIG. 9.24. Indirectly monitored system for a gyro-magnetic compass

An important variation of monitoring control is illustrated in Fig. 9.24. In this instance the control or monitoring signal is applied by electro-mechanical means to the transmission system rather than to the gyroscope. Thus by correction of the transmission which reproduces any azimuthal changes of the gyroscope's direction and communicates them to the rest of the equipment, the gyroscope is free to wander. The latter is no longer used as a primary direction indicator and is merely a stabilizer. Not only may it be sited in the most dynamically advantageous position in the craft, but it is unencumbered by precession coils or azimuth torque motor. Greater accuracy is possible and a larger and more refined instrument may be used since it is no longer necessary to install it in the steering position. Equipment of this kind is described as being *indirectly monitored* and is typical of the most recent practice in gyro-magnetic compasses. Variation and deviation corrections may conveniently be fed into the master indicator or the correcting device, so that true headings are indicated.

249

Fig. 9.25 illustrates the schematic arrangement of compass equipment, such as the Sperry Gyrosyn and the Kelvin-Hughes G5 compass, in which direct monitoring is employed. These are primarily aircraft compasses and use saturable inductors as the magnetically sensitive elements. Three different forms of magnetic detector and comparison means are given. At (*a*), is shown the Sperry Flux Valve, a three-legged device similar to the normal saturable inductor, with a single exciting coil energized at *xy* and three secondary windings, star-connected (see Chapter 4, p. 71). The ends of the secondaries are connected to a three-phase resolver, the rotor of which is driven by the follow-up motor. The Earth's magnetic field gives rise to three vectors of its horizontal component acting along the arms of the flux valve; hence corresponding electrical signals arise in the secondaries and are applied to the resolver windings. The resolver's resultant field, which, by threading the rotor

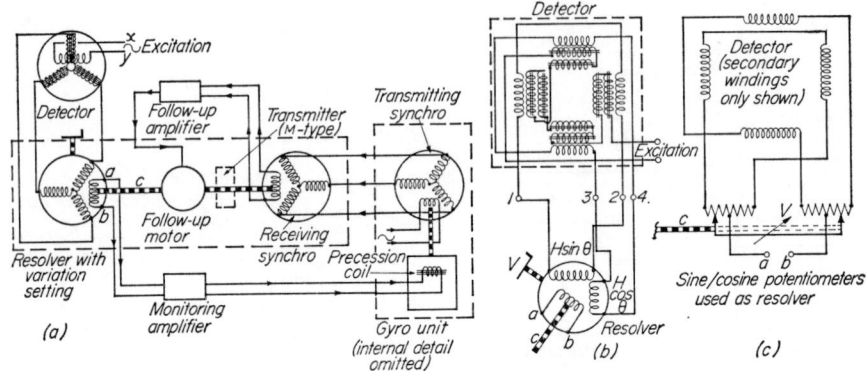

FIG. 9.25. Schematic arrangements of compasses with direct monitoring

winding, induces a voltage at the rotor terminals, is directed according to the angle which the horizontal component of the Earth's field makes with the arms of the flux valve. Thus the maximum and zero signal positions of the rotor are both related to the direction of the aircraft or ship at the time. Turning the craft will cause an apparent rotation of the zero signal position, which is the chosen reference or alignment condition. The comparison between the zero position of the rotor and the azimuthal direction of the gyroscope's axis is effected by coupling the rotor to the follow-up motor, which is controlled by the synchro-transmission link from the gyroscope.

At (*b*), a two-axis saturable inductor system is shown replacing the flux valve; four inductors are used, in pairs placed at right angles to each other. This system feeds a two-phase resolver exactly as the flux valve feeds the three-phase resolver, and comparison with the gyro-unit is as in (*a*).

At (*c*), the two-phase resolver is replaced by a pair of sine-cosine potentiometers, whose rotating arms are coupled to the follow-up motor. The use of multi-element inductors, resolvers and sine-cosine potentiometers is described in Chapter 4, p. 75.

In all three cases, the gyro follow-up system is identical and, in accordance with normal practice, has a coincidence transmitter electrically coupled to a receiver, the amplified output of which drives a follow-up motor mechanically coupled to the receiver rotor.

The gyroscope is monitored by amplifying the resolver or sine–cosine potentio-meter output, and using this voltage to energize precession coils on the gyroscope frame. A permanent magnet or armature attached to the gyroscope's inner gimbal passes through these coils, which, when energized, act on the magnet so that the appropriate precessional torque is applied to the gyroscope. When the gyroscope is precessed so that the follow-up motor brings the resolver output to zero, correct alignment between the gyroscope and the magnetic meridian is achieved and the monitoring signal ceases.

The Sperry Flux Valve, illustrated in Plate 9.2, is a typical detector unit, which would customarily be installed in the wing-tip of an aircraft.

Plate 9.3 shows a Sperry master indicator or primary navigational instrument, having a dial engraved in degrees over which moves a pointer, coupled to the main spindle of the resolver mechanism. The instrument contains, in addition, a receiv-ing synchro-element, follow-up motor and step-by-step or other form of transmitter. An alignment indicator giving 'dot' or 'cross' indication shows whether alignment between the gyroscope and the magnetic detector is complete. Fig. 9.26 shows the symbols for zero position and misalignment.

Variation can be applied, and true headings are obtained by rotating the stator of the resolver or the windings of the sine–cosine potentiometers [see V, in Fig. 9.25 (c)]. A control box is fitted when alternative gyroscopes are provided, and con-tains a change-over switch so that either may be used. A third position of the switch, usually known as the D.G. (directional gyro) position, allows the monitoring control to be taken off the gyroscopes.

The amplifier unit contains amplifiers for the gyro-transmission link follow-up system and for the monitoring signal, which may be either d.c. or a.c. In the unit there is sometimes a source of stable d.c. for compass correction. This correction is carried out by fore-and-aft and athwartship solenoids in the detector unit in place of the usual bar magnets.

The Sperry gyro-unit (Plate 9.4) comprises the gyroscope (a three-phase star-connected stator with a squirrel cage rotor, the latter being part of the gyroscope proper), its gimbals, a levelling switch and torque motor, the precession coils, a magnet or armature, a synchro-transmitter and a compass dial. A mechanical clutch disengages the dial and transmitter from the gyroscope, enabling quick alignment of the system. An alignment (dot and cross) indicator and a course-setting device complete the instrument, which provides heading information for the benefit of the pilot corresponding to that at the master indicator (see Fig. 9.26).

A variation of this system for small aircraft is shown in Fig. 9.27. The master indicator is omitted, and in the gyro-unit the synchro-transmitter is replaced by the resolver, so that the gyro-transmission follow-up system is no longer required. The gyro-unit is the only heading reference, and variation correction is not usually included in this type of equipment.

Indirectly monitored systems

In Fig. 9.28, the system shown in Fig. 9.24 is elaborated and illustrates the typical schematic arrangement of a group of compasses employing indirect monitoring. This group includes the Kearfott N.1, the Bendix Polar Path and the Admiralty gyro-magnetic compasses. The latter are described in some detail as they possess a number of special and unusual features.

A detector similar to the Sperry Flux Valve is commonly used, or, alternatively, two pairs of inductors as illustrated in the diagram. In the Bendix Polar Path compass an interesting type of compass-aided inductor is employed. This is a three-phase arrangement and is illustrated in Fig. 9.29.

The resolver and monitoring amplifier are similar to those previously described and the gyro follow-up system, its synchro-elements, amplifier and follow-up motor follow the same pattern as illustrated in Fig. 9.25. However, the receiving synchro is mounted in such a manner that not only may the rotor be turned with respect to the stator, but the stator may also be turned. This makes the element an electrical differential gear, whereby a monitoring signal may be added to the gyro-transmission instead of being applied directly to the gyroscope.

FIG. 9.26. Symbols for zero and misalignment in a Sperry master indicator

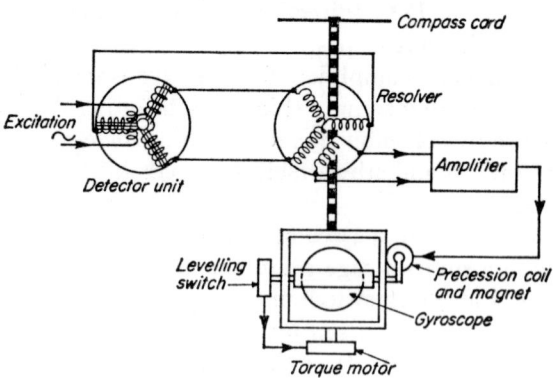

FIG. 9.27. Gyro-unit for small aircraft

The output from the monitoring amplifier drives the monitoring motor, which is mechanically coupled by a high-ratio reduction gear to the stator of the receiving synchro. This permits the appropriate rate of monitoring to be applied, and it will be appreciated that the correction of the gyroscope drift can be effected by rotating the stator windings so that the electrically and mechanically rotating fields set up in the synchro are mutually cancelled, the zero position of the rotor being unaffected. This system has several advantages, one being the simplification of the gyro-unit by comparison with that shown in Fig. 9.25 as no precession means are necessary. There is no indicating card or dial and the axis may point in any direction. An alternative to the differential receiving synchro in which a follow-through synchro is used is shown in Fig. 8.28.

As a rule, compass equipment of this type is fitted with high precision gyroscopes, with wander rates not exceeding 5° per hour. The gyro-unit is normally remotely situated from the pilotage or navigational position and can therefore be larger and more robust than a panel-mounted instrument; hence there is better possibility of finding a position unaffected by movement of the craft. The gyroscope is sealed in a container of helium and has a thermostatically controlled

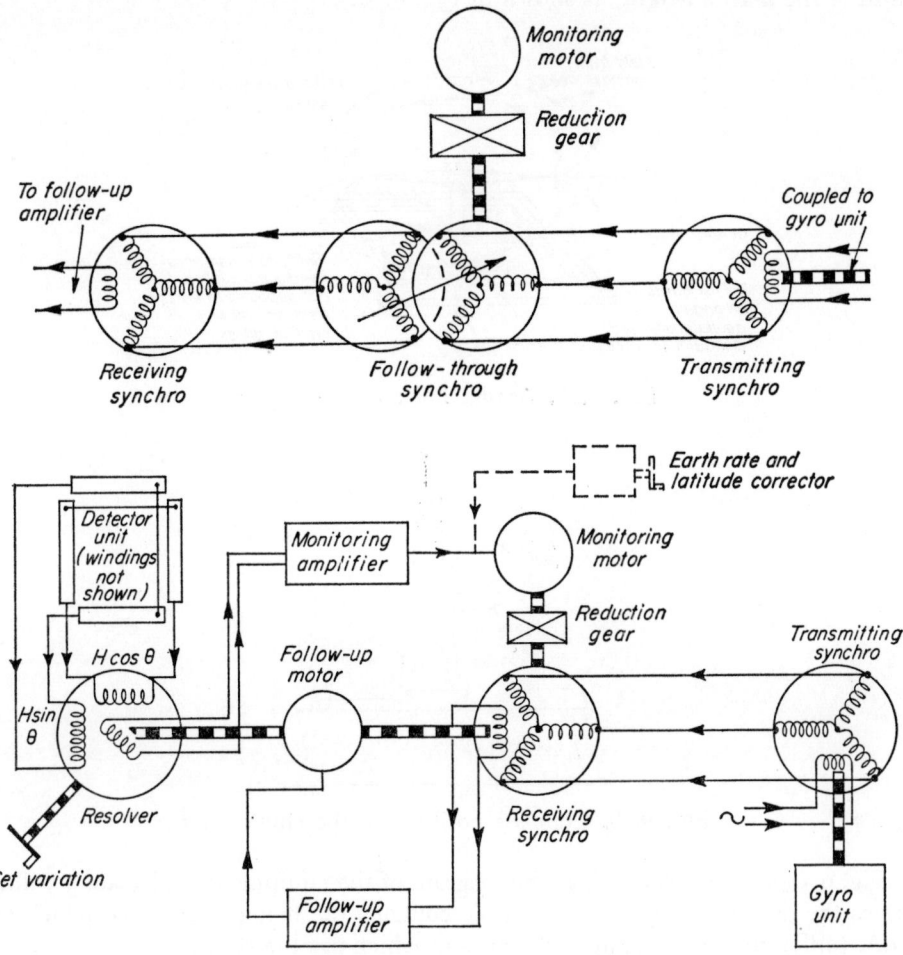

Fig. 9.28. Indirectly monitored systems

heater to avoid thermal effects that may cause changes of balance. Special care is taken in the choice of gimbal and rotor bearings. With gyroscopes that exhibit low rates of wander, the monitoring rate, as has already been shown, can be correspondingly low, resulting in increased accuracy of the compass system. This may be further enhanced by fitting an Earth rate or latitude corrector, taking the form of a balance weight or latitude rider on the gyroscope casing, or a motor-driven corrector, which applies an appropriate rate to the monitoring system through suitable differential means, to correct for the effect of the Earth's rotation.

253

Admiralty gyro-magnetic compass, Type 7

Whereas the compass systems already discussed are primarily aircraft instruments, the Admiralty gyro-magnetic compass was specifically designed for use in large naval vessels, such as cruisers and aircraft carriers, and as such, has certain interesting features. Owing to the difficulty of finding a good compass position on deck, the detector is mounted on a bracket extending from the mast, in the middle third of the mast's height, as shown in Fig. 9.30.

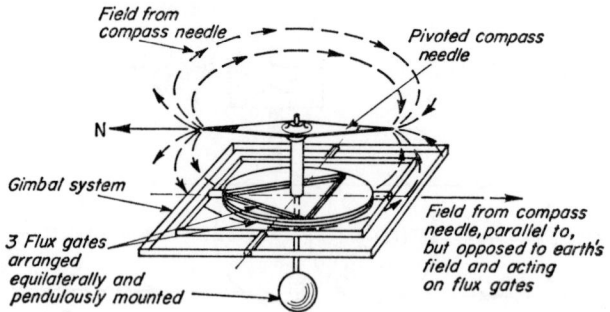

FIG. 9.29. Bendix Polar Path compass

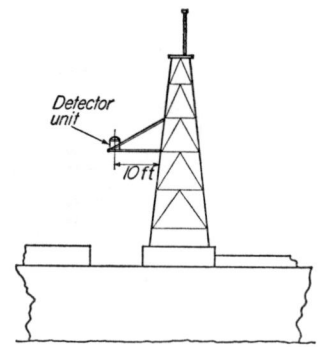

FIG. 9.30. Compass position on the ship's mast

Fig. 9.31 is a complete schematic diagram of the equipment, and Plate 9.5 illustrates the detector assembly. A console contains the meter and switch panel, the D.A. (differential and alignment) unit—in which are the receiving synchro of the gyro-transmission system—follow-up motor, monitoring control, resolver, and outgoing transmitters, and the necessary amplifiers and retransmission panels. Step-by-step and magslip outputs are provided. The complete console is a much elaborated master indicator with amplifiers, similar to that illustrated in Plate 9.8.

Since it is often an advantage to have a choice of monitoring rates, the speed of the monitoring motor is made variable to suit various conditions of use, sea and weather. The monitoring motor is provided with an eddy-current brake, and the braking effort is altered by varying the current in the brake-magnet coils (Fig. 9.32). Rates of 4°, 0°·8 and 0°·4 per minute are available, which can be multiplied

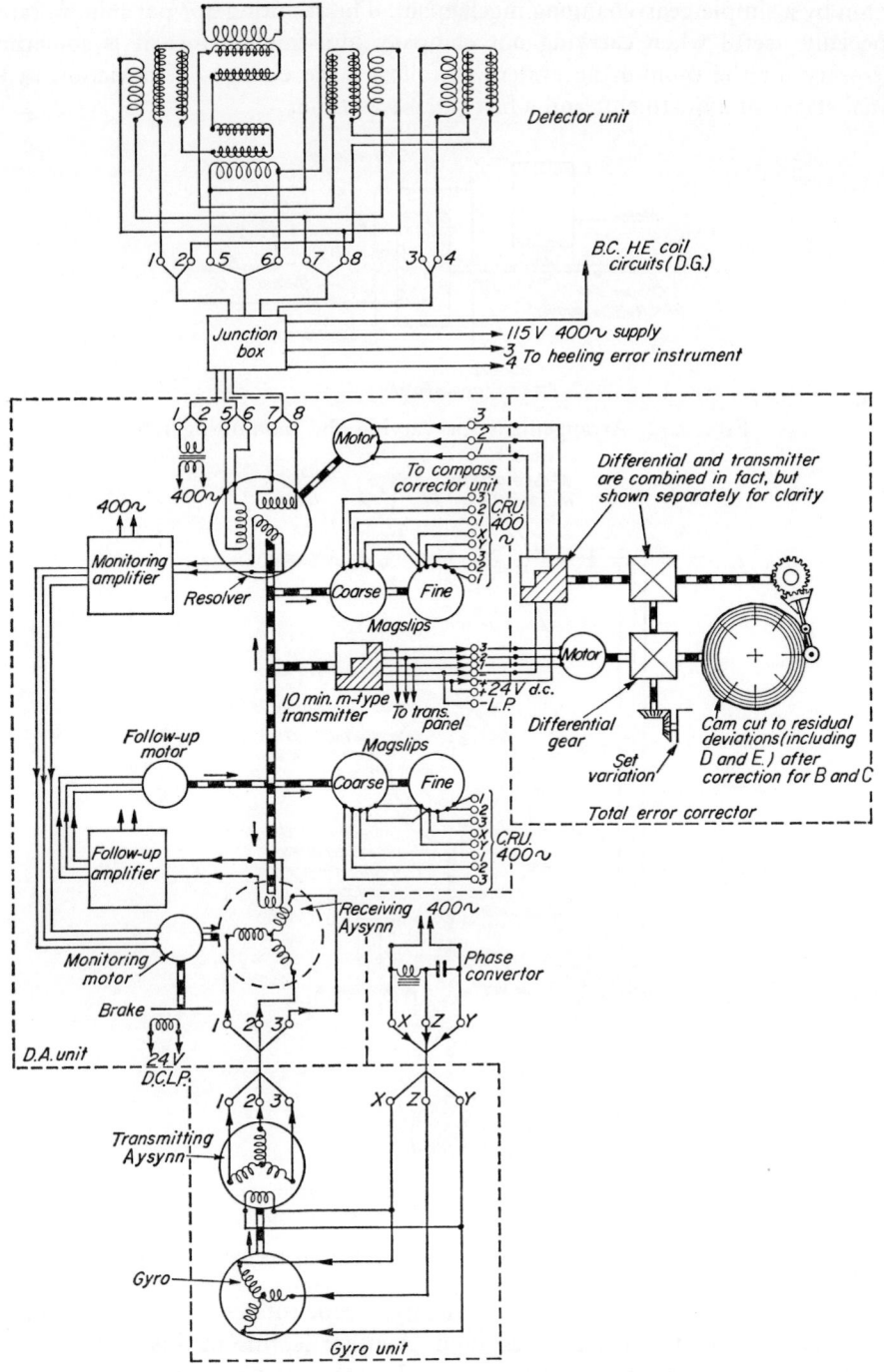

FIG. 9.31. Admiralty gyro-magnetic compass, Type 7
Schematic diagram

by ten by a simple gear-changing mechanism. The resulting 40° per minute rate is especially useful when carrying out compass adjustment, since it is sometimes necessary for the monitoring system to follow large changes of deviation in the initial stages of adjustment, and a fast rate saves time.

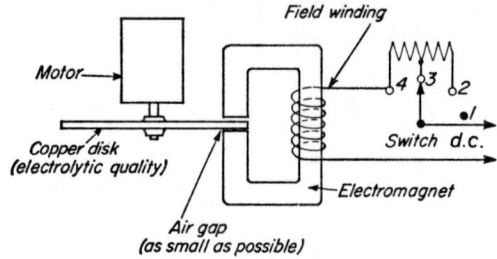

FIG. 9.32. Arrangement for varying the monitoring rate

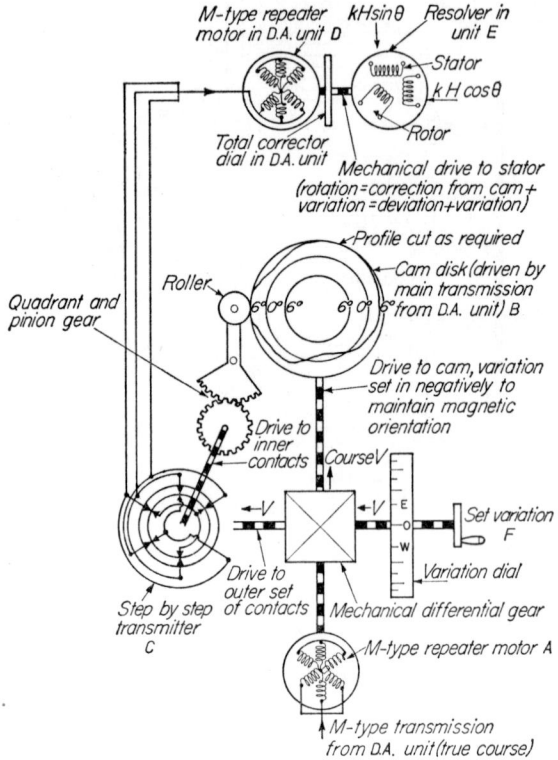

FIG. 9.33. Remote total error corrector

Plate 9.6 illustrates the gyro-unit, whose gyroscope rotates at 12 000 r.p.m., and has a wander rate of less than 3° an hour. Another feature of this compass is the remote total error corrector (Fig. 9.33), which enables variation and residual deviation corrections to be applied, giving true headings at the compass repeaters. The instrument, a photograph of which is shown in Plate 9.7, can be fitted in a position convenient to the bridge or chart-house.

The repeater motor A, which is driven by an M-type step-by-step link from the D.A. unit in the console, drives a cam B through a train of gears so that the cam does one revolution per degree of azimuth. The cam is cut to the polar diagram of the residual deviation curve. A bell-crank lever and quadrant gear are moved by the cam as it rotates, imparting a rotation, corresponding to the desired correction, to a finger-type step-by-step transmitter, C. This sends impulses to a receiving repeater motor D in the D.A. unit, geared to the stator of the resolver, E. The angular movement of the stator, corresponding to the deviation correction, initiates a monitoring signal, which rotates the transmission through the angle of correction.

To provide variation correction, the transmitter in the corrector is rotated through the appropriate angle by a manually-operated gear, F, which simultaneously turns the cam through the same angle in the *opposite direction*. This is to allow the cam to be related to magnetic headings after the variation correction has been 'monitored-in'. Otherwise the deviation curve would not be valid.

Compass adjustment is effected by a scissor-magnet type of permanent corrector operated by rods extending from the casing of the detector unit (or binnacle) to an operating unit at the mast end of the bracket upon which the detector is mounted.

Heeling error correction—the cancellation of the ship's vertical field—is also effected by the scissor-magnet corrector, but the actual measurement of the field to be corrected, and the indication that correction has been effected, is achieved by a third and vertical inductor mounted in the detector unit. This is energized from the 400 c/s supply and measures total vertical field (ship + Earth). A portable instrument, known as the electronic heeling error instrument, determines this value. Knowing the local undisturbed value of the vertical field of the Earth, the appropriate adjustment to the instrument is made, so that only the value of the ship's field is indicated. This is then cancelled by moving the vertical scissor-corrector until zero field is shown by the heeling error instrument. The process of correcting heeling error is further explained in Fig. 9.34. Since the vertical detector is further from the correctors than the rest of the detector unit, a proportionate correction of the measured field is made.

Admiralty gyro-inductor compass (A.G.I.C.)

The Admiralty gyro-inductor compass,* illustrated schematically in Fig. 9.35, was an experimental compass using indirect monitoring, and was, in fact, an earlier example of this type of control than that used in the compasses already described. Three similar modulated high-frequency second-harmonic inductors (see Chapter 4) excited at $5\frac{1}{2}$ kc/s were used, placed at right angles. The horizontal ones which measured $H \cos \zeta$ and $H \sin \zeta$ were fixed, i.e. not gimballed, fore-and-aft and athwartship, while the vertical one, which measured Z, was gimballed and associated with a 'Z' coil for the correction of tilt errors.

Sine-cosine potentiometers combined the sine and cosine signals after amplification and a high degree of negative feed-back was provided by tertiary windings

* Hine, A., and Hitchins, H. L., 'Improvements in and relating to azimuthal direction-indicating apparatus'. Brit. Pat. 624083, 1946 (Gimballed Z coil).

Hine, A., and Hitchins, H. L., 'Apparatus for measuring and detecting magnetic fields'. Brit. Pat. 619525, 1946 (Modulated H.F. saturable inductor).

National Research Development Corporation, 'Improvements in navigational compass systems'. Brit. Pat. 690011, 1951 (Admiralty gyro-magnetic compass—inventor, A. Hine).

on the inductors. The gyroscope was coupled to the potentiometers and outgoing transmission by a normal synchro follow-up system. Monitoring was effected by a motor (M), with an eddy-current brake enabling normal and fast rates (4° and 40° per min) to be obtained, acting on the transmission system through a mechanical differential gear (DF).

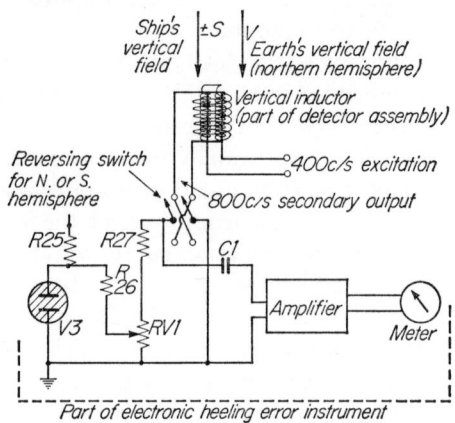

FIG. 9.34. Correction of heeling error

With H.E. magnets and RV1 set to zero, meter reading depends on $V+S$. RV1, which is a calibrated potentiometer, is rotated to obtain a zero reading on the meter, a direct current from the power supply to the H.E. instrument, of which V3, R25, R26 are components, being supplied to the inductor secondary, thereby opposing and cancelling $V+S$. The d.c. is kept from the H.E. amplifier by capacitor C1. The reading of the potentiometer dial is now an indication of $V+S = X$.
Add twice X to the local value of the Earth's vertical field and divide by 3. Set this value on the potentiometer dial.
Adjust the vertical corrector magnets in the binnacle until the meter reads zero, i.e. S is completely cancelled.
Heeling error is therefore corrected, H.E. instrument is now switched off, unplugged and removed.

The gimballed Z coil is in the feed-back circuit of the vertical inductor. Thus the vertical component of the Earth's magnetic field produces a direct current in the feed-back winding and the Z coil. By correctly proportioning the windings of the coil and the inductor, a field equal and opposite to the Earth's vertical field can be produced at the horizontal inductors, which are now virtually working in a horizontal field only. If the inductor system is tilted, the vertical inductor and coil remain vertical, and as zero vertical field is still maintained at the horizontal inductors, deviation due to tilt is avoided.

However, the effect of acceleration is to produce deviations similar to those of a pendulous detector system (Fig. 9.36). For instance, if, owing to a banked turn, the coil is tilted into the false vertical, its axis will be normal to the deck plane and therefore normal to the horizontal inductor array. The axial magnetic field from the coil will cancel the Earth's component $Z \cos \beta$, but an uncorrected component, $Z \sin \beta$, along the horizontal inductors will cause the usual deviation, given by

$$\tan \delta = \frac{\tan D \sin \beta \cos \alpha}{1 - \tan D \sin \beta \sin \alpha}$$

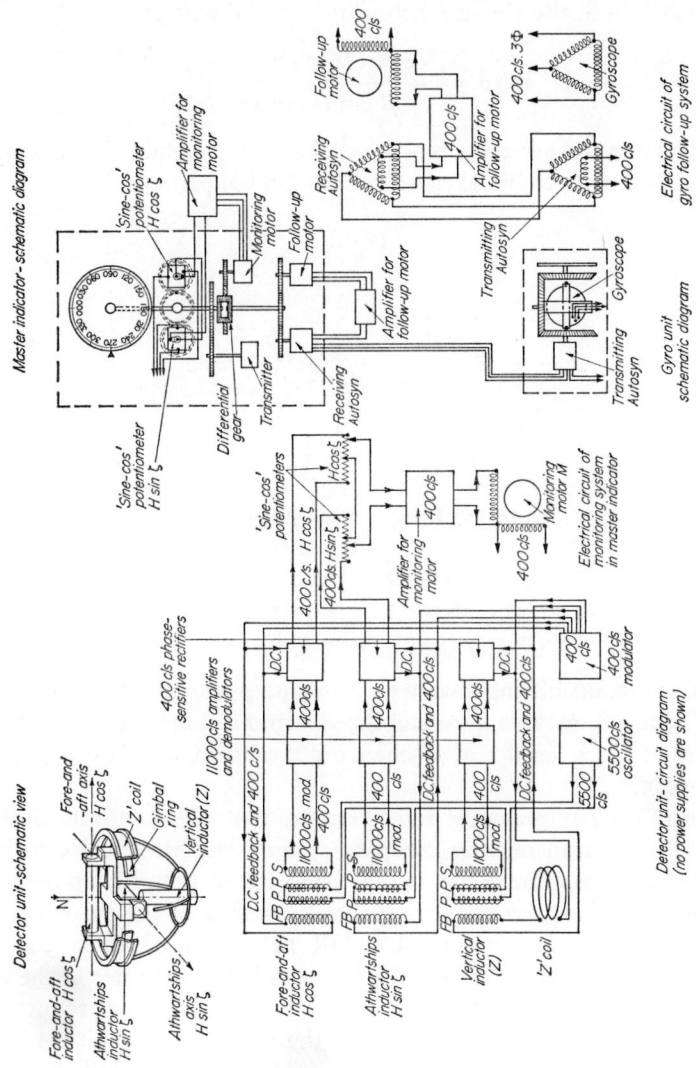

FIG. 9.35. Admiralty gyro-inductor compass
Schematic diagram

If the coil is tilted, leaving the horizontal detectors level, an error will be introduced by the component in the horizontal plane arising from the tilted coil. This is $(Z \sin 2\beta)/2$. Thus

$$\tan \delta = \frac{\frac{1}{2} \tan D \sin 2\beta \cos \alpha}{1 - \frac{1}{2} \tan D \sin 2\beta \sin \alpha}$$

When β is small, both the above expressions may be written

$$\tan \delta = \frac{\beta \tan D \cos \alpha}{1 - \beta \tan D \sin \alpha}$$

The method has certain advantages over other systems, notably in the suspension and in fewer leads having to be taken across the gimbal pivots.

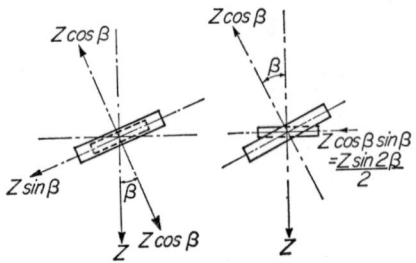

Fig. 9.36. Effect of acceleration

GYRO-MAGNETIC COMPASSES
EMBODYING A FOLLOW-UP TYPE OF MAGNETIC COMPASS
AND INDIRECT MONITORING

Almost any of the transmitting magnetic compasses using a follow-up system, described in Chapter 8, may be gyro-stabilized without difficulty. Indirect monitoring, as in Fig. 9.28, is used and a schematic arrangement of the equipment is shown in Fig. 9.37.

The compass unit C, amplifier A_1 and follow-up motor FM with transmitter TS form in themselves a transmitting compass system, the electrical circuit being indicated by dotted lines. The compass may have any of the pick-up devices described in Chapter 8. The gyro-stabilizing means consists of the gyroscope G and transmitter T feeding a differential receiver DR, the output of which feeds amplifier A_2. The resetting receiver (or motor) RS is coupled to the rotor of the receiver DR, while the monitoring motor MM is coupled through a large reduction gear to its stator, which is rotatable with respect to the rotor. The circuit is rearranged so that the output from A_2, instead of that from A_1, feeds the follow-up motor; the transmitter TS drives the resetter RS so that DR-A_2-FM-TS-RS is a closed follow-up loop controlled by the transmitter T in the gyro-unit. The follow-up member of the compass C, which is also part of the loop, moves in step with the transmitter T and so is orientated according to the angular position of the gyroscope's axis.

The output from the pick-up of the compass remains connected to the amplifier A_1, but the amplified signal now drives the monitoring motor MM. Thus any difference between the direction transmitted by the gyroscope and that indicated by the magnetic compass gives rise to a monitoring signal, which is amplified by A_1. This energizes the monitoring motor; hence the difference between the two directions is corrected by the rotation of the stator of DR.

Outgoing transmission to repeaters may be obtained from the mechanical link between RS and DR by means of transmitter TT. A feature of the system is that, by simple switching, the equipment may be operated as an unstabilized transmitting compass or as an unmonitored directional gyroscope in addition to its stabilized mode.

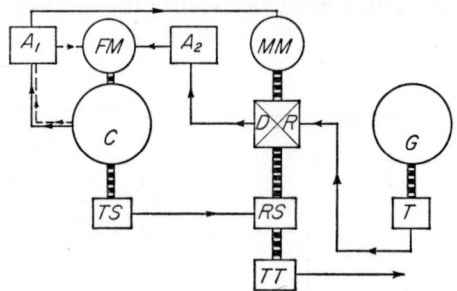

FIG. 9.37. Gyro-magnetic compass with follow-up magnetic compass

Admiralty gyro-magnetic compass, Type 5

A typical example of the system described above is the Admiralty gyro-magnetic compass, Type 5,* which is widely used in naval vessels. Its schematic arrangement is shown in Fig. 9.38. The master unit (see Fig. 9.39) contains a pivoted-needle magnetic compass with a Wheatstone bridge type of liquid-resistance network as a pick-up device (as in the A.T.M.C. system described in Chapter 8), a follow-up motor and a step-by-step transmitter. The unit is gimballed and mounted in a light-alloy metal binnacle containing the usual permanent magnet correctors and supporting soft-iron spheres and a Flinders bar.

The magnetic compass, which follows conventional practice, has a 5-in. card and a ring magnet with a moment of 750 c.g.s. units. The bowl rotates within the gimbals, which are fixed and stationary with respect to the binnacle. This differs from the arrangement of the A.T.M.C.2 (see Chapter 8) and avoids certain types of gimbal error. The bowl is carried on the azimuth gear wheel which is driven by the follow-up motor, its electrodes enter from underneath through ceramic insulating bushes, and a locking pin may be engaged to secure the bowl so that an internal lubber's line marks the fore-and-aft line of the ship, enabling the compass to be used as a simple non-transmitting instrument. An annular scale divided into 360° is attached to and encircles the upper part of the bowl, indicating compass heading when the equipment is operating as a transmitting compass, stabilized or otherwise. Below the bowl a small cylindrical coil is fitted, forming part of the residual deviation system described later.

The amplifiers A_1 and A_2 in Fig. 9.37 are mounted on a common chassis, which with the repeater panels, a step-by-step retransmission unit, a small a.c. motor

* *Handbook of the Admiralty Gyro-magnetic Compass Type* 5, Admiralty Compass Observatory, 1953, B.R. 1788.

generator and a DATEC unit are assembled in a console similar to that shown in Plate 9.8. The DATEC, which stands for 'Differential alignment unit and total error corrector' (see Fig. 9.40), embodies the monitoring motor and gearbox, differential receiver, resetter, transmitter, alignment indicator and variation and residual deviation corrector mechanism. The gyro-unit is separate, and is supported in gimbals. The gyroscope is $2\frac{1}{4}$ in. in diameter and rotates at 24 000 r.p.m.

To return to Fig. 9.38: the azimuth information from the gyro-unit is transmitted by the Aysynn (synchro) transmitter to the DATEC. The signal from the Aysynn receiver is amplified by the follow-up amplifier, whose output drives the follow-up motor in the master unit. This rotates the bowl of the compass and drives the M-type

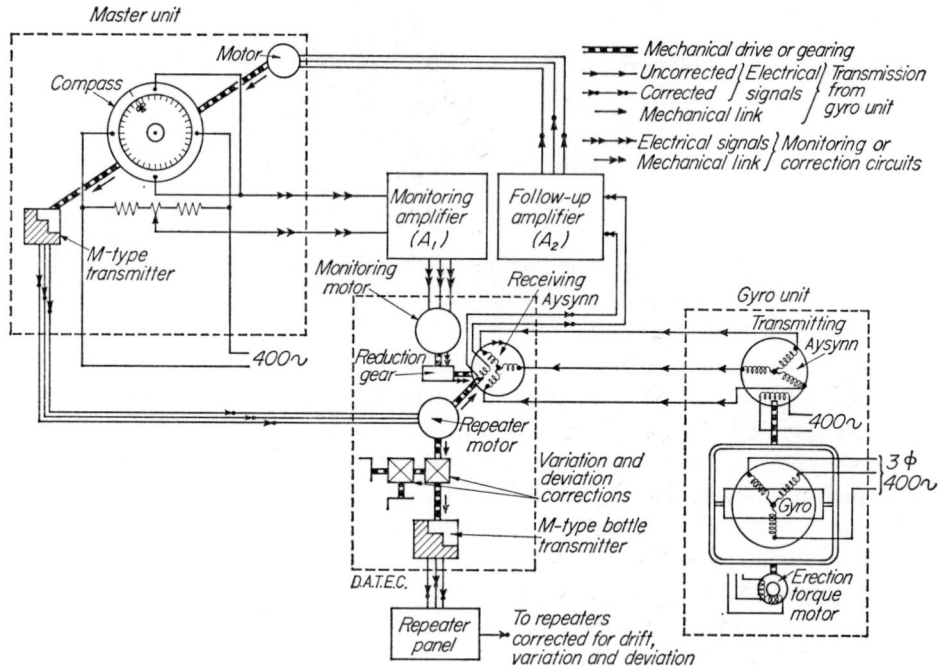

FIG. 9.38. Admiralty gyro-magnetic compass, Type 5
Schematic diagram

transmitter connected to the DATEC's repeater motor which, through mechanical gearing, 're-sets' the Aysynn receiver. The signal from the compass pick-up circuit is amplified by the monitoring amplifier and controls the monitoring motor. Correction for variation is applied through a differential gear in the main transmitter drive from the DATEC.

The residual deviation corrector (shown in Fig. 9.40) is described in more detail in Chapter 10 where compass correction methods are discussed. The two-speed (4° and 40° per min) monitoring control is similar to that illustrated in Fig. 9.32. Switching is provided so that in addition to its operation as a gyro-magnetic compass, the equipment may be used as an unstabilized (A.T.M.C.) system, or as a transmitting directional gyroscope.

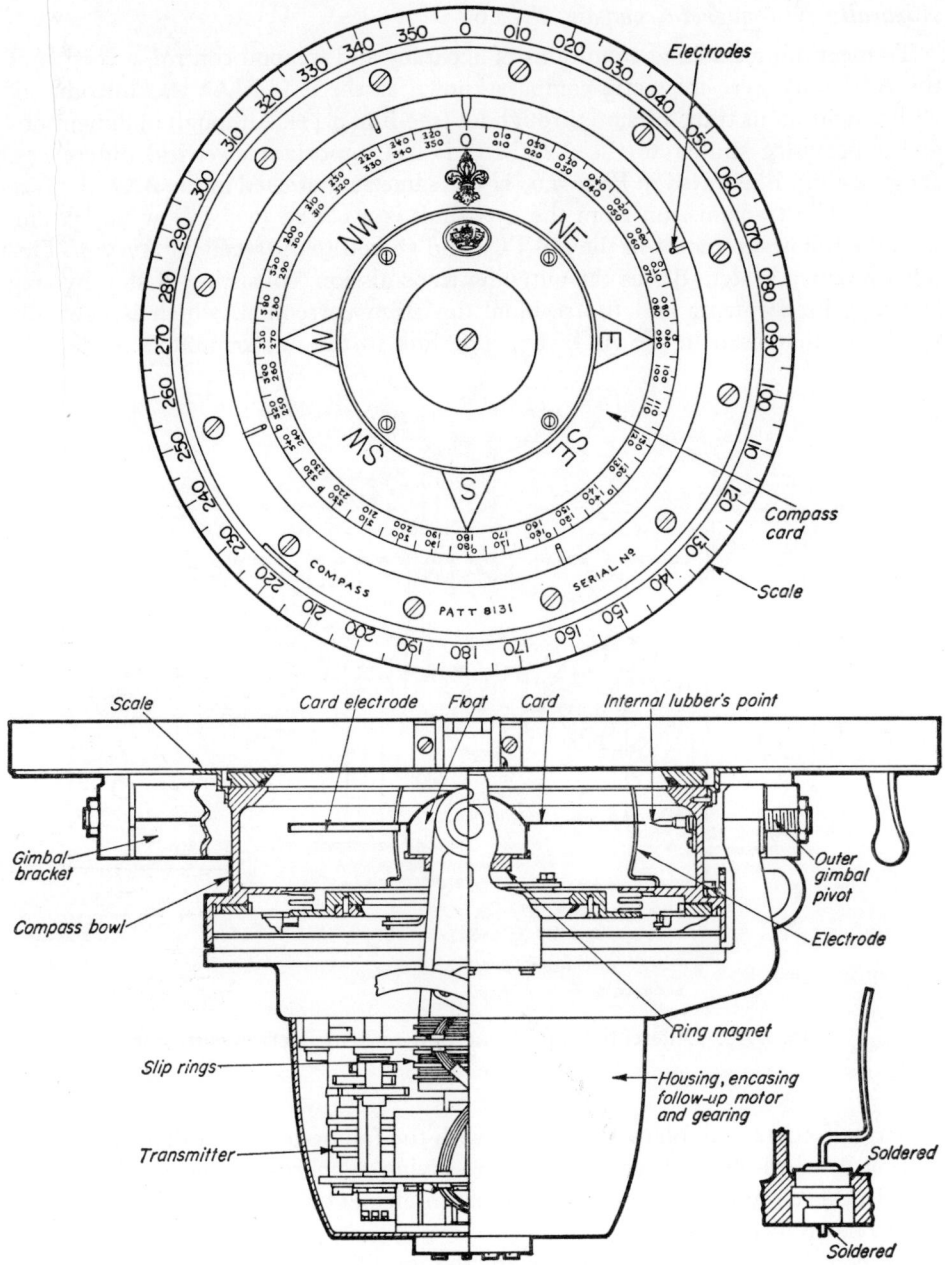

FIG. 9.39. Admiralty gyro-magnetic compass, Type 5
Master unit

In the event of a power failure, the master unit is used as a conventional magnetic compass. The small a.c. motor generator provides the a.c. supply for the amplifiers and the 3-phase supply for the gyroscope. The equipment is driven from the main 24-volt d.c. system of the ship and requires about 240 watts.

Admiralty gyro-magnetic compass, Type 6

To meet more exacting conditions of accuracy and weapon control, a version of the Admiralty gyro-magnetic compass known as the A.G.M.6* was introduced, with synchronous transmission throughout (see Fig. 9.41). Although in dimensions and appearance similar to the A.G.M.5, it has important electrical differences. Its gyro-unit, illustrated in Plate 9.6, is the same as that used in the A.G.M.7 (see p. 254). The transmission from the gyro-unit is a closed loop follow-up system, with the follow-up motor in the DATEC and geared to the receiving Aysynn. This self-contained system drives the outgoing transmission, consisting of step-by-step and magslip elements and the residual deviation correction, which is generally similar to that used in the A.G.M.5. The link to the master unit is made by a

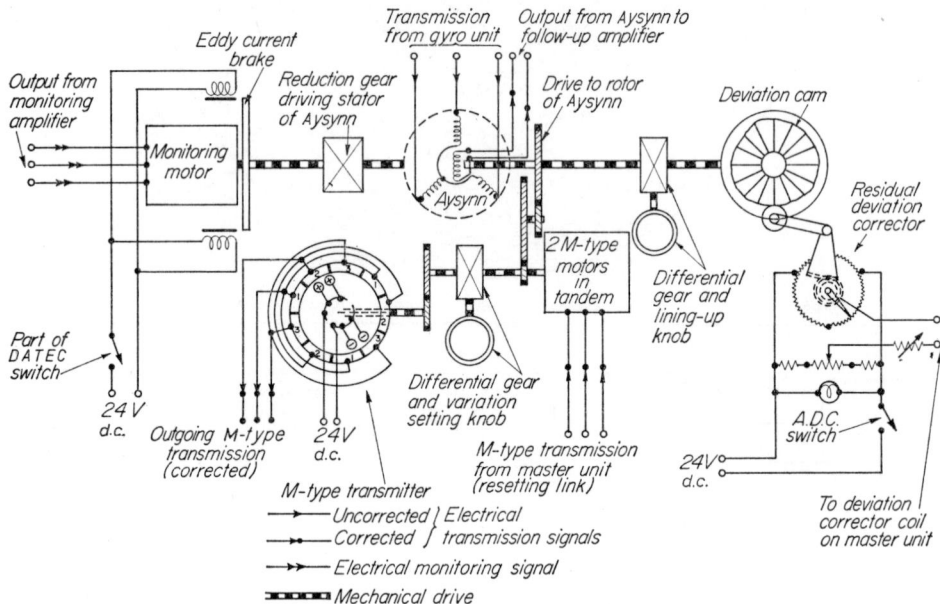

FIG. 9.40. Differential alignment unit and total error corrector
Schematic diagram

second self-contained follow-up system, with the follow-up motor driving a transmitting synchro-element, coupled to a receiving synchro-element at the master unit. In this circuit, a follow-through synchro-element allows for variation correction to be applied.

The compass follow-up is completed by the amplifier and compass motor coupled to the bowl. The bowl thus repeats the azimuth information transmitted by the gyro unit, effecting the appropriate monitoring control in the same way as in the A.G.M.5. In this case, however, three monitoring rates are provided, 4°, 0°·8 and

* *Handbook of the Admiralty Gyro-magnetic Compass Type* 6, Admiralty Compass Observatory, 1957, B.R. 109.

National Research Development Corporation, 'Improvements in navigational compass systems'. Brit. Pat. 690011, 1951 (Admiralty gyro-magnetic compass—inventor, A. Hine).

o°·4 per min. Provision is also made for using the equipment as an unstabilized compass or as a directional gyro-system.

The three identical amplifiers have a phase-advance section where the a.c. signal is demodulated and then applied as a 'gating' signal to the a.c.-driven output valves. This system is more stable and less liable to hunt than a system using a simple amplifier, as in the A.G.M.5. The equipment requires a power supply of 115-volt single-phase 400 c/s, the 3-phase supply for the gyroscope being obtained from a phase-convertor unit in the console. The power consumption is about 250 volt-amperes.

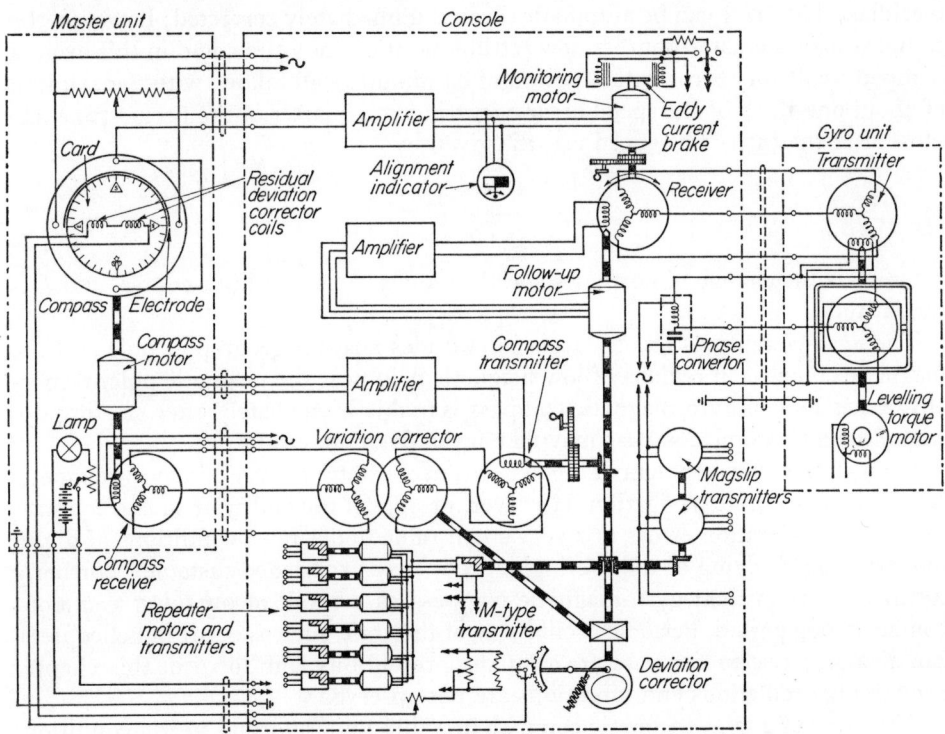

FIG. 9.41. Admiralty gyro-magnetic compass, Type 6

The console for the A.G.M.6, which is typical of those used in A.G.M. equipment, is shown in Plate 9.8. The magnetic compass and the binnacle (Plate 9.9) are identical with those used in the A.G.M.5.

PERFORMANCE OF GYRO-MAGNETIC COMPASSES IN SHIPS AND AIRCRAFT

The function of gyro-stabilization as applied to a magnetic compass is to reduce acceleration errors. It must be remembered that, on a stationary platform, no increase of accuracy results from stabilization; rather the reverse, since if any random drift of the gyroscope occurs, a stick-off error must arise.

Therefore in a large and stable ship in calm weather, little is to be gained by gyro-stabilization and a simple magnetic compass is likely to give a more correct definition of ship's head. However, the effect of rolling and pitching, manoeuvring at high speed and sundry vibrational disturbances will produce conditions in which stabilization would be advantageous.

The modern gyro-magnetic compass in a moving ship is as accurate as the gyroscopic compass, provided that care has been taken in adjustment and that variation is correctly applied. It has a further advantage in that after a turn it settles quickly and is free from an effect peculiar to certain gyroscopic compasses known as 'ballistic deflection', a result of northerly speed error. Discussion of this error is not within the scope of the present work, and it suffices to say that a northerly speed component may cause certain types of gyro-compass to settle to the west of the true meridian. The error can be automatically and immediately corrected; however, the compass may oscillate about its new settling position for a time, and in this event a damped oscillatory error with a period of 84 minutes will follow, with a maximum of about one-third of the speed error occurring at a quarter of the period (21 minutes) after the turn. The speed error is given by

$$\frac{V \cos \zeta_1}{5\pi \cos L}$$

where V is the change of northerly speed in knots, ζ_1 is the ship's course, and L is the latitude.

The gyro-compass is less accurate in latitudes near the geographic pole and the magnetic compass in regions of low magnetic field near the magnetic pole. Generally, a magnetic or gyro-magnetic compass is usable immediately after starting up, whereas a gyroscopic compass may take up to five hours to settle.

In an aircraft, the problem of high speeds necessitates the use of a gyro-magnetic compass for accurate navigation. However, in spite of the ability of such a compass to reduce acceleration errors to a very small value under most conditions of flight, modern aircraft flying at today's high speeds may experience sustained northerly turning errors in their gyro-magnetic compass as a result of pursuing a sinuous course of long period. Periodic oscillations of the craft, such as occur in helicopters, can also give rise to compass errors if the precautions laid down in this chapter concerning oscillations and vibrations are not observed.

The error of a gyro-magnetic compass in a ship under normal steady conditions is represented by a standard deviation, σ, of about ± 10 minutes of arc, and in an aircraft, of about ± 45 minutes of arc, always provided a good magnetic position has been chosen.

Chapter 10

The Correction of the Compass Position
in a
Ship or Aircraft

In the days of wooden ships, the magnetic compass gave a reasonably accurate indication of the ship's head. The proximity of cannon balls and other iron hardware (not to speak of the pistols that Capt. Bligh kept in the binnacle) doubtless caused unexpected errors or deviations of the compass, but it was not until iron and steel ships came into general use that these deviations assumed serious proportions. Notable tragedies, for instance the occasion when H.M.S. *Apollo* went ashore while escorting a convoy of merchantmen in the year 1804, and the loss of H.M.S. *Hero* and *Defence* in 1811, were attributed eventually to unknown compass errors; and the Admiralty Compass Committee, which was the progenitor of the present Admiralty Compass Division, was set up to inquire into the problem. G. B. Airy,* then Astronomer Royal, and Lord Kelvin, basing their methods on Poisson's† fundamental equations, devised a satisfactory means of correcting the *compass position* in a ship, so that the effects of the ship's magnetism were cancelled by local magnetic fields, leaving the Earth's field reasonably undisturbed to act on the compass.‡

It must be stressed that it is the compass position and not the compass that is corrected.

TYPES OF SHIP'S MAGNETISM

The ironwork of a ship may be magnetized whilst the ship is being built. The Earth's magnetic field, contact with electrical machinery or magnetic grabs, or the working on the material during construction, all have their effect.

Fig. 10.1 is an example of a ship built in the United Kingdom on a slip pointing magnetic north. The total field, being inclined at about 67° to the horizontal, passes through the ship as shown, with the result that the upper-works and stern of the ship receive 'blue' magnetic polarity, and the forepart and keel receive 'red'

---

* Airy, G. B., *Phil. Trans. roy. Soc.*, 1839, **3**; *Proc. I.N.A.*, 1859.

† Poisson, D., Memoir to the Institut de France, 1824.

‡ For a more comprehensive discussion of the theory of compass deviations and correction, the reader is referred to *The Theory of the Deviations of the Magnetic Compass* (1948), Admiralty Compass Observatory, BR.101(48).

magnetic polarity. A compass placed at X would thus be subject to disturbing fields as though a magnet were placed horizontally nearby with a red pole forward, and a further magnet placed vertically with a blue pole abaft the compass, as in Fig. 10.2.

Although the magnetism of a ship is very complex and not always stable, it is practical and convenient to divide it into two categories:

(*a*) Permanent magnetism or hard iron,
(*b*) Induced magnetism or soft iron.

An intermediate stage, sometimes called sub-permanent magnetism, is considered to be present, but this is probably mainly due to instability in the magnetic property of some of the ship's structure caused by a gradual magnetization or demagnetization of certain types of iron.

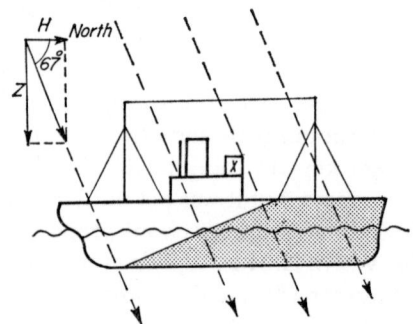

FIG. 10.1. Ship's magnetism induced by the Earth's field during shipbuilding

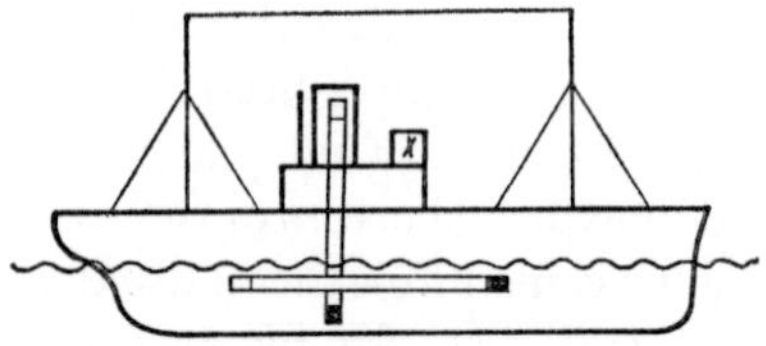

FIG. 10.2. Ship's magnetism represented by equivalent magnets

Permanent magnetism

Permanent magnetism is due to plates, stanchions, masts and other items which retain their magnetism and suffer no change as the ship alters heading. Slow changes may occur over a period of time owing to the 'working' or settling down of the vessel, but in general the magnetism may be considered equivalent to that of a bar magnet, and reasonably stable.

Induced magnetism

Some of the ship's structure will consist of iron that does not retain its magnetism and is magnetized only when lying in a magnetic field. Since the Earth's magnetic field is always passing through the ship's structure, the iron, soft though it is, is magnetized, but in a direction depending upon that of the ship's head.

Assume the ship in Fig. 10.1 to be built entirely of soft iron. When heading north, she would be magnetized as shown, but if she turned to south, her magnetism would be as indicated in Fig. 10.3, showing a reversal of horizontal magnetism though no reversal of vertical magnetism. The horizontal magnet in Fig. 10.2 would therefore have to be reversed. On east and west headings, the ship would be affected as shown in Fig. 10.4, showing a complete reversal of magnetism athwartships as the ship turns from east to west.

In the southern hemisphere the vertical magnetism would undergo reversal and the upper works and funnel would show red polarity uppermost.

268

At the compass position, the magnetic field due to the permanent magnetism of the ship may conveniently be divided into three orthogonal components as shown in Fig. 10.5. They are:

Component P in the fore-and-aft direction, positive when directed forward.

Component Q in the athwartship direction, positive when directed to starboard.

Component R perpendicular to the keel, positive when directed downward.

Fig. 10.2 illustrates components P and R.

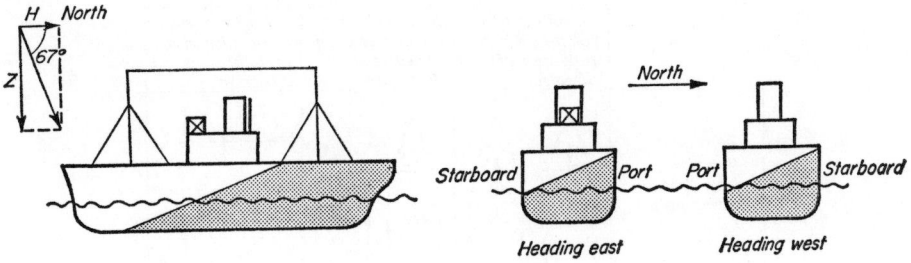

FIG. 10.3. Ship's 'soft-iron' magnetism reversed by reversal of ship's course

FIG. 10.4. Reversal of 'soft-iron' magnetism

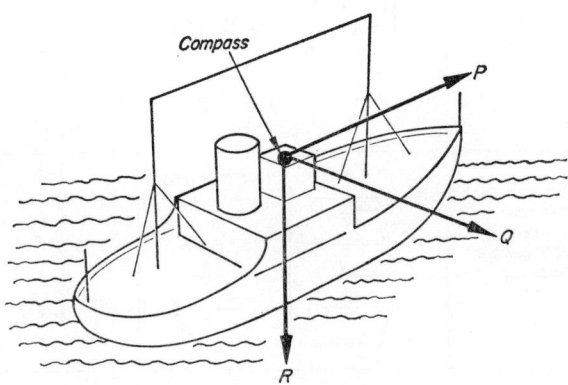

FIG. 10.5. Components of ship's magnetism

Components P, Q and R, as drawn, are all positive

The field due to the induced magnetism is more complex and needs to be divided into a greater number of components. The soft-iron distribution is represented by nine imaginary soft-iron 'rods' disposed orthogonally as shown in Fig. 10.6. These rods are magnetized by induction and take up the polarity corresponding to the direction of the ship with respect to the direction of the Earth's magnetic field. The vertical rods suffer reversal of magnetism only as the ship moves from one magnetic hemisphere to the other. The horizontal rods reverse their magnetism for a 180° change of heading, but their effect on the compass depends on the relationship of

269

the compass position to these imaginary rods. Whereas the strength of the permanent magnetism remains unaltered with change of intensity of the Earth's magnetic field, the strength of the induced magnetism depends on this intensity.

Poisson expressed the magnetic field at the compass position of a ship by the equations:

$$X_1 = X + aX + bY + cZ + P$$
$$Y_1 = Y + dX + eY + fZ + Q$$
$$Z_1 = Z + gX + hY + kZ + R$$

where X, Y and Z are three orthogonal components of the Earth's magnetic field;

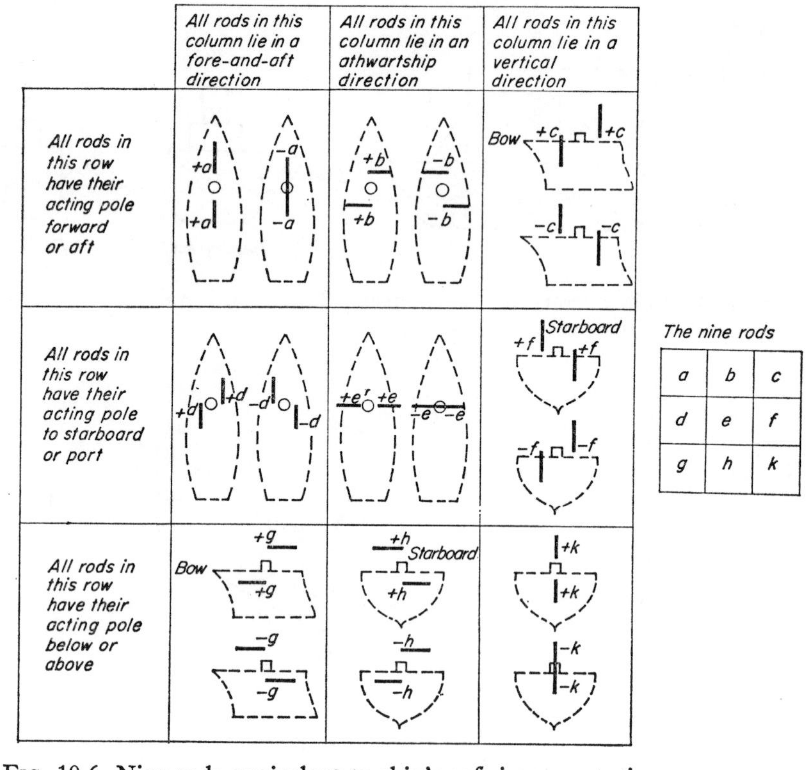

FIG. 10.6. Nine rods equivalent to ship's soft-iron magnetism

X_1, Y_1 and Z_1 are three components of the total field at the compass position, directed respectively to forward, to starboard and perpendicular to the keel; a to k represent the nine soft-iron rods, a mathematical conception, as the rods as such have no physical existence; and P, Q and R are the components of the permanent magnetic field.

A further consequence of the ship's magnetism is to alter the strength of the resultant field acting on the compass needle, known as the directive force. A constant λ is used to express the ratio of the resultant force to the free magnetic field of the Earth. The ship in Fig. 10.1 would, if permanently magnetized, cause an increase of directive force at the compass position when heading south, and a decrease when heading north. If the vessel were entirely of soft iron, the horizontal magnetism would tend to reduce the directive force on both headings.

Deviations of the compass due to permanent magnetism

Component P. The effect at the compass position of a magnetic field in a fore-and-aft direction (component P) is shown in Fig. 10.7. Here the ship has a red pole forward which gives rise to $-P$. She is shown in eight positions, corresponding to cardinal and intercardinal compass headings.

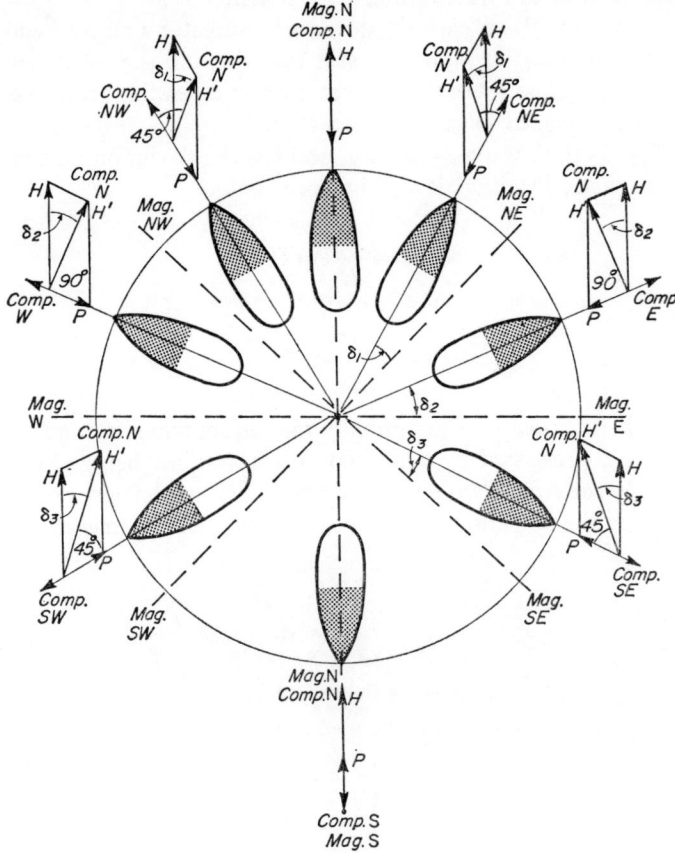

FIG. 10.7. Ship with $-P$ component

$$\lambda = \frac{H'}{H} = \frac{\sin(\delta + \zeta')}{\sin \zeta'}$$

For small angles,

$$\lambda = \frac{\delta}{\tan \zeta'} + 1$$

or

$$\lambda = \frac{P}{H} \cos \zeta' + 1$$

$$\sin \delta = \frac{P}{H} \sin \zeta'$$

For small angles,

$$\delta = \frac{P}{H} \sin \zeta'$$

On north, the ship and the Earth have magnetic fields in opposite directions but parallel to each other. The resultant field is still magnetic north and the compass exhibits no error or deviation, though the directive force is reduced. On south, once again the two fields are parallel but in the same direction, giving no deviation and an increase in directive force. On other headings, there will be deviations since the ship's field will be inclined to that of the Earth. The compass needle will lie along the resultant field and the deviation is said to be easterly or positive if the resultant is inclined to the east of magnetic north, and westerly or negative if the resultant is inclined to the west of magnetic north.

(*Note:* A compass with an easterly deviation indicates a ship's head lower than it is in fact, and the positive sign indicates that the addition of easterly deviation to ship's head by compass gives the correct magnetic heading; conversely, westerly deviation, having a negative sign, is subtracted.)

In Fig. 10.8, ζ' is the compass heading, and δ is the deviation, so that $\zeta (= \zeta' + \delta)$ is the ship's magnetic heading. It will be seen that

$$\sin \delta = \frac{P}{H} \sin \zeta'$$

or, if δ is small,

$$\delta = \frac{P}{H} \sin \zeta'$$

which is zero when ζ' is $0°$ or $180°$ and at a maximum when ζ' is $90°$ or $270°$. This deviation can be represented by a sine curve—deviations being plotted against ζ', the ship's head *by compass*. Two equal and opposite maxima occur on east and west and for this reason the deviation is known as semicircular. Its magnitude is inversely proportional to H.

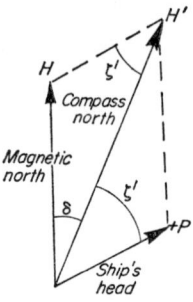

FIG. 10.8 Effect of component $+P$

The result can also be expressed in terms of ζ. Let H' (Fig. 10.8) be the resultant field acting on the compass needle:

$$\frac{H'}{\sin (\zeta' + \delta)} = \frac{H}{\sin \zeta'}$$

Thus

$$\sin \zeta' = \frac{H}{H'} \sin (\zeta' + \delta) = \frac{H}{H'} \sin \zeta$$

Therefore

$$\sin \delta = \frac{P}{H} \cdot \frac{H}{H'} \sin \zeta = \frac{P}{H'} \sin \zeta$$

or

$$\sin \delta = \frac{P}{\lambda H} \sin \zeta$$

in which H' has been set equal to λH, λ being the ratio of the strength of the horizontal field at the compass needle to the strength of the Earth's horizontal field in free space.

Component Q. Similarly, a magnetic field at the compass position in an athwartship direction (component Q) can be shown to produce a semicircular deviation also inversely proportional to H, but with maxima at north and south by compass. In this case

$$\sin \delta = \frac{Q}{H} \cos \zeta'$$

or, for small angles,

$$\delta = \frac{Q}{H} \cos \zeta'$$

Component R. A magnetic field at the compass position perpendicular to the keel (component R) produces no deviation as long as the ship is vertical.

Should, however, the ship list or heel through an angle, i, there will be a deviating force, $R \sin i$, to port or starboard, which, like component Q, produces a deviation dependent on $\cos \zeta'$ (see Fig. 10.9). If the ship is pitching, the deviating force is to forward or aft and depends on $\sin \zeta'$.

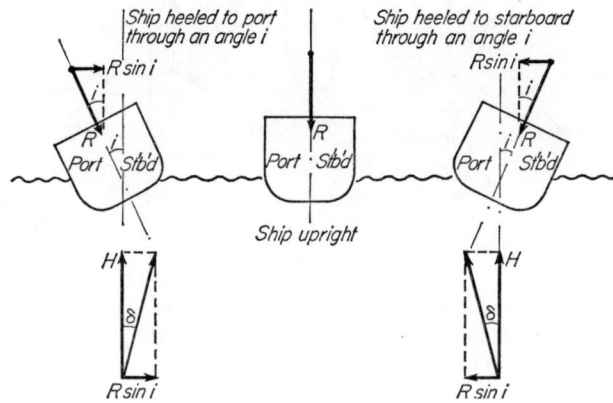

FIG. 10.9. Effect of component $+R$ with ship heading north

For rolling, when δ is small,

$$\delta = \frac{R}{H} i \cos \zeta'$$

For pitching, when δ is small,

$$\delta = \frac{R}{H} i \sin \zeta'$$

273

Deviations of the compass due to soft iron

The imaginary soft-iron rods, a, b, d and e, when lying in the Earth's magnetic field, produce a good approximation to the soft-iron magnetism of the ship in the horizontal plane when she is on an even keel. Rods a and e provide what is known as symmetrical soft iron, and b and d unsymmetrical soft iron.

When the a and e rods are lying in the direction of magnetic north, they are magnetized parallel to the Earth's horizontal component and so do not cause any deviation of the compass, though the directive force is reduced with $-a$ and $-e$

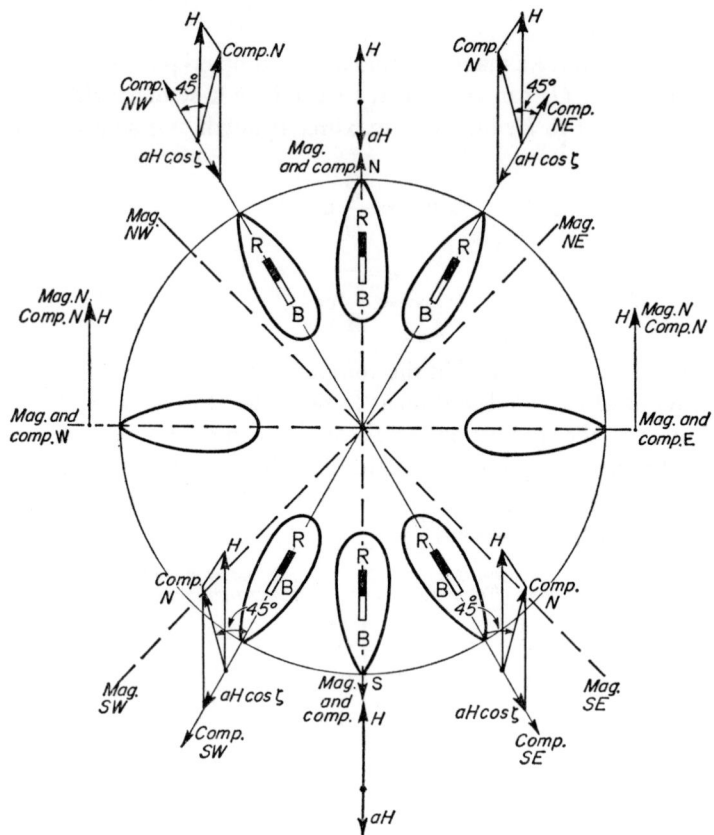

FIG. 10.10. Ship with $-a$ rod

A very unusual condition, it provides the best and simplest illustration of induced magnetism and quadrantal deviation.

rods and increased with $+a$ and $+e$ rods. On south there is also no deviation, but since the rods are still magnetized in the same sense, though reversed as far as the ship is concerned, the directive force is still reduced with $-a$ and $-e$ rods and increased with $+a$ and $+e$ rods. When lying east or west, the rods are not magnetized and so no deviation is produced on these headings, neither is the directive force affected. Thus there are four zero positions on compass north, east, south and west with four maxima in between. This form of deviation is known as quadrantal (see Fig. 10.10) and is independent of H.

274

Rods b and d also produce quadrantal deviations. In addition, a constant error may appear on all headings. Thus there are two positions of zero deviation, on north and south for rod b and on east and west for rod d. Maxima of similar sign occur at east and west with rod b and at north and south with rod d. The directive force is not affected with the ship heading north, east, south or west, but $+b$ and $+d$ rods increase the directive force on north-west and south-east and reduce it on north-east and south-west. With $-b$ and $-d$ rods, the increase occurs on north-east and south-west and the reduction on north-west and south-east.

It will be shown that

$$\sin \delta = \frac{(a-e)}{2\lambda} \sin 2\zeta$$

for symmetrical soft iron; or approximately

$$\delta = \frac{(a-e)}{2} \sin 2\zeta'$$

and

$$\sin \delta = \frac{(d-b)}{2\lambda} + \frac{(d+b)}{2\lambda} \cos 2\zeta$$

for unsymmetrical soft iron; or approximately

$$\delta = \frac{(d+b)}{2} \cos 2\zeta'$$

The two approximations apply only in a well-placed compass position when $(d-b)/2 \simeq 0$ and a and e are small.

Rods a and e. In Fig. 10.11,

$$H' \cos \zeta' = aH \cos \zeta + H \cos \zeta \qquad (10.1a)$$

$$-H' \sin \zeta' = - eH \sin \zeta - H \sin \zeta \qquad (10.2a)$$

Multiply (10.1a) by sin ζ, (10.1b) by cos ζ:

$$H' \sin \zeta \cos \zeta' = aH \sin \zeta \cos \zeta + H \sin \zeta \cos \zeta$$

$$-H' \sin \zeta' \cos \zeta = - eH \sin \zeta \cos \zeta - H \sin \zeta \cos \zeta$$

Adding,

$$H' \sin (\zeta - \zeta') = H \sin \zeta \cos \zeta (a-e) \qquad (10.3a)$$

or

$$\frac{H'}{H} \sin \delta = \frac{(a-e)}{2} \sin 2\zeta \qquad (10.4a)$$

Therefore

$$\sin \delta = \frac{(a-e)}{2\lambda} \sin 2\zeta$$

When δ is small,

$$\delta = \frac{(a-e)}{2\lambda} 2 \sin \zeta \cos \zeta$$

Since

$$\lambda = \frac{\sin \zeta}{\sin \zeta'}$$

$$\delta = (a-e) \sin \zeta' \cos \zeta$$
$$= (a-e) \sin \zeta' \cos (\zeta' + \delta)$$
$$= (a-e)(\sin \zeta' \cos \zeta' - \delta \sin^2 \zeta')$$

In a well-chosen compass position, a and e should be small; thus $\delta(a-e)$ is very small. Hence a good approximation in the case of small deviations is

$$\delta = \frac{(a-e)}{2} \sin 2\zeta' \qquad (10.5a)$$

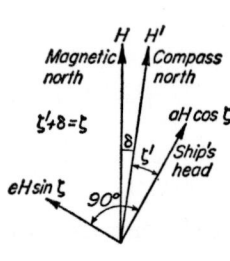

FIG. 10.11. Rods a and e
$\zeta' + \delta = \zeta$

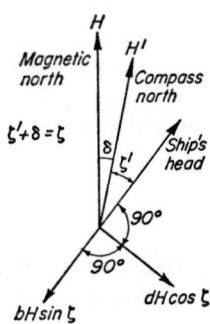

FIG. 10.12. Rods b and d
$\zeta' + \delta = \zeta$

Rods b and d. In Fig. 10.12,

$$H' \cos \zeta' = -bH \sin \zeta + H \cos \zeta \qquad (10.1b)$$
$$-H' \sin \zeta' = dH \cos \zeta - H \sin \zeta \qquad (10.2b)$$

Multiply (10.1b) by sin ζ and (10.2b) by cos ζ:

$$H' \sin \zeta \cos \zeta' = -bH \sin^2 \zeta + H \sin \zeta \cos \zeta$$
$$-H' \sin \zeta' \cos \zeta = dH \cos^2 \zeta - H \sin \zeta \cos \zeta$$

Adding:

$$H' \sin (\zeta - \zeta') = H(d \cos^2 \zeta - b \sin^2 \zeta) \qquad (10.3b)$$

or

$$\frac{H'}{H} \sin \delta = d \cos^2 \zeta - b(1 - \cos^2 \zeta)$$

$$= d \cos^2 \zeta - b + b \cos^2 \zeta$$

$$= -b + \cos^2 \zeta(d+b)$$

276

Therefore

$$\frac{H'}{H} \sin \delta = -b + \frac{(d+b)}{2}(\cos 2\zeta + 1)$$

$$= -b + \frac{d}{2} + \frac{b}{2} + \frac{\cos 2\zeta}{2}(d+b)$$

Therefore

$$\sin \delta = \frac{(d-b)}{2\lambda} + \frac{(d+b)}{2\lambda}\cos 2\zeta \qquad (10.4b)$$

For δ to be small, the compass must be well placed so that $(d-b)$ is negligible and $(d+b)$ is small. Then

$$\delta = \frac{(d+b)}{2\lambda}\cos 2\zeta$$

$$= \frac{(d+b)}{2\lambda}\cos 2(\zeta' + \delta)$$

or

$$\delta = \frac{(d+b)}{2\lambda}(\cos 2\zeta' - 2\delta \sin 2\zeta')$$

Since $\lambda = \sin \zeta / \sin \zeta'$,

$$\delta = \frac{(d+b)}{2} \cdot \frac{(\cos 2\zeta' \sin \zeta' - 2\delta \sin 2\zeta' \sin \zeta')}{\sin \zeta' + \delta \cos \zeta'}$$

Neglecting terms in δ^2 and in $(d+b) \times \delta$, a good approximation for a well-placed compass is

$$\delta = \frac{(d+b)}{2}\cos 2\zeta' \qquad (10.5b)$$

The remaining horizontal rods h and g cause heeling and pitching errors, but this effect is apparent to any appreciable extent only in a badly placed compass position. Rods b, d and f also cause heeling and pitching errors in a badly placed compass. Rod e contributes to the heeling error in most cases and produces a deviating force additional to that due to component R equal to

$$eZ \sin i \cos i \quad (= eZ \sin i, \text{ when } i \text{ is small})$$

Similarly, rod a contributes to the pitching error.

The remaining soft-iron rods, c, f and k are magnetized by the vertical component of the Earth's magnetic field.

Rods c and f produce poles proportional in strength to the vertical component of the Earth's field either forward or aft (rod c), or to port or starboard (rod f) of the compass. They therefore produce effects which are additional to the permanent iron components P and Q and are proportional to Z. Consequently, the total fore-and-aft deviating force is $P+cZ$ and the total athwartship deviating force is $Q+fZ$.

Thus the fore-and-aft deviation is given by

$$\sin \delta = \frac{(P+cZ)}{\lambda H} \sin \zeta$$

$$= \frac{1}{\lambda}\left(\frac{P}{H} + c \tan D\right)\sin \zeta \qquad (10.6a)$$

where D is the dip angle; or, for small angles

$$\delta = \left(\frac{P}{H} + c \tan D\right)\sin \zeta' \qquad (10.6b)$$

and the athwartship deviation, by

$$\sin \delta = \frac{(Q+fZ)}{\lambda H} \cos \zeta$$

$$= \frac{1}{\lambda}\left(\frac{Q}{H} + f \tan D\right)\cos \zeta \qquad (10.7a)$$

or, for small angles,

$$\delta = \left(\frac{Q}{H} + f \tan D\right)\cos \zeta' \qquad (10.7b)$$

Rod k produces similar effects to component R, positive in the northern hemisphere and negative in the southern, owing to its being magnetized by the vertical component Z of the Earth's magnetic field.

The deviating force is

$$F = kZ \sin i \cos i \qquad (10.8)$$

To summarize:

The combinations of P and c, and of Q and f give semicircular deviations.

The combinations of a, b, d and e give a constant deviation and quadrantal deviations.

Semicircular deviations are functions of ζ' or ζ and are inversely proportional to H. Quadrantal deviations are functions of $2\zeta'$ or 2ζ and are independent of H.

Exact coefficients

The total magnetic field to the east, which is also the easterly deviating field, is given by

$$H' \sin \delta = H\left[\frac{(d-b)}{2} + \left(\frac{P}{H} + c \tan D\right)\sin \zeta + \left(\frac{Q}{H} + f \tan D\right)\cos \zeta\right.$$

$$\left. + \frac{(a-e)}{2}\sin 2\zeta + \frac{(d+b)}{2}\cos 2\zeta\right] \qquad (10.9)$$

derived from equations (10.4a), (10.4b), (10.6a) and (10.7a). The same equations that were used to establish these relationships, viz: (10.1a), (10.2a), (10.1b), (10.2b)

(10.6*a*) and (10.7*a*) can be used to show that the total magnetic field to the north, which is a deviating field combined with H, is given by

$$H' \cos \delta = H\left[1 + \frac{(a+e)}{2} + \left(\frac{P}{H} + c \tan D\right)\cos \zeta - \left(\frac{Q}{H} + f \tan D\right)\sin \zeta \right.$$

$$\left. + \frac{(a+e)}{2} \cos 2\zeta - \frac{(d+b)}{2} \sin 2\zeta \right] \quad (10.10)$$

These last two equations may be written:

$$H' \sin \delta = \lambda H(\bar{A} + \bar{B} \sin \zeta - \bar{C} \cos \zeta + \bar{D} \sin 2\zeta + \bar{E} \cos 2\zeta) \quad (10.11)$$

$$H' \cos \delta = \lambda H(1 + \bar{B} \cos \zeta - \bar{C} \sin \zeta + \bar{D} \cos 2\zeta - \bar{E} \sin 2\zeta) \quad (10.12)$$

where $\lambda = 1 + (a+e)/2$ and

$$\bar{A} = \frac{(d-b)}{2\lambda}$$

$$\bar{B} = \frac{1}{\lambda}\left(\frac{P}{H} + c \tan D\right)$$

$$\bar{C} = \frac{1}{\lambda}\left(\frac{Q}{H} + f \tan D\right)$$

$$\bar{D} = \frac{(a-e)}{2\lambda}$$

$$\bar{E} = \frac{(d+b)}{2\lambda}$$

These symbols are known as the exact coefficients of functions of ζ and describe completely the magnetism of the ship when she is on an even keel for all values of deviation in terms of the hard-iron components and soft-iron rods.

Dividing (10.11) by (10.12) and writing $\sin \delta / \cos \delta$ for $\tan \delta$,

$$\sin \delta = \bar{A} \cos \delta + \bar{B} \sin \zeta' + \bar{C} \cos \zeta' + \bar{D} \sin (\zeta + \zeta') + \bar{E} \cos (\zeta + \zeta') \quad (10.13)$$

which gives a relationship between the exact coefficients and the deviation on any compass course: however, the deviation in terms of compass course and coefficient is not explicitly stated.

Approximate coefficients

For small angles, a Fourier series may be used to represent the deviation on equally spaced *compass* courses, the deviation being a cyclic function of ζ'. Thus

$$\delta = A + B \sin \zeta' + C \cos \zeta' + D \sin 2\zeta' + E \cos 2\zeta' + F \sin 3\zeta'$$

$$+ G \cos 3\zeta' + H \sin 4\zeta' + \ldots \quad (10.14a)$$

where A, B, C, D, E, etc. are the so-called approximate coefficients referring to the same magnetic forces as the corresponding exact coefficients, and ζ' is the *compass* course as opposed to ζ, the *magnetic* course. This formula is commonly used in determining the deviation of a ship's compass and has the obvious advantage of depending on compass course, the only information readily available.

TABLE 10.1 *The effect of soft iron*

Rod	Type of deviation	Deviation dependent (when small) on:
a	quadrantal pitching error	$\sin 2\zeta'$, $\sin \zeta'$, $\sin i$, $\cos i$, $\tan D$
b	quadrantal constant* heeling error*	$\cos 2\zeta'$ — $\sin i$, $\cos i$, $\tan D$
c	semicircular heeling error†	$\sin \zeta'$, $\tan D$ $\sin i$, $\tan D$
d	quadrantal constant* pitching error*	$\cos 2\zeta'$ — $\sin i$, $\cos i$, $\tan D$
e	quadrantal heeling error	$\sin 2\zeta'$ $\cos \zeta'$, $\sin i$, $\cos i$, $\tan D$
f	semicircular pitching and heeling errors*	$\cos \zeta'$, $\tan D$ $\sin i$, $\cos i$, $\tan D$
g ⎫ h ⎭	heeling and pitching errors*	$\sin \zeta'$, $\cos \zeta'$, $\sin i$, $\tan D$
k	heeling and pitching errors	$\sin \zeta'$, $\cos \zeta'$, $\sin i$, $\cos i$, $\tan D$

* occurs in a badly placed compass
† disappears when the correct amount of Flinders bar is used (see p. 287)

Although any cyclic function $f(x)$ can be written as a series,

$$f(x) = a_0 + a_1 \cos x + a_2 \cos 2x + a_3 \cos 3x + \ldots + a_n \cos nx$$
$$+ b_1 \sin x + b_2 \sin 2x + b_3 \sin 3x + \ldots + b_n \sin nx$$

there seems to be no justification for extending the formula for deviation beyond terms in $2\zeta'$, since, assuming that the approximate coefficients have the same meaning magnetically as the exact coefficients, the whole magnetic state of the ship is described not only by the coefficients $\bar{A}$ to $\bar{E}$, but also by the coefficients A to E.

Moreover, the exact and approximate coefficients are related as follows:

$$\bar{A} = \sin A$$
$$\bar{B} = \sin B(1 + \tfrac{1}{2} \sin D)$$
$$\bar{C} = \sin C(1 - \tfrac{1}{2} \sin D)$$
$$\bar{D} = \sin D$$
$$\bar{E} = \sin E$$

From these, the various hard- and soft-iron constants of the ship can be ascertained.

It can be argued that coefficients F, G and H have a real meaning magnetically, since for example

$$\sin F = \frac{\bar{D}\bar{B}}{2}, \quad \sin G = \frac{\bar{D}\bar{C}}{2}, \quad \sin H = \tfrac{1}{2}(\bar{D})^2$$

but as these are functions of what should, after compass adjustment, be small angles, they can usually be disregarded.

Obviously, many observations of deviations of the compass can be analysed to reveal coefficients beyond E. These may actually exist as errors of the compass system and be peculiar to the system, such as a coefficient H, giving a term in $\sin 4\zeta'$ which arises in certain ill-designed inductor compasses; or coefficients F and G, giving terms in $\sin 3\zeta'$ and $\cos 3\zeta'$, owing to incorrectly arranged deviation-correcting devices. As far as the ship is concerned, however, anything beyond E can fairly be regarded as observational error. The existence of residual errors of this kind enables the accuracy of a set of observations to be assessed.

Provided it is certain that the compass system and any method of deviation correction are beyond reproach, the expression

$$\delta = A + B \sin \zeta' + C \cos \zeta' + D \sin 2\zeta' + E \cos 2\zeta' \qquad (10.14b)$$

correctly describes the deviation of the compass when this does not exceed 20°. (In this expression, δ, A, B, C, D, E and ζ' are all in degrees.)

After analysing a table of deviations to determine the approximate coefficients, the necessary steps can be taken to deal with the various forms of ship magnetism. Given the coefficients, a synthesis of the deviations is possible, which, when compared with the observed deviations, reveals the observational or residual errors. These are normally of a random nature and capable of statistical definition. If the residual errors appear to have a cyclic character of their own, it is probable that higher coefficients exist, owing to an incorrectly designed or sited compass. The methods of deviation analysis and synthesis and their interpretations are described on p. 295.

To summarize the properties of the approximate coefficients:

A is due to a misalignment of the compass in the ship (apparent A) or to induced magnetism in the soft-iron rods b and d (magnetic A). It is constant, i.e. independent of course ζ'.

B is due to permanent magnetism fore-and-aft (component P) and to induced magnetism in soft-iron rod c. It is *semicircular* deviation, dependent on $\sin \zeta'$.

C is due to permanent magnetism athwartships (component Q) and to induced magnetism in soft-iron rod f. It is *semicircular* deviation, dependent on $\cos \zeta'$.

D is due to induced magnetism in soft-iron rods a and e and is a *quadrantal* deviation, dependent on $\sin 2\zeta'$. This mode of deviation is sometimes known as that due to 'symmetrical' soft iron.

E is due to induced magnetism in soft-iron rods b and d, and is a *quadrantal* deviation, dependent on $\cos 2\zeta'$. This mode of deviation is sometimes known as that due to 'unsymmetrical' soft iron.

Some further mention of coefficients F and G is advisable, since in certain circumstances they may have an appreciable magnitude. They are *sextantal* deviations—dependent on $\sin 3\zeta'$ and $\cos 3\zeta'$ respectively. Their existence is primarily due to incorrectly placed corrector magnets (see section on *Methods of deviation correction*) namely, having correctors at a distance compared with which the length of the compass needle is significant. Smith and Evans* (see Appendix 5, p. 355) showed that, with a single needle, if sextantal deviations were to be reduced to an insignificant amount, the corrector magnets needed to be six times the length of the needle away if in the same horizontal plane, or three times that length if in a horizontal plane above or below the compass.

If, however, two compass needles were used, placed parallel and with their poles subtending 60° at the pivot point, the sextantal deviation disappeared, even for comparatively short distances. The use of multiple needle compasses has already been mentioned.

A similar effect may arise in the correction of soft-iron effects, and appear as even higher coefficients, H and K, dependent on $\sin 4\zeta'$ and $\cos 4\zeta'$. In a compass position where very large deviations are found and before any correction is made, it is frequently noticed that these higher coefficients occur, but disappear after correction. Generally F is associated with B, G with C, H with D, and K with E.

Constant λ. The constant λ has already been defined as the ratio of the strength of the horizontal field at the compass needle to that of the Earth's horizontal field in free space (p. 270). Components P and Q alter λ on certain headings, but the mean value of λ, measured on a number of equally spaced headings, is unity. Soft-iron rods $+a$ and $+e$ will increase the mean value of λ because, as already stated on p. 279,

$$\lambda = 1 + \frac{(a+e)}{2}$$

Similarly $-a$ and $-e$ rods reduce the mean value of λ. A reduction in λ means increased deviations due to the various iron members of the ship.

Measurement of λ. It is shown in Chapter 3, p. 30, that the strength of a magnetic field is inversely proportional to the square of the period of oscillation of a magnetic needle placed in that field. The same method may be used to compare H', the strength of the total horizontal magnetic field at the compass position (after removing the compass), with H, that at a site ashore remote from any magnetic interference. If T_1 and T are the respective periods of oscillation of the needle,

$$\lambda = H'/H = T^2/T_1^2$$

Obviously the same needle must be used on board and ashore.

Constant μ (p. 283) may be measured similarly by using a vertical or dip needle. The deflector (p. 288) may also be used to determine λ.

Heeling error
In addition to those deviations dependent solely on the ship's course and for which a deviation or correction table can be prepared, there are other deviations which arise when the ship is inclined and which are known as heeling errors.

* Smith, A., and Evans, F. J., 'On the effect produced on the deviation of the compass by the length and arrangement of the compass needles and on a new mode of correcting the quadrantal deviation', *Phil. Trans. roy. Soc.*, 1861, **161**.

In a correctly placed compass and with horizontal correctors properly adjusted, the heeling error is caused by component R (see p. 269) and soft-iron rods k and e. The latter, owing to their inclination, effectively produce an apparent k rod (k_i), an apparent f rod (f_i), an apparent h rod (h_i) and an apparent e rod (e_i).

Since, in practice, the semi-circular part alone is considered, we may write for $\bar{C}$ in the inclined condition, $\bar{C}_i$, and from p. 279,

$$\bar{C}_i = \frac{1}{\lambda}\left(\frac{Q_i}{H} + f_i \tan D\right)$$

Q_i is the component due to component R and we can write

$$Q_i = -R \sin i$$

and

$$f_i = (e_2 - k) \sin i \cos i$$

Therefore

$$\bar{C}_i = \frac{1}{\lambda}\left[\frac{-R \sin i}{H} + (e_2 - k) \sin i \cos i \tan D\right] \qquad (10.15)$$

or, for small angles of heel,

$$\bar{C}_i = \frac{i}{\lambda}\left[(e_2 - k) \tan D - \frac{R}{H}\right] \qquad (10.16)$$

e_2 being the value of e after the compass has been corrected.

Thus the deviation is given by

$$\sin \delta_i = \bar{C}_i \cos \zeta'$$
$$= \frac{i}{\lambda}\left[(e_2 - k) \tan D - \frac{R}{H}\right] \cos \zeta' \qquad (10.17)$$

which may be written $i \times J \cos \zeta'$ where

$$J = \frac{1}{\lambda}\left[(e_2 - k) \tan D - \frac{R}{H}\right] \qquad (10.18)$$

This constant J is called the heeling error coefficient.

Constant μ. Corresponding to λ in the horizontal plane, μ is the mean ratio of the vertical magnetic force at the compass to that in free space. Since the compass is always above the upper pole owing to the component Z of the Earth's magnetic field, we have to compromise and write Z_0' for Z' when Z equals a particular value Z_0.

Thus the total mean vertical force at the compass is $Z_0 + kZ_0 + R$. So

$$\mu = \frac{Z_0 + kZ_0 + R}{Z_0}$$

From this,

$$\mu = 1 + k + \frac{R}{Z_0}$$

$$= 1 + k + \frac{R}{H \tan D}$$

or

$$k = \mu - 1 - \frac{R}{H \tan D}$$

Substituting in (10.18),

$$J = \frac{1}{\lambda}(e_2 - \mu + 1) \tan D$$

For $(e_2 + 1)$ write $-\lambda(\bar{D} - 1)$. Hence

$$J = -\left(\bar{D} + \frac{\mu}{\lambda} - 1\right) \tan D \qquad (10.19)$$

which is, perhaps, the more usual expression.

The heeling error coefficient is usually negative in the northern hemisphere and the coefficient $-J$ is defined as the change in deviation to windward (the high side of the ship) for one degree of heel with the ship's head north or south. Thus

$$-J = \left(\bar{D} + \frac{\mu}{\lambda} - 1\right) \tan D$$

Similarly, the pitching error coefficient may be found. The error is maximum on east and west and the magnitude of the coefficient is the same as that of J.

METHODS OF DEVIATION CORRECTION

The existence of large deviations in a magnetic compass, however accurately they are known, is undesirable. The compass may be very 'lively' in some sectors and very 'sluggish' in others, where the rate of change of deviation is in the same sense as or opposite to the movement of the ship respectively. Uncorrected heeling error will result in an unsteady compass when the ship rolls. Moreover, permanent magnetic effects (coefficients B and C) and certain soft-iron effects, depend on change of so-called magnetic latitude, i.e. on changing values of the components of the Earth's magnetic field.

Lord Kelvin and Sir George Airy were the pioneers in the field of practical compass correction and their methods are used to this day. The principle involved is the correction of like with like. That is to say, the field due to the permanent magnetism of the ship is cancelled *at the compass position* by small permanent magnets so placed and of such dimensions that the space occupied by the compass magnets is subjected only to a magnetic field parallel to that of the Earth in free space.

Thus in a compass binnacle, there are three sets of permanent magnets. One set is placed with their length fore-and-aft. These correct coefficient B, due to fore-and-aft permanent magnetism (component P), another set has their length athwartships, for the correction of coefficient C, due to athwartships permanent magnetism (component Q), and the third set is placed vertically under the compass for the correction of vertical permanent magnetism (component R).

Soft-iron effects, mostly coefficient D and sometimes coefficient E, depending on the horizontal rods a, b, d and e, are commonly corrected by soft-iron spheres placed on the binnacle athwartships, since the coefficient D found in ships is almost invariably positive, while the spheres in that position provide a $-D$ coefficient. Coefficient D, which is due to rods a and e is usually produced by an excess of $-e$ over $-a$, giving positive values of D. Since the spheres are, in fact, $+e$ rods, they correct the excess of $-e$ so as to make $D = 0$. (See Appendix 5, p. 355). The rare instance of a ship's having $-D$ would require the spheres to be fore-and-aft. The presence of coefficient E, usually quite small if it does exist, necessitates a slew of the spheres from the athwartships direction.

The amount of slew, θ, is given by

$$\tan 2\theta = E/D$$

and the total amount of soft-iron correction to be provided by the spheres is $(D^2 + E^2)^{1/2}$.

The Flinders bar, a vertical bar of soft iron, corrects for vertical soft iron due to soft-iron rod c in the majority of cases, though in a badly placed compass the effect of rod f has to be compensated. If this occurs, the Flinders bar, which is normally placed on the centre line of the compass and forward of it, needs to be slewed.

If ϕ is the angle of slew,

$$\tan \phi = f/c$$

and the force to be corrected is $Z(c^2 + f^2)^{1/2}$.

COMPASS ADJUSTMENT (CORRECTION)

The sequence of compass adjustment is usually carried out as follows:

(1) *Correction of coefficient D by soft-iron spheres*

Place the ship with her head successively to NE., SE., SW., and NW. by compass and observe the deviations of the compass by comparison of the compass bearings and the true bearing of a distant mark or of a celestial body. Coefficient D may now be determined and the spheres of the correct size placed at the appropriate distance from the compass in accordance with the correction table applicable to a particular compass and binnacle. The placing of the spheres is particularly related to the type of compass because the induction from the compass needles contributes to some extent to the magnetization of the spheres and therefore to the amount of correction provided. The sphere, when acted upon by the Earth's field, behaves as a magnetic dipole* and the field therefrom may effectively cancel the field due to soft iron in the ship. Since both the correcting field and the deviating field are proportional to the strength of the Earth's field, coefficient D is corrected at one place for all other places except for the effect of induction from the compass needles. By careful design of both compass and binnacle it is possible to keep to small values any error as the ship moves from place to place entailing change of magnetic latitude. Any residual coefficient D will be substantially constant.

If the spheres are too large or too close, it is possible to introduce deviations of the form $H \sin 4\zeta'$, especially if a single needle is used in the compass. The incidence of coefficient H is largely eliminated when two compass needles whose poles subtend 60° at the pivot point are used.*

Since a small amount of permanent magnetism may exist in even the best soft-iron spheres, hard-iron correction is carried out after the spheres are placed. Thus any small amount of hard iron in the spheres is subsequently corrected. Another reason for correcting with spheres first is that they correct part of the heeling error due to vertical soft iron. It is advisable to check the soft-iron correction after completing correction by permanent magnets and before 'swinging' for residual deviations.

(2) *Correction of component R by vertical magnets*

The compass is removed from the binnacle and in its place is put the heeling error instrument. This is a magnetic needle mounted on knife edges and balanced in a site where the Earth's field is undisturbed, so that when placed with its axis E.–W. (i.e. the needle lying in the meridian with the appropriate end to the north) it rests horizontally.

The heeling error instrument is then placed in the plane of the compass needles again with the needle in the meridian as before. If there is vertical magnetism present, the needle will tilt. Vertical magnets are placed in the binnacle at such a distance and of sufficient number to bring the needle horizontal. Thus the vertical magnetism of the ship, both hard and soft (rod k) is cancelled. Although the hard iron (component R) is corrected in one place for all parts of the Earth, the soft-iron correction holds good only for the place of correction. This is the one exception to the rule, where the principle of correcting like with like is not observed.

(3) *Correction of components P and Q by permanent magnets*

Place the ship with her head to east by compass or to west by compass and observe deviation. Insert fore-and-aft magnets as required until the compass indicates correctly.

Place the ship with her head to north (by compass) or to south (by compass) and observe deviation. Insert athwartship magnets as required until the compass indicates correctly.

Since there is always the possibility that small coefficients A and E are present, it is advisable to place the ship on south if correction was done on north and to halve any error that may appear by readjusting the athwartship magnets. (Conversely, place the ship on north if correction was done on south.) Similarly, if correction was done on east (or west), place the ship on west (or east) and halve any error by readjusting the fore-and-aft magnets.

Complete correction will remain so in all latitudes, but residual deviations will be inversely proportional to the strength of the Earth's horizontal field component.

* See Appendix 5, pp. 357–9.

(4) *The 'swing' for observation of residual deviations*

It now only remains to 'swing' the ship in order to prepare a table of residual deviations. This is done by placing the ship on a series of equally spaced *compass headings* and determining the deviation by observing the bearing of the sun or other heavenly body (whose elevation should not exceed 30°) or the bearing of a distant mark. Care must be taken, however, in the choice of a mark sufficiently distant, or parallax errors will result. Table 10.2 is of interest in this connection.

TABLE 10.2 *Parallax errors of bearings of distant objects*

Radius of swing	Relative bearing of object	Distance of object (miles)					
		1	2	3	4	5	6
100 ft	45°, 135°	0° 40′	0° 20′	0° 15′	0° 10′	0° 05′	0° 05′
	90°	1° 00′	0° 30′	0° 20′	0° 15′	0° 10′	0° 10′
150 ft	45°, 135°	1° 05′	0° 30′	0° 20′	0° 15′	0° 10′	0° 10′
	90°	1° 25′	0° 45′	0° 30′	0° 20′	0° 15′	0° 15′
200 ft	45°, 135°	1° 20′	0° 40′	0° 25′	0° 20′	0° 15′	0° 15′
	90°	1° 55′	1° 00′	0° 40′	0° 30′	0° 25′	0° 20′
250 ft	45°, 135°	1° 40′	0° 50′	0° 35′	0° 25′	0° 20′	0° 15′
	90°	2° 25′	1° 10′	0° 50′	0° 35′	0° 30′	0° 25′
300 ft	45°, 135°	2° 00′	1° 05′	0° 40′	0° 30′	0° 25′	0° 20′
	90°	2° 50′	1° 25′	1° 00′	0° 45′	0° 35′	0° 30′

Another method of obtaining deviations is by reciprocal bearings from a 'magnetic hut' ashore. These are signalled to the ship simultaneously with the taking of the bearing of the hut from the compass. Transits may also be used, where the bearing of two or more conspicuous objects in line is known or can be ascertained from a chart.

The practice of swinging 'by comparison' with another compass is not to be encouraged, except in an emergency. Especially is it inadvisable to use a gyro-compass repeater unless it is ascertained without doubt that the gyro-compass has been settled for at least an hour.

Swinging a ship should take at least 40 minutes, so as to avoid the so-called Gaussin error due to semi-permanent magnetic effects.

Flinders bar

So far no mention has been made of the procedure involved in placing the Flinders bar. This is because it is not possible to determine c or f from observation in one place and the amount of Flinders bar cannot be ascertained until the ship has been subject to an appreciable change of magnetic latitude. When correcting component P, for instance, by permanent magnets, it is impossible to avoid correcting for rod c at the same time.

It will be remembered that

$$\bar{B} = \frac{1}{\lambda}\left(\frac{P}{H} + c \tan D\right)$$

Similarly

$$\bar{C} = \frac{1}{\lambda}\left(\frac{Q}{H} + f \tan D\right)$$

Now for small deviations,

$$\lambda H \sin B = P + cZ$$

and

$$\lambda H \sin C = Q + fZ$$

Hence by determining coefficients B and C in two largely different magnetic latitudes, it is possible to find the amount of c or f that has to be corrected and to fit the requisite amount of Flinders bar or alter that already fitted. Obviously, none of the correctors must be altered between the two swings for finding c or f.[*] The amount of Flinders bar is found from appropriate tables, as with spheres.

A Flinders bar forward corrects $-c$ and aft corrects $+c$. A slew to port corrects $+f$ and a slew to starboard $-f$. The cross section of a Flinders bar produces small $+a$ and $+e$ rods and a small readjustment of the spheres may be necessary after fitting a Flinders bar.

The deflector

When it is not possible to take bearings of a distant mark, or of the sun, a compass may be corrected approximately by the use of the deflector (Fig. 10.13). A well-placed compass carefully corrected with this device should be within 2° of being correct, but no deviation table can be prepared and coefficients A and E cannot be determined.

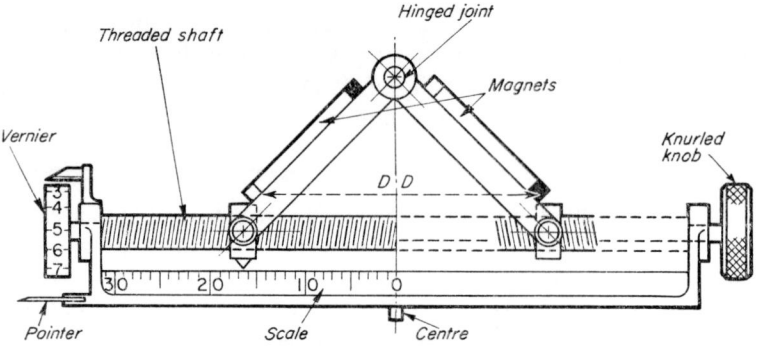

Fig. 10.13. Deflector

Briefly, the deflector consists of a magnet of variable strength, which is used to deflect the compass through 90°—the 'normal' deflection—when placed over E. by N. on the compass card. Its use is based on the principle that if the directive force at the compass is equal at all points, the deviations are zero. The strength of the variable magnet is a measure of the directive force and is indicated on a scale

*A special case occurs if a ship is swung at the magnetic equator, where Z is zero and so rods c and f are unmagnetized. Coefficients B and C will then be due to P and Q only and may be properly corrected by permanent magnets. Any subsequent change in coefficients B and C when the ship moves to where Z is other than zero may be attributed to C and f respectively.

288

attached to the instrument. The variable magnet is, in fact, a pair of hinged magnets which are opened and closed by the desired amount.

In use the deflector is placed centrally on the card, the ship being steered on a cardinal course, say north. The deflector magnet is adjusted by opening or closing the pair by means of the knurled knob until the card reads west at the lubber, with the deflector lying exactly over E. by N. The scale reading is noted. This process is repeated on east, by deflecting the card to north, on south by deflecting the card to east, and on west by deflecting the card to south.

Set the deflector to the mean of the scale readings for ship's head north and south, and with the ship steering on one of these courses obtain a normal deflection with the aid of athwartship corrector magnets. Similarly, by taking the mean readings for east and west, correct to normal deflection with the aid of fore-and-aft corrector magnets. If the readings on any two points alter by more than ten, the ship should be re-swung.

The spheres are placed by steering east or west, setting the deflector to the mean of all four readings on cardinal points and obtaining a normal deflection by moving the spheres in or out.

Theory of the deflector

When deflected, compass north lies on ON [Fig. 10.14 (*b*)] and $\hat{NOR} = 90°$. The deflector magnet lies along AOD, so that OD is east by north, or $\hat{EOD} = 11\frac{1}{4}°$. Thus in the equilibrium state,

$$\frac{OR}{OD} = \cos 11\frac{1}{4}°$$

Now

$$\lambda = \frac{OR}{OH}$$

Hence

$$\lambda.OH = OD \cos 11\frac{1}{4}°$$

or

$$\lambda = \frac{OD}{OH} \cos 11\frac{1}{4}°$$

Thus the field produced by the deflector is proportional to λ and, since the scale markings are a measure of the effective strength of the magnet DD (Fig. 10.13) and thus of the field it produces, the scale reading of the deflector can be considered as being proportional to λ, the directive force.

Other methods of quadrantal correction

The soft-iron spheres appear in different forms according to the fancy of various designers of compasses and binnacles. The theoretical form of soft-iron mass for correcting quadrantal deviations is an ellipsoid, and some foreign binnacles have appeared with ellipsoidal correctors instead of spherical. The use of soft-iron plates in racks, fitted in place of the sphere brackets, has been tried by the Japanese; soft-iron cylinders, soft-iron chains and sundry similar appliances are known in the art, but in general, the convenience and effectiveness of the sphere, albeit only an approximation to an ellipsoid, has proved the most satisfactory quadrantal corrector.

The Morel bar, a form of quadrantal corrector of French origin, is made of Mumetal, which has a very high degree of permeability. A bar placed close to the compass depends on the induction from the compass needles for its corrective properties. The high permeability permits the bar to be small in size while exhibiting a large degree of induced magnetism. If placed fore-and-aft, the bar is not magnetized when the ship is heading east or west, since it lies at right angles to the

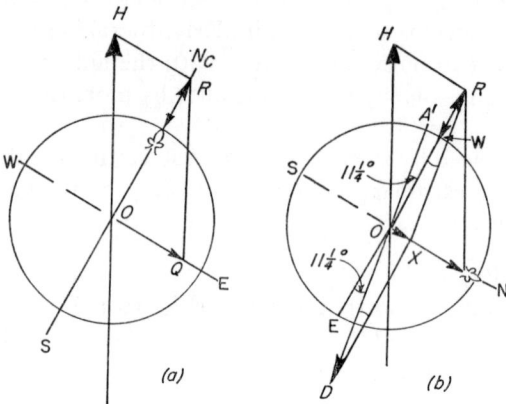

FIG. 10.14. Theory of the deflector

(a) Compass before deflection
  OH Earth's field
  N_M magnetic north
  N_C compass north (also direction of ship's head, as ship is steering north by compass)
  OQ component Q
  OR resultant magnetic field
(b) Compass after deflection
  OD field from deflector
  OX resultant of OD and OR
  $XOR = 90°$
  $OR = XD$
  $OD = XR$
  $\dfrac{OR}{XR} = \cos 11\tfrac{1}{4}° = \dfrac{OR}{OD}$
  $OR = \lambda OH = OD \cos 11\tfrac{1}{4}°$

or
$$\lambda = \frac{OD}{OH} \cos 11\tfrac{1}{4}°$$

compass needle. When the ship is heading north or south, the bar is magnetized in the same direction as the Earth's field. Thus, provided the remanence is nil, the Morel bar produces no deviation and therefore no correction on cardinal headings. However, on intercardinal headings the bar is magnetized along its length by the fore-and-aft component of the needle's field, producing a blue pole towards the north. Any soft iron in the fore-and-aft direction of the ship is magnetized by the fore-and-aft component of the Earth's field, but with a red pole forward.

Thus by adjusting the size of the Morel bar and its distance from the compass, the ship's soft iron can be corrected with this device. It must be remembered, however, that the correction is no longer constant when the value of the Earth's field is altered, and the bar needs to be readjusted on change of magnetic latitude.

Correction of quadrantal deviation by two compasses, a method devised by Capt. F. J. Evans* of the Admiralty Compass Department, is mainly of historical interest. It was known that two compasses placed near to each other had a mutual deviation effect, and Evans thought that this effect might be used to cancel the quadrantal errors arising from soft iron in the ship. The mutual effect of two similar needles is discussed on p. 136 and the existence of a quadrantal deviation can be demonstrated mathematically. Evans' two compasses were placed in a double binnacle at an adjustable distance apart, a scale being provided showing the amount of quadrantal correction. The scheme was tried in H.M.S. *Warrior* but was not very successful because the correction applied to only one magnetic latitude, and difficulties were experienced with the permanent magnet correctors.

Correction by coils

The use of compass-corrector coils to compensate for the effect of degaussing coils is a well-established practice. Briefly, if a compass is corrected without energizing the degaussing system of the ship the compass will be deviated when the degaussing system is switched on, owing to the change in the magnetic condition of the ship. The errors are usually of such a magnitude as to necessitate recorrection. To avoid this, fore-and-aft, athwartship and vertical coils are placed about the compass, as shown in Fig. 10.15, so that three orthogonal fields are produced, acting through the centre of the compass system.

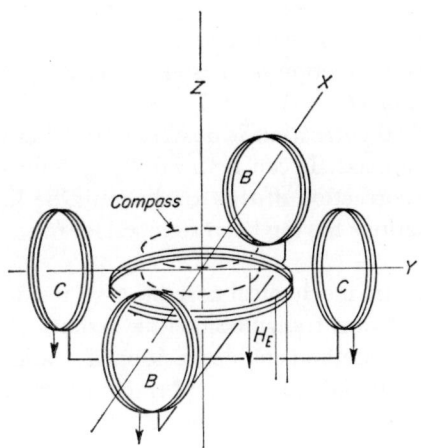

Fig. 10.15. Correction by coils

These coils are connected to the particular degaussing circuit of the ship according to the type of deviation being produced, and the current is set by means of variable resistances so that the change, due to degaussing, in the ship's field, at

* Smith, A., and Evans, F. J., 'On the effect produced on the deviation of the compass by the length and arrangement of the compass needles and on a new mode of correcting the quadrantal deviation', *Phil. Trans. roy. Soc.*, 1861, **161**.

the compass position, is counteracted automatically by the field set up by the corrector coils. Usually complete correction is not achieved and a separate deviation table is prepared. Multi-winding coils are often used, since each set of degaussing coils in the ship produces its own deviations.

A similar arrangement of coils can, of course, be used to replace the three sets of permanent magnets in the binnacle, but this is departing from the sound principle of correcting like with like and it must be remembered that a very stable source of electric current must be used, otherwise unexpected deviations may well arise. Such a system, however, has the great advantage that it is remotely controlled and convenient for correcting a transmitting compass that may be sited aloft or in some other awkward position. A similar system is adopted in certain aircraft compass systems, such as the Sperry Gyrosyn, in which the magnetic element is situated at the wing extremity.

METHODS OF COMPENSATING FOR RESIDUAL DEVIATIONS

Mechanical (non-magnetic) correction

In certain transmitting compasses a non-magnetic correction system is used, whereby a cam follower system injects differentially an angle of correction into the transmission system.

The cam is shaped, either by cutting a thin plate of metal, or by distorting a strip of metal by equally-spaced adjusting screws, to conform to the deviation curve. Such a system must be used with caution. As far as deviations due to soft iron are concerned, it has been said before that they are independent of the strength of the Earth's field, and so may quite legitimately be corrected mechanically. However, deviations due to permanent or hard iron depend on the strength of the Earth's field, and so mechanical correction of these is strictly applicable only to one given locality or to one value of H.

A system of mechanical correction is described in Chapter 9 in connection with the Gyro Flux Gate compass. It can be used safely only when the deviations are small, and mechanical correction of this kind should be limited to residual deviation, the principal deviations having been reduced by magnetic methods to as small a value as possible.

A form of non-automatic mechanical correction of residual deviations is simply a differential gear inserted in any transmission link so that any angle of correction may be applied manually. This method enables residual deviations, such as appear on the ship's deviation card, to be added to or subtracted from the transmitted heading angle by rotating the operating knob to the setting of deviation appropriate to the course being steered. Similarly, variation can be corrected differentially. The combination of variation and deviation known as total error is corrected in the total-error corrector used with the Admiralty Transmitting Magnetic Compass (Fig. 10.16 and Plate 10.1). If a ship is executing complicated manoeuvres and the transmitting compass is supplying heading information to an automatic plotting machine, quite appreciable errors will occur in the plot if the residual deviations are not continually applied as the ship alters course. For example, if a ship were steaming on a circular course and semicircular deviations were present, the plot would assume a spiral form, the direction of spiralling depending on the type and sign of the deviation. Quadrantal deviations would result in an elliptical plot. Hence the fitting of a total-error corrector was found to be desirable.

The indication of true direction by the correction of variation is an obvious advantage in plotting. This indication in all compass repeaters when fed through a total-error corrector system is a strong recommendation for the use of the system, particularly in the stabilization of a radar display.

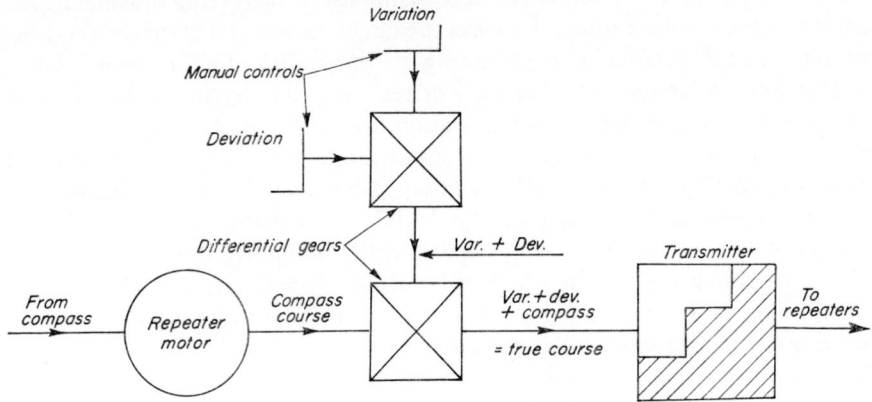

Fig. 10.16. Total-error corrector

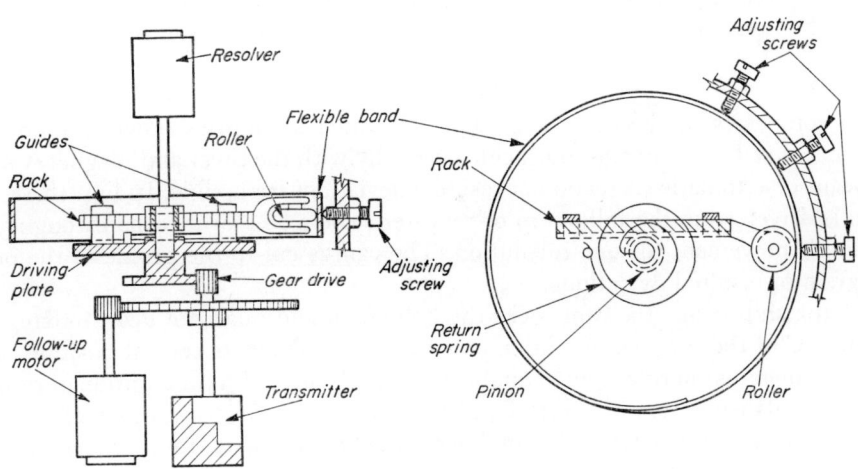

Fig. 10.17. Automatic residual-deviation correction

Fig. 10.17 shows another form of automatic residual-deviation correction, applied to a follow-up motor–resolver system (see p. 75). The motor drives a transmitter through gearing in the conventional way, but the resetting drive to the resolver is of a special nature. The motor is geared to a driving plate and on this is mounted a sliding rack which engages a pinion concentric with the motor drive.

If the rack is held in its guides, the pinion will rotate with the driving plate, but if the rack is moved to and fro in its guides as the driving plate rotates, an extra motion either retarding or advancing the angular position of the pinion with respect to the motor drive will be imparted to the pinion and to the resolver shaft connected to it.

The end of the rack carries a roller which bears against the inner surface of a stationary, but flexible, strip or band. The profile of this band can be altered at will by adjusting a number of equally-spaced screws, so as to impart a desired motion to the pinion in addition to the rotation of the motor. If the band is made to correspond to the residual deviation curve of the compass, a correcting movement will be imparted to the resolver rotor. This has the effect, as the motor drives the resolver rotor into a null position corresponding to the information received from the magnetic detector system, of advancing or retarding the driving system by a small angle corresponding to the residual deviation, so that the motor and transmitter are turned through the change of heading plus or minus any change in deviation. This system is applicable only to small deviations, since a slight error is introduced by the small movement of the roller on the flexible band as the correction takes place, the roller at last coming to rest in what is really a slightly ill-corrected position.

There are other systems in which the output from a cam or other mechanical means is used to rotate the stator of a resolver differentially with respect to the rotor in order to introduce a desired correction.

Mechanical methods of correcting deviations due to permanent magnetism are valid only for one value of H. On change of H, errors proportional to the change are introduced and re-correction is necessary. Such methods are of practical use only in correcting residual deviations of 2° or 3° at the most. Correction of variation, but not automatic correction, can be combined with these systems.

Automatic (electro-magnetic) correction of residual deviations.

In compasses of the A.T.M.C. type, in which a compass bowl or follow-up element is stabilized north–south concentrically with the pivot and magnet system, a means of automatic correction of residual deviations is possible. In Fig. 10.18, the cam is driven from the follow-up or repeater system so that it turns through 360° of ship's movement for one revolution. The cam is cut to the profile of the polar diagram of residual deviations.

As the ship turns, the cam rocks the bell crank and quadrant gear, moving the contact C of the toroidal potentiometer to and fro about its central position. The potentiometer is energized by a suitable d.c. supply and, with the addition of resistor RR and balancing potentiometer B, a bridge system is constructed, which delivers current of varying magnitudes in both directional senses to a corrector coil. This coil is attached to the compass bowl or follow-up element immediately under the centre of the magnet system and remains orientated east–west for all positions of ship's head. Through the agency of the cam and bell-crank systems, currents depending on the ship's head flow in the coil, and corresponding magnetic fields act upon the compass needle, at right angles to the magnetic meridian. The compass needle is, therefore, deviated according to the profile of the cam. Thus, if the cam is shaped in accordance with deviations that already exist, a reasonably good correction of small residual deviations is provided automatically.

The system is usually employed for correcting residual deviations of from 3° E. to 3° W. and is set up by fitting a 'zero' cam corresponding to no deviation. By adjusting the potentiometer B so that no current flows in the coil, no deviation is applied to the compass.

An uncut cam is then substituted for the zero cam, giving, say, 3° correction, and the potentiometer P adjusted to give 3° deviation of the compass needle. Allowing for the fact that, for small angles, tangents equal sines, the system is now correctly set up.

The application of this system is shown in the section on the Admiralty gyro-magnetic compass in Chapter 9. The correction is constant and it corresponds to a permanent magnet system. Thus only if residuals are due to coefficients *B* and *C*, is exact correction obtained as the ship moves about the seas. Soft-iron deviations corrected by this system will be exactly cancelled only at one value of *H*.

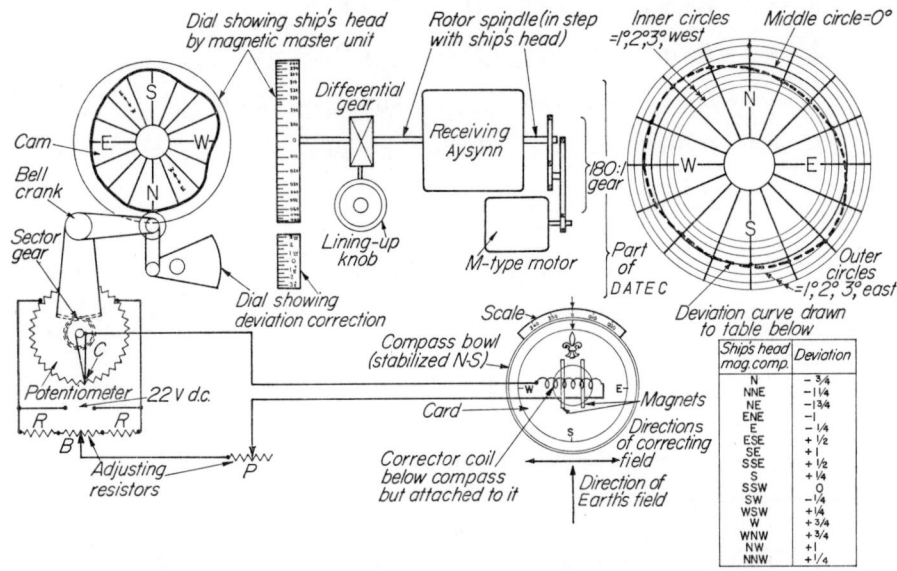

FIG. 10.18. Electro-magnetic residual-deviation correction

If any change of horizontal force occurs as the ship travels about, the potentiometer P may require adjustment to ensure that the correction provided by the cam is in fact 3° E. to 3° W. It will be readily appreciated that, if *H* increases, a greater deflecting field will be needed, and vice versa.

DEVIATION ANALYSIS

Having reduced the deviations of the compass to a minimum by correctors, it is customary to record the residual deviations on a deviation card; in fact, to produce a calibration table for the compass in that particular position.

Analysis of the residual deviations can be very informative in determining the success of the various modes of correction, the existence of high coefficients owing to unusual magnetic effects, the probable error of observation and the accuracy of the adjustment. Analysis consists in the determination of the coefficients from a deviation table, and since, in general, the deviations are small angles, equation

(10.14a) is the expression from which the approximate coefficients may be determined. Moreover, ζ' is the only heading that can conveniently be determined from a swing.

Uncorrected deviations, so long as they do not exceed about 20°, can also be treated in this way to find out the magnetic conditions of the compass position and the nature and distribution of the iron structure magnetically. The most effective method is based on that of 'least squares'.

From equation (10.14a):

$$\delta = A + B \sin \zeta' + C \cos \zeta' + D \sin 2\zeta' + E \cos 2\zeta' + \ldots$$

or for a complete table of equally spaced headings:

$$\delta_1 = A + B \sin \zeta'_1 + C \cos \zeta'_1 + D \sin 2\zeta'_1 + E \sin 2\zeta'_1 + \ldots$$

$$\delta_2 = A + B \sin \zeta'_2 + C \cos \zeta'_2 + D \sin 2\zeta'_2 + E \cos 2\zeta'_2 + \ldots$$

$$\delta_{n-1} = A + B \sin \zeta'_{n-1} + C \cos \zeta'_{n-1} + D \sin 2\zeta'_{n-1} + E \cos 2\zeta'_{n-1} + \ldots$$

$$\delta_n = A + B \sin \zeta'_n + C \cos \zeta'_n + D \sin 2\zeta'_n + E \cos 2\zeta'_n + \ldots$$

Thus

$$\sum_0^n \delta = nA + B \sum_0^{2\pi} \sin \zeta' + C \sum_0^{2\pi} \cos \zeta' + D \sum_0^{2\pi} \sin 2\zeta' + E \sum_0^{2\pi} \cos 2\zeta' + \ldots$$

Hence

$$A = \frac{\Sigma \delta}{n}$$

the constant error, which is the mean of the deviations on each heading. Multiplying by $\sin \zeta'$ and adding:

$$\sum_0^n \delta \sin \zeta' = A \sum_0^{2\pi} \sin \zeta' + B \sum_0^{2\pi} \sin^2 \zeta' + C \sum_0^{2\pi} \sin \zeta' \cos \zeta'$$

$$+ D \sum_0^{2\pi} \sin \zeta' \sin 2\zeta' + E \sum_0^{2\pi} \sin \zeta' \cos 2\zeta' + \ldots$$

from which,

$$\sum_0^n \delta \sin \zeta' = B \sum_0^{2\pi} \sin^2 \zeta'$$

All other terms $\equiv$ 0.

Thus

$$\sum_0^n \delta \sin \zeta' = B \sum_0^{2\pi} \frac{(1 - \cos 2\zeta')}{2}$$

$$= \frac{nB}{2} - \sum_0^{2\pi} \frac{\cos 2\zeta'}{2}.$$

Thus

$$B = \frac{2}{n} \sum_{0}^{n} \delta \sin \zeta'$$

Similarly

$$C = \frac{2}{n} \sum_{0}^{n} \delta \cos \zeta', \qquad D = \frac{2}{n} \sum_{0}^{n} \delta \sin 2\zeta', \qquad E = \frac{2}{n} \sum_{0}^{n} \delta \cos 2\zeta'$$

and so on.

For a 16-point swing, where deviations are measured on headings spaced by $22\frac{1}{2}°$,

$$A = \frac{\Sigma\delta}{16}, \qquad B = \frac{\Sigma\delta \sin \zeta'}{8}, \qquad C = \frac{\Sigma\delta \cos \zeta'}{8} \qquad \text{etc.}$$

With this spacing of headings, $\sin \zeta'$ and $\cos \zeta'$ can be given constant values, so that a form of tabular analysis can be devised and the coefficients found quickly by multiplying the observed deviations by the appropriate constants, summing each column, and dividing by 8. Thus, we have the constant 0·924 (or S_6) for $\sin 67\frac{1}{2}°$, $\cos 22\frac{1}{2}°$, and the sines and cosines of various multiple angles, 0·707 (or S_4) for $\sin 45°$, $\cos 45°$, etc., and 0·382 (or S_2) for $\sin 22\frac{1}{2}°$, $\cos 67\frac{1}{2}°$, etc.

Tables can, of course, be prepared for any number of equally-spaced deviations and the number of coefficients is limited by the number of equations, i.e. the number of headings at which the deviation is observed. Unless conditions are exceptional, it is seldom necessary to go beyond coefficient E, which requires five equations to determine the five unknowns. Examples of analysis forms are given in Appendix 4. Since a 16-point swing provides sixteen equations, eleven spare equations are left, which can be used to determine the accuracy of the observations by statistical methods.

A quick and somewhat rough-and-ready method of analysis is to calculate the amplitudes of the harmonic components from the deviations at their known peaks (Fig. 10.19). For instance, coefficient B is related to $\sin \zeta'$, thus maximum amplitudes are expected at 90° (E.) and 270° (W.). In case the deviation curve is displaced by other coefficients, the *mean* of the two deviations at E. and W., regardless of sign, is taken to give a rough value of coefficient B.

Thus

$$B = \frac{\text{dev. E.} - \text{dev. W.}}{2}$$

Similarly

$$C = \frac{\text{dev. N.} - \text{dev. S.}}{2}$$

Coefficient D, being related to $\sin 2\zeta'$, produces four peaks at NE., SW. and NW. and so

$$D = \frac{\text{dev. NE.} - \text{dev. SE.} + \text{dev. SW.} - \text{dev. NW.}}{4}$$

and similarly

$$E = \frac{\text{dev. N.} - \text{dev. E.} + \text{dev. S.} - \text{dev. W.}}{4}$$

It is sometimes dangerous to use this method unless it is certain that coefficients *F* and *G* are small or non-existent.

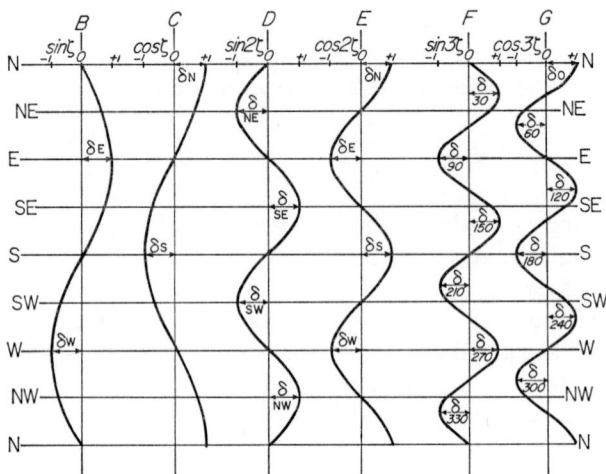

FIG. 10.19. Approximate coefficients

Although it will be clear that

$$F = \frac{\text{dev. } 30° - \text{dev. } 90° + \text{dev. } 150° - \text{dev. } 210° + \text{dev. } 270° - \text{dev. } 330°}{6}$$

and

$$G = \frac{\text{dev. } 0° - \text{dev. } 60° + \text{dev. } 120° - \text{dev. } 180° + \text{dev. } 240° - \text{dev. } 300°}{6}$$

a value given by (dev. E — dev. W.)/2 might just as well be due to coefficient *F* as to *B* and (dev. N. — dev. S.)/2 might similarly be due to *G* as well as to *C*. In fact:

$$\frac{\text{dev. E.} - \text{dev. W.}}{2} = (B - F)$$

and

$$\frac{\text{dev. N.} - \text{dev. S.}}{2} = (C - G)$$

Coefficients *H* and *K* can also be found from a 16-point swing if required.

$$H = \tfrac{1}{8}(\text{dev. NNE.} - \text{dev. ENE.} + \text{dev. ESE.} - \text{dev. SSE.}$$
$$+ \text{dev. SSW.} - \text{dev. WSW.} + \text{dev. WNW.} - \text{dev. NNW.})$$
$$K = \tfrac{1}{8}(\text{dev. N.} - \text{dev. NE.} + \text{dev. E.} - \text{dev. SE.} + \text{dev. S.}$$
$$- \text{dev. SW.} + \text{dev. W.} - \text{dev. NW.})$$

298

A study of Fig. 10.19 will clarify this method.

From a 4-point swing starting on north: A, B, C and E can be found.

From an 8-point swing starting on north: A, B, C, D, E, F, G and K can be found.

From a 12-point swing starting on north: A, B, C, D, E, F, G, H, K, L, M and O can be found.

From a 16-point swing, coefficients A to Q, with the exception of P, can be found.

It is possible, however, to deduce the values of coefficients B and C approximately from intercardinal values of the deviation. Considering the sine curves for coefficients B and C in Fig. 10.19, the deviations at 45°, 135°, 225°, and 315° for B are 0·707 multiplied by the full amplitude; similarly for the deviations at the same headings in the case of C.

Thus

$$B = \frac{1 \cdot 414}{4}(\text{dev. NE.} - \text{dev. SE.} + \text{dev. SW.} - \text{dev. NW.})$$

and

$$C = \frac{1 \cdot 414}{4}(\text{dev. NE.} - \text{dev. SE.} - \text{dev. SW.} + \text{dev. NW.})$$

Hence from observations made on NE., SE., SW., and NW. coefficients A, B, C and D can be found.

An extension of deviation analysis is the analysis of errors and other quantities observed in the testing of compasses where the magnitude of the error is written in the deviation expression for δ and an angle α, representing some reference system of angles for ζ'. Exactly similar reductions will produce a set of coefficients to which physical qualities may be assigned, indicative of definite properties of the compass. Coefficients B and C do not mean hard iron and D and E soft iron in such instances, since B or C may be due to eccentricity of the compass system in a pivoted-needle instrument. In this case higher coefficients than C are probably due to graduation errors and may or may not be truly harmonic, though from Fourier's theorem any cyclic phenomenon can be resolved into harmonic functions.

Coefficients D, E and H have a meaning for certain types of inductor compass and compasses using sine–cosine resolving systems. This has already been discussed.

DEVIATION SYNTHESIS

Once a set of coefficients is obtained from a table of deviations, it is possible to produce by synthesis a table of computed deviations. The latter represent the most probable values of the deviations and are preferable in practice to the 'observed' values.

The synthesis consists in solving a set of equations for δ thus:

$$\delta_{\text{N.}} = A + B \sin 0° + C \cos 0° + D \sin 0° + E \cos 0°$$

$$\delta_{\text{NNE.}} = A + B \sin 22\tfrac{1}{2}° + C \cos 22\tfrac{1}{2}° + D \sin 45° + E \cos 45°$$

$$\delta_{\text{NE.}} = A + B \sin 45° + C \cos 45° + D \sin 90° + E \cos 90°$$

$$\delta_{\text{NNW.}} = A + B \sin 337\tfrac{1}{2}° + C \cos 337\tfrac{1}{2}° + D \sin 315° + E \cos 315°$$

A table similar to that used for deviation analysis is very convenient for synthesiz-ing deviations and the terms of the equations are rapidly determined by multiplying the coefficients by the constants 1, 0·924, 0·707, and 0·382 for a 16-point swing, due attention being paid to the algebraic sign of the quantity. The analysis forms given as Appendix 4 are arranged to be used for synthesis also.

It has already been stated that it is seldom realistic to take analysis beyond coefficient E, which means that in a 16-point swing there are eleven more equations than are strictly necessary for determining the five coefficients. This, however, enables the accuracy of the swing to be determined by the implicit use of these spare equations. Having obtained the computed deviations, these are subtracted from the observed deviations, giving the errors of the latter, v, which should, of course, be random.

Σv^2 is then determined, and, in the case of a 16-point swing: $(\Sigma v^2/11)^{1/2}$ is the standard deviation of the error of a single observation (σ).

That is to say, for every 100 readings of the deviation on one heading, the error, by comparison with the true value of the deviation will lie in 68 cases within $\pm(\Sigma v^2/11)^{1/2}$. Also, for a single observation of deviation on a given heading, the error that may occur is not likely to be greater than $2\sigma = \pm(\Sigma v^2/11)^{1/2}$. This is sometimes called the 95 per cent confidence level.

The value of σ is a good guide to the validity of a swing. If σ is low, it is indica-tive of accurate observation and the verification of the existence of only the five normal coefficients.

If the number of points used in a swing is other than sixteen,

$$\sigma = \left(\frac{\Sigma v^2}{N-M}\right)^{1/2}$$

where N is the number of points, and M is the number of coefficients determined. The larger N is, the more accurate will be the result.

Further explanation of the statistical analysis of errors is given in Chapter 12.

GRAPHICAL METHODS

The dygogram

A form of deviation computation that is often useful in studying the magnetic state of a new ship or in assessing the probable magnitude of deviations is graphical in its construction and form (see Fig. 10.20).

The total horizontal magnetic force at the compass position is the vector sum of the forces whose magnitudes are λH, $\lambda H\bar{A}$, $\lambda H\bar{B}$, $\lambda H\bar{C}$, $\lambda H\bar{D}$ and $\lambda H\bar{E}$, their positive directions, measured clockwise from magnetic north, being given by the angles 0°, 90°, ζ, $\zeta+90°$, 2ζ and $2\zeta+90°$ respectively.

Let the magnetic course be north, and the vectors Oh, ha, db, bc, ed and ae. If the values of $\bar{A}$, $\bar{B}$, $\bar{C}$, $\bar{D}$ and $\bar{E}$ are known and if unit value is assigned to λH, a vectors diagram can be drawn, in which the vectors are drawn in the order Oh, ha, ae, ed, db, bc. The resultant vector Oc represents the total force in magnitude and direction when the magnetic course is north, and the angle $h\hat{O}c$ is the deviation on north.

To extend this method to any magnetic course, the points O, a, d and c are found as before (Fig. 10.21). With centre a and radius ad, draw a circle known as the generating circle. Let cd, produced if necessary, cut the generating circle at q, known as the pole of the diagram. From q draw a line qd_1c_1 so that $d\hat{q}d_1$ is ζ_1, the magnetic course, and d_1 is the point of intersection with the generating circle. Mark off d_1c_1 and d_1c_1' both equal to dc, then the vector Oc_1 represents the total magnetic force, and $h\hat{O}c_1$ the angle of deviation when the magnetic course is ζ_1. Oc_1 and $h\hat{O}c_1'$ represent the same quantities when the magnetic course is $(\zeta_1 + 180°)$. By marking off dc' equal to dc, Oc' and $h\hat{O}c$ represent the total magnetic force and the angle of deviation when the course is magnetic south.

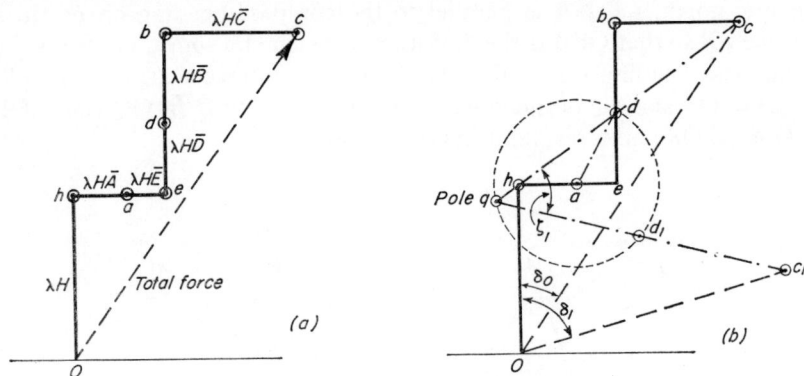

FIG. 10.20. Constructing the dygogram

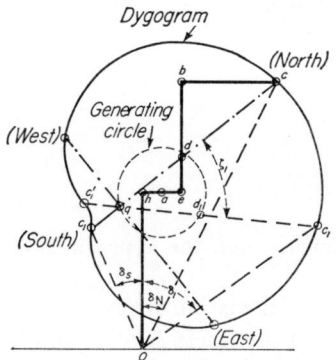

FIG. 10.21. Dygogram

It is important that the correct sense be given to the vectors; if $\bar{A}$, $\bar{E}$, and $\bar{C}$ are positive, ha, ae and bc must be directed towards magnetic east; if negative, to magnetic west. If $\bar{B}$ and $\bar{D}$ are positive, db and ed must be directed towards magnetic north; if negative, to magnetic south. If the construction is continued to give points c_1, c_2, c_3, ... for magnetic courses ζ_1, ζ_2, ζ_3 ... and c_1', c_2', c_3', for magnetic courses $(\zeta_1 + 180°)$, $(\zeta_2 + 180°)$, $(\zeta_3 + 180°)$ ... in increasing magnitude, points c_1', c etc. should read clockwise. The locus of c_1', c_1 is called a dygogram.

Given then this curve, the total magnetic force and deviation angle for any magnetic course may be found. To avoid actually drawing all the lines qd_1c_1 qd_2c_2, etc., a straight-edge of length cc' with the mid-point marked may be used to plot the dygogram. Place it so that the mid-point moves on the circumference of the generating circle, with its edge always passing through q. The points on the dygogram are found by marking the extremities of the straight-edge.

An approximate method of finding $\bar{B}$ and $\bar{C}$ (or P and Q) provided the soft-iron components are negligible

Given at least two deviations of the compass on magnetic headings (preferably north and south or east and west), draw the circle NWSE with centre O and radius representing λH [Fig. 10.22(a)]. Draw straight line NA so that $\hat{ONA}$ is the deviation on magnetic north, i.e. NA is parallel to the compass heading on north. Draw straight line SB so that $\hat{OSB}$ is the deviation on magnetic south, i.e. SB is parallel to the compass heading on south. Let these lines intersect at n. Draw nP perpendicular to OS and nQ perpendicular to OE. Then OP is $\bar{B}$, OQ is $\bar{C}$, λOP is P and λOQ is Q. On is the resultant deviating force.

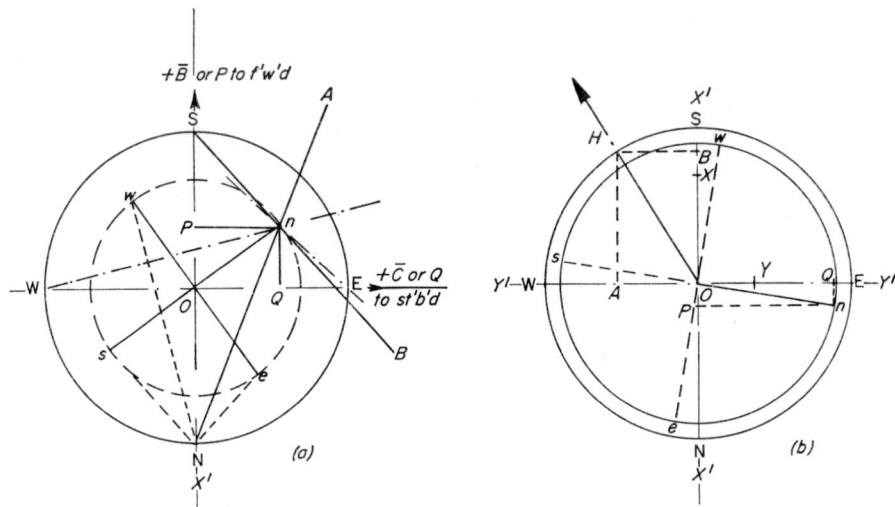

Fɪɢ. 10.22. Construction for approximation of $\bar{B}$ and $\bar{C}$

If more than the two values of deviation are available, the corresponding magnetic headings are marked off round the circle NWSE and lines parallel to the compass headings drawn through these points. The lines will intersect at n if there is no soft iron present, otherwise the point n may be estimated as the centre of the several intersections.

An approximate method of finding the deviations, given $\bar{B}$ and $\bar{C}$ or P and Q

The point n being obtained from the known values of $\bar{B}$ and $\bar{C}$ or P and Q, draw a circle of radius On. If now magnetic headings n, e, s, w, etc. are marked clockwise round this circle, deviations on the several magnetic headings may be found by drawing the lines Nn, Ne, Ns, Nw, etc., and measuring the angles $\hat{ONn}$, $\hat{ONe}$, $\hat{ONs}$, $\hat{ONw}$, etc., deviations to the right being easterly or positive and to the left westerly or negative. Alternatively, if the magnetic headings are marked off counterclockwise from N round the circle NWSE, lines drawn through these points and n are parallel to the compass headings.

Determination of P and Q (and hence $\bar{B}$ and $\bar{C}$) from field strength measurements

In Fig. 10.22(*b*), X'OX' and Y'OY' are the fore-and-aft and athwartship axes of the ship, OH is the magnitude of the horizontal component of the Earth's magnetic field, and HÔX' is the ship's heading (magnetic). Let OX and OY be the total field measured along OX' and OY' (by magnetometer, for instance).

Draw HB perpendicular to OX' and HA perpendicular to OY'. XB = P and AY = Q. Mark off OP = XB and OQ = AY, paying attention to the proper signs and directions, i.e. if XB is measured downwards from B, OP is measured downwards from O and vice versa. Similarly, if AY is measured to starboard from A, OQ is measured to starboard from O and vice versa. Draw P*n* and Q*n* pendicular to OX' and OY' respectively. O*n* is the resultant deviating force.

Draw the circle NWSE with centre O and radius ON, and the circle *nesw* with radius O*n*. These circles correspond to the similarly marked ones in the previous diagram and from which deviations and compass courses may be found.

If *n* falls outside the circle NWSE, the resultant deviating force is greater than λH and deviations of up to 180° will occur. The compass will appear to be 'locked' to a pole in the ship and will indicate only a limited range of heading.

Chapter 11

Other Methods for the Measurement
of
Magnetic Fields

The pivoted needle and the various forms of the saturable inductor, though probably the most versatile and best-known methods of studying the Earth's magnetic field and those most readily adaptable to the construction of compasses, are not the only devices sensitive to a magnetic field. In this chapter, a number of other ways of measuring a magnetic field will be discussed briefly.

'AE' effect

The 'AE' effect, sometimes called the skin effect, has already been referred to in the description of the Butterworth magnetograph on p. 152. To recapitulate, it was shown by Harrison, Turney, Rowe and Gollop* that the change in impedance in a wire of high permeability, along which an alternating current flows, is a function of the applied ambient magnetic field. The effect is proportional to $(\mu f)^{1/2}$, so that if μ is sufficiently large, quite low excitation frequencies can be used. The chief difficulties in using this system are those of instability of temperature, current, frequency and tension, and except in the Butterworth magnetograph, which provides stabilizing methods, it has never proved suitable for extensive use as a compass detector system.

The use of an 'AE' detector would follow lines already described. Two wires at right angles could be used for determining fore-and-aft and athwartship components of the Earth's field, and these would be combined in a resolver, or alternatively a single detector with a follow-up system could be used to seek zero field. The characteristics of a high-impedance ferro-magnetic wire subjected to a magnetic field are shown in Fig. 6.26.

Hall effect

A number of metals, when placed in a magnetic field, show what is known as the Hall effect.† This phenomenon was principally associated with the metal ger-

---

* Harrison, E. P., Turney, G. L. and Rowe, H., *Nature*, 1935, **135,** 961.
  Harrison, E. P., Turney, G. L., Rowe, H. and Gollop, H., *Proc. roy. Soc.*, 1936, **157**A, 451.
  Harrison, E. P., and Rowe, H., *Proc. phys. Soc. Lond.*, 1938, **50,** 176.
  British Pat. 562755, 1943.
  Harrison, E. P., and Smith (Miss) E. H., *Proc. phys. Soc. Lond.*, 1944, **56,** 31.
  † Torrey and Whitmer, *Crystal Rectifiers* (McGraw-Hill, New York and London, 1948).

manium, but recently several new materials have been discovered with greatly increased sensitivity. In Fig. 11.1, A is a thin plate of a single crystal of germanium. A battery or other source of d.c. applies a potential of a few volts at the points *bb*, opposite each another on the edge of the crystal. When a magnetic field *C* is applied as shown, normal to the plane of the crystal, a d.c. voltage proportional to the applied field will appear across the points *dd* in the centre of the two edges at right angles to those occupied by the electrodes *bb*.

Owing to the comparatively low sensitivity of germanium, the Hall effect was used principally for large fields, such as exist in the air gaps of electrical machines and between the poles of large magnets.

B.T.H. Ltd produced a convenient instrument embodying this principle, consisting of a probe, in the form of a very thin plate of germanium connected to a $4\frac{1}{2}$-volt dry battery and to a sensitive voltmeter. A calibration circuit balanced out stray effects.

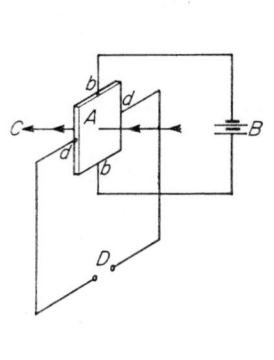

FIG. 11.1. Hall effect

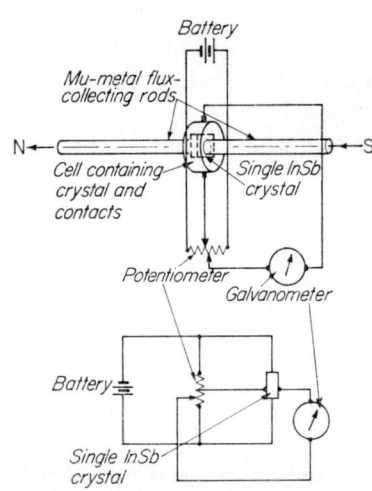

FIG. 11.2. Scheme for a compass using the Hall effect

Recently, however, work at the Services Electronics Research Laboratory* and by Siemens in Germany has shown that a number of combinations of elements exhibit a very pronounced Hall effect. The combinations are principally of those elements to be found in groups III and V of the periodic table. The elements concerned are:

Group III	Group IV	Group V
Ga	**Ge**	**As**
In	Sn	**Sb**
Tl	Pb	Bi

and the most practical combinations with high sensitivity are indium antimonide (InSb) and indium arsenide (InAs). The elements of Group IV are included to show the relationship of germanium to these newer combinations.

* Hilsum, C., 'Galvanometric effects and their application', *Brit. J. appl. Phys.*, 1961, **12**, 85.

The sensitivity of indium antimonide is said to be 300 times that of germanium and fields of the magnitude of that of the Earth can be detected without the use of amplifiers. The use of highly permeable 'flux-collecting' rods improves the sensitivity of the system and S.E.R.L. have demonstrated a compass or magneto-meter system on these lines. The instrument is a null-seeking device, the direction of the Earth's magnetic field being determined by finding the position of zero Hall voltage. The magnetic meridian is then at right angles to this direction. Lack of symmetry in the position of the contacts on the indium-antimonide crystal results in incorrect indications on the meter, and hence a balancing circuit is included.

Fig. 11.2 illustrates a compass system of this type and, although this appears to be very attractive, difficulties arise owing to temperature effects. If these could be corrected and the long-term stability of the system ensured, the use of Hall detector elements in the place of saturable inductors for either null-seeking follow-up systems or sine–cosine arrangements would be practicable. Among other features are the simplicity of the apparatus, the low power requirements and small weights and dimensions. The use of a.c. instead of d.c. suggests a method of obtaining a phase-sensitive output at any desired frequency, capable of simple amplification to any level required.

The magnitude of the Hall voltage is given by

$$V_H = \frac{RHi}{t} \times 10^{-8}$$

where R is the Hall constant in cm^3/coulomb, H the magnetic field in oersteds, i the current in amperes and t is the thickness of the specimen. R is determined from the expression

$$R = -\frac{3\pi}{8|e|} \cdot \frac{n\mu_n^2 - p\mu_p^2}{n\mu_n + p\mu_p} \text{ cm}^3/\text{coulomb}$$

where n and p are electron and hole concentrations, μ_n and μ_p electron and hole mobilities in cm^2/volt-second, and e the electron charge in coulombs. For germanium, $n \gg p$ and the expression reduces to

$$R = \frac{3\pi}{8ne}$$

For indium antimonide, $n = p$, but since $\mu_n \gg \mu_p$ the expression may again be written

$$R = \frac{3\pi}{8ne}$$

The carrier mobility of indium antimonide is 6×10^4 cm^2/volt-second, compared with $3 \cdot 6 \times 10^3$ cm^2/volt-second for germanium.

Cathode-ray compass

When a beam of electrons is acted upon by a magnetic field, it is deflected at right angles to the direction of the field. This deflection is comparatively small for fields of the order of magnitude of the Earth's, but the principle has been used in a recording magnetometer and can be applied to a compass (Fig. 11.3). A cathode-ray tube, mounted in gimbals, has a compass rose marked on its face. When the ship is steaming north, the electron stream is deflected so that the bright spot on the tube face is adjacent to the north point of the compass rose. A turn towards the east (ship and tube move clockwise) brings the bright spot adjacent to the east point, and so on.

R. T. Squier of the Minneapolis-Honeywell Regulator Co. enlarged upon this idea, using a modified cathode-ray tube having an electron gun with four anodes, a grid and centring electrodes (Fig. 11.4). At the end of the tube, remote from the gun, are four sector-shaped targets of 90° each. The electron beam is of appreciable cross-section, and when undeflected it strikes the centre of the target array, impinging equally on all four sectors.

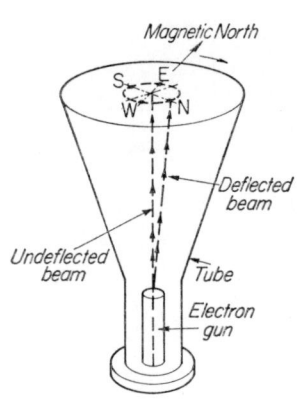

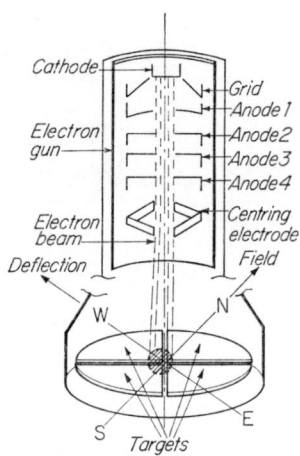

FIG. 11.3. Cathode-ray compass   FIG. 11.4. Squier's cathode-ray compass

The beam is modulated at 400 c/s by means of the grid. When the tube is mounted in gimbals, the electron beam is vertical until it is acted upon by the horizontal component of the Earth's magnetic field. If the beam is pointing downwards it is deflected to the west; if upwards, to the east. Thus the beam no longer impinges equally on all four targets and unequal amounts of current pass through the different sectors. As the tube turns owing to the movement of the vessel in which it is mounted, the beam remains stationary. However, the targets turn, thus altering the currents passing through the sectors.

Opposite sectors are connected, as shown in Fig. 11.5, to the primary of a transformer. The output of each transformer is the difference between the 400 c/s currents to opposite sectors and is proportional to the components of the magnetic field at right angles to a line through the centre of the sectors. The tube is set up so that the centre line of one pair of sectors lies in the fore-and-aft axis of the vessel and that of the other pair in the athwartship axis. The output of the two transformers is then proportional to the two components $H \sin \zeta$ and $H \cos \zeta$ of the Earth's horizontal field, where ζ is the angle between the fore-and-aft axis and magnetic north. As originally described, the outputs from the transformers are fed to sine–cosine potentiometers as described on p. 76 and shown in Figs. 4.39 and 4.40.

However, other means of combining these vectors could be used equally well. After combination, the resultant output is used to monitor a gyroscope, as in normal gyro-magnetic compass practice. An accuracy better than $\pm 1°$ and a sensitivity better than $\frac{1}{4}°$ are claimed for the instrument. Since the electron beam has no inertia, the system is free from some of the dynamic effects of a pendulous pivoted-needle system. Nevertheless, although an absence of northerly turning error is claimed for the system, there must be a deviation of the beam out of its correct position when the tube and electron gun are tilted into the false vertical owing to accelerations acting on the gimballed mass. The beam will then be acted upon by the resultant field of H and $Z \sin \beta$, where β is the angle of tilt.

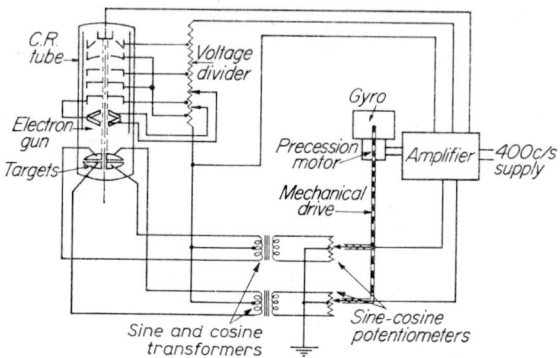

FIG. 11.5. Circuit for Squier's cathode-ray compass

Magnetic anomaly detector (M.A.D.)

On p. 66 reference was made to the unbalanced or peak difference inductor, which was described as being part of the M.A.D. system.

The magnetic anomaly detector is much in favour for airborne geological surveys and geophysical prospecting; and a high degree of sensitivity is claimed in determining magnetic anomalies. A second-harmonic inductor can also be used but it is considered by some authorities that either the peak difference method or the peak method is preferable in measuring field intensity.

A linear output in field strength measurement is not as essential as in the determination of direction, since feedback, backing-off and other corrections can be used and if necessary the whole equipment can be calibrated. Thus the most sensitive detector is to be preferred.

The purpose of M.A.D. is to find small irregularities in the general pattern of the Earth's magnetic field, which are associated with ferro-magnetic deposits of rock and oil-bearing strata. Thus the requirement is to measure the total intensity of the Earth's field. In principle, an inductor element is stabilized along the direction of the Earth's field by two subsidiary or orientating inductor elements. The stabilized element now measures the total intensity of the Earth's magnetic field and if the major part of this field is backed off, only its irregularities will be recorded. It is here that the high sensitivity of the equipment is important.

Thus the equipment consists of a detector head which is normally towed by the surveying aircraft in a streamlined body known as a 'bird'. This detector contains a gimballed framework in which the measuring inductor element is mounted. The attitude of the framework is determined by the signals derived from the orientating inductors which, by means of follow-up systems, seek to rest at right angles to the direction of the Earth's total magnetic field.

The scheme is shown diagrammatically in Fig. 11.6. The orthogonal axes XYZ are related to the Earth's total field vector X. The outer gimbal pivots about vertical bearings, and the inner gimbal, which supports the orientating inductors I_z and I_y, pivots about horizontal bearings carried in the outer gimbal. From these inductors, the output signals go to the orientating amplifiers A_z and A_y and

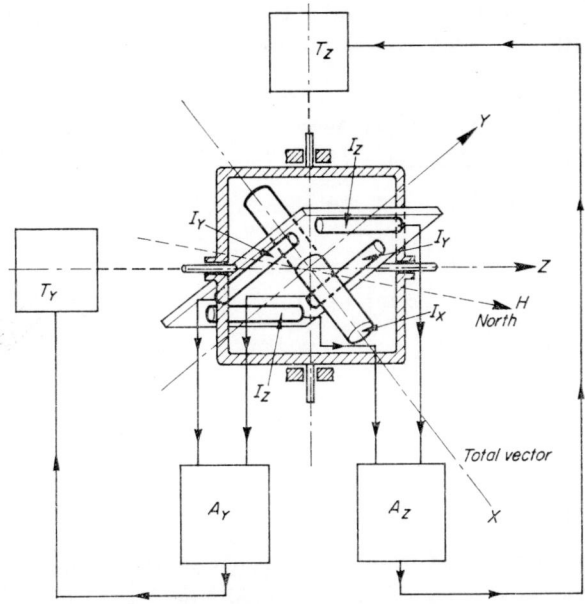

FIG. 11.6. Magnetic anomaly detector

the amplified signals energize the torque motors T_z and T_y which rotate the gimbal system until both inductors are in zero field. The plane of the inner gimbal is now normal to the total vector X and the measuring inductor element I_x lies along the total vector. Any change in the attitude of the framework will cause the necessary correcting signals to be applied to the torque motors, so maintaining I_x parallel to X.

The signal from I_x, in the form of a series of alternating large and small pulses (see p. 67), is amplified to produce a d.c. output corresponding to the magnitude of the anomaly as indicated by the difference in amplitude of the alternate pulses. The sense of the signal is determined by the phase relationship of the pulse pattern to a reference signal. For convenience, the diagram is reproduced in Fig. 11.7.

The range of the equipment is somewhat limited, being of the order of 200–1000 ft according to the size of the anomaly. The record is in the form of a line drawn by a recording voltmeter on a clock-driven chart, the zero line denoting the condition of an undisturbed magnetic field. Anomalies (Fig. 11.8) show up as irregularities in an otherwise straight line. In some versions of this instrument, a second-harmonic detector is used for field-strength measurement and peak-height detectors for orientation.

Transflux magnetometer

Variations of the saturable inductor dealt with in Chapter 4 have been described by D. O. Sproule, T. M. Palmer and others. Sproule demonstrated that if a specimen of highly permeable material M was made in the form of a rectangular plate (Fig. 11.9) with an exciting winding energized by a sinusoidal current, $i \sin \omega t$, and subjected to a magnetic field H at right angles to the axis of the excitation winding, an e.m.f. proportional to H and at twice the excitation frequency appeared across the extremities of a transverse winding.

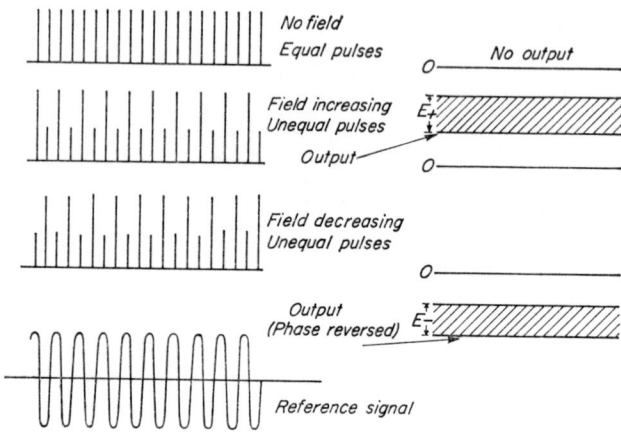

FIG. 11.7. Signal from the magnetic anomaly detector

Palmer's magnetometer

Another magnetometer (Fig. 11.10) is described by Palmer.* Here the excitation current is passed along a thin Mu-metal rod, the current being such as to produce a saturating circular field in the rod, around which is a pick-up winding. When the system is placed in an axial magnetic field H_X, an alternating e.m.f. at twice the excitation frequency appears at the output of the pick-up winding XY. Using a frequency of 5 kc/s, a wire rod of 42 S.W.G. 25 mm long, and a pick-up coil wire or rod of 42 S.W.G. 25 mm long, and a frequency of 5 kc/s, the pick-up coil having 235 turns of 39 S.W.G. copper wire and a length of 37 mm, sensitivities of 2 gamma have been achieved and fields up to 50 oersteds measured.

The principle of Palmer's magnetometer may be seen from Fig. 11.11. An alternating current $i \sin \omega t$ flows along the wire and produces an alternating circular saturating field H_Y. This combines with the axial field H_X to produce an oscillating resultant field H_R represented by the vector OP.

* Palmer, T. M., 'A small sensitive magnetometer,' *Proc. Instn. elect. Engrs.*, 1953, **100** (2), 545.

The induction in the wire at a given point may also be represented by a vector through O and provided hysteresis and eddy currents are neglected, this vector is in the same direction as the field vector and in Fig. 11.11 is represented by OQ. However, since the induction is not linearly related to the field, the inductor vector terminates on a curve AQYB instead of on a straight line as in the case of the field vector H_R.

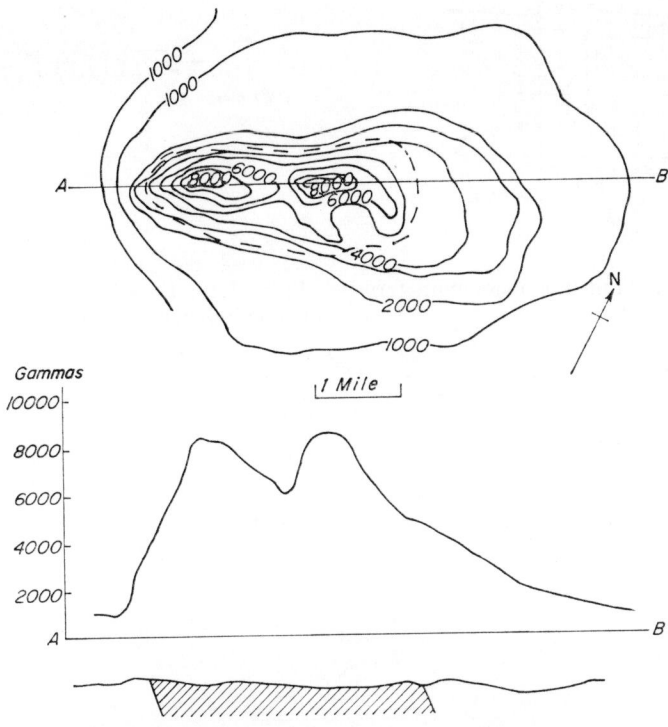

FIG. 11.8. A magnetic anomaly

At the bottom, a section through the country shows an outcrop of magnetic rock. The centre diagram shows the field strengths on a line from one side to the other of the country. The top diagram shows the magnetic anomaly as a 'contour' map, the lines joining points of equal magnetic force.

Thus the component of induction along the wire may be represented as alternating between the values OX and OY, thus generating an e.m.f. in the pick-up coil. Since the induction is in the same direction as the axial field, the phase of the induced e.m.f. depends on the direction of the applied field. Also, as the induction alternates between OX and OY twice in one cycle of the excitation, the e.m.f. is at twice the excitation frequency.

Palmer claims that since the use of a transverse field avoids mutual inductance between the excitation and output circuits, the amount of filtering customary in the second-harmonic inductor is considerably reduced.

For small values of H_X the output e.m.f. is proportional to H_X. The proposed circuit for this magnetometer is such that by applying d.c. to the pick-up winding, H_X can be neutralized and the system used as a null indicator. The value of H_X

is determined from the d.c. current flowing in the coil. Fig. 11.12 illustrates a block schematic circuit and Fig. 11.13 a section of the magnetometer head. A curve relating induced voltage to axial field is shown in Fig. 11.14.

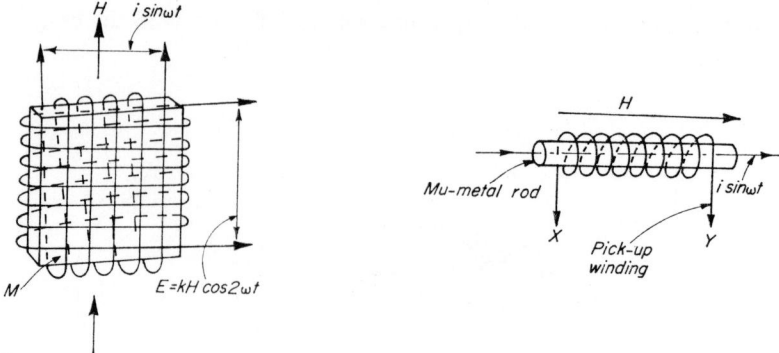

FIG. 11.9. Transflux magnetometer   FIG. 11.10. Fundamentals of Palmer's magnetometer

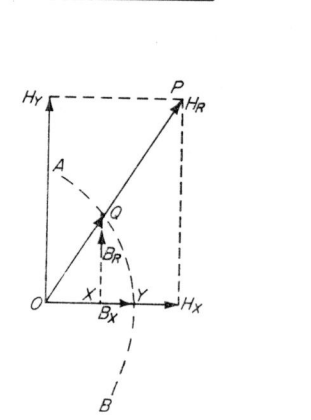

FIG. 11.11. Principle of Palmer's magnetometer

Nuclear induction and resonance

In 1954, Packard and Varian* described a means of measuring the total ambient magnetic field by what is known as the 'free precession' technique. This depends on the fact that the proton or nucleus of the hydrogen atom has a constant angular momentum and a constant magnetic moment. The atoms tend to act as small gyroscopes aligned in the ambient magnetic field.

Proton precession magnetometer. G. S. Waters† described a simple experiment in which the protons in a sample of water are acted on by a polarizing magnetic field

* Packard, H., and Varian, R., 'Free nuclear induction in the Earth's magnetic field,' *Bull. Amer. phys. Soc.*, 1953, **28**, No. 7, 7; *Phys. Rev.*, 1954, **93**, 941.
   † Waters, G. S., 'A measurement of the Earth's magnetic field by nuclear induction,' *Nature*, 1955, **176**, 691.

at right angles to that of the Earth. This field is switched off, and the protons that have been aligned in the field are now precessed back into alignment with the Earth's field. The frequency of precession is related to the ambient magnetic field as the precession frequency of a spinning top is related to the ambient gravitational field. The time during which the phenomenon is apparent is quite short, about five

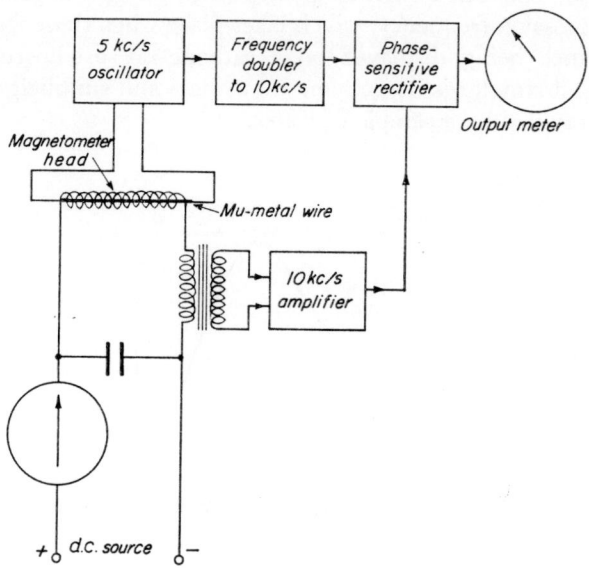

FIG. 11.12. Palmer's magnetometer

Block schematic circuit

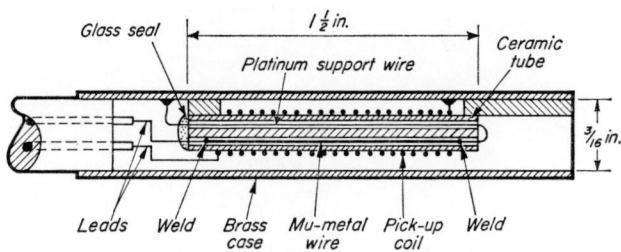

FIG. 11.13. Section of head of Palmer's magnetometer

seconds after switching off the polarizing field, but it is adequate for reliable results to be obtained. The precession frequency may be measured by beating it with a stable reference frequency and displaying the beat pattern on an oscilloscope.

The relation between precession frequency f, and ambient field T is given by

$$2f = \gamma_p T$$

where γ_p is the gyro-magnetic ratio for the proton.

The gyro-magnetic ratio, γ_p, is $26\ 753 \pm 0\cdot6$ sec^{-1}oersted^{-1} and, in a measurement quoted by Waters, the precession frequency of the protons after deflection

313

was 1997·22 c/s. This corresponds to a total ambient field of 0·46906 oersted ± 1 gamma. A schematic diagram of a proton precession magnetometer is shown in Fig. 11.15.

Nuclear resonance fluxmeter. A fluxmeter devised by the Admiralty Signal and Radar Establishment and demonstrated in 1957 at the Physical Society's Exhibition uses the resonance between a high-frequency field, at right angles to the Earth's field, and the precession frequency, which takes place when these frequencies coincide, the resonance being displayed on a cathode-ray oscilloscope. The great advantage of this fluxmeter is its extreme robustness and simplicity, the supply of protons being provided by a sample of water.

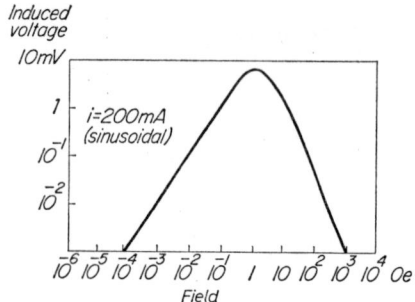

FIG. 11.14. Curve relating induced voltage to axial field

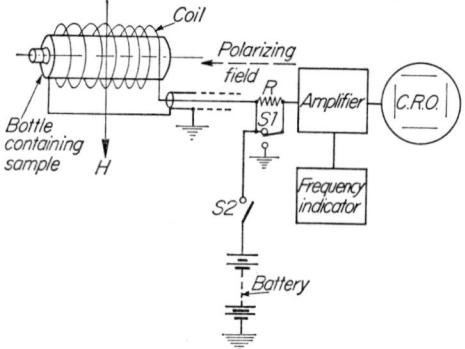

FIG. 11.15. Proton precession magnetometer
Schematic diagram

The range of the instrument described is from 1100 to 7500 oersted, but it is, of course, capable of detecting much higher fields.

Rubidium vapour magnetometer. The rubidium-vapour magnetometer, a schematic arrangement of which is shown in Fig. 11.16, utilizes the technique of optical pumping for absolute measurements of the Earth's total magnetic field by observing the radio-frequency absorption lines in the vapour of rubidium. Since the frequency and field are related by an accurately known atomic constant, precise results are obtained without reference to a standard magnetometer for calibration. Light from a rubidium lamp, which is excited at radio-frequency and so avoids the use of

314

electrodes, first passes through an interference filter, which transmits only the 7947 Å rubidium line. The light, passing through a polarizer and a quarter-wave plate, is thus circularly polarized. The light beam, which is orientated to be approximately parallel to the magnetic field, next passes through the rubidium vapour absorption cell. The circularly polarized light orientates the magnetic moments of the rubidium atoms along the direction of the ambient field. The photocell measures the light intensity. Surrounding the cell are Helmholtz coils, one pair of which is energized by an L.F. oscillator to modulate the ambient field through the cell. A further pair of coils is energized from a radio-frequency oscillator to provide an R.F. magnetic field at right angles to the ambient field. If this field has a frequency corresponding to the Zeeman transitions when energy changes occur, i.e. the Larmor frequency, the atoms return to a random orientation, causing a decrease in the intensity of the transmitted light. This frequency, which

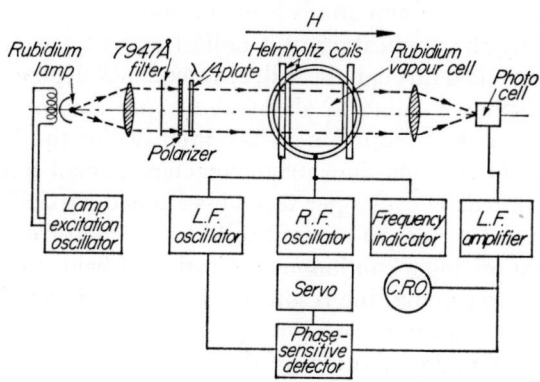

FIG. 11.16. Rubidium vapour magnetometer
Schematic diagram

is known accurately, is dependent on the strength of the magnetic field and is 466·7 kc/s per oersted. Thus, by sweeping the R.F. oscillator slowly through the resonance frequency, the point at which the minimum intensity of the light occurs may be observed and, by measuring the frequency, the field strength may be calculated. Modulation of the ambient field by the low-frequency oscillator results in the output from the photocell undergoing changes in intensity and phase on either side of the minimum signal as the R.F. field passes through resonance. Thus a servo-system may be used to lock the radio-frequency oscillator on to resonance, the frequency being displayed on a counter or other suitable instrument. As changes occur in the ambient field, the servo drives the oscillator to the resonance point, providing a continuous measurement of the ambient field. An accuracy of 0·05 gamma has been claimed for this method.

Metastable-helium magnetometer. The rubidium-vapour magnetometer has certain disadvantages. The nuclear spin possessed by all alkali atoms further splits the atomic levels and causes non-linearities between field strength and output. Moreover, with rubidium six resonance lines are present and these can be confused, especially in weak magnetic fields.

The method of optical pumping has been applied by Keyser, Rice and Schearer[*] to a magnetometer using helium atoms in the metastable state. This magnetometer is similar in construction and performance to the rubidium-vapour instrument, but the helium atom does not suffer from the disadvantages inherent in the use of alkali atoms.

As in the rubidium magnetometer, the helium atoms are orientated by passing a beam of circularly polarized 'resonance radiation' along the direction of the magnetic field to be measured, through an absorption cell containing helium excited to the metastable state. The excitation is provided by the application of a weak radio-frequency electric field. The resonance radiation, whose wavelength is 10 000 Å, is obtained from a helium lamp excited from a 50 Mc/s power supply, the light passing through a polarizer and a quarter-wave plate of polythene sheet before reaching the absorption cell. A photosensitive cell detects changes in the intensity of the light beam emerging from the cell.

The orientation of the helium atoms brought about by the resonance radiation can be destroyed by the application of a radio-frequency magnetic field at right angles to the ambient field through the cell, at a frequency depending on the strength of the field and having a value of about 28 c/s per gamma. This condition is accompanied by a marked reduction in the intensity of the light falling on the photocell. Also, as in the rubidium magnetometer, a small modulating magnetic field of several hundreds of cycles per second is applied coaxially with the ambient field, enabling a servo-system to be used to lock the radio-frequency oscillator on the resonance point, giving a continuous indication of field strength.

At a field of 50 000 gamma, the resonance frequency is about 1·4 Mc/s and a measurement to the nearest cycle represents a resolution of 0·03 gamma. Very satisfactory records of micropulsations of the Earth's field have been obtained with this instrument.

At the present stage of development, nuclear induction and resonance devices are applicable mainly to total field strength without determining direction.

[*] Keyser, A. R., Rice, J. A., and Schearer, L. D., 'A metastable helium magnetometer for observing small geomagnetic fluctuations', *J. geophys. Res.*, 1961, **66,** 4163–9.

Chapter 12

Compass Testing

Laboratory tests

The principal laboratory test of a compass ascertains the accuracy and reliability of the instrument in determining the direction of the magnetic meridian. It is followed by measurement of the constants of the compass—period of oscillation, damping, pendulousness, freedom of tilt, magnetic moment of the system, operational temperature and, perhaps, pressure range. In addition, the testing of transmitting compasses may involve measurement of the follow-up rate, the response of any servo-mechanisms used, accuracy of transmission and, in the case of gyromagnetic compasses, such parameters as monitoring rates, 'stick-off' errors, drift rate of the gyroscope and its various characteristics. Dynamic as well as static tests are often applied to determine the performance of the compass under accelerations.

Compass accuracy

Accuracy is the most important of the various properties of the compass, but it is necessary to ask what is meant by 'accuracy' and how it may be determined. Most definitions of accuracy, unless they are suitably qualified, are meaningless. For instance, to require a compass to be accurate to two degrees is an incomplete statement unless the conditions are given. It is therefore always advisable to quote accuracy figures on a statistical basis. It may reasonably be assumed that the errors of a magnetic compass in an undisturbed magnetic field are random. Since there is only one datum direction, namely the magnetic meridian, all courses or headings are equivalent as far as errors are concerned.

Random errors are considered to have a Gaussian distribution so that the square root of the sum of the squares of the errors divided by the number of observations minus one is what is known as the standard deviation of error, σ.

$$\sigma = \left(\frac{\Sigma \epsilon^2}{n-1} \right)^{1/2}$$

where ϵ is the error and n the number of observations. This is very nearly the r.m.s. value of errors.

In a Gaussian distribution, σ, which is a usual figure for expressing compass accuracy, represents the 68 per cent confidence level; that is to say, 68 per cent of the errors will be equal to or less than σ and 32 per cent will be greater than σ and could have any magnitude.

Frequently a knowledge of the greatest error that can occur is required. Theoretically this can be 180° and is represented by the 100 per cent confidence level; it is considered, however, that the 95 per cent confidence level is normally adequate, where only 5 per cent of the readings would give errors greater than the 95 per cent value, which is mathematically more readily related to σ than is the 100 per cent value. The 95 per cent value is 2σ, so that if σ = ± 1°, there is only a 5 per cent chance that any observations will be more in error than ± 2°. Another useful value is the 50 per cent confidence level or 'equal odds' chance, where half the errors will be within and half without the zone. The 50 per cent value is $\frac{2}{3}$σ or one-third of the 95 per cent value.

These confidence levels may be applied to all manner of observations, and the accuracy can be overall, limited to specified operating conditions, under dynamic or static conditions, and so forth. The standard deviation should thus properly be regarded as a coefficient from which various levels of error can be deduced rather than as an 'accuracy' figure.

An example of an accuracy test of a compass in which errors or departures from the meridian are observed will serve as a suitable demonstration of the computation and interpretation of the probability of error (see Tables 12.1 and 12.2).

TABLE 12.1 *Example of accuracy test on a compass—table of errors (degrees)*

$\zeta°$	I	II	III	IV	V	VI	VII	VIII	IX	X	Mean Error (ϵ)
0	+0·5	+0·3	0	−0·1	−0·2	+0·3	+0·3	−0·2	+0·5	+0·2	+0·16
22½	+0·5	+0·2	0	−0·1	−0·2	+0·3	+0·2	−0·2	+0·5	+0·2	+0·14
45	+0·5	+0·2	0	−0·1	−0·2	+0·2	+0·2	−0·1	+0·4	+0·4	+0·15
67½	+0·4	+0·4	+0·1	0	−0·2	+0·3	+0·3	−0·2	+0·3	+0·3	+0·17
90	+0·3	+0·3	+0·1	−0·2	−0·3	+0·4	+0·4	−0·4	+0·3	+0·3	+0·12
112½	+0·3	+0·2	0	−0·2	−0·3	+0·2	+0·2	−0·4	+0·4	+0·2	+0·06
135	+0·4	+0·3	+0·1	−0·2	−0·1	+0·3	+0·3	−0·3	+0·2	0	+0·10
157½	+0·2	+0·1	−0·1	−0·2	−0·4	+0·3	+0·1	−0·5	+0·2	0	−0.03
180	+0·2	+0·1	−0·2	−0·1	−0·4	+0·1	+0·1	−0·5	+0·4	0	−0.03
202½	+0·4	+0·2	0	0	+0·2	+0·2	+0·2	−0·3	+0·5	+0·2	+0·16
225	+0·5	+0·2	0	+0·1	+0·2	+0·2	+0·2	−0·2	+0·5	+0·1	+0·18
247½	+0·5	+0·3	0	+0·1	−0·2	+0·3	+0·3	−0·2	+0·4	+0·3	+0·18
270	+0·4	+0·1	−0·1	+0·1	−0·2	+0·1	+0·1	−0·3	+0·4	+0·2	+0·08
292½	+0·6	+0·3	0	+0·1	−0·2	+0·2	+0·3	−0·1	+0·4	+0·2	+0·18
315	+0·4	+0·2	0	0	−0·2	+0·3	+0·2	−0·3	+0·5	+0·1	+0.12
337½	+0·5	+0·2	−0·1	0	−0·2	+0·2	+0·2	−0·2	+0·5	+0·1	+0·12

$$\epsilon = A + B \sin \zeta + C \cos \zeta + D \sin 2\zeta + E \cos 2\zeta + F \sin 3\zeta + G \cos 3\zeta + H \sin 4\zeta + K \cos 4\zeta$$

$A = +0·116$ $\qquad$ $D = +0·043$ $\qquad$ $G = +0·019$

$B = −0·023$ $\qquad$ $E = −0·028$ $\qquad$ $H = +0·012$

$C = +0·043$ $\qquad$ $F = −0·021$ $\qquad$ $K = −0·028$

In spite of an initial assumption that all errors should be random, it is possible that a systematic error due to eccentricity or misalignment of the system could exist, which, being constant, is capable of being calibrated out; and it is therefore incorrect to include this. Consequently, the accuracy is usually defined with respect to a mean calibration (or deviation) curve which is obtained from a number of series of observations. Therefore δ, rather than being the departure from the magnetic meridian, now becomes the departure from the mean calibration curve. The

318

TABLE 12.2 *Example of accuracy test on a compass—table of departures from mean errors (∂), their squares (∂^2) and statistical analysis.*

ζ	\(16\)-point swings I	II	III	IV	V	VI	VII	VIII	IX	X	$\Sigma\partial^2$	$\sigma(\pm)$	$\sigma/\sqrt{10}$†	$(\partial_{max})^2$‡
0	+0·34 / 0·116	+0·14 / 0·020	−0·16 / 0·026	−0·26 / 0·068	**−0·36 / 0·130**	+0·14 / 0·020	+0·14 / 0·020	−0·36 / 0·130	+0·34 / 0·116	+0·04 / 0·002	0·648	0·255	0·081	0·130
22½	**+0·36 / 0·130**	+0·06 / 0·004	−0·14 / 0·020	−0·24 / 0·057	−0·34 / 0·116	+0·16 / 0·026	+0·06 / 0·004	−0·34 / 0·116	+0·36 / 0·130	+0·06 / 0·004	0·607	0·246	0·078	0·130
45	**+0·35 / 0·123**	+0·05 / 0·003	−0·15 / 0·022	−0·25 / 0·063	−0·35 / 0·123	+0·05 / 0·003	+0·05 / 0·003	−0·25 / 0·063	+0·25 / 0·063	+0·25 / 0·063	0·529	0·230	0·073	0·123
67½	+0·23 / 0·053	+0·23 / 0·053	−0·07 / 0·005	−0·17 / 0·029	**−0·37 / 0·137**	+0·13 / 0·017	+0·13 / 0·017	−0·37 / 0·137	+0·13 / 0·017	+0·13 / 0·017	0·482	0·220	0·070	0·137
90	+0·18 / 0·032	+0·18 / 0·032	−0·02 / 0	−0·32 / 0·102	−0·42 / 0·176	+0·28 / 0·078	+0·28 / 0·078	**−0·52 / 0·270**	+0·18 / 0·032	+0·18 / 0·032	0·832	0·288	0·091	0·270
112½	+0·24 / 0·058	+0·14 / 0·020	−0·06 / 0·004	−0·26 / 0·068	−0·36 / 0·130	+0·14 / 0·020	+0·14 / 0·020	**−0·46 / 0·212**	+0·34 / 0·116	+0·14 / 0·020	0·668	0·258	0·082	0·212
135	+0·30 / 0·090	+0·20 / 0·040	0 / 0	−0·30 / 0·090	−0·20 / 0·040	+0·20 / 0·040	+0·20 / 0·040	**−0·40 / 0·160**	+0·10 / 0·010	−0·10 / 0·010	0·520	0·228	0·072	0·160
157½	+0·23 / 0·053	+0·13 / 0·017	−0·07 / 0·005	−0·17 / 0·029	−0·37 / 0·137	+0·33 / 0·109	+0·13 / 0·17	**−0·47 / 0·221**	+0·23 / 0·053	+0·03 / 0·001	0·642	0·253	0·080	0·221
180	+0·23 / 0·053	+0·13 / 0·017	−0·17 / 0·029	−0·07 / 0·005	−0·37 / 0·137	+0·13 / 0·017	+0·13 / 0·017	**−0·47 / 0·221**	+0·23 / 0·053	+0·03 / 0·001	0·682	0·261	0·083	0·221
202½	+0·24 / 0·058	+0·04 / 0·002	−0·16 / 0·026	−0·16 / 0·026	+0·04 / 0·002	+0·04 / 0·002	+0·04 / 0·002	**−0·46 / 0·212**	+0·34 / 0·116	+0·04 / 0·002	0·448	0·212	0·067	0·212
225	+0·32 / 0·102	+0·02 / 0	−0·18 / 0·032	−0·08 / 0·006	+0·02 / 0	+0·02 / 0	+0·02 / 0	**−0·38 / 0·144**	+0·32 / 0·102	−0·08 / 0·006	0·392	0·198	0·063	0·144
247½	+0·32 / 0·102	+0·12 / 0·014	−0·18 / 0·032	−0·08 / 0·006	**−0·38 / 0·144**	+0·12 / 0·014	+0·12 / 0·014	−0·38 / 0·144	+0·22 / 0·048	+0·12 / 0·014	0·532	0·231	0·073	0·144
270	+0·32 / 0·102	+0·02 / 0	−0·18 / 0·032	+0·02 / 0	−0·28 / 0·078	+0·02 / 0	+0·02 / 0	**−0·38 / 0·144**	+0·32 / 0·102	+0·12 / 0·014	0·472	0·217	0·069	0·144
292½	**+0·42 / 0·176**	+0·12 / 0·014	−0·18 / 0·032	−0·08 / 0·006	−0·38 / 0·144	0 / 0	+0·12 / 0·014	−0·28 / 0·078	+0·22 / 0·048	0 / 0	0·512	0·226	0·072	0·176
315	+0·28 / 0·078	+0·08 / 0·006	−0·12 / 0·014	−0·12 / 0·014	−0·32 / 0·102	+0·18 / 0·032	+0·08 / 0·006	−0·32 / 0·102	**+0·38 / 0·144**	−0·02 / 0	0·498	0·223	0·071	0·144
337½	**+0·38 / 0·144**	+0·08 / 0·006	−0·22 / 0·048	−0·12 / 0·014	−0·32 / 0·102	+0·08 / 0·006	+0·08 / 0·006	−0·32 / 0·102	+0·38 / 0·144	−0·02 / 0·002	0·572	0·239	0·075	0·144
Sum Mean											9·036*		0·075	2·712§

* The sum of $\Sigma\partial^2$ may be written $\Sigma\Sigma\partial^2$. The overall $\sigma = \pm(\Sigma\Sigma\partial^2/160)^{\frac{1}{2}} = \pm0°\cdot238$ (14'). Overall 95% level $= \pm0°\cdot476$ (28').

† $\sigma/\sqrt{10}$ is the accuracy of the mean value of ∂; its average value is $0°\cdot075$ (4'·5).

‡ The several values of (∂_{max}) and $(\partial_{max})^2$ are shown under the 16-point swings in bold type.

§ The sum of $(\partial_{max})^2$ may be written $\Sigma\partial^2_{max}$. Approximate 95% level $= \pm[\Sigma\partial^2_{max}/16]^{\frac{1}{2}} = \pm0°\cdot41(25')$.

319

character of the mean calibration curve can be determined by a Fourier analysis as on p. 319.

Alternatively, the accuracy of any mean value may be stated as the standard deviation about the mean (σ) divided by the square root of the number of observations (n) i.e. as $\sigma/n^{1/2}$.

Where standard deviations are to be added,

$$\sigma_s = (\sigma_1^2 + \sigma_2^2 + \sigma_3^2 + \ldots + \sigma_n^2)^{1/2}$$

If it is preferred to calculate standard deviations from zero, the standard deviation about a mean error that is different from zero is

$$\sigma_m = (\sigma_0{}^2 + m^2)^{1/2}$$

where σ_m is the standard deviation about a mean, m the mean error and σ_0 the standard deviation about zero.

Assuming all headings to be equivalent, an overall standard deviation can be obtained using the sum of the squares of all departures from the mean, regardless of heading. Then n is the total number of observations—i.e. the number of headings multiplied by the number of swings or series of headings.

In the worked example (Tables 12.1 and 12.2), ten 16-point swings have been taken. The actual errors are shown in Table 12.1 and also the mean error, which may be analysed into its harmonic components or coefficients by using the forms shown in Appendix 4. These coefficients, of course, need not have any relation to the usual coefficients of deviation (Chapter 10). *A* could be incorrect initial alignment of the test gear; *B* and *C* are usually eccentric errors in the system; *D* and *E* are usually associated with certain defects in inductor compasses; *H* signifies lack of linearity in certain inductor systems. Table 12.2 shows the departures from the mean, with the squares, sum of squares and standard deviation about each mean. The accuracy of each mean is also given and the overall standard deviation. From these tables can be ascertained the systematic error of the compass, the accuracy of this error and the overall extreme error (95 per cent confidence level).

A short way, sufficiently precise for most purposes, of obtaining the last figure, is to determine the standard deviation of the maximum departures from the mean. These are in bold type in Table 12.2 and the resulting σ is approximately the overall error (95 per cent confidence level).

An inspection of the tables reveals any unusual features about the swings; e.g. swing VIII is systematically low, and the test conditions might well be examined in that particular case.

A conclusion that can be drawn from this example is that the calibration (mean) curve exhibits, primarily, a constant error of 7 minutes of arc. There are also small eccentric and quadrantal errors. About the mean curve there is a standard deviation of ± 14 minutes of arc, indicating that a single observation of error might be incorrect to the extent of ± 28 minutes of arc. The accuracy of the mean curve obtained from ten sets of observations is $\pm 4 \cdot 5$ minutes of arc.

The small variation of standard deviation about the mean value, reflected also in the accuracy of the mean on the several headings, justifies the assertion that in compass testing all headings are equivalent, since they are related to a single datum.

The approximate method of determining the 95 per cent zone gives 25 minutes, as against 28 minutes for the precise method.

Accuracy tests

Since the accuracy of a compass is generally determined by placing the instrument on a number of known headings and comparing the compass heading with the known, a device for orienting the compass at will is required. The simplest form is a graduated rotatable turntable provided with a vernier, and tangent screws for adjustment. Plate 12.1 illustrates such a turntable, which can be mounted on a stand, a table or, ideally, a concrete pillar. The surface of the turntable must be truly horizontal on all headings and should have, in addition to three grooves at 120° to take standard tribrach fittings, any necessary holes or sockets for the attachment of sighting vanes and other implements. In use, the table is set up so that the 0°–180° diameter of the turntable lies in the magnetic meridian. This is achieved either by placing a magnetic compass whose index error, if any, is known, on the turntable so that its lubber's line or index is on the 0°–180° diameter and then setting the table until the compass needle is aligned with the lubber's line; or by placing a theodolite or other sighting device on the turntable and training it on a mark whose bearing (either magnetic or true) is known, and then setting the table accordingly. Variation can be found by using a Kew or similar type of magnetometer.

Directional error. The directional error of a compass can now be determined on a series of headings by mounting the compass securely on the turntable so that its index mark is parallel to the 0°–180° datum, and rotating the turntable so that the compass lubber's line is in turn brought to the selected headings. The orientation of the turntable is noted and the difference between turntable and compass headings is the error of the compass.

Compass minus turntable = error

A table of errors (or deviations if the sign is changed) can be drawn up and analysed as already explained on p. 318. Such errors are directional errors and include:

 Card (or graduation) error;
 Eccentricity error of the pivot in the card;
 Index error due to misalignment of the magnetic axis of the magnet system with the 0°–180° diameter of the card.

These errors are separable and may need to be determined.

Card error is conveniently determined by setting the compass on the turntable so that compass heading 'zero' corresponds to turntable heading 'zero.' The table is turned to the required compass headings and the compass and turntable compared as before. This test is merely an angular check on the card graduations and is not related to the magnetic meridian.*

Included in these errors will be the eccentric error due to the incorrect placing of the pivot (see also sections on compass calibration pp. 328–30).

---

* Throughout this test, a careful watch must be kept for changes of variation. If the conditions are magnetically disturbed, testing for accuracy should not be continued.

Eccentricity error of the pivot is revealed by comparing card errors 180° apart by an analysis of semicircular errors corresponding to coefficients **B** and **C**. The phase of the semicircular error indicates the direction of displacement of the pivot.

By placing a sighting vane or wire directly over the compass, with the wire cutting the pivot point, eccentricity will be revealed by the card readings under the wire or vane being other than 180° apart. This test should be done in two positions at right angles to each other. If the divergences from 180° are A and B respectively, the semicircular error or eccentricity E is given by

$$E = \tfrac{1}{2}(A^2 + B^2)^{1/2} \quad \text{when the angles are small.}$$

Index error exists because the magnetic axis of the needle system is either misaligned with respect to the 0°–180° diameter of the card or, in simple needle instruments, different from the needle's geometric axis.

Index error is the difference between directional error and card error, but may be found independently by aiming the compass at a mark whose bearing is accurately known, observing the compass bearing of the mark and applying the card error. This corrected reading is then compared with the magnetic bearing of the mark to arrive at the index error. This test may be done either in the laboratory or at a previously surveyed base out of doors. Although one mark is adequate, the use of a number of approximately equally spaced marks around the base to determine the mean index error makes for improved accuracy (see also *Calibration of card compasses* for a more elaborate and precise method).

Lubber error. In compasses used for steering, the angular relationship of the lubber's line with the fore-and-aft direction of the ship is important. The lubber error is defined as the angle between the line joining the lubber's point or mark to the pivot point and the line parallel to the fore-and-aft line of the ship (defined by the outer gimbal axis), which passes through the pivot point.

To determine this error special jigs and mountings are required, and difficulties arise because the exact location of the pivot point is obscured in a completed compass. An accurate determination of lubber error may involve dismantling the compass.

Total error. The sum of the directional error and the lubber error is sometimes called total error. It has significance only in a steering compass. It is preferable, however, that the term total error be restricted in use to the difference between true north and compass north.

Bearing errors when using an azimuth circle

Bearing errors have been discussed on pp. 116–8 in connection with prismatic compasses. The two errors revealed by a rotatable circle are those due (*a*) to the pivot being eccentric to the rim of the compass bowl on which the azimuth circle is mounted, or (*b*) to the line of sight through the circle not passing through the centre of the card.

(*a*) will appear as a semicircular error and is measured by setting up the compass on the turntable and aiming the circle at a mark whose bearing is known. The turntable is then rotated 180° and the circle re-aimed. The difference is twice the eccentric error. This test should be done on four headings at right angles to each other, as in measuring the type of eccentric error described on p. 116.

(*b*) will appear as a constant error on all headings of the compass when tested in the same manner as that described under (*a*).

Other errors to be measured when testing azimuth circles are (*c*) ill-aiming, and (*d*) skew of prism. Ill-aiming error is determined by aligning the circle exactly on a mark, ensuring that the line of sight, as defined by a thread cutting the vee of the sight, the centre of the circle and the centre of the card, is aimed at a fixed mark.

The bearing is read and the circle 'ill-aimed' by the required angle, either by rotating it on the compass or by rotating the turntable. The mark should, of course, still be visible in the vee sight, although the eye may have to be moved to a different position, and its bearing is read again. The difference between this reading and the first is the ill-aiming error for the given amount of ill-aim.

The same principle applies in testing other circles where wire sights or optical images are used instead of a vee sight.

The prism skew is observed by aiming the circle at a plumb-line with the prism horizontal. As the prism is tilted, the bearing of the plumb-line should not alter more than an angle specified.

Testing transmitting compasses

The methods so far described appertain to direct-reading compasses, i.e. those in which there is a visible card or needle. Transmitting compasses are tested in a similar way by mounting the master unit on the turntable. Observations of compass reading are then taken at the repeating instrument. If the master unit has a visible card or needle, this is also observed as described in the tests for direct-reading compasses. Comparison between the master unit and repeater tests the accuracy of transmission, and both sets of readings can be subject to separate analyses.

Friction test

This test is to indicate the quality of the pivot and jewel suspension, and is a measure of the frictional torque that opposes the return of the needle into the meridian after deflection (see also p. 22).

The method applies to all types of instruments, whether read at the card of the master compass or at a repeater. The system is settled in the meridian and deflected through a given angle by a small magnet (see Fig. 12.1). It is then released and allowed to settle once again. The new settling position is noted, and the test repeated on the other side of the meridian. The compass must not be tapped or vibrated. The friction angle may be stated *either* as the included angle between the two settling positions after release on either side of the meridian, *or* as the mean departure from the meridian on settling after release first on one side and then on the other, *or* as the larger of these two angles.

The needle is deflected first to position OA and then released, settling in position OP. Observe PÔN. The needle is then deflected to position OB, so that NÔB = NÔA, and then released, to settle in position OQ. Observe QÔN.

The friction angle may be defined as: (*a*) PÔQ, (*b*) PÔQ/2, or (*c*) whichever is the greater of PÔN or QÔN.

Effect of swirl

The effect of swirl is the drag caused by any form of damping (pp. 30–4), especially in liquid compasses. The compass to be tested is set up on a turntable or other suitable stand, so that the needle system is settled in the meridian. The turntable is then rotated at a constant speed, preferably by a small motor, placed so as not to deviate the compass, through a given number of complete revolutions, the speed and number of revolutions being chosen to suit specific requirements. The table is then stopped, when an immediate observation is made of the deviation from the meridian of the compass or, in a transmitting compass, of the repeater.

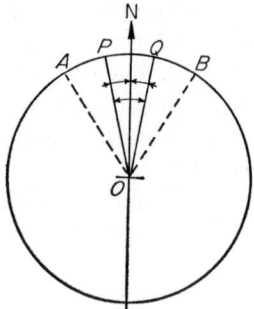

FIG. 12.1. Friction angle

Period of oscillation

The period of oscillation is a property of all magnet systems that are not critically damped or aperiodic. The degree of damping is the subject of a further test.

The magnet system is deflected, say 40°, after being settled in the meridian. It is kept deflected for about half a minute to allow any disturbance of the compass liquid to die away. The system is then released, and the times recorded of successive meridional passages of the needle or card. The interval between two such passages is half the period. As many passages as possible should be observed, subject to the degree of damping, and a mean value of the period calculated.

Time of swing

In aperiodic or critically damped systems, no period is measurable, so an arbitrary quantity known as the time of swing is measured.

The needle system, after being settled in the meridian, is deflected 90° and held there for about half a minute as before, and then released. Instead of the periodic time, the time is observed that the needle takes to return to 5° from the meridian.

Damping is measured in a periodic compass by observing the successive amplitudes of a number of swings past the meridian. This measurement can well be combined with a test of the period of oscillation.

If the needle is deflected to the *east*, the angle of overswing to the *west* is observed (θ_1). Then as the needle swings to the east again, its overswing to the *east* is observed (θ_2). Again the needle swings westward and its second westerly overswing is observed (θ_3). The damping factor is given by:

$$f = \frac{\theta_3 - \theta_2}{\theta_1 - \theta_2}$$

where θ_1, θ_2 and θ_3 are compass headings, or

$$f = \theta_2/\theta_1 = \theta_3/\theta_2$$

where θ_1, θ_2 and θ_3 are deflections. In an aperiodic or critically damped instrument, the overswing (probably only one can be noticed) is observed.

Pendulousness

Pendulousness depends on the balance of the compass needle system (p. 21). It is a requirement that the compass system should be substantially horizontal at all places on the Earth's surface. In the United Kingdom, for instance, the vertical field of the Earth is 0·43 oersted. It increases to about 0·6 oersted in navigable northern waters and to −0·6 oersted in southern waters; thus the maximum change of vertical field is about 1 oersted.

In the United Kingdom a change of vertical force of 1 oersted is applied by placing the compass within a Helmholz coil arranged to produce a vertical field. The requisite current is applied to the coil to produce 1 oersted in the appropriate direction and the change of tilt of the system is observed. Obviously, only those instruments with a visible card or systems can be tested in this way.

Freedom of tilt

Freedom of tilt is sometimes required to be measured. This is merely the angle through which the compass can be tilted before the system fouls any part of the bowl or internal fittings, however the compass is orientated. A tilting platform mounted on the turntable is necessary for this test.

Magnetic moment

The magnetic moment of the compass system is measured by offering up the complete compass or the separate magnet system to a suspension magnetometer of the type described on p. 127 and in Fig. 6.2.

Although the separate magnet system can be tested as previously described, with the magnet lying east-west, a complete compass requires a special rig, since its needle will, of course, lie north–south. This rig is shown diagrammatically in Fig. 12.2.

In a given horizontal field, the deflection of the magnetometer needle is equal to $\tan^{-1} F/H$, where F is the deflecting field. If the magnet under test is short compared with its distance from the magnetometer, F is proportional to M/d^3, d being the distance from the centre of the magnet to the magnetometer needle and M its moment. If deflections are limited to small angles, they will be proportional to M/d^3 and so a simple calibration of moment in terms of distance and deflection can be prepared.

Induction test (four-sphere test)*

The induction test is applied to determine the errors introduced by magnetic induction in soft-iron quadrantal deviation correctors owing to the magnet system of the compass.

The compass is mounted on a stand and two soft-iron spheres are placed diametrically opposite each other. The spheres are rotated about the compass and from deviations of the latter, coefficient D is determined. Two more identical spheres are then placed similarly but at right angles to the first pair and the spheres again

---

* See also Appendix 5, p. 364.

rotated. From the deflections of the compass that occur, coefficient H is found. The ratio of coefficient H to coefficient D is then determined.

All the foregoing tests need to be carried out in a specially constructed magnetic laboratory or test room. The materials used in the construction of the building must be non-magnetic and the site should be the subject of a magnetic survey to ensure that it is only the undisturbed magnetic field of the Earth that is experienced in the test area.

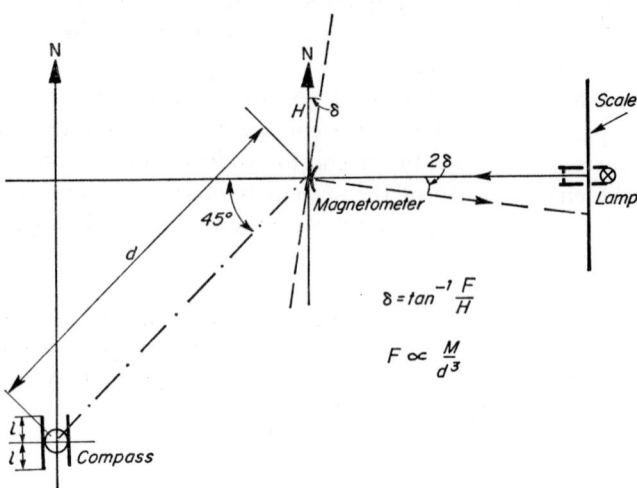

$$\delta = \tan^{-1} \frac{F}{H}$$

$$F \propto \frac{M}{d^3}$$

FIG. 12.2. Measuring magnetic moment

Clear lines of sight from the windows are necessary and the marks should be as far away as convenient. In poor visibility or at night, lights may be used or a collimator can be set up in the test room aligned in a chosen direction. The instruments required, other than special dimensional jigs, are few and simple and their use has already been described. They may be summarized as follows:

Graduated turntable, firmly mounted, level and commanding clear lines of sight;

Tilting platform for mounting on the turntable;

Sighting frame for aligning the turntable and any compass placed on it;

Magnetometer for determining variation;

Magnetometer for measuring magnetic moment;

Plumb-line for testing azimuth circles;

Sundry magnets, spirit-levels, stop-watches, sighting marks and, if possible, a collimator. Only the best quality of instrument should be used for test purposes.

A good theodolite is a useful accessory as it can be used for aligning turntables and test stands, and for approximate determinations of variation.

Dynamic tests

The behaviour of a compass on a moving platform, such as a ship at sea or an aircraft in flight, may be and frequently is very different from its behaviour in the laboratory or test room. The problem of acceleration will be appreciated from a study of pp. 34–46.

For dynamic testing, the essential piece of equipment is a platform which can be accelerated in any desired direction, or if this proves too complicated, in the magnetic east–west direction. The platform could conveniently be a trolley driven to and fro by some form of crank and link mechanism; a 'swing' comprising a parallel motion linkage or mounting platform, with some means of driving the linkage so that a variety of amplitudes and periods of oscillation can be selected; or a mast pivoted at its base, with the compass mounted at the top. The mast can be arranged to pivot as well as oscillate, so that a complete programme of yaw, pitch and roll experiments can be carried out. Otherwise, a platform of the Scoresby type may be used for roll, pitch and yaw tests. Great care needs to be paid to the non-magnetic character of swings, platforms and masts, and any electrical driving mechanism must be so situated that it does not cause any deviations of the compass. Photographic recording gear is useful in obtaining records of dynamic tests.

The instrumentation of dynamic testing depends largely on the type of compass under test and may range from simple direct visual or photographic observations to a complete recording system of the many variables that can be found in the more elaborate gyro-magnetic systems. The primary effects to be observed are changes in heading, perturbations of the system, oscillatory effects, and wanderings of the gyroscope.

There is, however, no land-based substitute for a practical test at sea or in the air, unless the effect of some specific dynamic influence only is to be recorded, when a test on land is simpler and more practicable.

Electrical and other tests

Other tests to which compasses may be subjected do not necessarily need a prepared site. The effects of pressure and temperature can be studied by the use of conventional equipment such as bell jars and vacuum pumps, refrigerators and hot chambers. There is considerable choice of such equipment and test methods and procedure can be varied to suit circumstances. Occasionally the performance of a compass needs to be studied under abnormal conditions of temperature and pressure, in which case the magnetic properties of the test gear and its surroundings need to be taken into consideration.

Transmitting compasses and gyro-magnetic compasses also require a variety of electrical tests, the specific nature depending upon the type of equipment. Normal measurements are:

> Current consumption;
> Operating voltage;
> Range of voltage and frequency over which equipment will work;
> Speed of follow-up and rate of transmission;
> Characteristics of any gyroscope fitted;
> Range of monitoring rates;
> Starting voltage of follow-up motors and the like;
> Backlash measurement;
> Lag in the servo-loops.

Most of these are covered by conventional electrical practice or, alternatively, may be studied by consulting text-books on gyroscopic instruments and servo-mechanisms.

Test rigs for the more elaborate transmitting and gyro-magnetic compasses are usually designed according to the nature of the equipment and specific requirements.

Calibration of card compasses

A calibration test sheet is illustrated in Fig. 12.3. The purpose of calibration is to determine the corrections that need to be applied to observed compass bearings to convert them to correct magnetic bearings. The corrections fall into two types: those necessitated by irregularities in the compass graduations, and cyclic effects due, for example, to eccentricity of the pivot; and those needed because of misalignment of the magnetic axis of the needle or other detecting device with the datum or north–south reference.

Incorrect angles are measured, say, from compass north or zero of the scale, because of errors of the first type, which are purely mechanical and independent of magnetic effects. Such errors can be determined by simple observation in the laboratory, comparing the angular movement of a turntable on which the compass is mounted with that of the index mark or lubber's point, with respect to the compass card.

The correct procedure is to set the compass index or lubber to predetermined values on the card or scale, by rotation of the turntable as necessary, and then to read the corresponding graduation on the turntable. Thus compass angle against turntable angle may be tabulated and a table of corrections prepared. To ensure accurate results, at least six sets of readings from 0° to 360° should be made at intervals according to the degree of precision required; every 15° is a practical subdivision, and the alignment of compass zero with turntable zero should be checked and, if necessary, reset at the start and finish of each cycle of angles. This is to prevent any change of magnetic variation being interpreted as part of the compass error.

The second type of error results in a constant difference between compass and magnetic bearings. Whilst the determination of such an error can be made in the laboratory, it is often more convenient and accurate to carry out the measurement out-of-doors on a specially prepared range or base. A series of roughly equally spaced marks is laid out, at upwards of 100 yds from and around the test point. For the highest accuracy, greater distances are advisable, though above one mile, heat-shimmer and poor visibility may prove a handicap. The geographic bearings of the several marks are determined astronomically.

By directing the compass at each of these marks in turn, the compass bearing, corrected by a previous calibration in the laboratory, may be compared with the true or geographic bearing. Subtracting the latter from the former gives variation plus compass index error (C.I.E.). At the commencement of each round of angles, the magnetic variation should be measured. Thus from each round, the mean value of variation plus C.I.E. can be ascertained. Hence, knowing the variation at the time, the index error is determined. The mean value from a number of rounds gives the required correction.

The combination of the laboratory or angular calibrations and the index error enables a complete calibration curve to be prepared. It will be apparent that a single datum mark could equally well be used, provided sufficient observations are

328

TEST SHEET FOR CALIBRATION
PRISMATIC COMPASSES

1	2	3 *
	Errors	
Hd'g	Lab calib$^{n.}$	Final calib$^{n.}$
0		
15		
30		
45		
60		
75		
90		
105		
120		
135		
150		
165		
180		
195		
210		
225		
240		
255		
270		
285		
300		
315		
330		
345		

Compass No.__________

Date of test Lab.__________

Field__________

Observer Lab.__________

Field__________

Variation at start of field test__________

Variation at end of field test __________

Mean variation__________

* + Errors are to be added to compass reading

− Errors are to be subtracted from compass reading

Col. 3 is col. 2 + constant error of compass

Field test

Mark	Mean compass bearing	Correc$^{n.}$ from col. 2	Corrected compass bearing	True bearing of mark	Total error	δ Dep. from mean	δ^2
I				338·67			
II				002·53			
III				019·67			
IV				013·77			
V				092·75			
VI				157·54			
VII				248·96			
VIII				281·05			
IX				336·76			

Mean total error________ Mean

Mean variation________ $\sigma^2 =$ ________

Constant error compass________ $\sigma =$ ________

FIG. 12.3. Calibration test sheet

taken, but a series of marks helps to average out any positional errors in placing the compass or systematic observational errors associated with any one mark.

Calibration of datum compasses

Now to be considered are instruments of the theodolite or bearing-plate type, in which the magnetic element is as illustrated in Figs. 7.2–7.4, and the Admiralty Pattern 1 compass. The distinguishing feature of these is that the compass needle is used to orientate a bearing plate or graduated azimuth circle with respect to magnetic north. Generally, the testing of the graduated circles is beyond normal test-room procedure and, in the case of well-made instruments, is unlikely to be necessary. It can, however, be carried out by setting up collimators on known bearings around a firm and rigid datum pillar, on which the instrument to be tested is set (as in the testing of sextants); or, if sufficiently careful astronomical observations are made, distant marks out-of-doors may be used.

Collimation and similar errors are best dealt with as laid down in works on theodolite design and construction, and, for the present purpose, it is assumed that the instrument is optically correct. Thus the magnetic problem is the determination of the index error of the needle, i.e. the correction to be applied when the bearing plate or circle is aligned to compass north to provide magnetic bearings.

The instrument should be set up on a test pillar or tripod, carefully levelled, and aimed at a fixed mark whose true bearing is known to at least five times the degree of accuracy expected from the compass. The best pivoted needles have accuracies of approximately ± 3 minutes of arc (standard deviation); hence the bearing of the mark must be known to 30 seconds of arc. The bearing plate may be set at zero or any convenient degree mark.

The sighting axis of the instrument is now turned into approximate alignment with the magnetic meridian and the needle is unclamped. By carefully rotating the instrument, the index marks of the compass element are brought into alignment with the needle. The circle or bearing plate reading is noted. The needle should be clamped and unclamped six to ten times and a mean of readings taken. Thus the angle between the mark and compass north is known, and consequently the compass bearing of the mark. Simultaneously with these readings, the variation is determined by a Kew-type magnetometer. If ten observations are taken, it is wise to take a variation reading at the beginning and end of the series, using the mean value. This, applied to the true bearing of the mark, gives its magnetic bearing, which, subtracted from the compass bearing, reveals the index error. It is advisable to take several such sets of observations, placing the theodolite or datum instrument in various positions, such as those suggested by the positions of the foot-screws in the grooves of a datum pillar, and ascertaining the mean index error.

PREPARING A TEST SITE

It is first of all essential to ensure that a test site is free from magnetic disturbances, anomalies or irregularities. The permissible amount of interference depends on the purpose for which the site is intended. As a rule, there are three kinds of site:

Magnetometer huts and calibrating sites;

Compass test rooms;

Swinging bases for the adjustment of aircraft and vehicle compasses.

Magnetic disturbances may arise owing to the proximity of power supplies, electric railways and tramways, passing traffic and welding plant, to name some of the more usual causes. Anomalies or irregularities may be due, for example, to iron-bearing inclusions in the ground, buried scrap-iron, drains, conduits, or neighbouring iron or steel buildings.

It is generally recommended that, when testing a standard reference instrument, the site should be distant at least 100 yards from any steel- or iron-framed building, power supplies, gas or water mains or large masses of iron or steel. For a magnetometer hut, 300 yards would be advisable. Electric railways or tramways should be no nearer than one mile, as even at that distance, transient disturbances are readily discernible.

For compass test rooms, provided the effect of neighbouring iron and steel work, pipes and electrical supplies can be determined and is unlikely to vary, a much closer distance can be allowed, perhaps as near as 100 feet. In such a case, however, frequent checks are necessary.

When siting a swinging base, no iron or steel buildings, pipes, drains or supply lines should be within the area of the base and no nearer than 200 feet to its perimeter.

A survey of the site is then carried out using field strength or direction measurements. For a reference instrument, the field strength must be stable to ± 2 gamma over an area of about 100 feet radius from the instrument (apart from recognizable changes in the Earth's field itself). If comparisons are being made with another site, it is sometimes necessary to apply a site correction for strength and direction measurements.

When testing compasses, much larger effects are permissible if constant reference is made to a standard instrument; site corrections are perfectly legitimate, but the movement of magnetic material within, say, 100 feet of the site, should be avoided. Uniformity of the field should be no worse than ± 10 gamma and of the field direction, ± 1 minute of arc. It should be stressed that a constant field, though differing in direction and strength from the Earth's undisturbed field, is to be preferred to one that is subject to random variations, and in this connection, fixed masses of iron and steel in the proximity of the site may sometimes be allowed. The quality of the site may be modified according to the type of instrument that is being tested. Datum compasses and surveying instruments, for example, need the highest grade of magnetic stability, when site corrections in excess of 5 minutes of arc should be avoided.

Bases for compass adjustment usually consist of a central area, on which the aircraft or vehicle is placed, and a concentric circular track, some 100 feet to 150 feet in radius, on which the datum compass is placed. Reciprocal measurements of the magnetic meridian are made over the central area and the datum compass circle between successive points 20 feet apart, so that the whole base may be regarded as divided up in the form of a 'grid'. For a 'Class One' base, the differences between successive points must not exceed 6 minutes of arc.

In addition to an accurate magnetic survey of the site, it is sometimes desirable to have surveyed lines of true bearing set up, so that magnetic variation may be measured directly; or, given the magnetic variation, compasses and other instruments may be set up in the magnetic meridian for testing.

The test 'point' or stand may be defined by a datum plate with a centre mark, the centre of a theodolite pillar as used by the Ordnance Survey, or a turntable mounted upon a rigid support. All such supports and mountings need to be of non-magnetic material. From this point the lines of true bearing radiate to distant marks. These should be as far distant as possible for accuracy, but heat-haze, atmospheric 'shimmer' or mist and fog sometimes compel the use of nearer marks. Collimators set up on adjacent rigid supports serve equally well, may be under cover and are unaffected by the weather.

Fig. 12.4 shows a simple test site with an adjacent magnetometer hut, all the materials of which need to be tested carefully to ensure that they are not magnetic. There are many variations of the lay-out shown in Fig. 12.4, which may be adapted to suit local requirements and terrain.

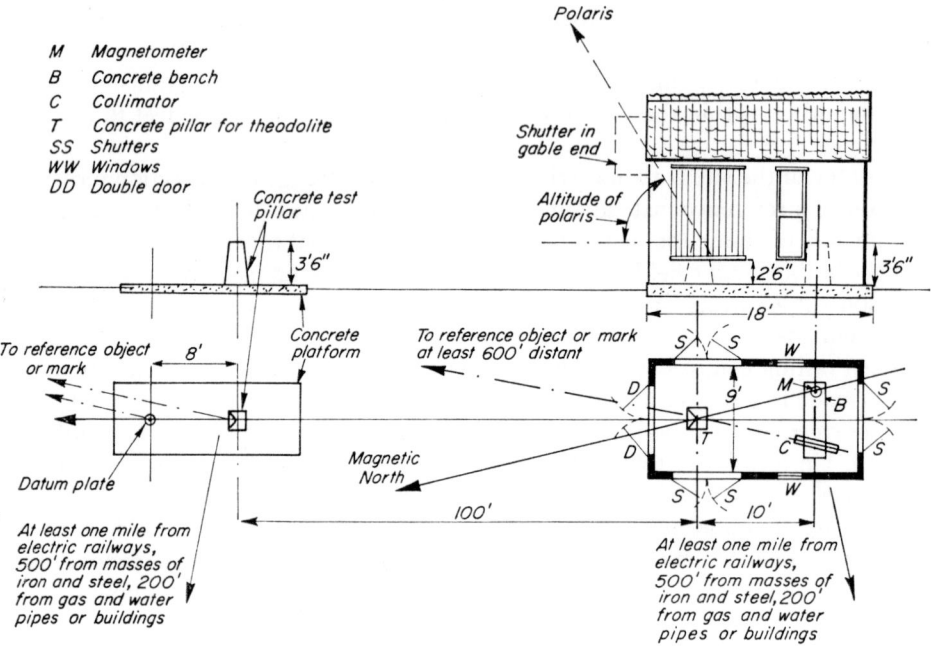

FIG. 12.4. Test site

The method of obtaining the true bearing of a mark is to observe the azimuth of certain convenient stars, or of the sun (or moon), by means of a high-grade theodolite and to relate this to the line of sight joining the test point to the mark. The most convenient computation of celestial azimuths is by means of the haversine formula, well-known to navigators, and which, for convenience, is condensed into the tabular form shown in Fig. 12.5.

Having determined the precise time of observation of the celestial body's bearing in relation to the mark, and remembering that as many observations as possible should be taken in the interests of accuracy, the hour angle and declination are found from the Nautical Almanac and entered on the form. Latitude and longitude are also entered as being known data for the site. Inman's Nautical Tables provide

CALCULATION OF TRUE BEARING OF SUN OR STAR

Sun/star _ _ _ _ _ _ _ _ _ _ _ _ _ _ _   Log hav. _ _ _ _ _ _ _ _ _ _ _ _ _ _ _

Date _ _ _ _ _ _ _ _ _ _ _ _ _ _ _   Log cos _ _ _ _ _ _ _ _ _ _ _ _ _ _

G.M.T. _ _ _ _ _ _ _ _ _ _ _ _ _ _

G.H.A. Υ _ _ _ _ _ _ _ _ _ _ _ _ _ _ _   Log cos _ _ _ _ _ _ _ _ _ _ _ _ _

Incr. _ _ _ _ _ _ _ _ _ _ _ _ _ _ _   Log hav. (e) ⎯⎯⎯⎯⎯⎯ Sum

_ _ _ _ _ _ _ _ _ _ _ _ _ _ _ _ _   Nat hav. (e)

S.H.A. ✱ _ _ _ _ _ _ _ _ _ _ _ _ _   Nat hav. (d) _ _ _ _ _ _ _ _ _ _ _

G.H.A. ✱ _ _ _ _ _ _ _ _ _ _ _ _ _

Incr. _ _ _ _ _ _ _ _ _ _ _ _ _ _   Nat hav. C.Z.D. ⎯⎯⎯⎯⎯ Sum

_ _ _ _ _ _ _ _ _ _ _ _ _ _ _ _   Alt = ⎯⎯⎯⎯⎯⎯

Long. _ _ _ _ _ _ _ _ _ _ _ _ _ _

L.H.A. _ _ _ _ _ _ _ _ _ _ _ _ _ _   Log cosec _ _ _ _ _ _ _ _ _ _ _ _

Lat. _ _ _ _ _ _ _ _ _ _ _ _ _ _ _   Log cosec _ _ _ _ _ _ _ _ _ _ _ _ _ _

Dec. _ _ _ _ _ _ _ _ _ _ _ _ _ _ _

Incr. _ _ _ _ _ _ _ _ _ _ _ _ _ _ $\frac{1}{2}$ log hav. _ _ _ _ _ _ _ _ _ _ _

_ _ _ _ _ _ _ _ _ _ _ _ _ _ _ _ _ $\frac{1}{2}$ log hav. _ _ _ _ _ _ _ _ _ _ _

    Log hav. AZ. ⎯⎯⎯⎯⎯⎯ Sum

(d)∗Lat.∼dec. _ _ _ _ _ _ _ _ _ _

C.Z.D. = ⎯⎯⎯⎯⎯⎯⎯⎯⎯⎯

Co lat. _ _ _ _ _ _ _ _ _ _ _ _ _ _

C.Z.D. _ _ _ _ _ _ _ _ _ _ _ _ _ _

C.Z.D.∼co lat. _ _ _ _ _ _ _ _ _ _

PX 90° dec. _ _ _ _ _ _ _ _ _ _ _   Azimuth _ _ _ _ _ _ _ _ _ _ _ _ _ _

Sum _ _ _ _ _ _ _ _ _ _ _ _ _ _

Diff. _ _ _ _ _ _ _ _ _ _ _ _ _ _   True bearing ⎯⎯⎯⎯⎯⎯

Notes

L.H.A. if over 180° subtract multiples of 360°   Long. add easterly
(d) Lat. and dec. same name subtract.   subtract westerly
    Lat. and dec. different names add.
Incr. applies to sun and Υ
PX = 90 − N. dec.
     90 + S. dec.

FIG. 12.5. Layout for computation of azimuth

the information for log cosines, log haversines, natural haversines and $\frac{1}{2}$ log haversines, and thus the azimuth of the body in question eventually appears at the end of the form.

The importance of accurate timekeeping cannot be overstressed, and some standard such as the B.B.C. time signals and a reliable chronometer or deck watch should be used.

The formula for the azimuth of a heavenly body is

$$\text{hav } PZX = \text{cosec } PZ \text{ cosec } ZX \left[\frac{\text{hav } (PX + \overline{PZ \sim ZX})X}{\text{hav } (PX - \overline{PZ \sim ZX})}\right]^{\frac{1}{2}}$$

where PZX is the azimuth, PZ the co-latitude $(90° - \text{Lat.})$, ZPX the hour angle, ZX the zenith distance, and PX the polar distance $(90° \pm \text{Dec.})$.

$$\text{hav } ZX = \text{hav } (PX \sim PZ) + \text{hav } \theta$$

$$\text{hav } \theta = \sin PX \sin PZ \text{ hav } ZPX$$

These expressions may be written thus:

$$\text{hav } ZX = \text{hav } (\text{Lat.} + \text{Dec.}) + \text{hav } \theta$$

$$\text{hav } \theta = \cos \text{Lat.} \cos \text{Dec.} \text{ hav } ZPX$$

The altitude of the body is $(90° - ZX)$.

An approximate azimuth for a celestial body can be determined from the alt-azimuth tables, in which altitude and azimuth are tabulated for values of latitude and declination against apparent time, or star's hour angle.

USE OF RECIPROCAL BEARINGS
IN MAGNETIC SURVEYS AND COMPASS CALIBRATION

Given two datum compasses whose index errors are known, the detection of magnetic anomalies across a site is a comparatively straightforward undertaking. The two instruments are set up at suitable points a convenient distance apart, say 100 feet, and reciprocal bearings of each other's telescope axis taken. These are related to the magnetic meridian as determined by each compass and, after applying any necessary correction, the magnetic bearings of the common line of sight are compared. Any difference would signify the presence of a magnetic anomaly; in an area free from magnetic interference the bearings would correspond exactly.

It is possible, nevertheless, to determine anomalies and site corrections even if the errors of the compass are not known. In Fig. 12.6 let A and B represent the two datum compasses. The magnetic bearing of B from A is given by θ_1, which is the correct bearing ϕ_1 + the error e_1. The magnetic bearing of A from B is given by θ_2, which is the correct bearing $(\phi_1 + 180°)$ + the error e_2 + the site difference Φ.

Thus

$$\phi_1 = \theta_1 - e = \theta_2 - e_2 - \Phi - 180°$$

The instruments are then changed over, B going to A and A to B, and the bearings again determined. In this instance,

$$\phi_2 = \theta_3 - e_2 = \theta_4 - e_1 - \Phi - 180°$$

(ϕ_2 is written since the direction of the meridian may have altered from ϕ_1.)
Thus

$$\theta_1 - e_1 = \theta_2 - e_2 - \Phi - 180°$$
$$\theta_3 - e_2 = \theta_4 - e_1 - \Phi - 180°$$

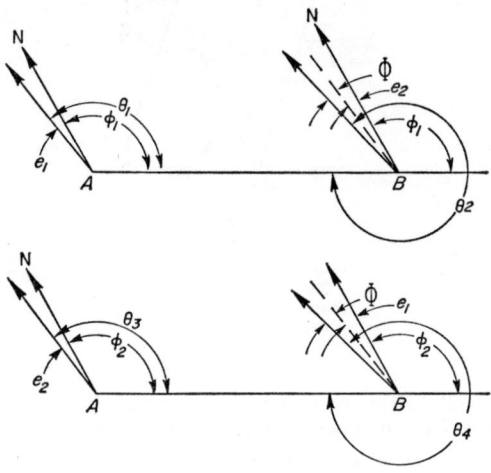

FIG. 12.6. Reciprocal bearings

Adding these two equations:

$$\theta_1 - \theta_3 - e_1 - e_2 = \theta_2 + \theta_4 - e_2 - e_1 - 2\Phi - 360°$$

Thus

$$\theta_1 + \theta_3 - \theta_2 - \theta_4 = -2\Phi$$

or

$$\Phi = \frac{\theta_2 + \theta_4 - \theta_1 - \theta_3}{2}$$

By subtracting the two equations,

$$\theta_1 - \theta_3 - e_1 + e_2 = \theta_2 - \theta_4 - e_2 + e_1$$

so that

$$\frac{\theta_1 - \theta_3 - \theta_2 + \theta_4}{2} = e_1 - e_2$$

Hence the *difference* of the errors of the two instruments can be obtained.

Calibration of datum compass by reciprocal bearings

If instead of two datum instruments, one instrument, whose error is required, and a Kew-type magnetometer are used at opposite ends of a base line, a method is available of finding the index error of the datum compass.

335

In Fig. 12.7, let O be the theodolite associated with the Kew magnetometer and C the datum compass. The telescopes of the two instruments are trained on each other, and the bearings are read of OC and CO. The datum compass is used to measure the apparent magnetic bearing of O from C, i.e. $\hat{PCO}$ (θ_1). The measurement of the variation by the magnetometer is effectively the determination of the angle

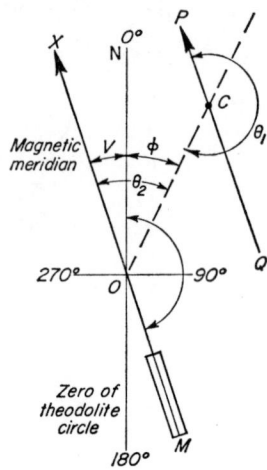

FIG. 12.7. Checking the accuracy of a compass by direct comparison with the Kew magnetometer, using reciprocal bearings

OC line of sight between theodolite and compass
Magnetometer
M magnetometer magnet
O theodolite
Angle between zero of circle and magnetic meridian = MON
$180° - MON$ = Variation V, if zero of circle = true north.
NOC = angle between zero of circle and line of sight OC = ϕ
NOC+V = magnetic bearing of C from O = XOC = θ_2 or if ON is *any* direction other than north–south, XOC is the angle between the magnetic meridian and OC.
Compass
C Compass
PCQ is the apparent meridian by compass and PCO is the apparent magnetic bearing of O from C, i.e. the angle between the apparent magnetic meridian and CO = θ_1.
Thus, for there to be no error, PCO−XOC = 180°.
All directions, including ON, can be arbitrary.

between the magnetic meridian XOM and a predetermined azimuth line ON through the theodolite's circle. Hence by measuring the prevailing variation and the angle $\hat{NOC}$ (ϕ), the angle between the magnetic meridian and OC, $\hat{XOC}$ (θ_2) is determined. Assuming no difference magnetically between O and C, the error of the datum compass is

$$e = (\theta_1 - 180) - \theta_2$$

$$= \theta_1 - (\phi + V + 180)$$

where θ_1 is the apparent magnetic bearing of O from C, ϕ the true bearing of C from O, and V the angle of magnetic variation.

336

The method of determining variation by means of a Kew-type magnetometer is described in Chapter 6, p. 128, but where high accuracy is not required, variation can be found by using a calibrated magnetic compass or a theodolite, the index error of whose magnetic compass is known.

Given a test site, the compass bearing of the fixed distant mark is observed, ten or more readings being taken and meaned. The necessary correction from the calibration table or curve, or the index error in the case of a theodolite, is applied to give magnetic bearing. From this is subtracted the true bearing of the mark, which gives the variation. Fig. 12.8 indicates a method of setting out the measurements.

Where a test site is not available, the magnetic bearing of the sun, moon or other celestial body can be found and compared with the true azimuth in order to find variation. The observations should be carried out when the altitude of the celestial body is below 30°, and the true azimuth of the body can be calculated from alt-azimuth tables or as described on p. 334. Care must be taken to select an area free from magnetic disturbances or irregularities.

SEA TRIALS

The testing of compasses at sea (or in the air) usually involves accuracy tests under steady conditions of course and speed, during and after turns, when rolling or executing other manoeuvres such as a 'broad weave' or changes of speed.

Before conducting sea trials, the deviations of the compass must be accurately known, and it is advisable to 'swing ship' immediately before and after the trial. The ship's position needs to be known accurately and also the local value of variation.

A careful programme of trials procedure (trials orders) is essential so that all taking part are aware of their particular duty or station, and so that the sequence of changes of course, speed and the like is prepared and, if necessary, rehearsed. A complete narrative of the conduct of the trial, including every manoeuvre, ship's position, variation, control settings (if any) in the compass equipment, times of starting and stopping each event, however trivial, speed, weather, course and so on, should be kept by an individual detailed to conduct the trial and to co-ordinate the observing team. The importance of the preparation and rehearsal of sea trials cannot be overemphasized.

In general, all these trials require that the error of compass or repeater be determined by comparison with an independent datum. In the simplest trial, the ship may be steered on a transit and the ship's head by compass compared with the actual course. Generally, however, a more refined method is required and the usual one is to compare the compass bearing of a heavenly body with the true bearing of that body. (In some cases it may be permissible to use a very distant mark.) The most direct method is to observe the heavenly body by means of an azimuth circle mounted on the compass (or repeater) and at pre-arranged times (say every 20 seconds) to note the compass bearing of the body. The times of observation, the compass bearings and the true bearings of the body corresponding to the time of observation are tabulated, revealing the error of the compass and permitting the determination, by statistical analysis, of the standard deviation, the mean error and the accuracy of the mean.

I *By compass*

Compass bearing of mark ° ′ ″ (At least 10 observations)	True bearing of mark * ° ′ ″ (2)	Site:
	Variation = (2)−(1) ° ′ ″	Date:
		Time:
(1) Mean	+ = east, − = west	Observer:

* Sun or other heavenly body can be used, given a suitable alidade or sight.

II *By theodolite and fixed mark*

Circle reading		Add 360° if (1)−(2) is negative	
Datum mark	Compass	Compass bearing of mark = (3) ° ′ ″ = (1)−(2)	Site:
Right face / Left face			Date:
° ′ ″ (5 readings) / ° ′ ″ (5 readings)	(At least 10 observations)	True bearing of mark = (4) ° ′ ″	Time:
			Observer:
(1) Mean ° ′ ″	(2) Mean ° ′ ″	Variation = (4)−(3) ° ′ ″ + = east, − = west	

III *By theodolite and sun*

Time G.M.T. h.m.s.	Circle reading			Add 360° if (1)−(2) is negative	
	Sun		Compass	Compass bearing of sun = (1)−(2)−(3) ° ′ ″	Site: (lat. and long.)
	R. face	L. face			
	☽ ☽ (5 readings) ☽ ☽ (5 readings)		(At least 10 observations)		Date:
	mean ☽ mean ☽ mean ⊙	mean ☽ mean ☽ mean ⊙		Calculated true bearing of sun at mean time = (4) ° ′ ″	Observer: ☽ = Sun's R. limb ☽ = Sun's L. limb ⊙ = Sun's centre
h.m.s. mean time	mean ⊙ (1) ° ′ ″		Mean (2) ° ′ ″	Variation = (4)−(3) ° ′ ″ + = east, − = west	

FIG. 12.8. Measurement of variation

For example:

H.M.S. Compass Rose* (Lat. 51° 20′ N., Long. 5° 12 W.)

Run 1

Course: NE.	Weather: Fine	Deck watch error: 4 sec fast
Speed: 10 knots	Sea: Calm	Observations on: Sun

Variation: 9° 40′ W.

G.M.T. = (D.W.T.)−(D.W. error)

Error = (True bearing of sun)−(Compass bearing of sun)

To obtain the mean error take a second reading.

There are obvious limitations to this method, the principal one being the observational accuracy possible with an azimuth circle, which, at best, is about $\pm\frac{1}{4}°$. Errors in timing and recording, and the impossibility of rechecking an observation are further handicaps to direct observation of a datum mark or body.

Only automatic methods using photography ensure accuracy and reliable records, which may be re-read if any doubt arises during the reduction of the results. The methods, of which there are several, all compare the ship's head by compass with the ship's course determined by some reliable and independent means. For instance, over short runs, a precisely balanced directional gyroscope may be used as the datum instrument. Alternatively, if due allowance is made for its errors, the gyro compass may be used as a datum with which the magnetic compass under test is compared.

The appropriate repeater dials are arranged on a panel displaying counter index numbers, clock time, trial number and other relevant information, and the whole assembly is photographed continuously by a cine-camera or by single shots at appropriate intervals. Whilst manual control is often adequate, an automatic system is always to be preferred.

After processing, the record films are read and the data tabulated on suitable forms to enable the compass error to be determined. There is, however, no better datum than the ship's heading as determined from the observation of a heavenly body with which to compare the ship's heading by compass. Its use removes many instrumental errors and avoids the risks involved in using other directional devices which are often only a small amount better than the compass under test. However, it is rather more difficult to record photographically the relative bearing of a heavenly body than to photograph a repeater dial.

Azimuth datum instrument

One form of the azimuth datum instrument that was developed at the Admiralty Compass Observatory has been used with considerable success in determining ship's head by sun. In principle, the instrument consists of a transparent bearing plate, mounted in gimbals and graduated in tenths of degrees, with the 0°–180° diameter accurately aligned in the fore-and-aft line of the ship. (Alignment is carried out by reference to the datum training marks when in dry dock.) A rotatable head carrying a prism and lens projects the image of the sun on the bearing plate, the graduation on which this image falls being the apparent relative bearing of the sun, recorded photographically with a single-shot 35 mm camera. Photographs of

* With acknowledgement to Nicholas Monsarrat.

the compass repeater are taken simultaneously. Given the precise time of each photograph, the sun's true bearing is ascertainable and hence the relative bearing of the sun may be converted to ship's head by sun.

The instrument has numerous refinements. For instance, it may not always be possible to have the prism aimed exactly at the sun, and the optical system, therefore, is designed to correct for ill-aiming errors. A fiducial line or lubber's line, which lies in the optical plane of the solar prism and lens, is provided in the rotating head. When the system is aimed exactly, the sun's image falls on this line, and the position of the line on the bearing plate is the sun's relative bearing. Ill-aim is evidenced if the sun's image on the bearing plate is displaced from the fiducial line. If the altitude of the sun were zero, no correction for this displacement would be necessary, and the relative bearing could be read directly from the sun's image.

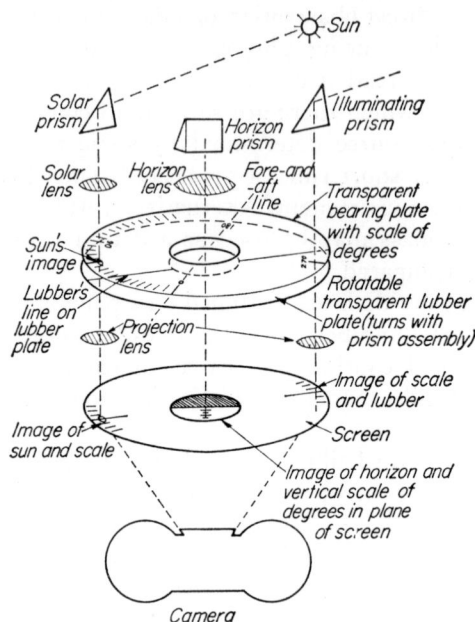

FIG. 12.9. Azimuth datum instrument

Optical principles

However, a correction for this apparent ill-aim must be applied according to the sun's altitude, namely (apparent ill-aim × secant altitude) which gives the correct value of ill-aim to be applied to the angular position of the fiducial line.

Also necessary is the 'cross-level' correction, to correct the effect of tilting across the line of sight. This causes an error in the apparent relative bearing of the sun and is given by (R sin altitude) where R is the angle of tilt. The tilt is measured by photographing the horizon through the axis of the instrument, the displacement of the horizon up or down a diametral scale indicating the degree of tilt. A counter is included in the field of view of the camera so that each 'frame' is correctly identified.

Fig. 12.9 illustrates the optical principles of the instrument, and Fig. 12.10 is a *pro forma* for the reduction of the results.

340

Ship _____

Analysis of A.D.I. data

Watch start _____

G.M.T. / B.S.T.

White to red / White to green

Trial No. _____

Starboard / Port

Run No. _____

Date _____

(Cols 1, 4, 5, 6 and 9 are observed directly on A.D.I.)
(Cols 10 and 18 are calculated for given G.M.T.)

Sun scale filter _____   Sun light value _____
Reciprocal scale filter _____   Horizon light value _____
Horizon scale filter _____

1	2	3	4	Lubber (e =)			8	9	10	11	12	13	14	15	16	17	18	19
				5	6	7												
Counter No.	D.W.T. h.m.s.	G.M.T. h.m.s.	Sun's app. bg	S	X	Mean	S' (S + e)	R	Alt.	Sin alt.	R sin alt. (9 x 11)	Corr'd sun's b'g (4 − 12)	App. ill-aim (13 − 8)	Corr'n factor sec alt.	Corr ill-aim (14 x 15)	Sun's rel. bg (7 x 16)	Sun's true bg	Ship's head (18 x 17)

Fig. 12.10(a)

1	2	3	4	5	6	7	8	9	10
Counter No.	Ship's head by A.D.I.	Ships head by compass					Compass error		

Ship _____ Trial No. _____ Run No. _____

Trials recording room record (cols 3,4,5 and 6)
Compass error record (cols 7,8,9 and 10)

FIG. 12.10(*b*). Pro forma for reduction of results obtained by azimuth datum instrument

The conduct of trials in aircraft follows similar lines to the procedure at sea, except that the manoeuvres are likely to be more violent and the observations of a heavenly body more difficult. Although an azimuth datum instrument can be used in the air, horizon observations for tilt are impracticable and difficulties can arise on this account.

Measurement of variation at sea

The determination of variation is frequently carried out at sea by using the ship's standard compass, by a method very similar to that used in 'swinging ship' to obtain a deviation table.

The ship is placed on each of sixteen equally spaced compass headings in turn and the compass bearing of the sun observed. Given the correct time, the sun's true bearing is known and hence the *total error* of the compass.

The total error is equal to *deviation+variation*. The sum of the harmonic or cyclic deviations (coefficients B, C, D and E) is zero over 360°. If there is no coefficient A, the sum of the deviations divided by 16 is the *magnetic* variation.

$$E_t = \text{total error} = \text{var.} + \text{dev.}$$

$$\text{dev.} = A + B \sin \zeta' + C \cos \zeta' + D \sin 2\zeta' + E \cos 2\zeta'$$

$$\sum_0^{2\pi} \text{dev.} = \sum_0^{2\pi} A + \sum_0^{2\pi} B \sin \zeta' + \sum_0^{2\pi} C \cos \zeta' + \sum_0^{2\pi} D \sin 2\zeta' + \sum_0^{2\pi} E \cos 2\zeta' = \sum_0^{2\pi} A$$

$$\sum_0^{2\pi} E_t = n(\text{var.}) + \sum_0^{2\pi} A$$

$$\text{var.} = \frac{1}{n}\left[\sum_0^{2\pi} E_t - \sum_0^{2\pi} A\right]$$

If there is a coefficient A whose value is known, the sum of the deviations divided by 16 less A is the magnetic variation.

This procedure is carried out by clockwise and counter-clockwise swings, the average of the two variation values obtained being regarded as the actual value of variation. The swing should be carried out slowly and carefully and when dealing with transmitting magnetic compasses, any total-error corrector mechanism must have its variation setting at zero.

APPENDIX 1

Extract from the Admiralty Specification for Compass Liquid

The compass liquid shall be a solution of alcohol and pure distilled water of specific gravity of 0·93 at 15°C. The compass shall be de-aerated in an exhaustion chamber immediately before it is finally sealed.

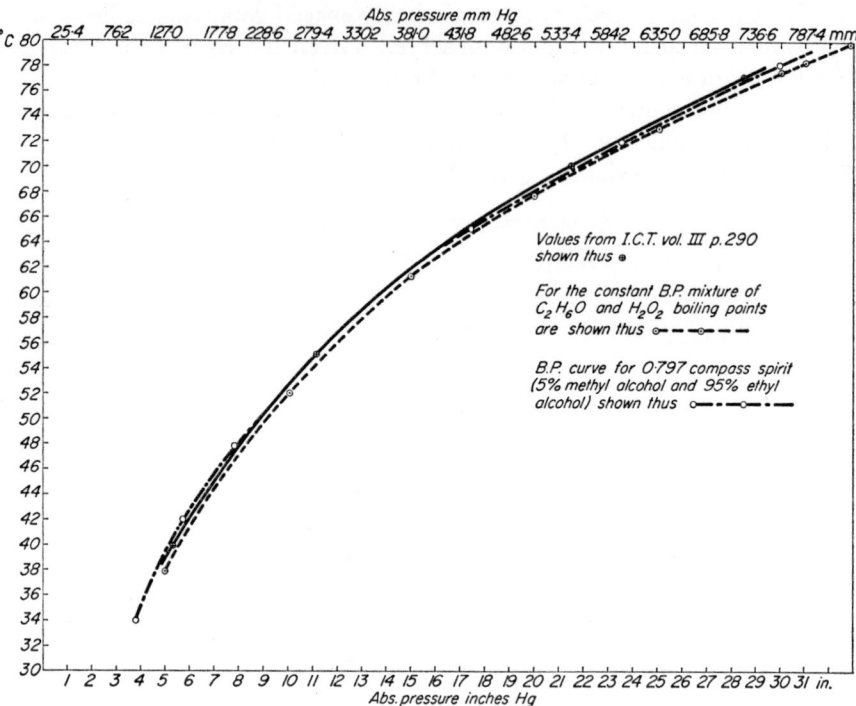

FIG. 1. Boiling point of compass liquid and its dependence on pressure
The ordinates are degrees Celsius.

$$\Delta t = \frac{273 \cdot 1 + t}{\phi}(2 \cdot 8808 - \log_{10}\rho_{mm})$$

where Δt is the change of B.P., t the B.P. in °C, and ρ_{mm} the absolute pressure in mm Hg.
For ethyl alcohol

$$\Delta t = 78 \cdot 32 - t$$

$$t = \frac{78 \cdot 32 - 273 \cdot 1 \times k}{1 + k}$$

where $k = \dfrac{2 \cdot 8808 - \log_{10}\rho_{mm}}{\phi}$

Normal B.P. taken as 78·32°C.

The alcohol used shall consist only of pure absolute ethyl alcohol de-natured with 5 per cent pure methanol. It shall contain no materials likely to cause discoloration of the liquid itself or of the paint etc. inside the compass bowl. It shall conform to the following requirements:

(i) The liquid shall be colourless, transparent, mobile and volatile, shall burn with a blue flame and shall have a characteristic odour.

(ii) *Specific gravity*. The specific gravity at 15°C shall not exceed 0·797.

(iii) *Residue*. The amount of non-volatile material remaining from 100 cc of the liquid after evaporation and drying for 1 hour at 100°C shall not be greater than 0·01 per cent w/w.

(iv) *Free acid*. The acidity shall not be greater than will allow a decided pink colour to result from the addition of two drops N/10 caustic soda to 10 cc of the material in the presence of two drops of phenolphthalein solution as indicator.

(v) *Free bases*. The amount of basic substances calculated as caustic soda shall not exceed 0·01 per cent by weight when determined by titration with N/100 sulphuric acid using methyl red as indicator.

(vi) *Esters*. The spirit shall contain no esters. The estimation is carried out in the following manner:

10 cc of the material is carefully neutralized with N/10 caustic soda after addition of two drops of phenolphthalein indicator; then a further quantity of 20 cc of caustic soda solution is added and the mixture boiled for two hours in a flask under a reflux condenser. The excess of alkali is then titrated with N/10 sulphuric acid. Each cubic centimetre of the N/10 caustic soda used up in the hydrolyzing of the esters in the sample represents 0·0074 gram of esters, calculated.

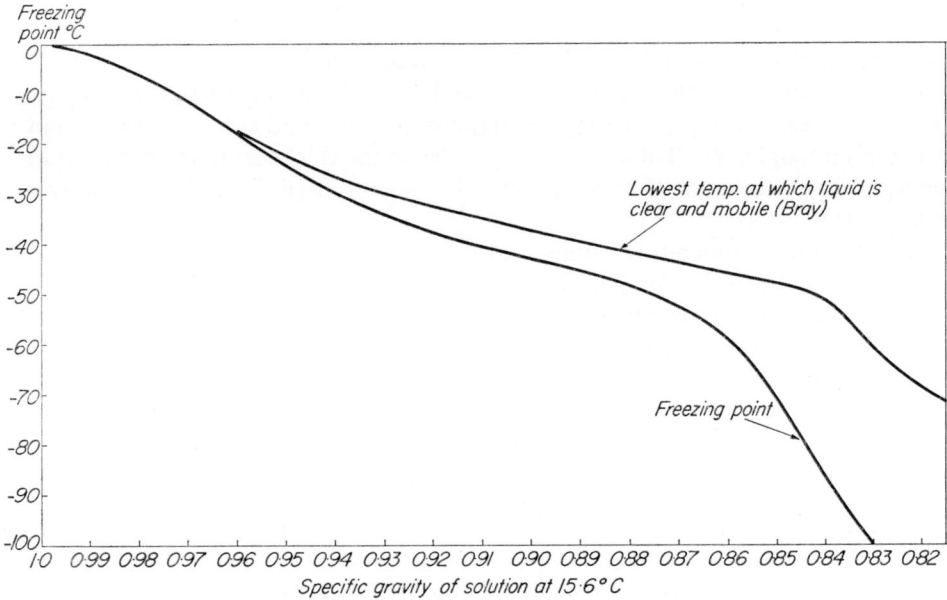

FIG. 2. Dependence of clarity and mobility of compass liquid on temperature and density

APPENDIX 2

Extract from British Standard 1699: Part 1: 1966.
Specification for Magnetic Compasses and Binnacles
Part 1, Class A, for use in merchant ships

SECTION TWO: MAGNETIC COMPASSES

Construction and materials

4. (*a*) The magnets used in the directional systems of magnetic compasses shall be of a suitable magnetic material having a high remanence and a high coercivity. All other materials used in magnetic compasses, other than transmitting compasses, shall be of non-magnetic material.

(*b*) The distance between the lubber mark and the outer edge of the card shall be between 1·5 mm and 3·0 mm (0·06 in. and 0·12 in.). The width of the lubber mark shall be not greater than 0°·5 of the graduation of the card. The lubber mark shall be of such design as to allow the compass to be read when the bowl is tilted 10° in the case of a gimballed compass, or 30° in the case of a hemispherical compass.

(*c*) When the verge ring and the seating for the azimuth reading device are both horizontal, the graduated edge of the card, the lubber mark if a point, the pivot point and the outer gimbal axis shall all lie within 1 mm (0·040 in.) of the horizontal plane passing through the gimbal axis fixed to the bowl. In dry card compasses only, the graduated edge of the card may be below the horizontal plane passing through the pivot point by a vertical distance not exceeding 15 mm (0·6 in.).

(*d*) The gimbal axes shall be mutually perpendicular within a tolerance of 1°. The outer gimbal axis shall be in the fore and aft direction.

(*e*) The thickness of the top glass cover and of the bottom glass of the compass shall be not less than 4·5 mm (0.175 in.) if non-toughened and not less than 3·0 mm (0·125 in.) if toughened. These values apply also to the thickness of the top glass in hemispherical compasses. If material other than glass is used, it shall be of equivalent strength.

(*f*) Within the temperature range −30°C to +60°C:

    (i) the liquid in the compass bowl shall remain clear and free from bubbles, and neither emulsify nor freeze;

    (ii) there shall be neither inward leak of air nor outward leak of liquid;

    (iii) paint shall not blister, crack or discolour appreciably.

(*g*) The balance of the compass bowl shall not be disturbed by any magnifying glass or by any azimuth reading device provided for use with the compass.

Mounting

5. (*a*) The bowl of the compass shall be mounted so that the verge ring remains horizontal when the binnacle is tilted 40° in any direction and so that the compass cannot be dislodged under any conditions of sea or weather.

(*b*) In compasses of the hemispherical type, in which no supporting gimbal is provided, the freedom of the card shall be 30° in all directions.

6. (*a*) *Moment of inertia*. The moment of inertia of the directional system shall be

appreciably the same about all horizontal axes passing through the bearing surface of the pivot jewel.

(*b*) *Suspension.* The directional system shall be retained in position by suitable means and remain free when the bowl is tilted 10° in any direction.

(*c*) *Supporting force.* (i) When measured at a temperature of 20 ± 3°C, the force exerted by the directional system on the pivot in the liquid used shall be between 4 gf and 10 gf when the diameter of the card is 166 mm (6·5 in.) or less, and shall be between 4 gf and 14 gf when the diameter of the card is greater than 165 mm (6·5 in.).

(ii) In a dry card compass the force exerted on the pivot bearing by the directional system shall be between 10 gf and 20 gf.

(*d*) *Magnetic moment.* The magnetic moment of the magnets in the directional system of liquid compasses shall not be less than the value given in Fig. 5.2. [See p. 90.]

(*e*) *Period.* When measured at a temperature of 20 ± 3°C the half period of the directional system shall not be less than 12 seconds following an initial deflection of the card of 40° from the magnetic meridian, in a horizontal magnetic field of strength 0·18 oersted.

NOTE. After the initial deflection has been given, the half period is measured between the first two consecutive passings of the original indication of the course.

Alternatively, the compass may be aperiodic or heavily damped. In this case, the time taken to return to within 1° of the magnetic meridian, following an initial deflection of the card of 90°, shall not be more than 56 seconds at the above temperature.

(*f*) *Tilt of the card.* (i) When the directional system is in the bowl, the card shall not incline more than 0·5° from the horizontal plane when the vertical field is zero. (ii) The tilt of the directional system shall not change more than 3° when the vertical field strength changes by 1 oersted.

Graduation

7. (*a*) *Compass card.* The compass card is graduated in 360 single degrees starting from north in a clockwise direction as viewed from above. The cardinal points are indicated by the capital letters N., S., E. and W.; the intermediate points may also be marked. The north point may alternatively be indicated by a suitable emblem.

(*b*) *Readability by the helmsman.* If a steering compass is provided for the helmsman, it shall be possible for a person with normal vision to read at a distance of 1·4 m (55 in.), in both daylight and artificial light, the lubber mark and those graduations on the card which are contained within a sector whose width is not less than 15° to each side of the lubber mark. The use of a magnifying glass is permitted.

For reflecting and projecting compasses, the lubber mark shall be visible and the 30° sector of the card shall be readable by a person with normal vision at a distance of 1 m (40 in.) from the periscope tube.

(*c*) *Standard compass.* The standard compass shall be provided with a scale for the measurement of bearings relative to the ship's head, the scale being graduated in 360 degrees in a clockwise direction, zero, as seen through the azimuth reading device, indicating the direction of the ship's head.

Accuracy

8. (*a*) *Constructional errors*. In compasses other than transmitting compasses, the directional error, as defined in 2(*v*), shall not exceed 0°·5 on any heading.

In transmitting compasses, whether the transmitting system is energized or not, the directional error shall not exceed 0°·6.

(*b*) When the lubber mark is fixed, the lubber error, as defined in 2(*w*), shall not exceed 0°·3.

(*c*) *Error due to friction*. With the compass at a temperature of 20 ± 3°C, the card when given an initial deflection of 2° first on one side of the magnetic meridian and then on the other, shall return to within 0·17 degree of arc of its original position, in a horizontal magnetic field of strength 0·18 oersted.

(*d*) *Swirl error*. When the compass is at a temperature of 20 ± 3°C, and in a horizontal magnetic field of strength 0·18 oersted, rotation of the compass bowl through 360° in the horizontal plane at a uniform speed of 1°·5 per second shall not cause a deflection of the card from the magnetic meridian of more than 2° for cards of less than 200 mm diameter and of 3° for cards of 200 mm diameter or more.

(*e*) *Induction error*. When coefficient *D* is corrected by spheres or some similar conventional device, the error introduced by magnetic induction in these correctors, due to the magnets in the directional system, shall be such that the value of the ratio of coefficient *H* to coefficient *D*, (i.e. *H*/*D*), obtained from the 4 corrector (Meldau) test, will not exceed 0·08.

NOTE. In this test, the compass is mounted on a stand and two soft-iron correctors are placed diametrically opposite to each other and symmetrical to the centre of rotation. The device with the two soft-iron correctors is then rotated round the fixed compass and co-efficient *D* calculated.

To cancel out the quadrantal deviation, two additional exactly similar correctors are placed at the same distance from the centre with their line of connection at right angles to that of the original pair. The arrangement of the four soft-iron correctors is then again rotated around the compass and coefficient *H* calculated.

From these values, the ratio of coefficient *H* to coefficient *D* is obtained.

(*f*) *Mounting error of azimuth reading device*. Where the azimuth reading device is mounted on the compass bowl, the vertical axis of the device shall be within 0·5 mm (0·02 in.) of the pivot point.

(*g*) *Error due to eccentricity of the verge ring*. If the verge ring is graduated, the perpendicular to the plane of this ring through the centre of the graduations shall be within 0·5 mm (0·02 in.) of the pivot point.

SECTION THREE: BINNACLES

Construction and materials

9. (*a*) Only high quality non-magnetic materials shall be used for the construction of the binnacle, brackets and holding down bolts. All sheet metal parts shall be not less than 0·8 mm (0·032 in.) in thickness.

(*b*) Provision shall be made in the binnacle to allow correction for any misalignment thereof in respect of the fore-and-aft line of the ship, by an angle of not less than 4° and not more than 6°.

(*c*) The compass shall be mounted in a binnacle in such a way that the centre line of the magnets of the directional system is between 1·07 m and 1·24 m (42 in. and 49 in.) above the under surface of the binnacle deck fittings.

Provision for correction of deviation

10. (*a*) *Material*. Where corrector magnets are used they shall be of a suitable magnet material with coercivity not less than 140 oersted. Material used for correcting induced fields shall have a coercivity not higher than 2 oersted.

(*b*) *Compensation for horizontal permanent magnetism*. The binnacle shall contain a device for correcting the deviation due to the horizontal components of the ship's permanent magnetism. This device shall be capable of correcting a coefficient *B* of at least 40° and a coefficient *C* of at least 40°.

Provision shall be made in the binnacle so that no magnet of the correcting system comes nearer than twice its own length to the magnets of the directional system. Magnets used in the device shall be 200 mm (8 in.) long and either 10 mm or 5 mm ($\frac{3}{8}$ in. or $\frac{3}{16}$ in.) in diameter.

(*c*) *Correction for heeling error*. The binnacle shall contain a device for correcting heeling error. This device shall be adjustable and capable of providing a vertical field at the magnets of the directional system over the range +0·75 to −0·75 oersted.

The magnets of the correcting device shall not be nearer than twice their own length from the magnets of the directional system. When heeling error magnets are used for this purpose they shall be 10 mm ($\frac{3}{8}$ in.) in diameter.

NOTE. The magnetic fields produced by the devices referred to in 10(*b*) and (*c*) above should be as uniform as possible in the space swept by the directional system, and should in no case introduce a measurable sextantal error.

(*d*) *Compensation for horizontal induced fields due to the horizontal component of the Earth's magnetic field in the soft iron in the ship*. The binnacle shall be provided with a device for correcting the deviation due to the horizontal components of fields induced in the soft iron in the ship. It shall be capable of correcting a coefficient *D* of up to +7° in dry card compasses and up to +10° in liquid compasses. When soft-iron spheres are used for this purpose, they shall be fitted with bolts 16 mm ($\frac{5}{8}$ in.) diameter and the centres of the spheres shall be 127 mm (5 in.) above the securing brackets.

When the binnacle is vertical, the centre of this correcting device shall not be further than 15 mm (0·6 in.) from the horizontal plane passing through the magnetic element of the directional system.

(*e*) *Compensation for horizontal induced fields due to the vertical component of the Earth's magnetic field in the soft iron in the ship*. The binnacle shall be provided with a device for compensating the horizontal magnetic fields due to induction caused by the *vertical* component of the Earth's magnetic field in the soft iron in the ship. Where a Flinders bar is used for this purpose, it shall be of soft iron 76 mm (3 in.) in diameter, either solid, or hollow provided that the diameter of the hole does not exceed 30 mm ($1\frac{1}{4}$ in.). It shall be supplied in lengths of 38 mm, 76 mm, 150 mm and 305 mm ($1\frac{1}{2}$ in., 3 in., 6 in. and 12 in. respectively).

When the binnacle is vertical, the magnetic pole of this correcting device shall lie in the same horizontal plane as the centres of the magnets of the directional system.

Where a Flinders bar is used its magnetic pole shall be taken to be $\frac{1}{12}$ of the length from the end.

(*f*) *Positions and attachment of correcting devices.* Provision shall be made in the binnacle for recording the positions of the correcting devices referred to in sub-clauses 10(*b*), (*c*) and (*d*). Numbers are used for this purpose in the case of sub-clauses 10(*b*) and (*c*) above, and they shall read from the bottom upwards.

Provision shall be made for all correcting devices to be satisfactorily secured after adjustment.

(*g*) Provision may be made for the fitting of corrector coils to provide compensation if the ship is fitted with degaussing coils.

SECTION FOUR: AZIMUTH READING DEVICE

Field of view

15. The standard compass shall be equipped with an appropriate azimuth reading device. The field of view shall be 5° on each side of the line of sight and it shall be possible to measure the azimuth of a distant object, either celestial or terrestrial, whose altitude lies between 5° below and 60° above the horizontal.

16. (*a*) When the azimuth reading device is correctly aimed, the error on any reading shall not exceed the values in column 2 of Table 2.

(*b*) However, if the device allows for incorrect aiming, the error in reading shall not exceed the values in column 3 of Table 2 when the device is 5° incorrectly aimed.

TABLE 2 *Azimuth reading error*

1. Altitude of observed object	2. Maximum permissible error, Condition 16(*a*)	3. Maximum permissible error, Condition 16(*b*)
Between 5° below and 45° above the horizontal	0°·3	1°
More than 45° above the horizontal	0°·5	1°·5
27° above the horizontal	0°·3	0°·5

APPENDIX 3

Pivots and Jewels

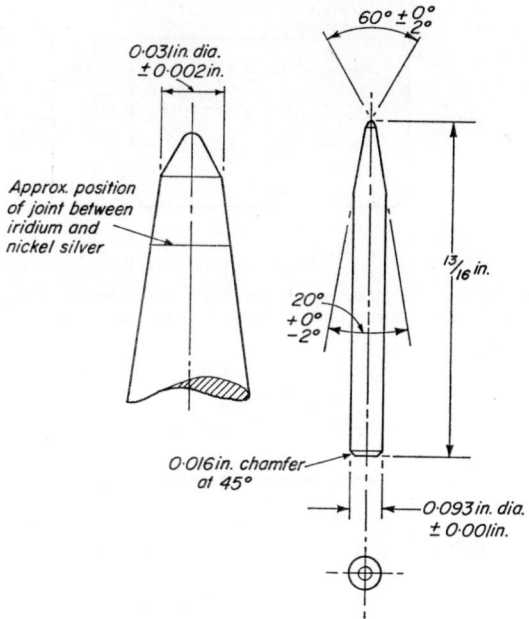

FIG. 1. Dimensions of pivot for a liquid compass

The material is 20 per cent nickel-silver and osmium-iridium, left bright, with cone and point ground and polished. The radius of the tip is 0·005 in. ± 0·0005 in. The flanks of the 90-degree cone are to be ground tangential to the curved surface of the tip junction of iridium and nickel-silver.

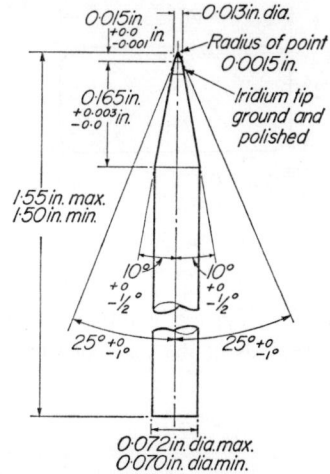

FIG. 2. Dimensions of pivot of Kelvin & Hughes 10-inch dry-card compass

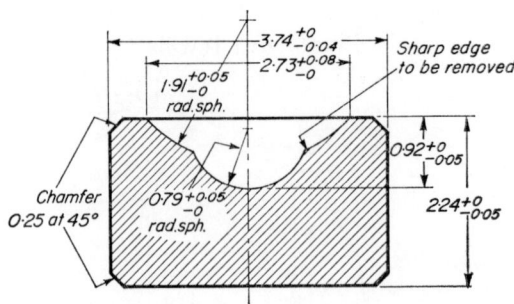

FIG. 3. Type 1 compass jewel

The eccentricity of the axis of recess shall not exceed 0·025 mm. The dimensions in this illustration and in Figs. 4 and 5 are all in millimetres.

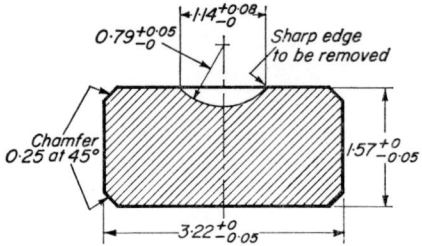

FIG. 4. Type 2 compass jewel

The eccentricity of the axis of recess shall not exceed 0·025 mm.

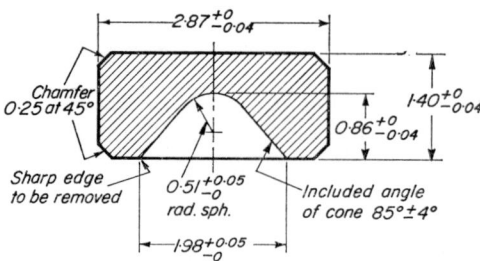

All dimensions are in millimetres

FIG. 5. Type 3 compass jewel

The eccentricity of the axis of the cone shall not exceed 0·005 mm.

Examples of Forms for the Analysis and Synthesis of Deviations of the Compass

DEVIATION ANALYSIS FORM

This form [Fig. 1] is designed to facilitate the calculation of the Deviation Coefficients A, B, C, D, E, F, G, H, K of the formula

$$\delta = A + B \sin \zeta' + C \cos \zeta' + D \sin 2\zeta' + E \cos 2\zeta'$$
$$+ F \sin 3\zeta' + G \cos 3\zeta' + H \sin 4\zeta' + K \cos 4\zeta' + \ldots$$

from a set of small deviations observed on 16 equidistant points.

The terms in F and G, H and K, giving the sextantal and octantal components, frequently have sensible values, and their omission makes the usual analysis imperfect.

H.M.S._____   Date_____

Points	Deviation curve	points	1	2	3	4	5	6	7	8	9	10	11	12	13	14	15	16	17	18
		points	Observed deviations	sin ζ'	Col 1 x col 2	cos ζ'	Col 1 x col 4	sin 2ζ'	Col 1 x col 6	cos 2ζ'	Col 1 x col 8	sin 3ζ'	Col 1 x col 10	cos 3ζ'	Col 1 x col 12	sin 4ζ'	Col 1 x col 14	cos 4ζ'	Col 1 x col 16	Totals of rows
N		0		0		1		0		1		0		1		0		1	
NNE		2		S_2		S_6		S_4		S_4		S_6		S_2		1		0	
NE		4		S_4		S_4		1		0		S_4		$-S_4$		0		-1	
ENE		6		S_6		S_2		S_4		$-S_4$		$-S_2$		$-S_6$		-1		0	
E		8		1		0		0		-1		-1		0		0		1	
ESE		10		S_6		$-S_2$		$-S_4$		$-S_4$		$-S_2$		S_6		1		0	
SE		12		S_4		$-S_4$		-1		0		S_4		S_4		0		-1	
SSE		14		S_2		$-S_6$		$-S_4$		S_4		S_6		$-S_2$		-1		0	
S		16		0		-1		0		1		0		-1		0		1	
SSW		18		$-S_2$		$-S_6$		S_4		S_4		$-S_6$		$-S_2$		1		0	
SW		20		$-S_4$		$-S_4$		1		0		$-S_4$		S_4		0		-1	
WSW		22		$-S_6$		$-S_2$		S_4		$-S_4$		S_2		S_6		-1		0	
W		24		-1		0		0		-1		1		0		0		1	
WNW		26		$-S_6$		S_2		$-S_4$		$-S_4$		S_2		$-S_6$		1		0	
NW		28		$-S_4$		S_4		-1		0		$-S_4$		$-S_4$		0		-1	
NNW		30		$-S_2$		S_6		$-S_4$		S_4		$-S_6$		S_2		-1		0	
Algebraic totals of columns			16A=		8B=		8C=		8D=		8E=		8F=		8G=		8H=		8K=	
Deviation coefficients			A=		B=		C=		D=		E=		F=		G=		H=		K=	Computed deviations
Products for computing deviations			A x 1		B x col.2		C x col.4		D x col.6		E x col.8		F x col.10		G x col.12		H x col.14		K x col.16	

FIG. 1. Form for analysis and synthesis of compass deviations

Analysis. The rules are as follows: (i) To find A take the algebraic mean of all the deviations. (ii) To find B, multiply the deviation on each point by the corresponding value of the factor of B in the above formula, and take double the mean; similarly for C, D, E, F, G, H, K.

The form is arranged so as to carry out these rules conveniently. The symbols S_2, S_4, S_6 represent the sines of the angles of 2, 4, and 6 points, and the necessary products can be taken from Table I., p. 181, Admiralty Manual, and entered in the appropriate compartments of the Form. (It is a saving of labour to take each deviation in turn and enter the products along the row in which the deviation stands; for the products repeat themselves.)

The coefficients thus found are the best values (according to the principle of least squares) derivable from the observed deviations. Any attempt to amend the observed deviations by drawing a 'flowing curve' past the points plotted in the diagram will be less successful than the orthodox process described above.

Synthesis. The same Form (or another copy) may also be used to compute the deviations corresponding to the coefficients. The necessary products are noted at the foot of the Form, in the bottom row; these are to be entered from the Table into the columns of the Form (noting again the repetitions which occur), and the additions are then made horizontally, by rows, and set in the last column, 18. If the observed deviations are of normal type they should differ very little from the computed deviations. The synthesis thus confirms the results of the analysis.

Note. The arcs given by the Table are at five-minute intervals. If a coefficient falls between two successive entries, interpolation may conveniently be avoided by taking 10 times the arc and dividing the '*S*' products by 10.

Fig. 2 shows another form suitable for deviation analysis.

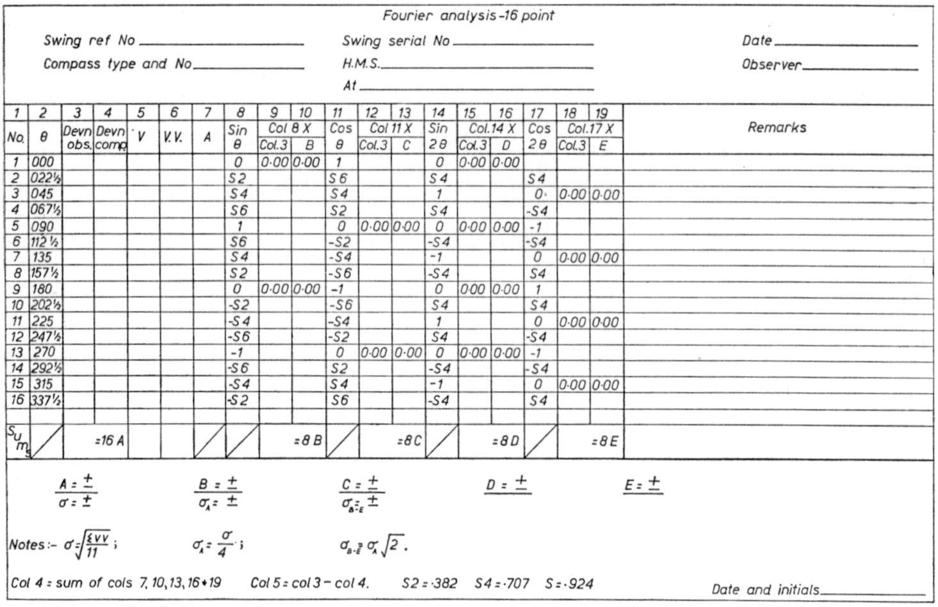

Fourier analysis-16 point

Swing ref No ______ Swing serial No ______ Date ______

Compass type and No ______ H.M.S. ______ Observer ______

At ______

1	2	3	4	5	6	7	8	9	10	11	12	13	14	15	16	17	18	19	
No.	θ	Devn obs.	Devn comp.	V	V.V.	A	Sin θ	Col 8 X		Cos θ	Col 11 X		Sin 2θ	Col.14 X		Cos 2θ	Col.17 X		Remarks
								Col.3	B		Col.3	C		Col.3	D		Col.3	E	
1	000						0	0·00	0·00	1			0	0·00	0·00				
2	022½						S2			S6			S4			S4			
3	045						S4			S4			1			0	0·00	0·00	
4	067½						S6			S2			S4			-S4			
5	090						1			0	0·00	0·00	0	0·00	0·00	-1			
6	112½						S6			-S2			-S4			-S4			
7	135						S4			-S4			-1			0	0·00	0·00	
8	157½						S2			-S6			-S4			S4			
9	180						0	0·00	0·00	-1			0	0·00	0·00	1			
10	202½						-S2			-S6			S4			S4			
11	225						-S4			-S4			1			0	0·00	0·00	
12	247½						-S6			-S2			S4			-S4			
13	270						-1			0	0·00	0·00	0	0·00	0·00	-1			
14	292½						-S6			S2			-S4			-S4			
15	315						-S4			S4			-1			0	0·00	0·00	
16	337½						-S2			S6			-S4			S4			
Sums			=16 A						=8 B			=8 C			=8 D			=8 E	

$A = \pm$ / $\sigma = \pm$ $B = \pm$ / $\sigma_A = \pm$ $C = \pm$ / $\sigma_{B-E} = \pm$ $D = \pm$ $E = \pm$

Notes:- $\sigma = \sqrt{\frac{\Sigma vv}{11}}$; $\sigma_A = \frac{\sigma}{4}$; $\sigma_{B-E} = \sigma_A\sqrt{2}$.

Col 4 = sum of cols 7, 10, 13, 16 + 19 Col 5 = col 3 − col 4. S2 = ·382 S4 = ·707 S = ·924 Date and initials ______

FIG. 2. Alternative form for analysis and synthesis of compass deviations

APPENDIX 5

The Compass Position and Correctors
in Relation to the Compass Magnet System

The effect of placing a compass needle in the compass position

The fundamental equations discussed in Chapter 10 which describe the magnetism of a ship and its effect at the compass position may not necessarily be strictly applicable when a compass needle or magnet system is placed in this position. This is particularly the case if a single needle is used whose length is comparable with the distance from its centre to neighbouring masses of hard or soft iron or to the correcting devices in the binnacle. If, however, the construction of the magnet system is such that it approximates to an infinitely short needle or conforms to certain geometric arrangements of pairs of magnets, the occurrence of spurious deviations, usually of a sextantal or octantal nature, owing to the proximity of the neighbouring iron masses to the compass needle, can be avoided.

Smith and Evans* showed that such deviations vanish if a pair of equal parallel needles are used whose like extremities subtend 60° at the centre of the system. Similarly, two pairs of magnets symmetrically disposed about a mean angle of 60° (30° on either side of the north-south axis of the system) are equally effective. This was the pattern adopted in the Admiralty Standard Compass (see Chapter 5, p. 115) primarily to ensure that the moments of inertia about all diameters of the card and system were the same, and it was entirely fortuitous that, by experiment, it was found that such a system was not subject to the sextantal and octantal deviations hitherto experienced with compasses with long single needles. Experiments carried out by Evans aboard the *Great Eastern* are of special interest;† in this notable vessel the two compass needles, placed close together, were $11\frac{1}{2}$ in. in length.

The mathematical investigation of the problem readily confirms the experiments.

The effect of a nearby magnetic pole.‡ Consider now the effect on a single compass needle of a nearby *permanent* magnetic pole. In Fig. 1, AB is a needle, centre C and length between poles equal to $2r$. It has a moment $M = 2mr$. Let the pole P be of unit strength. The force acting along PA on the pole A is m/AP^2, where m is the pole strength of the magnet AB. Owing to this force, there is a turning moment about C of R sin APC (m/AP^2). Similarly, there is a force along BP equal to m/BP^2, which gives rise to a turning moment about C of R sin BPC (m/BP^2).

---

* Smith, A., and Evans, F. J. 'On the effect produced on the deviation of the compass by the length and arrangement of the compass needles and on a new mode of correcting the quadrantal deviation.' *Phil. Trans. roy. Soc.*, 1861, 161.

† Evans, F. J. 'Reduction and discussion of the deviations of the compass observed on board of all the iron-built ships, and a selection of the wood-built steam-ships in H.M. Navy, and the iron steamship *Great Eastern*,' *Phil. Trans. roy. Soc.*, 1860, 337.

‡ *The Theory of the Deviations of the Magnetic Compass*, Admiralty Compass Observatory BR(101) 48, Part II, 25–29.

Now sin APC $= r \sin \phi/AP$ and sin BPC $= r \sin \phi/BP$, so that the total turning moment about C due to the unit pole at P is

$$L = Rmr \sin \phi \, (AP^{-3} + BP^{-3})$$

Since $AP^2 = R^2 + r^2 - 2Rr \cos \phi$ and $BP^2 = R^2 + r^2 + 2Rr \cos \phi$, we have

$$L = Rmr \sin \phi \, [(R^2 + r^2 - 2Rr \cos \phi)^{-3/2} + (R^2 + r^2 + 2Rr \cos \phi)^{-3/2}]$$

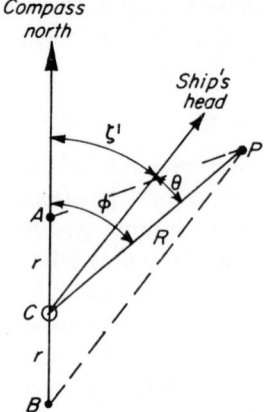

FIG. 1. The effect of a nearby pole

On expansion, excluding terms beyond r^2/R^2,

$$L = \frac{M}{R^2}\left[\left(1 + \frac{3}{8}\frac{r^2}{R^2}\right)\sin \phi + \frac{15}{8}\frac{r^2}{R^2}\sin 3\phi\right] \tag{1}$$

Now $\phi = \zeta' + \theta$; thus

$$L = \frac{M}{R^2}\left[\left(1 + \frac{3}{8}\frac{r^2}{R^2}\right)\sin (\zeta' + \theta) + \frac{15}{8}\frac{r^2}{R^2}\sin 3(\zeta' + \theta)\right]$$

$$= \frac{M}{R^2}\left[\left(1 + \frac{3}{8}\frac{r^2}{R^2}\right)(\sin \zeta' \cos \theta + \cos \zeta' \sin \theta)\right.$$

$$\left. + \frac{15}{8}\frac{r^2}{R^2}(\sin 3\zeta' \cos 3\theta + \cos 3\zeta' \sin 3\theta)\right] \tag{2}$$

If the deviation due to the turning moment is δ, then

$$HM \sin \delta = \frac{M}{R^2}\left[\left(1 + \frac{3}{8}\frac{r^2}{R^2}\right)\cos \theta \sin \zeta' + \left(1 + \frac{3}{8}\frac{r^2}{R^2}\right)\sin \theta \cos \zeta'\right.$$

$$\left. + \frac{15}{8}\frac{r^2}{R^2}\cos 3\theta \sin 3\zeta' + \frac{15}{8}\frac{r^2}{R^2}\sin 3\theta \cos 3\zeta'\right]$$

and

$$\sin \delta = \frac{1}{HR^2}(p \sin \zeta' + q \cos \zeta' + s \sin 3\zeta' + t \cos 3\zeta') \tag{3}$$

where p, q, s and t are constants; this may also be written

$$\sin \delta = B \sin \zeta' + C \cos \zeta' + F \sin 3\zeta' + G \cos 3\zeta' \tag{4}$$

showing sextantal components as well as semicircular.

356

If a pair of equal parallel needles are used whose poles subtend an angle of 2α at the centre of the system, the turning moment on it may be written

$$L = \frac{M}{R^2}\left\{\left(1 + \frac{3}{8}\frac{r^2}{R^2}\right)[\sin(\phi-\alpha)+\sin(\phi+\alpha)]\right.$$

$$\left.+ \frac{15}{8}\frac{r^2}{R^2}[\sin(3\phi-3\alpha)+\sin(3\phi+3\alpha)]\right\} \tag{5}$$

since such a system is equivalent to two crossed magnets inclined to each other at an angle 2α (Fig. 2). Thus

$$L = \frac{M}{R^2}\left\{\left(1 + \frac{3}{8}\frac{r^2}{R^2}\right)[\sin\theta\cos\alpha-\cos\phi\sin\alpha+\sin\phi\cos\alpha+\cos\phi\sin\alpha]\right.$$

$$\left.+ \frac{15}{8}\frac{r^2}{R^2}[\sin 3\phi\cos 3\alpha-\cos 3\phi\sin 3\alpha+\sin 3\phi\cos 3\alpha+\cos 3\phi\sin 3\alpha]\right\}$$

$$= \frac{M}{R^2}\left[\left(1 + \frac{3}{8}\frac{r^2}{R^2}\right)2\sin\phi\cos\alpha + \frac{15}{8}\frac{r^2}{R^2}2\sin 3\phi\cos 3\alpha\right]$$

$$= \frac{2M}{R^2}\left[\left(1 + \frac{3}{8}\frac{r^2}{R^2}\right)\sin\phi\cos\alpha + \frac{15}{8}\frac{r^2}{R^2}\sin 3\phi\cos 3\alpha\right] \tag{6}$$

For the sextantal term to disappear, $\cos 3\alpha$ must equal zero, i.e. $3\alpha = 90°$, or $\alpha = 30°$. Thus a pair of equal parallel magnets whose like poles subtend $60°$ at the centre of the system is not subject to sextantal deviations owing to the proximity of the permanent magnetic pole P.

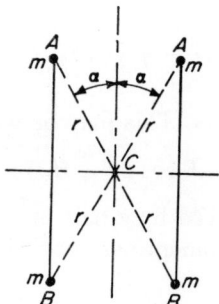

FIG. 2. Parallel needles are equivalent to crossed needles for the calculation of turning moments

Since the poles of a magnet are usually regarded as being one-twelfth of the length from each extremity, the disposition of magnets in which the like poles subtend $60°$ at the centre has the magnets a little too close to give exact equality of the moments of inertia about all diameters: but the approximation is sufficiently good for all practical purposes.

Now if the permanent magnet pole is replaced by a *transient induced* pole which normally gives rise to quadrantal deviations, an octantal deviation will occur if the pole is too close to a compass needle which is not equivalent to an infinitely short magnet.

Let the pole have a strength equal to unity $\times \cos \phi$; the turning moment is now given by

$$L = \frac{M}{R^2}\left[\left(1 + \frac{3}{8}\frac{r^2}{R^2}\right)\sin\phi\cos\phi + \frac{15}{8}\frac{r^2}{R^2}\sin 3\phi\cos\phi\right] \tag{7}$$

$$= \frac{M}{R^2}\left[\left(1 + \frac{3}{8}\frac{r^2}{R^2}\right)\frac{\sin 2\phi}{2} + \frac{15}{8}\frac{r^2}{R^2}\frac{(\sin 4\phi + \sin 2\phi)}{2}\right]$$

$$= \frac{M}{2R^2}\left[\left(1 + \frac{18}{8}\frac{r^2}{R^2}\right)\sin 2\phi + \frac{15}{8}\frac{r^2}{R^2}\sin 4\phi\right]$$

$$= \frac{M}{2R^2}\left[\left(1 + \frac{9}{4}\frac{r^2}{R^2}\right)\sin 2(\zeta' + \theta) + \frac{15}{8}\frac{r^2}{R^2}\sin 4(\zeta' + \theta)\right]$$

$$= \frac{M}{2R^2}\left[\left(1 + \frac{9}{4}\frac{r^2}{R^2}\right)(\sin 2\zeta'\cos 2\theta + \cos 2\zeta'\sin 2\theta)\right.$$

$$\left. + \frac{15}{8}\frac{r^2}{R^2}(\sin 4\zeta'\cos 4\theta + \cos 4\zeta'\sin 4\theta)\right] \tag{8}$$

If the deviation due to this turning moment is δ, then:

$$HM\sin\delta = \frac{M}{2R^2}\left[\left(1 + \frac{9}{4}\frac{r^2}{R^2}\right)\cos 2\theta\sin 2\zeta' + \left(1 + \frac{9}{4}\frac{r^2}{R^2}\right)\sin 2\theta\cos 2\zeta'\right.$$

$$\left. + \frac{15}{8}\frac{r^2}{R^2}\cos 4\theta\sin 4\zeta' + \frac{15}{8}\frac{r^2}{R^2}\sin 4\theta\cos 4\zeta'\right]$$

and

$$\sin\delta = \frac{1}{2HR^2}(p'\sin 2\zeta' + q'\cos 2\zeta' + s'\sin 4\zeta' + t'\cos 4\delta') \tag{9}$$

where p', q', s' and t' are constants. This may be written

$$\sin\delta = D\sin 2\zeta' + E\cos 2\zeta' + H\sin 4\zeta' + K\cos 4\zeta' \tag{10}$$

If now the single needle is replaced by a pair of equal parallel magnets as in the previous example, the turning moment may be written

$$L = \frac{M\cos\phi}{R^2}\left\{\left(1 + \frac{3}{8}\frac{r^2}{R^2}\right)[\sin(\phi - \alpha) + \sin(\phi + \alpha)]\right.$$

$$\left. + \frac{15}{8}\frac{r^2}{R^2}[\sin(3\phi - 3\alpha) + \sin(3\phi + 3\alpha)]\right\} \tag{11}$$

$$= \frac{M\cos\phi}{R^2}\left[\left(1 + \frac{3}{8}\frac{r^2}{R^2}\right)2\sin\phi\cos\alpha + \frac{15}{8}\frac{r^2}{R^2}2\sin 3\phi\cos 3\alpha\right]$$

$$= \frac{M}{R^2}\left[\left(1 + \frac{3}{8}\frac{r^2}{R^2}\right)\sin 2\phi\cos\alpha + \frac{15}{8}\frac{r^2}{R^2}2\sin 3\phi\cos\phi\cos 3\alpha\right]$$

$$= \frac{M}{R^2}\left[\left(1 + \frac{3}{8}\frac{r^2}{R^2}\right)\sin 2\phi\cos\alpha + \frac{15}{8}\frac{r^2}{R^2}(\sin 4\phi + \sin 2\phi)\cos 3\alpha\right] \tag{12}$$

The octantal term and part of the quadrantal term vanish when $\alpha = 30°$, so that the same magnet arrangement that avoids sextantal deviations due to the proximity of hard iron also avoids the octantal deviations due to the proximity of soft iron.

Operation of the correctors*

The correction of semicircular deviations by permanent magnets is simple and straightforward and consists of the cancellation of the deviating field at the compass position by the field from suitably placed magnets.

In the case of soft iron, the operation is complicated because the correcting property of the spheres is due in part to induction from the compass needle as well as to induction from the Earth's magnetic field.

Consider first the induction by the Earth's field. In Fig. 3, S is a sphere of soft iron whose centre is distant R from the centre C of the compass position, and placed to starboard of the ship's head by the angle θ. Let ζ be the magnetic heading of the ship.

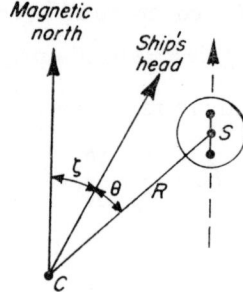

FIG. 3. Earth-induced magnetism of soft-iron correcting sphere

The sphere, when magnetized by the horizontal component H of the Earth's field, acts as a dipole of moment pH directed towards magnetic north. The forces at C due to this dipole are, in the direction CS, $2pH \cos (\zeta+\theta)/R_3$, and, in the direction $90°$ to clockwise from CS, $pH \sin (\zeta+\theta)/R_3$.

The components of these two forces towards magnetic east are

$$\frac{2pH}{R^3} \cos (\zeta+\theta) \sin (\zeta+\theta)$$

and

$$\frac{pH}{R^3} \sin (\zeta+\theta) \cos (\zeta+\theta)$$

The total force to magnetic east is:

$$\frac{3pH}{2R^3} \sin 2(\zeta+\theta) = \frac{3pH}{R^3} (\sin 2\zeta \cos 2\theta + \cos 2\zeta \sin 2\theta)$$

* *The Theory of the Deviations of the Magnetic Compass*, Admiralty Compass Observatory BR(101)48, Part IV, 76–9.

359

The components of the two forces in the direction of magnetic north are

$$\frac{2pH}{R^3}\cos^2(\zeta+\theta)-\frac{pH}{R^3}\sin^2(\zeta+\theta)$$

and

$$\frac{pH}{2R^3}+\frac{3pH}{2R^3}(\cos 2\zeta\cos 2\theta-\sin 2\zeta\cos 2\theta)$$

The total force to magnetic north is

$$H+\frac{pH}{2R^3}+\frac{3pH}{2R^3}(\cos 2\zeta\cos 2\theta-\sin 2\zeta\sin 2\theta)$$

$$=H+\frac{pH}{2R^3}+\frac{3pH}{2R^3}\cos 2\theta\cos 2\zeta-\frac{3pH}{2R^3}\sin 2\theta\sin 2\zeta$$

Therefore

$$\tan\delta=\frac{(3/2)(p/R^3)\cos 2\theta\sin 2\zeta+(3p/2R^3)\sin 2\theta\cos 2\zeta}{1+(p/2R^3)-(3p/2R^3)\sin 2\theta\sin 2\zeta+(3p/2R^3)\cos 2\theta\cos 2\zeta}\quad(13)$$

and from equations (10.11) and (10.12) we may write:

$$\bar{D}_e=\frac{1}{\lambda_e}.\frac{3p}{2R^3}\cos 2\theta\tag{14}$$

$$\bar{E}_e=\frac{1}{\lambda_e}.\frac{3p}{2R^3}\sin 2\theta\tag{15}$$

$$\lambda_e=1+\frac{p}{2R^3}\tag{16}$$

When, as is normal in ships, $\theta=90°$,

$$\bar{D}_e=-\frac{1}{\lambda_e}.\frac{3p}{2R^3}$$

$$\bar{E}_e=0$$

$$\lambda_e=1+\frac{p}{2R^3}$$

Consider now the effect of induction from a compass needle at C whose length is small compared with R and having a moment M (Fig. 4).

The forces due to the needle system at S are:

$$\frac{2M\cos(\zeta+\theta)}{R^3}\qquad\text{in the direction CS}$$

and

$$\frac{M}{R^3}\sin(\zeta+\theta)\qquad 90°\text{ clockwise from CS}$$

The sphere, being magnetized by these forces, gives equivalent dipoles of moment

$$\frac{2pM \cos(\zeta+\theta)}{R^3} \quad \text{in the direction CS}$$

and

$$\frac{pM \sin(\zeta+\theta)}{R^3} \quad \text{90° clockwise from CS}$$

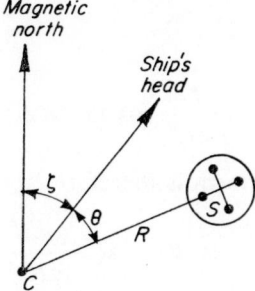

FIG. 4. Compass-induced magnetism of soft-iron correcting sphere

These dipoles give forces at the compass

$$\frac{4pM}{R^6} \cos(\zeta+\theta) \quad \text{in the direction CS}$$

and

$$\frac{pM}{R^6} \sin(\zeta+\theta) \quad \text{90° anti-clockwise from CS}$$

The easterly components of these forces are

$$\frac{4pM}{R^6} \cos(\zeta+\theta) \sin(\zeta+\theta)$$

and

$$-\frac{pM}{R^6} \sin(\zeta+\theta) \cos(\zeta+\theta)$$

giving a total force to magnetic east of

$$\frac{3pM}{R^6} \sin 2(\zeta+\theta)$$

$$= \frac{3pM}{2R^6}(\sin 2\zeta \cos 2\theta + \cos 2\zeta \sin 2\theta)$$

$$= \frac{3pM}{2R^6}(\cos 2\theta \sin 2\zeta) + \frac{3pM}{2R^6}(\sin 2\theta \cos 2\zeta)$$

The northerly components of the forces are

$$\frac{4pM}{R^6} \cos^2 (\zeta + \theta)$$

and

$$\frac{pM}{R^6} \sin^2 (\zeta + \theta)$$

giving a force to magnetic north of

$$\frac{5pM}{2R^6} + \frac{3pM}{2R^6} \cos 2(\zeta + \theta)$$

$$= \frac{5pM}{2R^6} + \frac{3pM}{2R^6} \cos 2\theta \cos 2\zeta - \frac{3pM}{2R^6} \sin 2\theta \sin 2\zeta$$

The total force to magnetic north is therefore

$$H + \frac{5pM}{2R^6} + \frac{3pM}{2R^6} \cos 2\theta \cos 2\zeta - \frac{3pM}{2R^6} \sin 2\theta \sin 2\zeta$$

and

$$\tan \delta = \frac{(3pM/2HR^6) \cos 2\theta \sin 2\zeta + (3pM/2HR^6) \sin 2\theta \cos 2\zeta}{1 + (5pM/2HR^6) - (3pM/2HR^6) \sin 2\theta \sin 2\zeta + (3pM/2HR^6) \cos 2\theta \cos 2\zeta} \tag{17}$$

and thus

$$\bar{D}_m = \frac{1}{\lambda_m} \times \frac{3pM}{2HR^6} \cos 2\theta \tag{18}$$

$$\bar{E}_m = \frac{1}{\lambda_m} \times \frac{3pM}{2HR^6} \sin 2\theta \tag{19}$$

$$\lambda_m = 1 + \frac{5pM}{2HR^6} \tag{20}$$

When $\theta = 90°$

$$\bar{D}_m = -\frac{1}{\lambda_m} \times \frac{3pM}{2HR^6}$$

$$\bar{E}_m = 0$$

$$\lambda_m = 1 + \frac{5pM}{2HR^6}$$

Consequently, a pair of spheres placed athwartships produce coefficients $\bar{D}_e$ and $\bar{D}_m$ in opposition to coefficient $\bar{D}_s$ of the ship. That is to say:

$$\frac{1}{\lambda}\left(\frac{-3p}{2R^3} - \frac{3pM}{2HR^6}\right) = \frac{a - e}{2\lambda}$$

or

$$-\frac{3p}{R^3}\left(1 + \frac{M}{HR^3}\right) = a - e$$

The correction is therefore due partly to the effect of the Earth's induction and partly to that of the needle's induction, the former being independent of H and the latter being inversely proportional to H. This gives incomplete correction of the compass position if H alters as the ship moves from one magnetic latitude to another, leaving a residual $\bar{D}$. The permissible amount of correction due to the induction from the compass needle system is governed by the appropriate tables showing the placing of the spheres for various types of compass and binnacle, which is the result of experiment.

In terms of compass heading ζ', coefficient $\bar{D}$ may be written $\sin D°$; moreover, $\frac{1}{2}(\bar{D})^2 = \sin H°$, which is an octantal deviation. Therefore the deviation due to the soft iron of the ship represented by $\bar{D}_s$ and the corrective effect of the spheres represented by $\bar{D}_e$ and $\bar{D}_m$ contain both quadrantal and octantal components in terms of compass heading. Nevertheless if correction is complete, $\bar{D} = 0$ and therefore $\sin D° = 0$ and $\sin H° = 0$.

The effect of the proximity of the correctors

The foregoing argument on the effect of nearby magnetic poles, whether permanent or induced, on a compass needle or magnet system would also be expected to apply to the effect of the permanent magnets and soft-iron masses used in compass correction.

The application of the theory to the permanent magnet correctors is straight-forward and, as would be expected, the form of magnet system that avoids sextantal deviations due to the proximity of a permanent magnetic pole is equally effective for a neighbouring permanent magnet.

The proximity of the soft-iron spheres is not so simple a matter. Equations (8) and (12) apply to a single induced pole near the compass, whereas the magnetization of the spheres described on pp. 359–61 is clearly a more complex state of affairs; in fact, although the compass needle may indeed induce a transient pole as it comes near the sphere, there is in addition the effect of a nearby dipole induced by the Earth's magnetic field and even perhaps by the field from the compass needle as well.

Divergent opinions have been held as to the form of compass magnet system that would avoid the octantal deviations due to the proximity of the soft-iron correctors, values of $45°$ and $60°$ being suggested as the correct angle to be subtended by the like poles of two equal parallel magnets at the centre of the array. Since the mathematical treatment is complicated and needs certain theoretical and arbitrary assumptions, Meldau investigated the matter practically, using an array of four soft-iron spheres spaced equally around the compass.

It has been shown that a single sphere, or two spheres diametrically opposite, when acted upon either by the Earth's field or the induction from an *infinitely short magnet* provide a coefficient $\bar{D}$ in terms of magnetic heading, which in terms of compass heading appears as a coefficient D (quadrantal) and a coefficient H (octantal). The latter is present even with an ideal magnet system and it was in order to remove it and to reveal only the effect of the spheres on an incorrectly designed system that Meldau used four spheres placed athwartships and fore-and-aft.

363

Referring to equations (14) and (18), if $\theta = 90°$ and $270°$

$$\bar{D}_e = -\frac{1}{\lambda} \cdot \frac{3p}{2R^3}$$

$$\bar{D}_m = -\frac{1}{\lambda} \cdot \frac{3pM}{2HR^6}$$

and if $\theta = 0°$ and $180°$

$$\bar{D}_e = \frac{1}{\lambda} \cdot \frac{3p}{2R^3}$$

$$\bar{D}_m = \frac{1}{\lambda} \cdot \frac{3pM}{2HR^6}$$

Thus the total $(\bar{D}_e + \bar{D}_m) = 0$; consequently, D and H equal zero.

If a needle system that is neither infinitely short nor in conformity with the pattern indicated by Smith and Evans is placed at the centre of the array of the four spheres, octantal deviations will arise. This was demonstrated by Meldau,[*] who also showed that when the needle system was of the correct form, the spurious octantal deviations disappeared. He concluded that the two-needle system with the poles subtending 60° at the centre avoids both the sextantal deviations due to permanent magnet correctors and the octantal deviations due to induced poles. Meldau further showed that a residual octantal deviation due to terms of the order of $(r/R)^4$ occurs where $2r$ is the equivalent length of the needle and R the distance from the centre of the needle system to the centre of a sphere. Terms of a higher order than the 4th power give rise to a dodecantal deviation which is almost $2\frac{1}{2}$ times the octantal effect, but both are very small when R is at least $2\frac{1}{3}$ times r.

The four-sphere test

The four-sphere array is therefore a simple means of ascertaining whether a compass magnet system is of the correct form, since all deviations due to the array are zero, except when the compass magnet system is incorrectly designed, in which case the octantal deviations are readily observed, their magnitude revealing the extent by which the system departs from the correct disposition, namely one in which the like magnet poles subtend a mean angle of 60° at the centre.

In the simple case of a transient induced pole, we see in equation (8) that

$$L = \frac{M}{2R^2}\left[\left(1 + \frac{9}{4}\frac{r^2}{R^2}\right)(\sin 2\zeta' \cos 2\theta + \cos 2\zeta' \sin 2\theta)\right.$$

$$\left. + \frac{15}{8}\frac{r^2}{R^2}(\sin 4\zeta' \cos 4\theta + \cos 4\zeta' \sin 4\theta)\right]$$

If $\theta = 0°$ or $180°$,

$$L = \frac{M}{2R^2}\left[\left(1 + \frac{9}{4}\frac{r^2}{R^2}\right)\sin 2\zeta' + \frac{15}{8}\frac{r^2}{R^2}\sin 4\zeta'\right]$$

* Meldau, H., 'Die Nadelanordnung der Kompassrose mit Rücksicht auf den D-Korrectoren', *Ann. Hydrogr. Berl.*, 1907, **35**, 17–25.
The Four-sphere Method of Testing Compasses, German Hydrographic Institute pamphlet, 1947.

If $\theta = 90°$ or $270°$,

$$L = \frac{M}{2R^2}\left[-\left(1 + \frac{9}{4}\frac{r^2}{R^2}\right)\sin 2\zeta' \times \frac{15}{8}\frac{r^2}{R^2}\sin 4\zeta'\right]$$

Consequently for a four-sphere array,

$$L = \frac{M}{R^2}\cdot\frac{15}{2}\cdot\frac{r^2}{R^2}\sin 4\zeta'$$

an expression containing only an octantal term.

With a pair of equal parallel magnets subtending 2α at the centre and also for four spheres, from equation (12):

$$L = \frac{M}{2R^2}\cdot\frac{15}{2}\cdot\frac{r^2}{R^2}\sin 4\zeta' \cos 3\alpha$$

an octantal term only which vanishes when $3\alpha = 90°$ or when $\alpha = 30°$.

APPENDIX 6

Multiple-Needle Systems and Ring Magnets

Multiple-needle system

In addition to the conventional two-needle system discussed in Appendix 5, other arrays of parallel pairs of magnets may be used. Fig. 1 shows a system with an odd number of pairs and Fig. 2 one with an even number. In both cases all the needles have equal pole strength and the poles lie on a circle about the centre C, the lines joining the poles being chords of the circle. The needles are symmetrically disposed about chords whose extremities subtend an angle, at C, of 2α. In the case of an odd number of pairs, the middle pair lies along this chord, whilst with an even number the chord lies midway between the two middle magnets on either side.

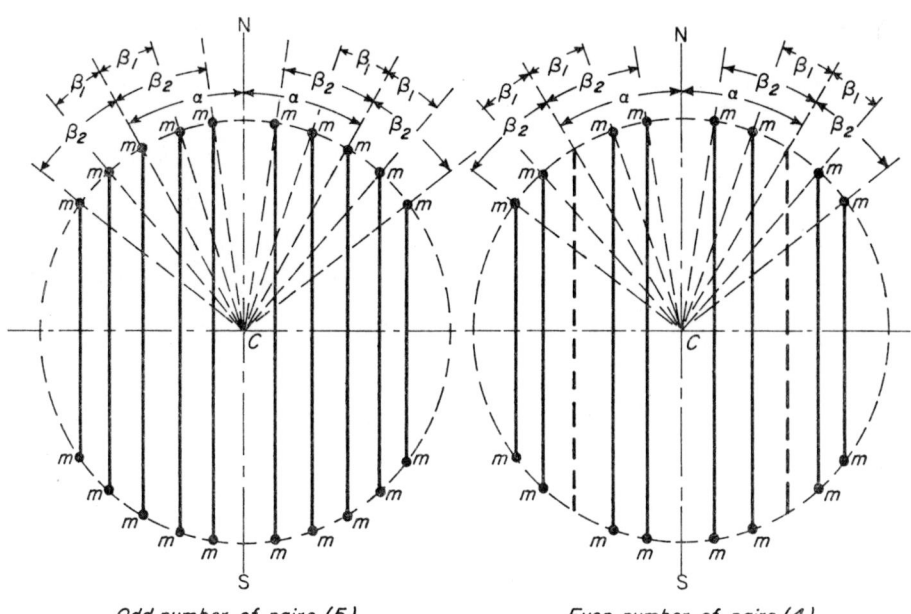

Odd number of pairs (5)   Even number of pairs (4)

FIG. 1. Multiple-needle system with an odd number of pairs of magnets

FIG. 2. Multiple-needle system with an even number of pairs of magnets

Both these systems can be replaced by arrays of crossed magnets of moment $M (= 2mr)$ intersecting at C, with their poles lying on the circumference of a circle of radius r, as in Figs. 3 and 4.

For an odd number of pairs of magnets, the turning moment L due to a unit pole at a point distant R from C and in a direction ϕ clockwise from compass north

$(\phi = \zeta' + \theta)$ is:

$$L = \frac{M}{R^2}\left\{\left(1 + \frac{3}{8}\frac{r^2}{R^2}\right)[\sin(\phi - \alpha) + \sin(\phi + \alpha) + \sin(\phi - \overline{\alpha - \beta_1})\right.$$

$$+ \sin(\phi + \overline{\alpha - \beta_1}) + \sin(\phi - \overline{\alpha + \beta_1}) + \sin(\phi + \overline{\alpha + \beta_1}) + \ldots$$

$$+ \sin(\phi - \overline{\alpha - \beta_n}) + \sin(\phi + \overline{\alpha - \beta_n}) + \sin(\phi - \overline{\alpha + \beta_n}) + \sin(\phi + \overline{\alpha + \beta_n}]$$

$$+ \frac{15}{8}\frac{r^2}{R^2}[\sin 3(\phi - \alpha) + \sin 3(\phi + \alpha) + \sin 3(\phi - \overline{\alpha - \beta_1}) + \sin 3(\phi + \overline{\alpha - \beta_1})$$

$$+ \sin 3(\phi - \overline{\alpha + \beta_1}) + \sin 3(\phi + \overline{\alpha + \beta_1}) + \ldots + \sin 3(\phi - \overline{\alpha - \beta_n})$$

$$\left. + \sin 3(\phi + \overline{\alpha - \beta_n}) + \sin 3(\phi - \overline{\alpha + \beta_n}) + \sin 3(\phi + \overline{\alpha + \beta_n})]\right\} \tag{1}$$

where $n = \frac{1}{2}(p-1)$, p being the number of pairs. Thus:

$$L = \frac{M}{R^2}\left\{\left(1 + \frac{3}{8}\frac{r^2}{R^2}\right)\left[2\sin\phi\cos\alpha + 2\sin\phi\cos(\alpha - \beta_1) + 2\sin\phi\cos(\alpha + \beta_1)\right.\right.$$

$$\left. + \ldots + 2\sin\phi\cos(\alpha - \beta_n) + 2\sin\phi\cos(\alpha + \beta_n)\right]$$

$$+ \frac{15}{8}\frac{r^2}{R^2}\left[2\sin 3\phi\cos 3\alpha + 2\sin 3\phi\cos 3(\alpha - \beta_1) + 2\sin 3\phi\cos 3(\alpha + \beta_1)\right.$$

$$\left.\left. + \ldots + 2\sin 3\phi\cos 3(\alpha - \beta_n) + 2\sin 3\phi\cos 3(\alpha + \beta_n)\right]\right\}$$

$$= \frac{M}{R^2}\left[\left(1 + \frac{3}{8}\frac{r^2}{R^2}\right)(2\sin\phi\cos\alpha + 4\sin\phi\cos\alpha\cos\beta_1 + \ldots\right.$$

$$+ 4\sin\phi\cos\alpha\cos\beta_n) + \frac{15}{8}\frac{r^2}{R^2}(2\sin 3\phi\cos 3\alpha + 4\sin 3\phi\cos 3\alpha\cos 3\beta_1$$

$$\left. + \ldots + 4\sin 3\phi\cos 3\alpha\cos 3\beta_n)\right]$$

$$= \frac{M}{R^2}\left\{\left(1 + \frac{3}{8}\frac{r^2}{R^2}\right)\left[2\sin\phi\cos\alpha\left(1 + 2\sum_1^{k=n}\cos\beta_k\right)\right]\right.$$

$$\left. + \frac{15}{8}\frac{r^2}{R^2}\left[2\sin 3\phi\cos 3\alpha\left(1 + 2\sum_1^{k=n}\cos 3\beta_k\right)\right]\right\} \tag{2}$$

For the sextantal term to vanish, $\cos 3\alpha = 0$, i.e. $\alpha = 30°$. Thus for an array of an odd number of pairs of needles placed symmetrically along chords of a circle with the poles lying on that circle, the mean angle subtended by like poles at the centre of the circle is 60° if sectantal deviations are to be avoided. As demonstrated in equations (7) to (12) in Appendix 5, the octantal deviations due to a neighbouring induced pole are also avoided.

For an array of an even number of pairs of magnets:

$$L = \frac{M}{R^2}\left\{\left(1 + \frac{3}{8}\frac{r^2}{R^2}\right)[\sin(\phi - \alpha - \beta_1) + \sin(\phi + \alpha - \beta_1) + \sin(\phi - \alpha + \beta_1)\right.$$

$$+ \sin(\phi + \alpha + \beta_1) + \sin(\phi - \alpha - \beta_2) + \sin(\phi + \alpha - \beta_2) + \sin(\phi - \alpha + \beta_2)$$

$$+ \sin(\phi + \alpha + \beta_2) + \dots + \sin(\phi - \alpha - \beta_n) + \sin(\phi + \alpha - \beta_n)$$

$$+ \sin(\phi - \alpha + \beta_n) + \sin(\phi + \alpha + \beta_n)]$$

$$+ \frac{15}{8}\frac{r^2}{R^2}[\sin 3(\phi - \alpha - \beta_1) + \sin 3(\phi + \alpha - \beta_1) + \sin 3(\phi - \alpha + \beta_1)$$

$$+ \sin 3(\phi + \alpha + \beta_1) + \sin 3(\phi - \alpha - \beta_2) + \sin 3(\phi + \alpha - \beta_2) + \sin 3(\phi - \alpha + \beta_2)$$

$$+ \sin 3(\phi + \alpha + \beta_2) + \dots + \sin 3(\phi - \alpha - \beta_n) + \sin 3(\phi + \alpha - \beta_n)$$

$$\left. + \sin 3(\phi - \alpha + \beta_n) + \sin 3(\phi + \alpha + \beta_n)]\right\} \tag{3}$$

where $n = p/2$, p being the number of pairs. So

$$L = \frac{M}{R^2}\left\{\left(1 + \frac{3}{8}\frac{r^2}{R^2}\right)[4\sin\phi\cos\alpha\cos\beta_1 + 4\sin\phi\cos\alpha\cos\beta_2\right.$$

$$+ \dots + 4\sin\phi\cos\alpha\cos\beta_n] + \frac{15}{8}\frac{r^2}{R^2}[4\sin 3\phi\cos 3\alpha\cos 3\beta_1$$

$$\left. + 4\sin 3\phi\cos 3\alpha\cos 3\beta_2 + \dots + 4\sin 3\phi\cos 3\alpha\cos 3\beta_n]\right\}$$

$$= \frac{4M}{R^2}\left[\left(1 + \frac{3}{8}\frac{r^2}{R^2}\right)\left(\sin\phi\cos\alpha\sum_1^n\cos\beta + \sin 3\phi\cos 3\alpha\sum_1^{k=n}\cos 3\beta_k\right)\right] \tag{4}$$

So once again for the sextantal term to vanish, 3α must equal zero or $\alpha = 30°$.

Therefore for any array of pairs of parallel magnets of equal pole strength placed along chords of a circle with their poles lying on its circumference, the mean angle subtended at the centre of the system by like poles must be 60° if sextantal and octantal deviations are to be avoided.

A further important consideration is that the moment of inertia of the array of magnets about every diameter must be the same. Consider a magnet in the form of a thin rod, ACB, placed as the chord of a circle of radius a with centre at O (Fig. 5). Let ZZ be any diameter and inclined at an angle θ to AB. Let the angle AÔN $= \alpha$ and let AC $=$ CB $= r$. The moment of inertia about ZZ of an element of the rod of mass dm is $y^2\,dm$. If the rod has a cross-sectional area, q, and density ρ and the element has thickness dx,

$$dI = q\rho y^2\,dx$$

Now $y/x = \sin\theta$

Thus

$$dI = q\rho\,x^2\sin^2\theta dx$$

368

and

$$I = q\rho \sin^2\theta \int_0^{r+CX} x^2 \, dx + q\rho \sin^2\theta \int_0^{r-CX} x^2 \, dx$$

Since CX = CO cot θ and CO = AO sin α, we have CX = a sin α cot θ.

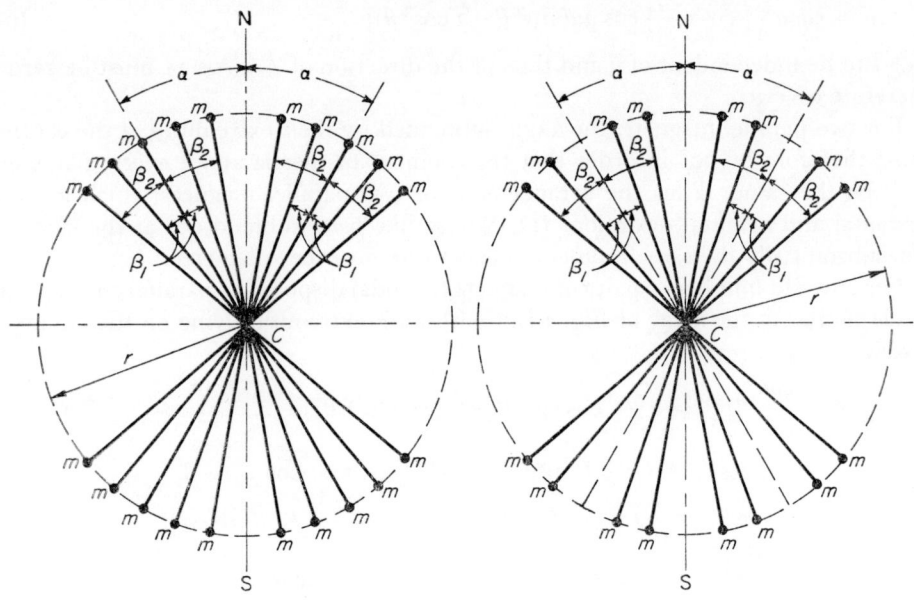

Odd number of pairs (5)

FIG. 3. Array of crossed magnets equivalent to Fig. 1

Even number of pairs (4)

FIG. 4. Array of crossed magnets equivalent to Fig. 2

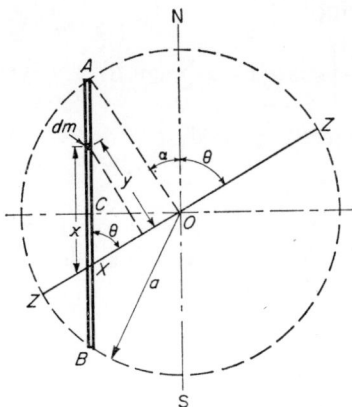

FIG. 5. Moment of inertia

Therefore

$$I = q\rho \sin^2\theta \left(\int_0^{r+a \sin \alpha \cot \theta} x^2 \, dx + \int_0^{r-a \sin \alpha \cot \theta} x^2 \, dx \right)$$

$$= \tfrac{2}{3}q\rho \sin^2\theta (r^3 + 3a^2 r \sin^2 \alpha \cot^2 \theta) \qquad (5)$$

369

But $r = a \cos \alpha$. Therefore

$$I = \tfrac{2}{3}q\rho \sin^2\theta(a^3 \cos^3\alpha + 3a^3 \sin^2\alpha \cos\alpha \cot^2\theta)$$

$$= \tfrac{2}{3}q\rho a^3(\cos^3\alpha \sin^2\theta + 3\sin^2\alpha \cos\alpha \cos^2\theta)$$

$$= \tfrac{2}{3}q\rho a^3(\tfrac{3}{4}\cos\alpha \sin^2\theta + \tfrac{1}{4}\cos 3\alpha \sin^2\theta + \tfrac{3}{4}\cos\alpha \cos^2\theta - \tfrac{3}{4}\cos 3\alpha \cos^2\theta)$$

$$= \tfrac{2}{3}q\rho a^3[\tfrac{3}{4}\cos\alpha + \tfrac{1}{4}\cos 3\alpha(\sin^2\theta - 3\cos^2\theta)] \tag{6}$$

For I to be independent of θ and thus of the direction of ZZ, $\cos 3\alpha$ must be zero; therefore $\alpha = 30°$.

For two parallel magnets, the angle subtended by their extremities at the centre must therefore be $60°$ in order that the moment of inertia about every diameter may be the same. Thus the arrangement of two parallel magnets which avoids sextantal and octantal deviations (i.e. whose like *poles* subtend $60°$ at the centre) has substantially the same moment of inertia about every diameter.

For any odd number of pairs of magnets (or rods) disposed as parallel chords of a circle similar to the array of Fig. 1 but with their extremities lying on the circumference:

$$\sum I = \frac{q\rho a^3}{6}\{[3\cos\alpha + 3\cos(\alpha - \beta_1) + 3\cos(\alpha + \beta_1) + \ldots + 3\cos(\alpha - \beta_n)$$

$$+ 3\cos(\alpha + \beta_n)] + (\sin^2\theta - 3\cos^2\theta)[\cos 3\alpha + \cos 3(\alpha - \beta_1)$$

$$+ \cos 3(\alpha + \beta_1) + \ldots + \cos 3(\alpha - \beta_n) + \cos 3(\alpha - \beta_n)]\}$$

$$= \frac{q\rho a^3}{6}[3\cos\alpha + 6\cos\alpha \cos\beta_1 + \ldots + 6\cos\alpha \cos\beta_n$$

$$+ (\sin^2\theta - 3\cos^2\theta)(\cos 3\alpha + 2\cos 3\alpha \cos 3\beta + \ldots$$

$$+ 2\cos 3\alpha \cos 3\beta_n)] \tag{7}$$

where $n = \tfrac{1}{2}(p - 1)$. Therefore

$$\sum I = \frac{q\rho a^3}{2}\left[\cos\alpha\left(1 + 2\sum_1^n \cos\beta\right)\right.$$

$$\left. + \left(\frac{\sin^2\theta - 3\cos^2\theta}{3}\right)\cos 3\alpha\left(1 + 2\sum_1^n \cos 3\beta\right)\right] \tag{8}$$

and so if $\alpha = 30°$, $\cos 3\alpha = 0$ and the term dependent on θ vanishes.

For an even number of pairs disposed as parallel chords of a circle similar to Fig. 2 but with their extremities lying on the circumference:

$$\sum I = \frac{q\rho a^3}{6}[(6\cos\alpha \cos\beta_1 + 6\cos\alpha \cos\beta_2 + \ldots + 6\cos\alpha \cos\beta_n)$$

$$+ (\sin^2\theta - 3\cos^2\theta)(2\cos 3\alpha \cos 3\beta_1 + 2\cos 3\alpha \cos 3\beta_2 + \ldots$$

$$+ 2\cos 3\alpha \cos 3\beta_n)] \tag{9}$$

where $n = \tfrac{1}{2}p$; and

$$\sum I = q\rho a^3\left[\cos\alpha \sum_1^{k=n} \cos\beta_k + \left(\frac{\sin^2\theta - 3\cos^2\theta}{3}\right)\cos 3\alpha \sum_1^{k=n} \cos 3\beta_k\right] \tag{10}$$

So once again if $\alpha = 30°$, $\cos 3\alpha = 0$ and the term dependent on θ vanishes. Therefore arrays of parallel magnets disposed as chords of a circle and whose like extremities subtend a mean angle of 60° at the centre of the array have the same moment of inertia about every diameter. Thus a magnet system composed of parallel pairs of needles and constructed so as to avoid sextantal and octantal deviations will have substantially the same moment of inertia about any diameter.

Disk and ring magnets

A disk or ring of magnetic material is assumed to be magnetized parallel to a diameter so as to have a uniform distribution of pole strength along each semi-circumference, as shown in Fig. 6. The disk or ring may therefore be regarded as being made up of an infinite number of pairs of parallel, chordal elementary magnets of pole strength dm with their poles lying on the circumference of the disk or ring (Fig. 7). This approximation is adopted in the interests of simplicity

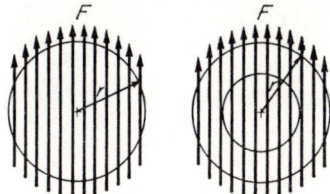

FIG. 6. Disk and ring magnets

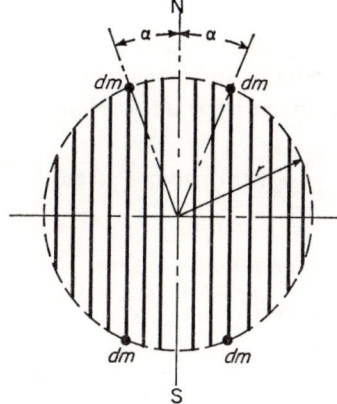

FIG. 7. Infinitesimal elements of a ring magnet

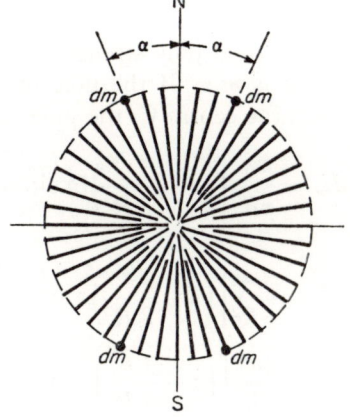

FIG. 8. Crossed elements equivalent to Fig. 7

in the following mathematical argument, since, strictly speaking, the poles of these elementary magnets should be regarded as being situated at one-twelfth of their length from either extremity and therefore lying on the circumference of an oblate disk or ring. The chordal magnets are equivalent to an infinite number of crossed pairs of elementary magnets intersecting at the centre of the disk or ring, of length equal to its diameter and with poles of strength dm lying on the circumference (Fig. 8).

Consider a typical pair of magnets acted upon by a unit pole (Fig. 9). The turning moment on the pair will be

$$dL = \frac{2r\,dm}{R^2}\left\{\left(\mathbf{1}+\frac{3}{8}\frac{r^2}{R^2}\right)\left[\sin(\phi-\alpha)+\sin(\phi+\alpha)\right]\right.$$

$$\left.+\frac{15}{8}\frac{r^2}{R^2}\left[\sin 3(\phi-\alpha)+\sin 3(\phi+\alpha)\right]\right\} \tag{11}$$

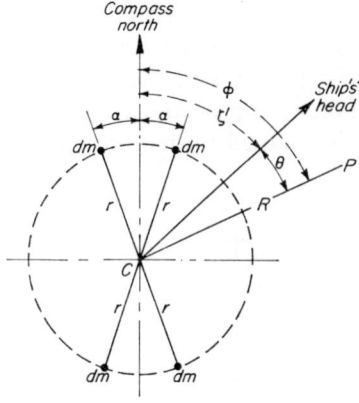

FIG. 9. Action of a unit pole on a pair of magnets

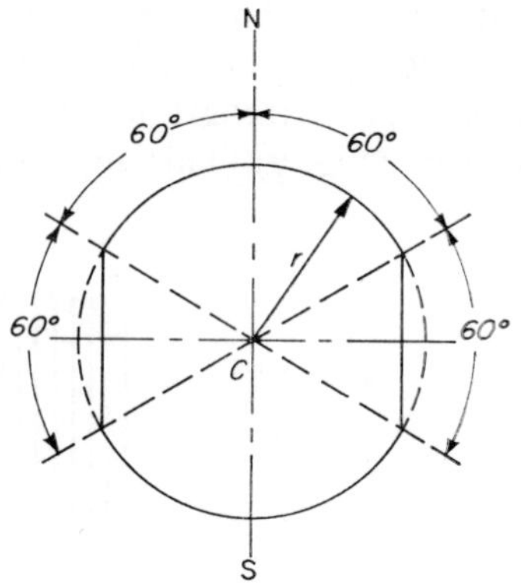

FIG. 10. Avoidance of sextantal and octantal deviations

Now if the total pole strength is m and is equally distributed, the pole strength per radian is m/π. If the pole of each infinitesimally small magnet subtends $d\alpha$ radians at the centre C, the pole strength dm is given by $m\,d\alpha/\pi$. Hence

$$dL = \frac{2mr}{\pi R^2}\left[\left(\mathbf{1}+\frac{3}{8}\frac{r^2}{R^2}\right)(2\sin\phi\cos\alpha\,d\alpha)+\frac{15}{8}\frac{r^2}{R^2}(2\sin 3\phi\cos 3\alpha\,d\alpha)\right]$$

and

$$L = \frac{2mr}{\pi R^2}\left[\left(\mathbf{1}+\frac{3}{8}\frac{r^2}{R^2}\right)2\sin\phi\int_0^{\pi/2}\cos\alpha\,d\alpha+\frac{15}{8}\frac{r^2}{R^2}2\sin 3\phi\int_0^{\pi/2}\cos 3\alpha\,d\alpha\right]$$

$$= \frac{4mr}{\pi R^2}\left[\left(\mathbf{1}+\frac{3}{8}\frac{r^2}{R^2}\right)\sin\phi-\frac{15}{8}\frac{r^2}{R^2}\sin 3\phi\right] \tag{12}$$

revealing a sextantal term as well as a semicircular one, but of less magnitude than and opposite in sign to that present in the case of a single needle. For the sextantal term to disappear,

$$\int_0^x \cos 3\alpha\,d\alpha = 0$$

i.e.

$$\sin 3x = 0$$

or

$$x = \frac{\pi}{3} = 60°$$

Therefore in order to avoid sextantal (and octantal) deviations the disk or ring needs to be modified by removing diametrically opposite segments which subtend 60° at the centre, as in Fig. 10. It will be noted that the mean angle subtended by the distributed pole of this magnet is 60°, as with multiple magnet systems.

Nevertheless, the complete disk or ring has a dynamic advantage that is not possessed by the modified magnet shown in Fig. 10, namely the moment of inertia is the same about every diameter, thereby fulfilling an important requirement of a compass magnet system. A compromise might be effected by replacing the unwanted segments of the disk or ring by appropriately shaped pieces of non-magnetic material.

In practice, a complete disk or ring magnet performs remarkably well, and experiments show it to have similar characteristics to the theoretical array of multiple needles. This is probably owing to the possibility of the elementary magnets comprising the segments of the disk or ring, diametrically opposite on the east–west diameter, becoming strongly demagnetized on account of their short length. Thus these segments may, fortuitously, be ineffective magnetically.

BIBLIOGRAPHY

HISTORICAL AND GENERAL

1. AIRY, G. B. Account of experiments on iron-built ships instituted for the purpose of discovering a correction for the deviation of the compass produced by the iron of the ship. *Phil. Trans. roy. Soc.*, 1839, **3**, 167–214.
2. AIRY, G. B. On the connection between the mode of building iron ships and the ultimate correction of their compasses. *Trans. Instn. nav. Archit.*, 1860, **1**, 105.
3. AIRY, G. B. *Syllabus of a Course of Three Lectures on Magnetical Errors, Compensations and Corrections with Special Reference to Iron Ships and their Compasses.* 1864, pp. 14, 20.
4. AIRY, G. B. *Syllabus of Lectures on Magnetism.* 1869, p. 23.
5. BASSNETT, T. British Patent 642 1854.
6. BERTELLI, T. Sopra Pietro Peregrino de Maricourt e la sua epistola 'De Magnete'. *Boll. bibliogr. sci. mat. Roma*, 1868, **1**, 1–32.
7. FINCH, H. F. Royal Greenwich Observatory, Hartland. *Nature*, 1957, **179(2)**, 994.
8. GILBERT, W. *De Magnete*, London, 1600. English translations: P. Fleury Mottelay, John Wiley and Sons, New York, 1893: with notes by S. P. Thompson. The Gilbert Club, London, 1900.
9. HARRIS, W. S. *Rudimentary Magnetism*, 1852, part III, p. 166.
10. HINE, A. Modern compasses, *Research, Lond.*, 1957, **10**, 97–107.
11. HITCHINS, H. L., and MAY, W. E. *From Lodestone to Gyrocompass.* Hutchinson, London, 1952 (2nd Edn 1955).
12. JOHNSON, E. *Practical Illustrations of the Necessity for Ascertaining the Deviations of the Compass.* 2nd Edn 1852, p. 33.
13. KALMER, A. British Patent 5, 1873.
14. MAY, W. E. The compass makers of Deptford dockyard, *Naut. Mag.*, 1950, **163**, 386–90.
15. MAY. W. E. Corrected compass cards, *Naut. Mag.*, 1948, **160**, 269–71.
16. MAY, W. E. Ring-shaped compass needles. *Naut. Mag.*, 1950, **163**, 78–80.
17. MAY, W. E. Alexander Neckham and the pivoted compass needle. *J. Inst. Navig.*, 1955, **8**, 283–4.
18. MAY, W. E. Floating compasses. *Naut. Mag.*, 1953, **169**, 1–5.
19. MAY, W. E., BRODIE, L. S., HINE, A. The Compass. *Encyclopaedia Britannica*, 1963, Vol. 6, p. 225.
20. POISSON, D. *Memoirs de l'Institut de France.* 1824.
21. SCHUCK, A. *Der Kompass.* Hamburg, 1911, 1915, and 1918.
22. SMITH, A., and EVANS, F. J. On the effect produced on the deviation of the compass by the length and arrangement of the compass needles and on a new mode of correcting the quadrantal deviation. *Phil. Trans. roy. Soc.*, 1861, 161.
23. SMITH, A., and EVANS, F. J. *Admiralty Manual for Ascertaining and Applying the Deviations of the Compass caused by the Iron in a Ship.* 1862 (reprint 1920).
24. STEBBING, J. R. Compasses of iron ships, *The Artisan*, 1850, 178–181.
25. THOMPSON. S. P. *William Gilbert and Terrestrial Magnetism in the Time of Queen Elizabeth.* Charles Whittingham, London, 1903, 16 pp.
26. WALKER, W. *The Magnetism of ships and the Mariner's Compass.* London, 1853.
27. WOLKENHAUER, A. Der Schiffskompass im 16. Jahrhundert und die Ausgleichung der magnetischen Deklination. *Ann. Hydrogr. Berl.*, 1905, **33**, 29–37.

TERRESTRIAL MAGNETISM

28. BAUER, L. A. Halley's earliest variation chart. *Terr. Mag.*, 1896, **1**, 28–31.
29. CHAPMAN, S. Theories of terrestrial and solar magnetism. *Dictionary of Applied Physics*, ed. Sir Richard Glazebrook, 1922, **2**, 543–61.
30. CHAPMAN, S. *The Earth's Magnetism.* Methuen, London, 1936, xi + 116 pp.
31. CHAPMAN, S., and BARTELS, J. *Geomagnetism.* O.U.P., London, 1940.
32. CHREE, C. An inquiry into the nature of the relationship between sunspot frequency and terrestrial magnetism. *Phil. Trans. roy. Soc.*, 1904, **203A**, 151–87.
33. CHREE, C. Terrestrial Magnetism. *Encyclopaedia Britannica*, 11th Edn, 1911. Vol. 17, 353–85.
34. CHREE, C. *Studies in Terrestrial Magnetism.* Macmillan, London, 1912, xii + 206 pp.

35. CHREE, C. The 27-day period in magnetic phenomena. *Proc. roy. Soc.*, 1904, **90A**, 583–99.

36. CHREE, C. The difference between the magnetic diurnal variations on ordinary and quiet days at Kew Observatory. *Proc. roy. Soc.*, 1915, **91A**, 370–81.

37. CHREE, C. Observational methods in terrestrial magnetism. *Dictionary of Applied Physics*, ed. Sir Richard Glazebrook, 2, 532–45.

38. CHREE, C., and STAGG, J. M. Recurrence phenomena in terrestrial magnetism. *Phil. Trans. roy. Soc.*, 1927, **227A**, 21–62.

39. HAZARD, D. L. The Earth's magnetism. Dept. Comm. *US Coast Geod. Survey, Spec. Publ. No.* 117, 1925, 52pp.

40. HINE, A. Some aspects of terrestrial magnetism phenomena. *Research, Lond.*, 1961, **14**, 143–6.

41. HOLMES, E. L. *Improvements in course recording or like apparatus for use in ships and the like.* British Patent 345362.

42. VON PEICHL, J. *Electric compasses and course recorders.* British Patent 1734, 1892. See also many and varied contributions to the *Journal of Geophysical Research*, Carnegie Institute, Washington, D.C., and *Journal of Atmospheric and Terrestrial Physics*, Pergamon Press, London. See also refs. 8 and 25.

COMPASS DESIGN AND THEORY

43. FIELD, M. B. The navigational magnetic compass considered as an instrument of precision. *J. Instn. elect. Engrs.*, 1919, **57**, 349–86.

44. HINE, A. and HITCHINS, H. L. *Improvements in and relating to azimuthal direction-indicating apparatus.* British Patent 624085, 1946, (Gimballed two-core).

45. HINE, A., and HITCHINS, H. L. *Apparatus for measuring and detecting magnetic fields.* British Patent 619525, 1946. (MMF inductor).

46. National Research Development Corporation. Improvements in navigational compass systems. British Patent 690011, 1951. (AGM system).

47. STARLING, S. G. The equilibrium of the magnetic compass in aeroplanes. *Phil. Mag.*, 1916, **32**, 461–76. See also refs. 10, 16, 21, and 22.

MAGNETOMETERS AND ALLIED OBSERVATORY AND SURVEYING INSTRUMENTS

48. BARNETT, S. J. A sine galvanometer for determining in absolute measure the horizontal intensity of the Earth's magnetic field. *Res. Dept. terr. Mag., Carnegie Inst., Washington, D.C., Pub.* **175**, 1921, **4**, 373–94.

49. BAUER, L. A., and FLEMING, J. A. The C.I.W. deflector in use on the 'Carnegie' for determining the magnetic horizontal intensity and magnetic declination at sea. *Terr. Mag.*, 1913, **18**, 57–62.

50. BAUER, L. A. W., PETERS, W. J., and FLEMING, J. A. *Compass variometer.* US Patent 1701603, 1929; also *Res. Dept. Terr. Mag., Carnegie Inst., Washington, D.C. Pub.* **175**, 1926, **5**, 341–57.

51. BILDINGMAREN, F. *Der Doppelkompass, seine Theorie und Praxis* (Sonderabdruck aus 'Deutsche Sudpolar-Expedition 1901–1903'. 5. Erdmagnetismus 1. Heft 1) Georg Reimer, Berlin, 1907, 104 pp., 14 figs.

52. FAUSELAU, G. Der Doppelkompass als magnetisches Universalinstrument. *Veröff. met. Inst., Univ. Berl.*, 1933, **395** 132–6.

53. FAUSELAU, G. Uber Messungen, mit dem Doppelkompass. *Veröff. met. Inst. Univ. Berl.*, 1931, **380**, 186–93.

54. ESCHENHAGEN, M. Ueber erdmagnetische Intensitätvariometer. *Verh. dt. phys. Ges.*, **1**, 147–52.

55. FLEMING, J. A. Two new types of magnetometers made by the Department of Terrestrial Magnetism of the Carnegie Institution of Washington. *Terr. Mag.*, 1911, **16**, 1–12.

56. FLEMING, J. A. Description of C.I.W. marine earth conductor. *Terr. Mag.*, 1913, **18**, 39–45.

57. FLEMING, J. A., and WIDNER, J. A. Description of the C.I.W. combined magnetometer and earth inductor.

58. FRASER, H. A. D. The unifilar magnetometer of the magnetic survey of India. *Terr. Mag.*, 1901, **6**, 65–9.

59. HARTNELL, G. The vertical intensity variometer. *Terr. Mag.*, 1919, **24**, 49–64.

60. HARTNELL, G. Horizontal-intensity variometers. *US Coast Geod. Surv. Spec. Pub. No. 89*, Washington, D.C., 62pp., 1 pl.

61. HAUSMANN, K. Der Magnettheodolit von Eschenhagen-Tesdorpf. *Z. Instrumkde*, 1906, **26**, 1–15.

62. HEILAND, C. A. Theory of Adolf Schmidt's horizontal field balance. *Geophys. Pros., Trans. Am. Inst. Min. metall. Petrol. Eng.*, 1929, **81**, 261–314; also *Z. angew. Geophys.*, **1**, 289, 329.

63. HEILAND, C. A., and PUGH, W. E. Theory and experiments concerning a new compensated magnetometer system (1934). *Geophys. Prospect., Trans. Am. Inst. Min. metall. Petrol. Engrs.*, 110, 334–72.

64. JOHNSON, E. A. Application of alternating current methods of detection to earth-inductors for marine and land observations. *Terr. Mag.*, 1936, 41, 251–60.

65. LA COUR, D. A vertical intensity magnetometer. *Terr. Mag.*, 1927, 32, 16.

66. LA COUR, D. La balance de Godhaun: Balance magnetique à l'aimant monade dans l'air rarefié. *Met. Inst. Comm. Mag.*, Copenhagen, 1930, 8, 28pp.

67. LA COUR, D. Le quartz magnetometre QHM (quartz horizontal force magnetometer). *Met. Inst. Comm. Mag.*, Copenhagen, 1936, 15, 22pp.

68. LLOYD-CREAK. The Lloyd-Creak dip circle for observations at sea. *Terr. Mag.*, 1901, 6, 119–21.

69. McNISH, A. G. A new type of vertical intensity induction variometer. *Terr. Mag.*, 1936, 41, 161–72; also *Rev. scient. Instrum.*, 7, 336–8.

70. ORDNANCE SURVEY, SOUTHAMPTON. *A portable magnetometer of the null type.* H.M.S.O., London, 1930, 4pp. 3pls.

71. ROMAN, I., and SERMON, T. O. A magnetic gradiometer. *Geophys. Prospect., Trans. Am. Inst. Min. metall. Petrol. Engrs.*, 1934, 110, 373–87.

72. SMITH, F. E. On an electromagnetic method for the measurement of the horizontal intensity of the Earth's magnetic field. *Phil. Trans. roy. Soc.*, 1922, 223A, 175–200; also Schuster, *Terr. Mag.*, 1914, 19, 19–22.

73. STEARN, N. H. Practical geomagnetic exploration with the Hotchkiss superdip. *Geophys. Prospect. Trans. Am. Inst. Min. metall. Petrol. Engrs.*, 1932, 97, 169–97.

74. SUCKSMITH, W. An improved magnetometer. *J. scient. Instrum.*, 1945, 22, 7.

75. WATSON, W. A quartz thread version force magnetograph. *Phil. Mag.*, 1904, 7, 393–9; *Terr. Mag.*, 9, 62–8.

76. WILD, H. Theodolit für magnetische Landesaufnahmen. *Vjschr. naturf. Ges. Zurich*, 1896, 41, 1–25.

PIVOTED-NEEDLE COMPASSES

77. ADMIRALTY COMPASS OBSERVATORY. *Handbook of the Admiralty Transmitting Magnetic Compass.* B.R. 1795, 1948; B.R. 1795(1), 1952.

78. ADMIRALTY COMPASS OBSERVATORY. *Handbook of the Admiralty Gyro-magnetic Compass Type 5.* B.R. 1788, 1953.

79. ADMIRALTY COMPASS OBSERVATORY. *Handbook of the Admiralty Gyro-magnetic Compass Type 6.* B.R. 109, 1957.

80. WILLIAMS, E. B., and BRANCH, W. J. V. *Air Navigation, Theory and Practice.* Brown, Son & Ferguson, 1952.
   See also refs 6, 9, 10, 13–16, 18, 20, 25, 26, 42, 46,

INDUCTOR AND SATURABLE INDUCTOR MAGNETOMETERS

81. ANTRANIKIAN, H. *Magnetic field direction and intensity finder.* US Patent 2047609, 1936.

82. ARMSTRONG, L. D. The use of high-permeability materials in magnetometers. The application of a saturated core type magnetometer to an automatic steering control. *Can. J. Res.*, 1947, 25(3), 124.

83. ASCHENBRENNER, H., and GOUBAU, G. Eine Anordnung zur Registrierung rascher magnetischer Störungen. *Hochfreq. Tech. Electroakust.*, 1936, 47, 177.

84. FELDTKELLER, R. The measurement of magnetic fields by using Forster Probes. (Eng. trans. in A.C.O. library).

85. GERARD, V. B. The airborne magnetometer. *Radio and Electronics (N.Z.)*, 1952, 6, 11–14.

86. GEYGER W. A. Fluxgate magnetometer— Toroidal core with semicircularly wound second harmonic detector winding acts as a field sensitive element; used with switching transistors in a battery operated small size magnetometer. *Electronics*, June, 1962, 48; also *J. app. Phys.* supp. 33(3).

87. GREGG, F. C. An alternating current probe for measurement of magnetic fields. *Rev. scient. Instrum.*, 1947, 18, 77.

88. KELLER, F., BAKER, L. S., and BYRNES, B. C. Airborne magnetic operation. *Trans. Am. geophys. Un.*, 1951, 32, 321.

89. McNISH, A. G. An induction magnetometer. *Trans. Am. geophys. Un.*, 1946, 27, 49.

90. MEE, C. D., and STREET, R. An improved precision magnetometer. *Proc. Instn. elect. Engrs.*, 1954. (II), 639.

91. PALMER, T. M. A small sensitive magnetometer. *Proc. Instn. elect. Engrs.*, 1953, 100 (II), 545.

92. RUMBAUGH, L. H. and ALLDREDGE L. R. Airborne equipment for geomagnetic measurements. *Trans. Am. geophys. Un.*, 1949, 30, 836–48.

93. RUMBAUGH, L. H., and TICKNER, A. J. Airborne magnetometers. *Electl. Engng, N.Y.*, 1947, 66, 680.

94. SCHONSTEDT, E. O., and IRONS, H. R. N.O.L. vector airborne magnetometer, Type 2A. *Trans. Am. geophys. Un.*, 1952, 33, 319.

95. SEISM, P. M., and HANNAFORD, W. L. A portable electrical magnetometer. *Can. J. Technol.*, 1956, **34**, 232–43.

96. STOCKARD, H. P. Airborne geomagnetic surveys by the United States Hydrographic Office. *Navigation, Los Ang.*, 1955, **4** (8).

97. STOCKARD, H. P., and WOODCOCK, F. B. A new magnetic survey aircraft for the United States Navy Hydrographic Office. *Inst. Hydr. Rev.*, 1960, **37**, 117–22.

98. WILLIAMS, F. C., and NOBLE, S. W. The fundamental limitations of the second harmonic type of magnetic modulator as applied to the amplification of small d.c. signals. *J. Instn. elect. Engrs.*, 1950, **97**(II), 445.

99. WYCKOFF, R. D. The Gulf airborne magnetometer. *Geophysics*, April 1948, 182.

100. AGA-BALTIC AKTIEBOLAG. *An electrical course indicating navigation instrument.* British Patent 569839, 1943.

101. GUNN, R. US Patent 2054318.

102. HEY, I. P. R., and BRIGGS, L. J. The Earth-inductor compass. *Proc. Am. phil. Soc.*, 1922, **61**, 15–32.

103. HINE, A. The inductor compass. *Proc. Instn. elect. Engrs.*, 1951, **98**(II), **64**, 485–98.

104. HUDSON, C. S. *Improvements in or relating to remote indicating compass.* British Patent 591019, 1945.

105. SIEMENS-APPARATE UND -MASCHINEN GmbH Berlin. *Improvements in or relating to compasses.* British Patent 520826.

See also refs. 10, 11, 19, 46.

MAGNETOMETERS

1. Hall Effect Magnetometers

106. HILSUM, C. Galvanometric effects and their application. *Brit. J. app. Phys.*, 1961, **12**, 85.

107. ROSHON, D. D., JR. Microprobe for measuring magnetic fields. *Rev. sci. Inst.*, 1962, **33**, 201–6.

108. TORREY, H. C., and WHITMER, C. A. Crystal Rectifiers, Chapter III. McGraw-Hill, New York, 1948.

109. SCHWARTZ, S. A. Stroboscopic Earth inductor compass. *Electl. Engng.* (*N.Y.*), 1951, **70**, 1001–3.

110. WEIDER, H. H. Sampling magnetometer based on the Hall effect. *J. app. Phys.*, 1962, **33**, Supp. No. 3, 1287.

111. WOOD, C. Hall effect transducers. *Measurement* (Full ref. unknown.—Publisher)

2. Nuclear Precession and Resonance Magnetometers

112. ANDREW, E. R. *Nuclear magnetic resonance.* C.U.P., London, 1955.

113. BELL, W. E., and BLOOM, A. L. Optical detection of magnetic resonance in alkali vapor. *Phys. Rev.*, 1957, **107**, 1559.

114. BLOOM, A. L. and JOHNSON, L. E. A magnetometer for the satellite. *Electron. Inds. Tele-Tech.*, 1957, **16**, 76–8, 148, 150, 152, 154, 158, 165. (The characteristic frequency of precessing protons in a weak magnetic field can serve as a measure of Earth's field in space. Data from rocket-borne magnetometers have been extrapolated to produce a tentative design for a satellite magnetometer.)

115. BURROWS, K. A rocket-borne magnetometer. *J. Br. Instn. Radio Engrs.*, Dec. 1959, 769.

116. CAHILL, L. J., and van ALLEN, J. S. High altitude measurements of the Earth's magnetic field with a proton precession magnetometer. *J. geophys. Res.*, 1956, **61**, 547–59.

117. COLEGROVE, F. D., and FRANKEN, P. A. Optical pumping of helium in the metastable state. *Phys. Rev.*, 1960, **119**, 680.

118. DEHMELT, H. G. Slow spin relaxation of optically polarized sodium atoms. *Phys. Rev.*, 1957, **105**, 1487.

119. FRANKEN, P. A. Magnetometer for space-exploring rockets. *J. Franklin Inst.*, Feb. 1959, 184.

120. HAWKINS, W. E. Orientation and alignment of sodium atoms by means of polarized resonance radiation. *Phys. Rev.*, 1955, **98**, 478.

121. HILL, M. N. A ship-borne nuclear-spin magnetometer. *Deep Sea Res.*, 1959, **5**, 309–11.

122. Instruments for satellites: A description of the Varian magnetometer. *J. Franklin Inst.*, 1957, **264**, 258–9.

123. KASTLER, A. Les méthodes optiques d'orientation atomiques et leurs applications. *Proc. phys. Soc.*, 1954, **67A**, 853.

124. KEYSER, A. R., RICE, J. A. and SCHEARER, L. D. A metastable-helium magnetometer for observing small geomagnetic fluctuations. *J. geophys. Res.* 1961, **66**(12), 6163.

125. LOWE, G. C. Measurement of magnetic fields by nuclear resonance. *Electronic Engineering*, March, 1959.

126. PACKARD, H., and VARIAN, R. Free nuclear induction in the Earth's magnetic field. *Bull. Amer. phys. Soc.*, 1953, **28**(7), 7; also *Phys. Rev.*, 1954, 93, 941.

127. RAMSEY, N. F. *Nuclear Moments.* John Wiley, New York, 1953.

128. SCHLUMBERGER WELL SURVEYING CORPORATION. *Surveying corporation improvements in the magnetic resonance apparatus.* British Patent 791866.

129. SKILLMAN, T. L., and BENDER, P. L. Measurement of the Earth's magnetic field with a rubidium-vapour magnetometer. *J. geophys. Res.*, 1958, **63**, 513.

130. UNTERBERGER, R. R. Direct recording of small geomagnetic fluctuations. *J. geophys. Res.*, 1960, **65**, 4213–5.

131. VARIAN ASSOCIATES OF THE USA. *Gyromagnetic compass*. British Patent 806702.

132. WATERS, G. S. A measurement of the Earth's magnetic field by nuclear induction. *Nature*, 1955, **176**, 691.

133. WATERS, G. S., and PHILLIPS, G. A new method of measuring the Earth's magnetic field. *Geophys. Prospect.*, 1956. **4**, 1–9.

3. Miscellaneous Types of Magnetometer

134. BUTTERWORTH, A. A sensitive recording magnetometer. *J. Instn. elect. Engrs.*, 1944, **94**(II), 325.

135. HARRISON, E. P. *Improvements in and relating to means for ascertaining and measuring the direction of a magnetic field*. British Patent 563522.

136. HARRISON, E. P. *Improvements in and relating to magnetic surveying apparatus*. British Patent 532333.

137. HARRISON, E. P., and HAMILTON-SMITH, E. *Improvements in and relating to means for ascertaining and indicating the direction of the Earth's or other magnetic field*. British Patent 584521.

138. HARRISON, E. P., and ROWE, H. An impedance magnetometer, *Proc. phys. Soc.*, 1938, **50**, 176.

139. HARRISON, E. P., TURNEY, G. L., ROWE, H., and GOLLOP, H. The AE effect. *Nature*, 1935, **135**, 961.

140. HARRISON, E. P., TURNEY, G. L., ROWE, H., and GOLLOP, H. The electrical properties of high permeability wires carrying alternating current. *Proc. roy. Soc.*, 1936, **157**, 451–79.

141. HARRISON, E. P., and HAMILTON-SMITH, E. A new method of measuring the inclination of the Earth's magnetic field. *Proc. phys. Soc.*, 1944, **56**, 31.

142. MARTON, L., LEADER, L. B., COLEMAN, J. W., and SCHUBERT, D. C. Electron-beam magnetometer. *J. Res. nat. Bur. Stand.*, *C. Engineering and Instrument*, 1959, **63c** 69.

143. TURNEY, G. L., and COUSINS, G. E. A portable direct-reading magnetometer. *J. scient. Instrum.*, 1938, **15**, 360–7.

144. TURNEY, G. L., HARRISON, E. P., and SERBY, J. E. A magnetic repeater compass. British Patent 458034.

SHIP'S MAGNETISM—COMPASS CORRECTION

145. ADMIRALTY COMPASS OBSERVATORY. *Notes on the Correction of the Magnetic Compass in Ships*. 1943.

146. ADMIRALTY COMPASS OBSERVATORY. *The Theory of the Deviations of the Magnetic Compass*. B. B. 101 (48), H.M.S.O., 1948.

147. *Admiralty Manual of Navigation*, Vol. 1. H.M.S.O., 1938 and 1954.

148. *Air Navigation*, Vol. I. H.M.S.O., 1944.

149. EVANS, F. J. Reduction and discussion of the deviations of the compass observed on board of all the iron-built ships and a selection of the wood-built steam-ships in H.M. Navy and the iron steam-ship *Great Eastern*. *Phil. Trans. roy. Soc.*, 1860, 337.

150. EVANS, F. J. *Elementary Manual for the Deviations of the Compass*. 1870.

151. HECK, H. N., and PARKER, W. E. Instructions for the compensation of the magnetic compass. *US Coast geod. Surv., Spec. Pub. No. 96*, Washington, D.C., 1923, iv + 49pp.

152. MAY, W. E. Compass Adjustment. Hutchinson, London, 1951.
See also refs 1, 3, 12, 20, 22, 23, 24, and 26.

SURVEYING, FIELD WORK, ETC.

153. FLEMING, J. A. Construction of non-magnetic experiment building of the Department of Terrestrial Magnetism. *Res. Dept. terr. Mag., Carnegie Inst. Wash. Pub.* **175**, 1921, **4**, 351–8.

154. HAZARD, D. L. Directions for magnetic measurements. 3rd Edn. *US Coast geod. Surv. No. 166*, Washington, D.C., 1930, vi + 129pp.

155. JOYCE, J. W. Manual on geophysical prospecting with the magnetometer. *US Dept. Int. Bur. Mines, Washington, D.C.*, 1937, xi + 129pp.

156. JOYCE, J. W. Airborne magnetic surveys. *Nature*, 1954, 173, 281–3. (Report of Geophysical discussion at R. astr. Soc., 27th November, 1953.)
See also refs 37 and 72.

COMPASS TESTING

157. MATEZKY, F. Zur Frage einer 'dynamischen' Prüfung von Magnetkompassen. *Dt. hydrogr. Z.* 1955, **86**, 242.

158. MELDAU, H. The four-sphere test. *Annln. Hydrogr., Berl.*, 1907, **35**, 17–25.

159. ADMIRALTY COMPASS OBSERVATORY. *Manual of the Admiralty gyro-compass (Sperry type)*, Chapter 7. B.R. 9, H.M.S.O., 1941.

160. BELL, J. Data transmission systems. *J. Instn. elect. Engrs.*, 1947, **94** (IIA), 222.

161. GARVEY, R. J. Electrical remote position-indicating systems as applied to aircraft. *J. Instn. elect. Engrs.*, 1947, **94** (IIA), 283.

162. HELPS, F. G. Data transmission by synchros. *Electron. Engng.*, 1956, **28**, 438–45. See also refs 77, 78, and 79.

REFERENCE BOOKS AND TEXTBOOKS

163. AIRY, G. B. *A Treatise on Magnetism: Designed for the Use of Students in the University.* Macmillan, London, 1870, xv + 220pp.

164. BATES, L. P. *Modern Magnetism.* C.U.P., 1961.

165. DAVIDSON, M. (Ed.) *The Gyroscope and its Action.* Hutchinson, London, 1946.

166. DUNCAN, J., and STARLING, S. G. *Text-Book of Physics: Magnetism and Electricity.* Macmillan, London, 1920.

167. FLEMING, F. A. (Ed.). *Terrestrial Magnetism and Electricity*—Physics of the Earth, Vol. VIII. McGraw-Hill, New York, 1939. (Has a very full bibliography.)

168. MOTTELAY, P. F. *Bibliographical History of Electricity and Magnetism.* Griffin, London, 1922.

169. STARLING, S. G. *Electricity and Magnetism for Degree Students.* Longmans, London, 1953.

170. VIGOUREUX, P., and WEBB, C. E. *Electric and Magnetic Measurements*, Blackie, London, 1936.

INDEX